American Geophysical Union

ANTARCTIC
RESEARCH
SERIES

Antarctic Research Series Volumes

Volume 72 | ANTARCTIC RESEARCH SERIES

Ecosystem Dynamics in a Polar Desert
The McMurdo Dry Valleys, Antarctica

John C. Priscu
Editor

American Geophysical Union
Washington, D.C.
1998

	ANTARCTIC
Volume 72	RESEARCH
	SERIES

ECOSYSTEM DYNAMICS IN A POLAR DESERT
John C. Priscu, Editor

Library of Congress Cataloging-in-Publication Data

Ecosystem dynamics in a polar desert : the McMurdo Dry Valleys,
     Antarctica / John C. Priscu, editor.
                p. cm. -- (Antarctic research series ; v. 72)
                Includes bibliographical references.
                ISBN 0-87590-899-3
                1. Desert ecology--Antarctica--McMurdo Dry Valleys. I. Priscu,
    John Charles. II. Series
     QH84.2.E276 1998
   557.54'098--dc21   97-46526
                                     CIP

ISBN 0-87590-884-5
ISSN 0066-4634

Published by
American Geophysical Union
2000 Florida Avenue, N.W.
Washington, D.C. 20009
With the aid of grant OPP-9414962
from the National Science
Foundation

Printed in the United States of America.

CONTENTS

THE ANTARCTIC RESEARCH SERIES

The Antarctic Research Series, published since 1963 by the American Geophysical Union, now comprises more than 70 volumes of authoritative original results of scientific work in the high latitudes of the southern hemisphere. Series volumes are typically thematic, concentrating on a particular topic or region, and may contain maps and lengthy papers with large volumes of data in tabular or digital format. Antarctic studies are often interdisciplinary or international, and build upon earlier observations to address issues of natural variability and global change. The standards of scientific excellence expected for the Series are maintained by editors following review criteria established for the AGU publications program. Publication of the Series is aided by a grant from the National Science Foundation, which supports much of the underlying field work. Priorities for publication are set by the Board of Associate Editors. Inquiries about published volumes, work in progress or new proposals may be sent to Antarctic Research Series, AGU, 2000 Florida Avenue NW, Washington, DC 20009 (http:/www.agu.org), or to a member of the Board.

PREFACE

The McMurdo Dry Valleys of southern Victoria Land comprise the largest ice-free expanse (about 4000 km²) on the Antarctic continent. Research in this region began during British expeditions of the early 1900's and has yielded much information on specific physical, chemical and biological features of the area. Only recently have scientists begun to view the region as an integrated system which includes dynamic interactions among biotic and abiotic components of the environment. The McMurdo Dry Valleys represents the coldest and driest desert on this planet. Photoautotrophic and heterotrophic microorganisms that are intimately linked with the presence of liquid water and nutrients dominate the biological assemblages. Owing to the low average temperature (-20° C) in the region, liquid water is a rare commodity that often exists for a short period only and occurs in many inconspicuous places. It is now clear that the presence of liquid water produces a cascade of tightly coupled events that ultimately leads to the biological production and cycling of organic carbon and related elements. It also is clear that an integrated knowledge of biological, chemical, and physical factors is required to understand biogeochemical dynamics within the cold desert ecosystem of the McMurdo Dry Valleys. While various aspects of this ecosystem have formed the basis of several excellent publications, the compendium of manuscripts published within this volume represents a first attempt to compile complementary information on the abiotic and biotic components of the McMurdo Dry Valleys and link them in a final synthesis chapter.

I hope that the information contained in this volume will help lay the foundation for future research directions of the current NSF funded Long-Term Ecological Research (LTER) program now underway in the McMurdo Dry Valleys. The McMurdo Dry Valleys LTER, which began in 1994, is the newest member of the LTER network of sites that now numbers 18. These sites range from the tropical rain forests of Puerto Rico to the northern Chihuahuan Desert of New Mexico and include forest, prairie, and tundra ecosystems.

The current over-arching hypotheses of the McMurdo LTER are (1) the structure and function of the McMurdo Dry Valleys are differentially constrained by physical and biological factors and (2) their structure and function is modified by material transport. While such relationships are fundamental to all ecological systems, their importance is more obviously demonstrated in extreme environments. For example, the McMurdo Dry Valleys environment generates low and episodic biological productivity, inextricably linked to the presence of liquid water resulting from ice melt. Despite relatively low biological production, biomass accrues in the area because loss rates are often equally low. The paucity of liquid water and the delicate balance that exists between gains and losses of organic carbon makes the McMurdo Dry Valleys ecosystem one of the most sensitive indicators of environmental change on our planet.

The manuscripts in this volume are arranged in six parts. The first part presents information on the primary abiotic driving forces and conditions defining the system. This section also demonstrates how liquid water links the geochemistry of glaciers, streams, and lakes in the region. The second and third parts present information on the hydrology, biogeochemistry, and physics of the perennial streams and permanently ice-covered lakes that are the major sites of organic carbon production in the McMurdo Dry Valleys. Part four focuses on the physical, chemical, and biological properties of the dry valley soils. Aeolian transport of soils may be the major mode of distributing organic matter among system components within the dry valleys. Part five defines and formalizes ecosystem organization and linkages in the McMurdo Dry Valleys. This section also discusses the role of environmental management in maintaining scientific and environmental values, which are critically important for long-term ecological studies. Environmental management is becoming increasingly important given the escalating levels of anthropogenic stress and disturbance from tourists and the scientific community. The impacts of human activities only now are being explored. Finally, the CDROM and accompanying text, which comprises part six of this volume, presents thematic data too voluminous to be included within specific papers. The CDROM also includes a detailed set of geospatial data that supports all manuscripts in this book. The information contained on the CDROM provides details and scale that should help the reader visualize the McMurdo Dry Valleys from an ecosystem perspective.

John C. Priscu
Montana State University
Bozeman, Montana

THE COMPOSITE GLACIAL EROSIONAL LANDSCAPE OF THE NORTHERN MCMURDO DRY VALLEYS: IMPLICATIONS FOR ANTARCTIC TERTIARY GLACIAL HISTORY

Michael L. Prentice

Climate Change Research Center, Institute for the Study of Earth, Oceans, and Space and Department of Earth Sciences, University of New Hampshire Durham, New Hampshire

Johan Kleman and Arjen P. Stroeven

Department of Physical Geography, Stockholm University,10691, Stockholm, Sweden

We reassessed glacial versus non-glacial hypotheses for the excavation of the McMurdo Dry Valleys using bedrock geomorphologic evidence. We find three glacial erosional landscapes, namely, high, intermediate, and low, within the Wright-Victoria Valley system below an elevation of 1300 m. The principal evidence for glacial erosion is the molded asymmetry of paired corners on tributary-valley spurs at trunk-valley intersections. This reflects confluence of wet-based alpine and trunk-valley glaciers flowing east to southeast. Hanging glacial benches at two different elevations in numerous valley locations coupled with valley floors that exhibit classic glacial morphology delineate the three glacial landscapes. We propose progressive glacial incision by wet-based ice in separate high, intermediate, and low phases to cut the stepped glacial landscapes. Apparent continuity of each glacial surface for more than 50 km from Transantarctic Mountain (TAM) crest toward the sea requires large ice flux from either local mountain and/or continental ice sheets. The lowest elevation of definitive fluvial morphology is c. 1300 m. Erosion of trunk valleys below this level has been by glacial processes. This implies that the TAM crest was significantly broader during the high and intermediate glacial phases than at present. It follows that a linear TAM ice sheet could account for high- and intermediate-phase glaciation. The East Antarctic Ice Sheet probably dominated during the low phase. The presence of wet-based alpine glaciers during each phase indicates that the coeval climate was significantly warmer and wetter than at present.

INTRODUCTION

To understand how the climate in the Southern Hemisphere, as well as globally, has evolved over the Cenozoic, the long-term history of the Antarctic ice sheet and of the Antarctic climate are of fundamental importance. Antarctic ice extent and surface climate strongly influence global sea level [e.g., *Bentley and Giovinetto*, 1991] as well as the temperature and circulation of the Southern Hemisphere atmosphere

[e.g., *Van Loon and Shea*, 1988; *Oglesby*, 1989] and global ocean [e.g., *Gordon et al.*, 1993]. Of the variety of records and data that record Antarctic glacial history, the more continuous are from the far field, in other words, from off of the Antarctic continent and its margin. It is difficult to discern Antarctic climate change in these records unambiguously. The near-field record of geomorphology and surficial geology on the other hand, directly reflects the history of different subaerial and glacial processes and spatial-temporal

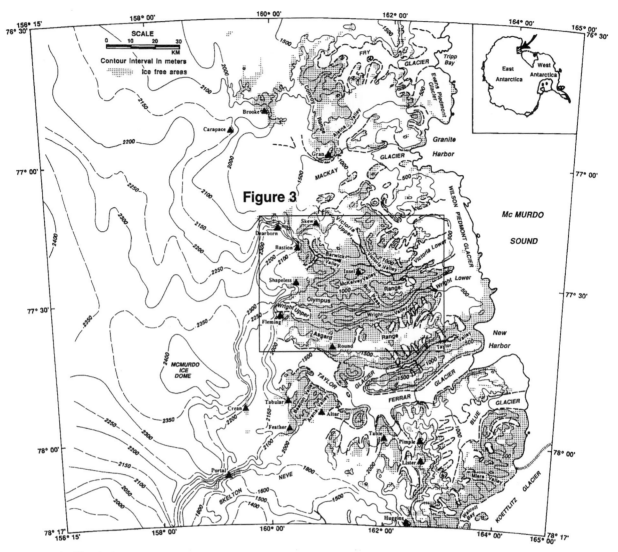

Fig. 1. Generalized topographic map of the McMurdo Dry Valleys region, Antarctica. Sources are: U.S. Geological Survey topographic map, McMurdo Sound, Antarctica at scale 1:1,000,000 (1974) and *Pyne et al.* [1985].

patterns. Hence our knowledge about Antarctic climate can be enhanced by inversion of the geomorphologic record, whereby a sequence of landform-creating climatic events is reconstructed through the use of established landform-process relationships and relative-age determinations of landscape elements.

There is an exceptional bedrock morphologic record exposed in the McMurdo Dry Valleys on the seaward flank of the Transantarctic Mountains (TAM) across from McMurdo Sound with nearly 3000 m of relief (Figure 1). The McMurdo Dry Valleys have long been considered a classic composite glacial landscape reflecting millions of years of glacial erosion and deposition [*Taylor*, 1922; *Bull et al.*, 1962; *Gunn and Warren*, 1962; *Nichols*, 1962; *Denton et al.*, 1991]. Recognizing recent evidence for minimal erosion of slopes and valley floors in the last 15 million years (m.y.) and applying geomorphologic concepts from partly analogue areas, *Denton et al.* [1993] reassessed the geomorphologic evolution of the dry valleys. They discerned surfaces arranged in a stepped fashion and attributed most of the valley cutting to subaerial processes operating under semi-arid conditions. They suggested that the main valleys were fluvially cut to

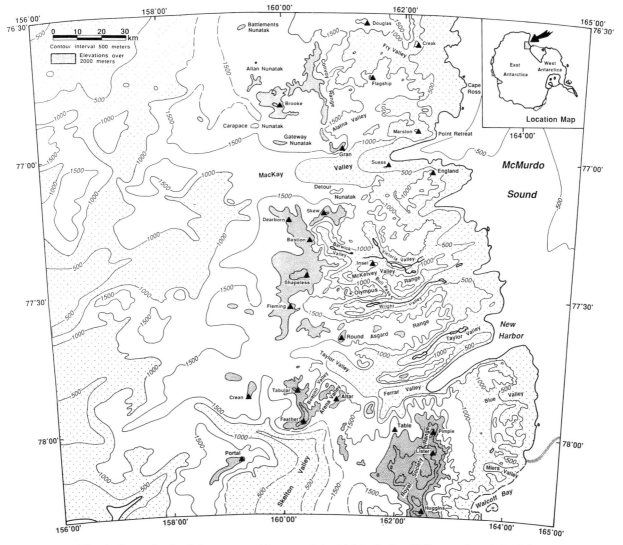

Fig. 2. Generalized sub-ice topographic map of the McMurdo Dry Valleys region, Antarctrica. Adapted from U.S. Geological Survey topographic map, McMurdo Sound, Antarctica at scale 1:1,000,000 (1974). Supplemental data are from *Drewry* [1982] and *Calkin* [1974].

their present depths, with minor overdeepening and sharpening by ice.

The scheme of *Denton et al.* [1993] was a major step forward in emphasizing the inherited elements of the landscape and relating the geomorphic history to the tectonic evolution. We are, however, critical regarding the relative roles attributed to fluvial and glacial erosion in the dry valleys landscape and here describe evidence that indicates some important revisions to the scheme put forward by *Denton et al.* [1993]. In particular, we recognize wet-bed glacial erosion in the Tertiary as a process that was very important in shaping the dry valleys morphology.

Physiography and Tectonic History

Three large valleys extending from the TAM crest to the ocean make up the core of the McMurdo Dry Valleys, namely, from south to north, Taylor Valley, Wright Valley, and the Victoria Valley system (Figure 1). The valleys are about 80–100 km in length, 5–10 km in width, and up to 3 km deep. Mountain ranges flanking the trunk valleys, notably the Kukri, Quatermain, Asgard, Olympus, and St. John's Ranges, exhibit a second-order system of highland valleys. Higher still is the rolling topography of the TAM crest, as represented by Mt. Fleming and Shapeless Mountain at the west end of the valleys (Figure 2). Our study area within the dry valleys includes Wright Valley and the Victoria Valley system and stretches from the Asgard Range on the south to the St. John's Range on the north (Figure 1).

The bedrock of the McMurdo Dry Valleys consists of the Precambrian metamorphic rocks of the Koettlitz Group and the Ordovician to Precambrian granitoid rocks of the Granite Harbor Intrusives [*Gunn and Warren*, 1962; *Grindley and Warren*, 1964; *Findlay et al.*, 1984]. The basement rocks are truncated regionally by the Kukri Erosion Surface [*McKelvey et al.*, 1977], formerly Kukri Peneplain [*Gunn and Warren*, 1962], which has a relief of c. 77 m in the area [*Turnbull et al.*, 1994]. The sedimentary sequences of the Devonian Beacon Supergroup, a collection of sandstones, shales, conglomerates, coal, and volcanic rock, were deposited on the peneplain and slope gently (2–5°) westward [*Webb*, 1963; *Barrett*, 1991]. The Jurassic Ferrar Dolerite Group, igneous rock of tholeiitic affinity, occurs as sills and dikes intruding Beacon and basement rocks [*Kyle et al.*, 1981]. Of geomorphologic significance are the basement [*Gunn and Warren*, 1962] or lower [*Turnbull et al.*, 1994] sill, intruded into the basement complex, the peneplain or upper sill, intruded between the Kukri Erosion Surface and the overlying sandstone, and a group of sills intruded into the higher sandstone sequence. The uppermost volcanic unit, the subaerially erupted Kirkpatrick basalts, are only preserved at inland high-elevation locations, such as the Allan Nunatak.

Uplift and inland tilting of the TAM shoulder was initiated by rifting about 50–55 million years before present (Ma) [*Fitzgerald*, 1992]. Apatite fission-track analyses indicate surface uplift of 5 km since the start of mountain formation [*Gleadow and Fitzgerald*, 1987]. Major faults paralleling the coast separate the uplifted mountain shoulder from the subsided Ross Sea embayment. Major faults under Ferrar and MacKay glaciers define the boundaries of the dry valleys block toward the south and north, respectively.

Glacial Hypothesis

Taylor [1922] suggested that the deep trunk valleys of the dry valleys were excavated by headward erosion of cirque glaciers. *Bull et al.* [1962, 1964] pointed out a patchwork of hanging structurally and non-structurally controlled benches on the walls of the trunk valleys that defined glacial valley floors older and higher than the present one. Their glacial interpretation was based on their inference that the old valleys have parabolic transverse profiles as indicated by curvilinear floor remnants and slopes separating benches. *Bull et al.* [1962] proposed that the trunk valleys were excavated by erosive East Antarctic outlet glaciers flowing over the TAM crest during two glacial episodes, their First and Second Glaciations. The hanging benches are characterized by abrupt breaks-of-slope. These were interpreted to indicate that younger glacial maxima involved lower ice surfaces.

Gunn and Warren [1962] proposed that the glacial origin of the trunk valleys resulted from headward migration of "pseudo-cirques." Pseudo-cirques refer to the large steps in the longitudinal profiles of the trunk valleys, the tread of one step and the riser of the next producing a cirque-like basin. The Airdevronsix Icefalls in Wright Valley is an example. According to Gunn and Warren, pseudo-cirques migrate upvalley because ice calves down the headwall onto horizontally stratified rocks of differing resistance. In response, thickened ice at the foot of the icefalls rotates, removing weaker rock and undermining resistant rock which crumbles away at the edge.

Nichols [1962, 1971] preferred subglacial erosion of Wright Valley by wet-based ice during his Vanda Glacial Episode, which corresponds to the Second Glaciation of *Bull et al.* [1962]. *Nichols* [1971] cited: 1) The overdeepened longitudinal profile of Wright Valley, reflecting thicker and faster-flowing upvalley ice; 2) hanging tributary valleys; 3) asymmetric U-shaped transverse cross-sections; 4) truncated spurs; and 5) "rock bastions" in front of alpine glacier valleys. In the Victoria Valley system, *Calkin* [1971] inferred a similar episode of glaciation, the Insel glaciation.

Numerous researchers [e.g., *Calkin*, 1964; *Selby and Wilson*, 1971a; *Aniya and Welch*, 1981) interpreted the amphitheater-shaped basins in the highlands of the dry valleys as cirques and cirque-headed alpine valleys formed following TAM uplift in the early Tertiary.

Denton et al. [1984] proposed two episodes of thick ice-sheet overriding of the dry valleys in the Tertiary. They took the dominant NE-SW trend of highland valley morphology as evidence for heavy erosion underneath a thick wet-based ice sheet that flowed northeastward obliquely across the grain of the trunk valleys because of blocking ice in the Ross Sea. Stoss/lee topography exhibited by ridge-lines and peaks in the Asgard Range was assigned to sub-ice sheet processes as was the near obliteration of such features in the western Olympus Range. After *Shaw and Healy* [1977], *Denton et al.* [1984] explained the irregularly channeled dolerite landscapes, of which the Labyrinth is the most spectacular example, as reflecting glacial erosion.

Non-Glacial Hypothesis

Denton et al. [1993] presented the first comprehensive non-glacial hypothesis for the evolution of the dry valleys landscape. A number of workers had previously suggested that isolated parts of the landscape were significantly impacted by polar non-glacial processes. For instance, sandstone slope retreat in high alpine valleys had been ascribed to salt weathering and deflation [*Selby*, 1971] as well as nivation [*Shaw and Healy*, 1977]. It has also been proposed that the dolerite channels of the Labyrinth in upper Wright Valley formed by salt-weathering along joint planes [*Selby and Wilson*, 1971] and catastrophic flooding [*Smith*, 1965]. On the other hand, *Denton et al.* [1993] suggested that the major elements of the dry valleys landscape are inherited from an early Tertiary warm semi-arid climatic regime. This view was partly based on their discovery of volcanic ash 15 m.y. in age on the slopes of the high dry valleys escarpment landscape that implied great antiquity and minimal damage due to glacial processes [*Marchant et al.,* 1993a, 1993b]. Additionally, *Denton et al.* [1993] recognized strong similarities between the tabular landscape of the western dry valleys and the cuestaform landscape of warm semi-arid regions such as the southwestern United States, principally the rectilinear slopes and the escarpments.

Denton et al. [1993] extended this non-glacial theory by attributing the cutting of the deep trunk valleys to fluvial downcutting. The principal basis was their interpretation that fluvial spurs are preserved on the floors of central-eastern Wright and Taylor Valleys. Erosion achieved by wet-bed glaciers was seen as minor. The main valley floors were considered to be unmodified since the middle Miocene, c. 15 Ma. The impact on landscape relief of cold-based glaciation and polar-desert subaerial processes since that time was regarded as negligible. By inferring that denudation pulses were driven by tectonically induced base-level changes, *Denton et al.* [1993] and *Sugden et al.* [1995] linked denudational history and large-scale tectonic events. Because the trunk valleys were fluvially cut and graded to sea level, *Denton et al.* [1993] and *Sugden et al.* [1995] invoked a period of tectonic subsidence to account for fjord occupation of Wright Valley during the late Neogene as long ago as 9 ± 1.5 Ma.

Scope of the Paper

The glacial and non-glacial hypotheses for dry valleys landscape evolution have significantly different implications for the Tertiary history of the Antarctic climate, the Antarctic Ice Sheet, and Antarctic tectonics, all of which, in turn, have important implications for global change. Both deserve careful examination. In this paper, we present the results of our test of both theories using bedrock geomorphology. Our study area encompasses the northern half of the McMurdo Dry Valleys. Though this is less than the area examined as basis for the non-glacial hypothesis, we consider it sufficiently representative to constitute a significant test. Before presenting our results, we provide some geomorphic background relevant to them. The crux of our analysis is distinction between major bedrock morphologic features produced glacially, fluvially, and through subaerial weathering. Identification of these features in the study area depends critically on concepts for these features and knowledge of process developed from non-Antarctic landscapes known to have been affected by glacial erosion, fluvial erosion, and various types of subaerial weathering.

GEOMORPHIC BACKGROUND AND METHODS

Given the evidence for the antiquity of the dry valleys landscape and the large probable variation in climate from early Tertiary warmth to present-day polar desert [*Mercer*, 1983; *Prentice and Matthews*, 1991], a wide range of denudation processes are likely to be reflected in this landscape. It seems safe to assume that the uplifting TAM experienced an increasingly wet climate, fed by the opening Ross Sea, and that fluvial denudation occurred [e.g., *David and Priestly*, 1914; *Taylor*, 1922]. As the snowline lowered in the TAM, glaciers and upland ice sheets formed. As the snowline continued to lower over the Tertiary, wet-based glaciation reached deeper and deeper into the dry valleys. With further climate cooling, dry-based glaciation, which trailed the wet-bed glaciation front, reached to sea level. Hence, a priori, the fundamental geomorphic processes are glacial and subaerial.

Glacial Processes/Features

Basal-ice thermal regime exerts a fundamental control on glacial processes and the efficiency of glacier erosion and deposition. Wet-based ice is at the pressure melting point. Hence basal sliding, abrasion, and plucking can occur. Dry-based ice is below the pressure melting point and bond strength between ice and substratum exceeds the yield strength of pure ice [e.g., *Paterson*, 1994]. Hence ice flow is by internal deformation alone. Erosion by cold-based ice is generally small [e.g., *Mercer*, 1971; *Drewry*, 1986]. In the subglacial environment, frozen and thawed beds can be patchy and thermal boundaries between them can be sharp [*Hughes*, 1981; *Kleman* 1992; *Dyke*, 1993; *Kleman and Börgstrom*, 1994]. In high-relief terrain, ice-surface slope and ice thickness largely govern basal ice temperature and, hence, location of erosion zones. The expected as well as documented pattern [*Glasser*, 1995; *Kleman and Stroeven*, 1997] is frozen ice-bed on main interfluves and thawed ice-bed in main valleys, in line with the concept of "selective linear erosion" [*Sugden and John*, 1976]. The dry-bed to wet-bed transition is a threshold separating erosional regimes that differ by many orders of magnitude in efficiency.

Because wet-bed conditions critically depend on topography that dictates ice thickness and influences ice-surface slope, we see wet-bed glacial erosion primarily as an exploiter of pre-existing topographical patterns. Hence valley patterns in a glaciated landscape may reflect processes other than glacial, whereas the bulk of the excavation represents glacial erosion. Glaciers can erode over any valley width transverse to flow direction. In high-relief terrain, wet-bed cirque and valley glaciation create distinctive bedrock features such as glacial walls, troughs, cirques, asymmetrically molded spurs and corners, enclosed basins, and hanging valleys.

A glacial trough is typically characterized by a parabolic transverse cross section and a distinct break-of-slope marking the limit to bordering uplands displaying a subaerially developed morphology, or an older glacial generation. In formerly glaciated areas, it is common that the glacial imprint is highly variable along a valley, such that both typical fluvial and glacial cross-sections coexist within short distances. Truncated spurs from older fluvial-valley generations may or may not occur. In glaciated mountains it is common that only one side of a main valley has been glacially eroded to such an extent that a glacial wall is formed. Steep rock walls can be considered diagnostic of glacial erosion if they either appear as "scars" in a terrain otherwise characterized by moderate slope angles or form a spatial pattern likely reflecting the lateral boundary of fast ice flow. Trough heads are characterized by abrupt breaks-of-slope separating high ground

with little evidence of glaciation but for small fluvioglacial gorges from precipitous, stepped headwalls to the trough floor [*Holtedahl*, 1967].

Glacial deepening and/or widening of the trunk valley commonly outpaces that in its tributary valleys by virtue of greater ice-flux in the trunk valley resulting in discordance between trunk and tributary valley floor at their junction. The tributary-valley floor hangs above the trunk-valley floor, with relief on the order of a few hundreds of meters [*Flint*, 1971]. Hence, in a single glacial phase, accordant trunk and tributary ice surfaces can mask discordant valley floors.

The break-of-slope or brink indicates the location of a shear zone between fast ice stream-flow in the trough and slow ice sheet-flow on adjoining uplands. Many examples exist where relict surfaces above the break-of-slope indicate that cold-based conditions on the uplands coexisted with vigorous wet-based flow in the troughs [*Kleman and Stroeven*, 1997]. The break-of-slope in a trough may vary considerably in elevation. An important point for paleo-glacier reconstruction is that the process boundary for glacial erosion need not coincide with the boundary between glacier and ice-free terrain. Hence features such as trough edges provide only minimum elevations for coeval ice-surfaces. The thickness of overlying cold-based ice becomes the issue.

Glacial troughs can exhibit stacked valley-in-valley profiles [*Flint*, 1971, p. 130; *Cotton*, 1942]. This refers to one or more sets of benches interpreted as remnants of older valley floors above an inner narrow glacial valley. Where the benches can be shown to be glacially eroded, they are generally considered to reflect older glaciations [*Hobbs*, 1911; *Cotton*, 1942]. An alternative glacial interpretation is that the benches were cut by cirque glaciers. As bedrock structure can explain the occurrence of many benches, the relationship of structure to benches should be understood. Benches exhibiting particularly sharp slope-breaks indicate that valley glaciers reflecting younger glacial phases did not overflow their banks.

A key geomorphic feature that we use to substantiate glacial erosion in the dry valleys, not previously recognized there, is the molded corner. Molded corners typify the junction of tributary and trunk valleys where wet-based alpine glaciers merge with wet-based valley glaciers [e.g., *Sharp*, 1988, p. 30]. The molding of the corners is commonly asymmetric. The up-flow corner is smoothly concave as it is in a pressure shadow. The down-flow corner under relatively high pressure is smoothly convex. Narrow interfluves separating adjacent tributaries develop a pronounced hook.

Occasionally, an offset exists between a pair of corners where tributary valleys contribute an appreciable ice volume to the trunk-valley glaciers.

Overdeepenings provide a minimum estimate of vertical glacial erosion. Another more crude measure is the altitude of hanging valleys after allowance for glacial erosion of the tributaries.

Subaerial Processes/Features

Under subaerial conditions, comminution of bedrock to grain sizes that can be efficiently transported by fluvial or eolian processes is the necessary first step in the denudation process. Under the present climate, chemical weathering is negligible in the dry valleys area. Given deep permafrost and the absence of an active layer, frost weathering is regionally inhibited, but locally active where dark rock surfaces of favorable aspect warm sufficiently to melt snow in sub-freezing air temperature [McKay et al., 1993]. During warmer conditions in the past, the full range of weathering processes have most likely been operational. Fluvial and eolian processes are the two non-glacial transport processes capable of transporting material out of the local system, whereas mass-wasting can be regarded as a short-range "commuter-transport" system feeding the long-range fluvial system.

In high-relief terrain, fluvial valleys are typically V-shaped and show a monotonic upstream decrease in valley cross-sectional area. In contrast to glaciers, which can erode over any width transverse to flow direction, streams directly erode narrow corridors [e.g., Harbor, 1992]. Fluvial denudation of a landscape therefore implies the presence of slopes at an angle adequate for transport of debris to streams through mass-wasting. Fluvial erosion is most efficient on steep gradients, but minimal on low gradients. Hence fluvial valleys tend to become graded to a base level, local or regional, depending on available time. Intact spurs are diagnostic of fluvial erosion. The plan-view pattern is one of interlocking spurs and winding rivers. Valley-in-valley profiles are initiated by local base-level lowering [Kleman and Stroeven, 1997].

The eolian transport system is restricted to sand and finer grain-sizes, but can transport material uphill and hence erode basins. It represents transport that is extremely variable in direction and magnitude. If the terminal grade from weathering of a particular rock type is sand or smaller, eolian processes can transport the material efficiently out from a given area. If weathering produces larger particles, lag deposits will result.

A number of features are diagnostic of subaerial weathering. Weathering and deflation basins are mainly enclosed, 2–10 m deep, and occasionally bounded by upstanding and less weathered rock. Their relief provides estimates of subaerial weathering and denudation. Tor-like features are isolated residual rock masses that project through scree or colluvium in the upper part of slopes, while no exposed bedrock is visible on the lower slopes. Such features indicate that a once steeper valley wall has been weathered, with an accompanying slope reduction. Although a precise reconstruction of the original wall is not possible, these features are diagnostic of older steeper slope profiles.

The fundamental control on the morphology of terrain with horizontally layered rocks is their relative rock-mass strength [Augustinus and Selby, 1990]. Following accelerated down-cutting to the top of a resistant rock layer, steep stream-parallel scarps retreat. Less resistant rock immediately above the expanding planation surface is removed by a combination of mass-wastage via freeze/thaw, salt-weathering, groundwater sapping and eolian/fluvial transport undercutting the cliff-forming rock. The rock face retreats by rockfall. The typical transverse profile of a retreating scarp is rectilinear or bounded by straight lines with constant slope angles from top to base of cap rock. Accelerated retreat in restricted areas results in formation of box-canyons. Isolated scarps shrink to buttes and mesas.

Methods

We mapped bedrock morphology in the study area based on stereoscopic viewing of aerial photographs.

Fig. 3. Map of selected major bedrock morphologic features in Wright Valley and the Victoria Valley system, McMurdo Dry Valleys, Antarctica. Background is from U.S. Geological Survey Satellite Image Map, McMurdo Dry Valleys, Antarctica (1995). The satellite scene is from Landsat 4 Thematic Mapper, Path 57 (shifted), Row 115. The projection is Lambert conformalconic. Symbols are explained in the inset. Topographic information is from U.S. Geological Survey topographic maps at scale 1:50,000 (1977). This figure is on the enclosed CDROM.

Fig. 3.

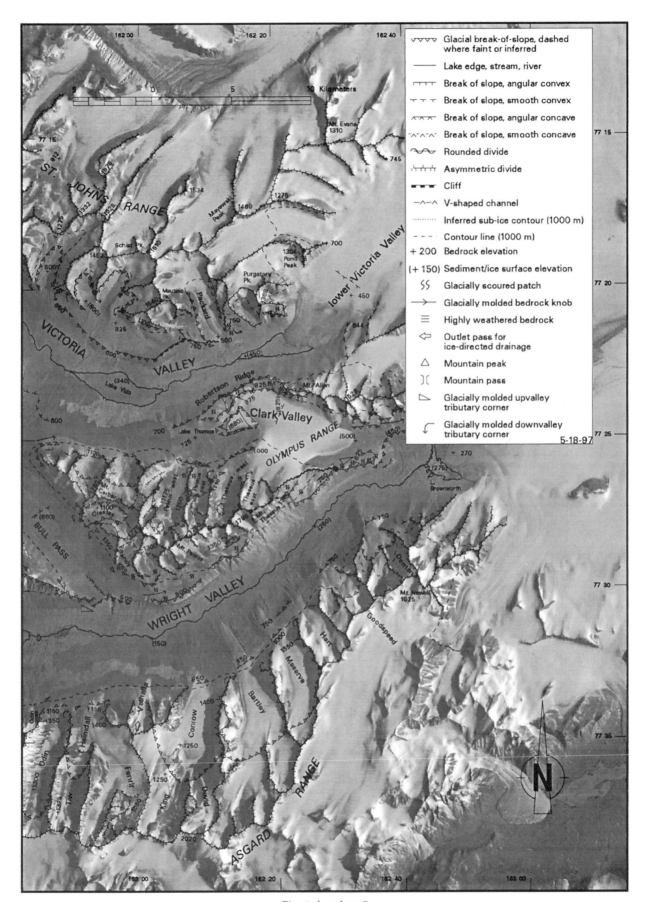

Fig. 3. (continued)

The scale of the photographs ranges between 1:25,000 and 1:5,000. We compiled the resulting unrectified map patterns onto a photographic print of a geo-corrected Landsat 4 scene (Path 57, Row 115) printed at 1:50,000. This bedrock morphologic map was digitized in ArcInfo. Topography was estimated from the 1:50,000 USGS topographic map series.

We have conducted detailed field studies in central Wright Valley and the TAM crest region between Mt. Fleming and Shapeless Mountain. In addition, we conducted reconnaissance field work in the western Olympus Range, McKelvey Valley, eastern Victoria Valley, and Bull Pass.

RESULTS

In our analysis of the northern dry valleys bedrock geomorphology, we proceed from upstream to downstream valley sections (Figure 3). This permits consideration of valley sections in the context of potential influx of ice or water from upstream. We first present the upstream sections: upper Wright Valley, central-western McKelvey Valley, Balham Valley, Barwick Valley, upper Victoria Valley, and Bull Pass-eastern McKelvey Valley. Then we consider the downstream sections: central Victoria Valley with the Clark Valley and central-to-lower Wright Valley. For each valley section, we proceed from high to low elevation through the principal physiographic units, namely, plateau, upper highlands, lower highlands, valley-side benches and tributaries, and trunk valleys. We recognize a variety of glacial features. The glacial features are compatible with three different glacial erosional landscapes applicable across the northern dry valleys, namely, the high, intermediate, and low landscapes. The basis for each landscape is that it reflects a separate phase of glaciation. Here we tentatively assign the glacial morphologies to one of the landscapes on the assumption of comparable denudation of the different valleys of the study area.

Upper Wright Valley

The plateau. West of the Asgard and Olympus Ranges, the terrain rises above the highest Ferrar Dolerite sill and on to the sandstone of the TAM crestal

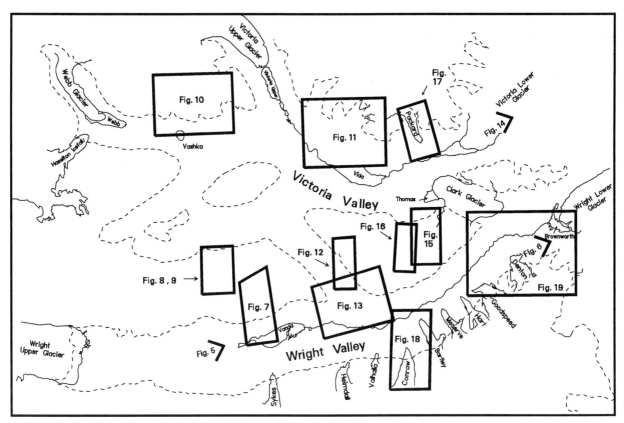

Fig. 4. Location map of aerial photographs used in figures. Dashed line is the 1000 m contour line from the USGS 1:50,000 topographic map series.

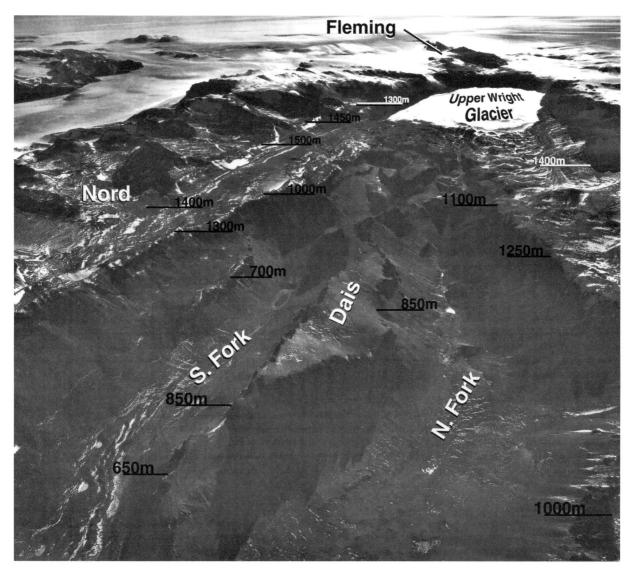

Fig. 5. Westward-directed view of the upper Wright Valley region showing selected features and elevations of slope breaks. Terrain below 1300 m a.s.l. is divisible into three different glacial erosional landscapes: the high landscape (c. 1300–1000 m a.s.l.), the intermediate landscape (c. 1000–850 m a.s.l.), and the low landscape (below c. 850 m a.s.l.). (Source: TMA 541, # 81).

plateau from which rise Mt. Fleming on the south and Shapeless Mountain on the north (Figures 3, 4, and 5). With one or two exceptions, the plateau does not exhibit morphology diagnostic of erosion by wet-based glaciers. One exception is the southern side of the Mt. Fleming ridge beneath the Mt. Fleming peak that exhibits a truncated spur. Below this glacial facet is a moderately inclined, glacially molded surface from 2200 to 2000 meters (m) above sea level (a.s.l.) that is directly overlain by basal till from wet-based ice belonging to the Sirius Group [*Stroeven et al., 1996*;

Stroeven and Prentice, in press]. Another possible exception is whaleback forms on the dolerite plateau west of Tyrol Valley at c. 1800 m a.s.l. [*Marchant et al.*, 1993a]. The glacial interpretation for spur truncation and bench molding is consistent with the southern surface of Mt. Fleming being a low TAM threshold along the inland perimeter of the dry valleys [*Drewry*, 1982].

Cut into the upper Fleming surface is a basin with a floor at c. 1700 m a.s.l. and relief of 250 m referred to as lower Fleming (Figure 3). The lower Fleming basin presents steep lateral walls, a steep backwall, and a parabolic transverse profile. The backwall and floor of lower Fleming show numerous 10–20 meter-high convex bumps, the cumulative effect being a large convexity in the middle of the cavity that is lower Fleming. The upper surfaces of the convexities are commonly dip-slopes of the sandstone. One of the lowest of these surfaces is molded, striated, and directly covered by basal till deposited from wet-based ice [*Stroeven and Prentice*, in press]. Not knowing the difference in age between this till and the latest valley carving, we cannot use the drift to infer that the lower Fleming basin was excavated by ice. Overall, we do not regard lower Fleming morphology as definitively glacial.

Upper highlands. The upper highland valleys of the western Olympus and Asgard Ranges were interpreted as formed through wet-based alpine-glacial erosion in a significantly warmer-than-present climate [*Selby and Wilson*, 1971; *Wilson*, 1973; *Campbell and Claridge*, 1987] (Figures 3 and 5). These valleys were glaciated for till with striated stones crops out in Koenig, Sessrumnir, Nord, and Nibelungen Valleys [*Ackert*, 1988; *Marchant et al.*, 1993a]. Dates on in situ ash overlying the till indicate that the valley floors are older than 15 Ma [*Marchant et al.*, 1993a]. Taking this evidence for great antiquity into account, *Denton et al.* [1993] reinterpreted the valleys as box canyons in the image of the cuestaform landscape of semi-arid platform deserts. *Denton et al.* [1993] suggested that denudation occurred under semiarid conditions in which groundwater and surface water flow were important and sufficiently extensive to supply water necessary to cut the deep trunk valleys. *Gunn and Warren* [1962] had previously suggested that the valleys were "pseudo-cirques" formed in a polar-desert climate by headward-migrating cliff erosion through undercutting at dry-based icefalls. Retaining only a little of the evidence used by *Denton et al.* [1984] to infer glacial overriding, *Denton et al.* [1993] also inferred that this escarpment landscape was completely submerged beneath an overriding ice sheet with near-negligible impact.

The bedrock morphology of the upper highland valleys in the western Olympus and Asgard Ranges is to us largely ambiguous (Figures 3 and 5). We agree that the valley floors are very old and were glaciated. A key question is the extent to which older generation morphologies such as wet-glacial have been eroded by subaerial polar-desert processes. Because sandstone nanoclimate is sufficiently warmer and wetter than the macroclimate that it is life-sustaining [*McKay et al.*,

1993], sandstone surfaces undergo numerous freeze/thaw cycles and exfoliate. We show below that deflation of Olympus Range sandstone interfluves facing central McKelvey Valley has been considerable. We suspect a similar situation in the western Asgard Range. For example, the eastern interfluve in lower Sessrumnir Valley is missing and has apparently been eroded away. In its place is a linear zone of heavily pitted sandstone that separates lower Sessrumnir Valley from Folkvanger Valley. Our preliminary interpretation of the pitted sandstone is that it represents extreme subaerial erosion of the former interfluve. An alternative interpretation for the pitted sandstone terrain is as a glacial meltwater channel system associated with an overriding ice sheet [*Sugden et al.*, 1991].

Lower highlands. East of the upper highland valleys, the lower highland valleys are cut through the Beacon sandstones well into the peneplain sill of Ferrar dolerite and are floored at elevations of 1000–1200 m. We consider Hercules valley in the western Olympus Range and Odin, Heimdall, and J. Sykes valleys in the western Asgard Range as exhibiting morphology that is definitively glacial (Figures 3 and 6).

The corners at the intersection of Hercules and Wright Valley are asymmetrically molded to the east (Figure 7). The western upvalley corner is sharp below a break-of-slope at an elevation of c. 1350 m. The eastern downvalley corner is rounded and offset inside the western corner. Hercules valley itself is parabolic in transverse cross-section. As an assemblage, these morphologic elements are glacial. The glacier configuration indicated is one of wet-based Hercules alpine ice merging with eastward-flowing ice in Wright Valley. The alpine glacier was steered eastward by the trunk glacier and eroded the eastern corner of Hercules valley in order to fit into the trunk valley. In this scenario, the downvalley corner was subjected to relatively high subglacial pressure; the upvalley corner was in a subglacial pressure shadow.

We do not think that any combination of subaerial processes could produce this morphology. The molded pair of corners are made of resistant dolerite. Fluvial erosion of Hercules valley could not produce the asymmetry of the corners, nor the parabolic cross-section. Whereas salt weathering, nivation, and deflation have probably shaped sandstone outcrops in the Olympus Range [e.g., *Selby*, 1971; *Shaw and Healy*, 1977], it is unlikely they are important for dolerite cliff morphology. Besides, the predominant southwesterly wind should have facilitated erosion of the trunk-side of the western corner and sharpening of the wind-aligned but rounded eastern corner.

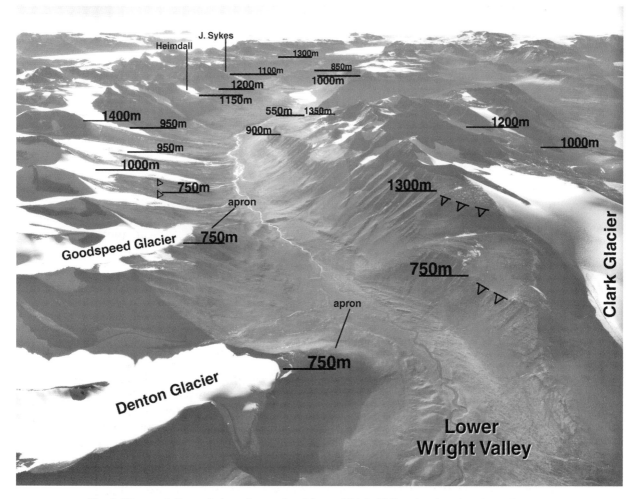

Fig. 6. Westward-directed view of central and lower Wright Valley showing elevations of stepped glacial landscapes. Elevated benches on the north wall in the right foreground (Clark ridge) reflect two separate glacial landscapes above the valley-floor glacial landscape. Elevated benches are also apparent on the north wall in the middle distance (Peleus bench and, on the Clark Glacier side, Theseus bench). The southern wall of Wright Valley features hanging valleys with thresholds that decrease in elevation from 1100 m a.s.l. in the west to 750 m at Denton Glacier in the east. Tributary truncated and molded corners can be traced to an elevation of c. 1400 m. The Denton and Goodspeed aprons are in the left foreground. (Source: TMA 353, #191).

J. Sykes, Odin, and Heimdall valleys, on the opposite side of Wright Valley from Hercules valley, exhibit glacial erosional morphology (Figures 3 and 6). The corners where all three join Wright Valley are asymmetrically molded toward the east. The effect is prominent on a narrow interfluve such as between Odin and Heimdall valleys. Odin valley presents a uniform parabolic cross profile throughout its 6-km length. Both the Heimdall and J. Sykes valleys feature parabolic cross profiles in their lowest sections. Supplemental evidence is provided by till tongues extending from the mouths of both Odin and Heimdall valleys eastward out onto the main Wright Valley slope with upper edges decreasing in elevation to the east. These tills demonstrate that wet-based alpine ice was steered eastward by a trunk glacier, though not necessarily contemporaneous with valley carving.

There is good evidence for fluvial erosion in the upper reaches of the J. Sykes and Heimdall valleys indicating that the lower highland valleys predate glaciation coeval with the molded corners. The Fenrir Valley, a tributary of the Heimdall Valley, for in-

Fig. 7. Stereogram of glacially molded corners at the junction of Hercules valley, western Olympus Range, and Wright Valley. The pair of corners is asymmetric in that the upvalley corner (left) is sharp and hooks east; the downvalley corner (right) is rounded and cut back further from the trunk valley. The transverse profile of lower Hercules valley is parabolic. The explanation of symbols is on Figure 3. Figure 4 shows the location. (Source: TMA 2473, #30, 31).

stance, shows a winding V-shaped channel as well as spurs typical of fluvial excavation (Figure 3). We regard the presence of the small tributary valleys in the J. Sykes and Heimdall drainages and their size increase from head to the valley outlets as a fluvial pattern.

West of J. Sykes and Hercules valleys, the strikingly continuous sandstone slopes that rise above and ring the Labyrinth are difficult to interpret (Figure 3). The western steep slopes truncate the floors of the Asgard

and Olympus upper highland valleys from the Airdevronsix Icefalls east to Sessrumnir Valley at elevations as high as 1550 m (Figure 5). This steep slope might be glacially carved by ice flowing into the North and South Forks of Wright Valley as suggested by *Bull et al.* [1962]. It also might be a cuesta-like escarpment dating to the early Tertiary [*Denton et al.,* 1993].

A variety of glacial hypotheses [e.g., *Gunn and*

Warren, 1962; *Cotton,* 1966; *Shaw and Healy,* 1977; *Denton et al.,* 1984] and non-glacial hypotheses [*Smith,* 1965; *Selby and Wilson,* 1971] have been proposed for the Labyrinth. We regard the two 150 meter-high troughs at the eastern end of the Labyrinth as glacial because of the parabolic transverse cross-profiles. Genetic interpretation for the lower relief channels and depressions in the western Labyrinth is beyond the scope of this paper.

Trunk valley. We agree with *Bull et al.* [1962] that the principal surface of Dais, the elongate mountain fragment separating the North and South Forks, is a remnant of an old glacial valley floor (Figure 3). We agree with *Denton et al.* [1993] that the smooth undulating surface is glacially scoured (Figure 5). Below the principal surface at c. 850 m a.s.l. are three isolated benches at elevations between 650 and 700 m. One is on the south wall of the South Fork; the other two are off the southwestern and southeastern ends of Dais. We think that all were carved glacially given their scoured surfaces.

The North and South Forks of upper Wright Valley exhibit strong glacial character (Figures 3 and 5). Both exhibit steep parabolic transverse profiles. The breaks-of-slope above these narrow valleys are formed in both granite and dolerite and are sharp between the obvious gullies. The headwalls of the North and South Forks exhibit steps upvalley into the Labyrinth (Figure 5).

East of Dais, Wright Valley doubles in width and exhibits glacial character (Figures 3 and 6). One glacial trait is the overdeepened longitudinal profile with the lowest portion being attained in the Lake Vanda basin (Figure 3). Another glacial trait is hanging tributary valleys. Both the Hercules and J. Sykes valleys are hanging with a sharp break-of-slope at an elevation of 1000 m.

Glacial landscapes. All of the glacial features could not be produced in a single glacial phase. We tentatively subdivide the glacial morphology into three different glacial landscapes, namely, high, intermediate, and low. Their age decreases with elevation. The bench with accordant tributaries between 1350 m and 1000 m a.s.l. on both valley sides constitutes a single high glacial landscape formed during a high glacial phase. Glacial incision into the trunk valley below 1000 m reflects a separate younger glaciation. It is unlikely that the 1000 m bench could have been molded at the same time as a glacial wall was cut into it. In other words, the 1000 m a.s.l. bench is not likely to survive thick wet-based glaciation floored well below 1000 m. In the latter

situation, obstructions such as the bench would be streamlined, if not removed, bringing the trough profile closer to the equilibrium parabolic shape [*Harbor,* 1992]. Incision into part of the bench would be unlikely during deglaciation from the thicker ice conditions that molded the top of the bench and tributary corners. The valley-glacier configuration must have changed considerably between the time it scoured the elevated bench to the time it incised the adjacent wall. This is explained by a separate phase of glaciation.

Similarly, we tentatively infer that the surface of Dais represents an intermediate glacial landscape formed during an intermediate glacial phase. The floor of the North and South Forks constitutes a low glacial landscape cut during a low glacial phase. We tentatively assign the benches at 650–700 m a.s.l. to another low landscape. As these benches are associated with the South Fork and were not observed elsewhere, their incision could reflect ice-drainage capture by the N Fork. Hence they are consistent with a very local change in valley-glacier configuration. We revisit all the landscapes and associated glacial phases in the discussion.

Central-Western McKelvey Valley

Highlands. We regard the two lower highland valleys in the western Olympus Range, referred to here as Aeolus and Jason valleys, as largely glacial in character (Figure 3). Both valleys exhibit molded corners at their junctions with McKelvey Valley indicating eastward ice-flow in McKelvey Valley (Figure 8). The upper limit of the molded corners is at an elevation of c. 1350 m. Both valleys exhibit parabolic transverse cross sections. The floor of Jason valley continues out onto a bench, the Jason bench, at 1150 m a.s.l. that extends around the corner of the Olympus Range into Bull Pass.

On the northeastern shoulder of Mt. Aeolus, two dolerite dikes cut the Beacon sandstone and reveal extensive sandstone weathering and eolian transport (Figures 8 and 9). The dolerite dikes rise high over the less resistant sandstone treads and less so over the more resistant sandstone risers. The sandstone surface therefore exhibits a stepped morphology with 5–20 m of local relief. We infer weathering in the current polar-desert environment. That this weathering is not older is illustrated by a patch of colluvium on the northern shoulder of Mt. Aeolus which has a smooth surface. The colluvium is much thinner than the depth of the depression in the adjacent exposed sandstone

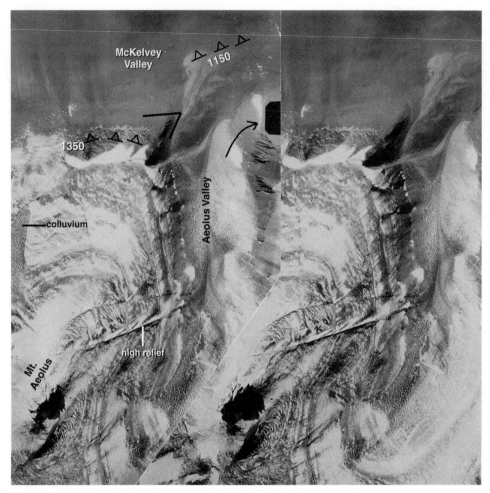

Fig. 8. Stereogram showing glacially molded corners of Aeolus valley at the intersection with McKelvey Valley. The upvalley (left) corner is angular; the downvalley corner is rounded and offset inside the upvalley corner. Erosion of the sandstone surface on the northeastern shoulder of Mt. Aeolus relative to dolerite dikes is apparent and diagrammed in Figure 9. (Source: TMA 2472, #228, 229).

area to the east. Hence the local sandstone relief is much lower under the colluvium than in the bare-sandstone areas on the mountain. Similar relationships are found at other locations in the western Olympus Range, such as on the slopes of Mt. Electra. The implication is that the weathering responsible for the relief is post-colluvium in age. This does not negate the conclusion by *Sugden et al.* [1995] that the large-scale cuestaform morphology, on the scale of the planation surfaces recognized by these authors, may be very old and related to landscape development prior to ice-sheet development.

The morphology of the Olympus Range west of Mt. Aeolus above 1300 m a.s.l., which includes Boreas, Dido, and Circe-Rude valleys, is not distinctly glacial

with one exception. At the western end of the Olympus Range in the broad valley referred to here as the Circe-Rude valley, we observed a 10 meter-high arcuate moraine traceable to a backwall that bordered an overdeepened basin in bedrock. This assemblage is a remnant of wet-based alpine glaciation. Additionally, the far western embayment of Circe-Rude valley exhibits left- and right-lateral glacial slopes indicating ice-flow from over the headwall fronting the Shapeless Mountain plateau down into western McKelvey Valley (Figure 3).

We regard the dolerite-capped Rude Spur and Insel Range fragments as glacial surfaces. Their planed morphology is consistent with glacial scouring. Further, till with numerous striated stones indicating

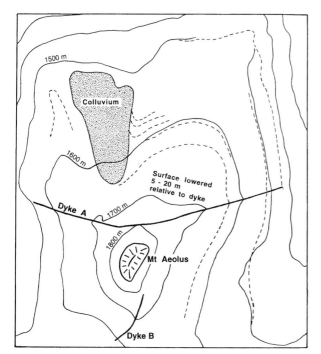

Fig. 9. Sketch of Mt. Aeolus, western Olympus Range, showing sandstone erosion relative to dolerite dikes. Most of the area is visible in Figure 8. Dashed lines provide supplementary contours that reveal 5–20 m of sandstone deflation in a polar-desert environment. Dashed contours on the Aeolus valley wall (right) reveal a significant concavity on the sandstone slope visible in Figure 8. Dashed contours east of the colluvium patch show a significant depression in the sandstone surface that is not present in the surface of the thin colluvium layer.

deposition from wet-based ice crops out on the eastern block of the Insel Range [*Calkin*, 1971] and also on Rude Spur at c. 1100 m a.s.l. (Figure 3). We suggest that the surfaces of these mountain blocks were glacially carved and reflect old glacial-valley floors.

Trunk valley. McKelvey Valley is a glacial valley for three reasons. The tributary valleys on the north side of the Olympus Range hang c. 400–600 m above the floor of McKelvey Valley. McKelvey Valley features a broad parabolic transverse cross section with matching breaks-of-slope on north and south walls at elevations between 1100 and 1250 m. The valley is likely overdeepened in its western end (Figure 3). McKelvey Valley is also in line with a tributary, Circe-Rude valley, cut by inflowing TAM ice.

Glacial landscapes. The glacial features are not compatible with a single glacial phase. We tentatively subdivide the glacial morphology into high and intermediate glacial landscapes. The Jason bench with accordant lower highland valleys, the alpine glacial features in Circe-Rude valley, and the tops of Rude Spur and the Insel Range are remnants of the high glacial landscape. The floor of McKelvey Valley reflects a separate younger phase of glacial erosion and is tentatively assigned to the intermediate glacial landscape.

Balham Valley

The plateau. Little is known about the extent of glacial erosion on the Shapeless Mountain plateau between elevations of 2500 and 1900 m. We mapped glacially scoured terrain under Sirius Group till over some undulations in this surface as well as small-scale alpine glacial erosion northeast of the Shapeless peak (Figure 3). Hence the high plateau was glaciated but the erosional impact of that glaciation is largely unknown.

Highlands. Two highland tributary valleys connect the TAM crest to the western end of Balham Valley (Figure 3). The northern, broader, and longer tributary heads into the saddle between Shapeless Mountain and Mt. Bastion. The southern tributary plunges from 1700 m a.s.l. on the Shapeless plateau west of Rude Spur to the bottom of Balham Valley at 800 m. Both tributary valleys are cut into the hummocky surface of a dolerite sill with a sharp break-of-slope at the intersection. The dolerite surface is extensive around Shapeless Mountain; small remnants of it were also observed along the divide in the Apocalypse Peaks north of the northern tributary.

Both tributary valleys exhibit a valley-in-valley transverse profile (Figure 3). The upper edge of the outer, older valley drops from c. 2050 m a.s.l. northeast of Shapeless Mountain to 1750 m a.s.l. on the east. The break-of-slope at the upper edge of the inner, younger valley lowers from 1700 m a.s.l. on the west to 1400 m a.s.l. on the east. From the curvilinear valley-side remnants, we infer that the older valley had a parabolic transverse profile with over 350 m of relief. It is likely to have been carved glacially. The inner valley of the northern tributary features 200–300 m of relief, a parabolic transverse section, a stepped longitudinal profile and streamlined bedrock on its floor. This assemblage of northern tributary features is of glacial origin.

The inner valley of the northern tributary is truncated at 1250 m a.s.l. by the western wall of Balham Valley leaving it about 400 m above the western end of Balham Valley. The longitudinal axis of the bottom of

Balham Valley is aligned with the southern tributary valley that leads back up to the plateau between Shapeless Mountain and Wright Upper Glacier. We infer that the southern tributary valley was glacially eroded based on its similarity with the northern tributary. Given that the southern tributary crosscuts the northern tributary, we infer that the last significant ice-flow into western Balham Valley from the plateau region came through the southern tributary.

Trunk valley. The bottom few hundred meters of Balham Valley exhibit a parabolic transverse profile that increases in width from west to east. The lateral walls of Balham Valley, particularly the northern wall, exhibit distinct breaks-of-slope between 1000 and 1200 m a.s.l. If the valley-floor sediment surface parallels the underlying bedrock topography, Balham Valley is also overdeepened by 150 m, the deepest segment being just downvalley from the junction of the tributaries.

Glacial landscapes. We tentatively postulate two to three separate phases of glacial downcutting for Balham Valley. In the trunk valley, elevated glacial surfaces at c. 1200 m a.s.l. are widespread, very likely contemporaneous, and so tentatively grouped into the high glacial landscape. Both tributary valleys in the Balham highlands might correlate to this high landscape. Alternatively, only the lower of the two tributaries might correlate to the high landscape. This alternative implies that the higher tributary is an older glacial landscape.

We tentatively regard the landscape incised below 1100 m in the trunk valley as representing younger glacial incision. Our rationale is the same as used in upper Wright Valley. The blocky remnants of the 1300–1100 meter-high glacial landscape could not be shaped and incised during a single phase of wet-based glaciation.

Barwick Valley

Highlands. The three large upper highland valleys in the Apocalypse Peaks on the southern slope of Barwick Valley exhibit a parabolic transverse profile (Figure 3). The western valley is typical with its broad floor at 1400–1600 m a.s.l. and stepped longitudinal profile. The corners where these valleys intersect the slope of Barwick Valley do not exhibit molding. On the basis of gross morphology, we infer that these valleys were glacially excavated. The high Apocalypse valleys are all hanging with distinct thresholds at elevations of c. 1300 m.

Below and downvalley from the upper highland valleys is a prominent bench in the peneplain dolerite sill at an elevation of c. 1200 m. Other peneplain-sill benches at c 1200 m a.s.l. occur north of Lake Vashka (Figure 10). We think that these surfaces are remnants of an older valley floor. Further, this floor was glacially eroded based on vertical alignment with the top of the eastern Insel Range inferred above to be glacially eroded (Figure 3).

Valley-side benches/tributaries. Barwick Valley features a glacially produced valley-side bench from above Lake Vashka on the west to the eastern end of the valley, hereafter, the Vashka bench (Figures 3 and 10). Two tributary valleys that are accordant with the Vashka bench incorporate glacially streamlined dolerite knobs (Figure 10). The eastern tributary exhibits glacially molded corners where it joins the trunk valley. Both valleys head into passes over the glacially molded drainage divide between Barwick Valley and the Victoria Upper Glacier drainage. The Vashka bench geometry is aligned with the tributary valley confluences. Downvalley from Lake Vashka, the bench widens considerably and the single slope-break splits into two that lower to 800 m and 700 m a.s.l. at the mouth of Barwick Valley (Figure 10). The Vashka bench features all imply that considerable wet-based ice flowed south over the divide with the Victoria Upper Glacier and joined eastward-flowing trunk-valley ice at the molded corners.

The Vashka-bench morphology is replicated upvalley north of Webb Lake where short valleys, 200 m high, are carved into a sandstone wall, the Fortress (Figure 3). The valleys exhibit parabolic profiles and irregular floors at about 1100 m a.s.l. The interfluves of two of the valleys exhibit eastward molded corners. Because the divide between the short valleys and the Victoria Upper Glacier drainage is glacially rounded, we infer considerable ice influx from the Victoria Upper Glacier drainage. These valleys are accordant with the valley-side Fortress bench at an elevation of c. 1050 m.

Across from the Fortress bench is the Bastion tributary valley floored at 1400 m a.s.l. below Mt. Bastion of the Willett Range. Bastion valley exhibits corners glacially molded to the east and a parabolic cross profile indicating erosion by wet-based ice (Figure 3). Ice flux into Bastion valley was likely from the Willet Range. Adjacent to Bastion valley on the east is a shallow valley occupied by the Haselton Icefall. Both the Bastion and Haselton tributary valleys hang c. 300 m above the floor of Barwick Valley with thresholds at

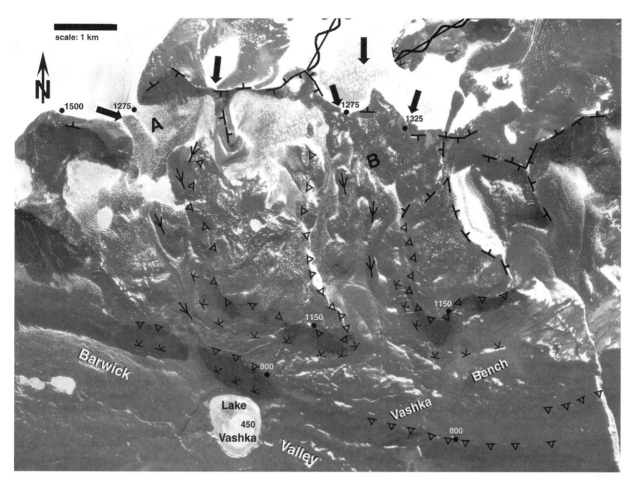

Fig. 10. Vertical view of ice-carved tributary valleys merging into the valley-side Vashka bench along the northern wall of lower Barwick Valley. Location is shown in Figure 4. Symbol explanation is on Figure 3. The two tributaries, A and B, head at gaps in the divide above the bench and contain streamlined dolerite knobs. The bench widens from west to east in alignment with the channel confluences. The bench heads into tributary A and widens at the confluence with tributary B. Tributary B features glacially molded asymmetric corners. We infer that ice from the accumulation area of the Victoria Upper Glacier overtopped the 1300 m a.s.l. divide and merged with eastward-flowing ice in Barwick Valley. The bench was formed in order to accommodate the influx of tributary ice. (Source: TMA 3056, #227).

c. 1000 m a.s.l. There is a distinct break-of-slope at 1350 m above the molded corner of eastern Bastion valley that wraps around into the trunk valley. We consider it an upper bound for glacial erosion associated with this tributary.

Trunk valley. We infer that the trunk valley below both the Vashka and Fortress benches is glacially eroded. The wall which truncates those benches is steep and straight at Lake Vashka implying glacial erosion. The break-of-slope at the edge of the benches is nearly continuous from the Vashka bench upvalley to the Fortress bench. The hanging Bastion and Haselton valleys on the south complement the hanging bench on the north. Such features result in a parabolic transverse profile.

Glacial landscapes. We tentatively suggest that Barwick Valley exhibits three different glacial landscapes separated by two erosional boundaries. It is unlikely that thick wet-based ice could simultaneously erode the Vashka and Fortress benches as well as the Barwick valley floor. Hence we assign the benches to the intermediate glacial landscape and the trunk-valley

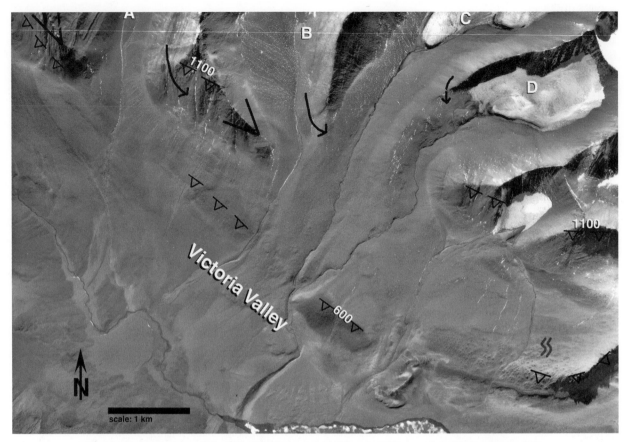

Fig. 11. Vertical view of the Vida bench in central Victoria Valley and molded corners of St. John's Range tributary valleys. Figure 4 shows the location. Symbols explained in Figure 3. The bench widens from west to east. The junctions between alpine valleys A-D and Victoria Valley are asymmetrically molded eastward. The western (upvalley) corner of the valley A junction is sharp; the eastern corner is cut back further from Victoria Valley and rounded. Alpine valley C has asymmetric corners with offset. The bench terminates at an elevation of 650 m. We infer that bench carving and tributary corner asymmetry reflect eastward steering of tributary ice from the four alpine valleys as it merged with trunk-valley ice. *Calkin* [1971, p.399] inferred coalesence between cirque-fed glaciers from the tributary valleys and Victoria Upper Glacier. (Source: TMA 2470, # 167).

floor to the low glacial landscape. As the break-of-slope between those valley-side benches is nearly continuous, we favor correlation of the benches. We place the Bastion and Haselton valleys into the intermediate landscape given comparable floor elevations.

The peneplain-sill bench in the eastern valley with its brink at 1150 m a.s.l. is sufficiently prominent that we assign it to a high glacial landscape. This means that the 1150 m break-of-slope in the eastern valley is the upper erosional boundary for the intermediate landscape. We see no difficulty in correlating this high glacial landscape with the upper highland Apocalypse

valleys. As trunk valleys are characterized by hanging tributaries, the 1350-m brink of the Apocalypse upper highland valleys is consistent with one phase of glaciation.

Upper Victoria Valley

Valley-side benches/tributaries. In the upper Victoria Valley at its junction with Barwick Valley, there are three valley-side benches with tributaries that exhibit glacial features. One, north of Lake Vida at c. 700 m a.s.l. and referred to here as the Vida bench, is glacially cut as demonstrated by the glacially molded corners of the accordant tributary valleys (Figures 3 and 11). Another, across the valley floor on the eastern slope of the Insel Range at c. 650 m, extends west into Barwick Valley and eastward down Victoria Valley. The third bench is continuous with the Vashka bench and is part of the glacially molded upvalley corner at the intersection of Barwick and upper Victoria Valley (Figure 3). The southern downvalley corner of the Barwick-Victoria Valley junction, above the 600 m bench of the eastern Insel Range, is rounded and offset inside the upvalley corner.

Upper Victoria Valley north of Victoria Upper Lake features hanging valleys with molded corners as well as a parabolic transverse profile (Figure 3). A tributary valley in the St. John's Range between the snout of Victoria Upper Glacier and Lanyon Peak exhibits corners glacially molded in the downvalley direction. The eastern wall of Victoria Valley shows scouring and straight facets surmounted by a distinct break-of-slope just below 1300 m a.s.l. The tributary on the western side of Victoria Valley, southeast of Mt. Leland and north of Sponsor's Peak, shows glacially molded corners in the downvalley direction. The back wall of this tributary valley is breached and glacially molded along a major saddle. Glacial facets cut into this saddle indicate that wet-based ice altered it. Both tributaries hang above the trunk valley with brinks at 750 m.

Trunk valley. The trunk valley below the valley-side benches appears strongly glacial (Figure 3). Evidence is provided by the glacially molded corners at the junction of Barwick and Victoria Valleys below the valley-side benches. The upvalley corner is a spur that projects below the c. 700 m a.s.l. benches. Additionally, the Vida bench is truncated and hangs c. 200 m above the trunk-valley floor.

Further upvalley, two major tributary valleys merge to form the upper Victoria Valley. The western tributary is the more incised and drains a larger accumulation area than the eastern tributary. The corners at the junction between the western tributary valley and the trunk valley are asymmetrically molded in the downvalley direction. Further evidence for the glacial carving of the western tributary includes its steep, linear northern wall and asymmetric parabolic transverse profile.

Glacial walls are not apparent along the sides of the major eastern tributary that flows out of the Mt. Mahoney region of the Clare Range. However, signs of glacial smoothing and scouring were observed along the headwall that drains ice from the Clare Range into the Victoria Upper Glacier. We estimate that the main pass through the Clare Range above the eastern tributary is at an elevation of 1600 m a.s.l. which is about 400 m above the MacKay Glacier surface to the north (Figure 1).

Glacial landscapes. We subdivide glacial features within upper Victoria Valley north of its junction with Barwick Valley into two different glacial landscapes. The intermediate glacial landscape incorporates valley terrain from the molded corners of the highest tributary valley at 1250 m a.s.l. to the hanging valley-side benches. Preservation of the benches and their incision are not compatible with a single phase of wet-based glaciation. Hence we propose that the trunk valley constitutes a low glacial landscape below the major break-of-slope at an elevation of 700–800 m.

Bull Pass-Eastern McKelvey Valley

Lower highlands. There is good morphologic evidence for wet-based glaciation throughout Bull Pass in the lower highlands between elevations of 1000 and 1300 m. The primary evidence for glacial shaping is molded corners at the junctions between Bull Pass and its tributary valleys indicative of ice-flow to the south toward Wright Valley. The tributaries are Orestes Valley and Sandy (Glacier) valley on the east side as well as Jason valley on the northwestern corner of Bull Pass (Figure 3). The northern corner of the Orestes Valley-Bull Pass junction is sharp; the southern corner is smooth and rounded (Figure 12). Orestes Valley also exhibits a parabolic transverse profile. The corners of Sandy and Jason valleys are similarly molded. We infer from the combined evidence that wet-based alpine glaciers merged with trunk-valley ice flowing south through Bull Pass from central McKelvey Valley into Wright Valley.

Trunk valley. Evidence for the glacial deepening of the floor of Bull Pass is provided by the tributary valleys hanging above it by c. 400 m. Additionally, the corners of the Bull Pass-Wright Valley junction are strongly molded below 1000 m a.s.l. (Figure 13). Those molded corners indicate that Bull Pass ice merged with Wright Valley ice flowing eastward down Wright Valley. Bull Pass also has a parabolic transverse profile.

Topographically, eastern McKelvey Valley is more closely associated with Bull Pass than with central McKelvey Valley (Figure 3). The floor of eastern McKelvey Valley, at 650 m a.s.l. is about 150 m below that of central McKelvey Valley but at about the same elevation as that of Bull Pass. The western glacial wall of Bull Pass continues into McKelvey

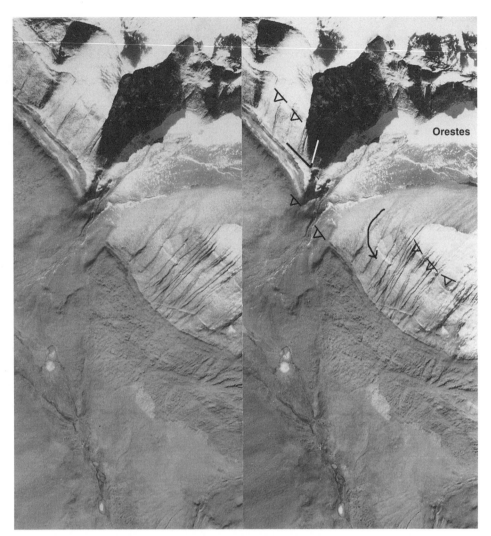

Fig. 12. Stereogram of glacially molded corners at the intersection of Orestes Valley and Bull Pass. Molding is asymmetric with the upvalley corner (upper left) sharp and the downvalley corner (lower right) rounded indicating confluent alpine and trunk-valley ice flowing downvalley to the southeast. Symbols are shown in Figure 3. Location is shown in Figure 4. (Source: TMA 2472, #224, 225).

Valley crosscutting the southern less-steep wall of central McKelvey Valley below c. 1000 m a.s.l. essentially dividing central from eastern McKelvey Valley. Coupled with the strong evidence for ice-flow to the south in Bull Pass, we infer that the western slope of eastern McKelvey Valley and Bull Pass below c. 800 m was carved by ice from Victoria Valley.

Glacial landscapes. We tentatively divide glacial evidence within Bull Pass and eastern McKelvey Valley into two separate landscapes. The high landscape incorporates the surface of the Insel Range, the Jason valley bench, and the high rounded shoulder southeast of Sandy valley. The floor and lower slope of

Bull Pass-eastern McKelvey Valley is assigned to the intermediate glacial landscape.

A critical question is the flow-pattern of trunk-valley ice that carved the intermediate landscape. Previously we inferred glacial carving of the intermediate landscape of central McKelvey Valley by eastward-flowing ice presumably from the Shapeless Mountain region. This ice is expected to have eroded much of eastern McKelvey Valley and Bull Pass. However eastern McKelvey Valley morphology also indicates ice-flow from Victoria Valley through and into Bull Pass crosscutting the central McKelvey trough. We tentatively explain the crosscutting mor-

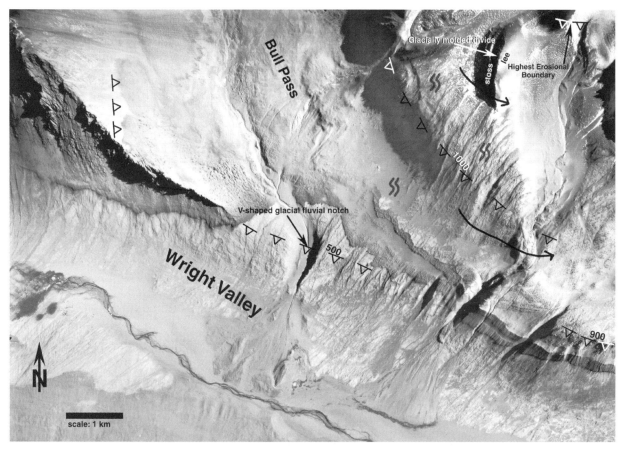

Fig. 13. Oblique view of glacially molded Bull Pass corners at the junction with Wright Valley. Between 1000 m a.s.l. and 500 m, the western (upvalley) corner is sharp and hooks east. The eastern (downvalley) corner is rounded and scoured. We infer that wet-based Bull Pass ice merged with trunk-valley ice flowing east in Wright Valley. In the upper right, the interfluve between Sandy valley and the adjacent unnamed valley is molded on both top and sides such that it hooks eastward, consistent with eastward turning ice flow. This interfluve is associated with a bench (Peleus bench) that is glacially scoured down to the break-of-slope at an elevation of 800–900 m. It is possible that the Peleus bench shows two separate landscapes divisible around the break-of-slope at c. 1000 m. Bull Pass hangs above Wright Valley with a threshold at 500 m. Symbols are shown in Figure 3. (Source: TMA 2194, # 18).

phology of the intermediate landscape as reflecting a shift in ice-surface height during this phase of glaciation. Earlier in the intermediate glacial phase, the ice surface over central McKelvey had greater height than that over central Victoria Valley. Later in the phase, ice over central Victoria Valley became relatively thick driving ice flow down Bull Pass.

Central Victoria and Clark Valleys

Lower highlands. The lower highland valleys in the eastern Olympus Range east of Mt. Cerberus exhibit eastward molded corners (Figures 3, 14, 15, and

16). *Calkin* [1971] previously noted that the alpine valley spurs were truncated and indicated ice-flow to the southeast. Additionally, these valleys have parabolic transverse profiles. The Theseus-west valley (Figure 15) also features a till tongue emanating from it that swings east out onto the southern slope of Clark Valley. The floors of three Olympus Range glacial valleys, Peleus-east, Peleus-west, and Theseus-west, project out into Clark Valley forming a significant bench at c. 1100 m a.s.l. referred to as the Clark bench. Together, the evidence indicates confluent wet-based alpine and trunk-valley ice flowing eastward through central Victoria and Clark Valleys. On the north side of

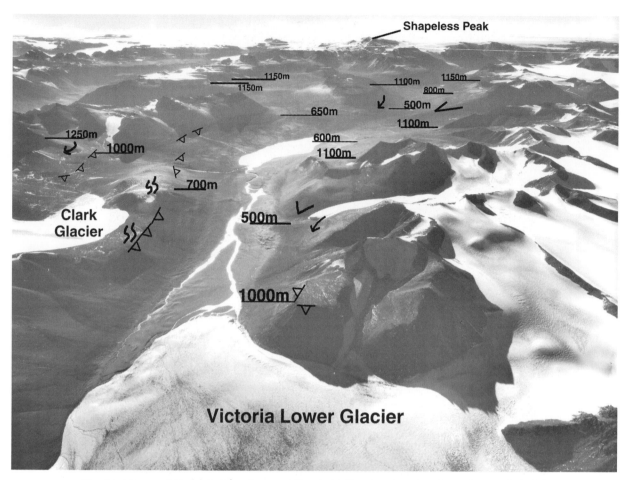

Fig. 14. Westward-directed view of central Victoria Valley showing elevations of stepped glacial landscapes. The Olympus Range which is cut through by Clark Valley and Clark Glacier is on the left. The St. John's Range is to the right (north) of Victoria Valley. In the distance is the flat-topped Insel Range that separates McKelvey Valley on the south from Balham and Barwick Valleys on the north. The high glacial landscape is above 1000 m a.s.l. and represented by: the smooth top of Purgatory Peak in the center foreground, the Theseus bench located above Clark Glacier, and the top of the Insel Range. The intermediate glacial landscape features the floor of Clark Valley, sections of the spur, Robertson Ridge, separating Clark Valley from Victoria Valley, the Vida bench north of Lake Vida (right of photo center), and the floor of McKelvey Valley. The low glacial landscape is the valley bottom up to breaks-of-slope at elevations of 600–700 m. (Source: TMA 353, # 197).

lower Victoria Valley, both Purgatory and Pond Peak feature glacially scoured surfaces at an elevation of c. 1000 m.

Above their glacial sections, the Peleus high valleys exhibit fluvial character. These valley sections exhibit V-shaped transverse profiles and projecting spurs (Figure 16). The fluvial-glacial transition occurs at an elevation of c. 1300 m.

Valley-side benches/tributaries. Clark Valley is parabolic-shaped and opens to both Victoria Valley and Wright Valley (Figure 3). At the west end facing

Victoria Valley, Clark Valley is overdeepened and rises to a valley-mouth threshold at 725 m. The eastern half of the valley apparently slopes gently toward Wright Valley and has a threshold at c. 450 m. We regard Clark Valley and benches at its ends, Robertson Ridge on the west and Clark ridge (our term) on the east as a valley-side feature relative to Victoria and Wright Valley. We regard the major features of this valley-side tributary and its benches as glacially produced.

Clark Valley features hanging glacial tributaries with thresholds at c. 1000 m a.s.l. In the east, its

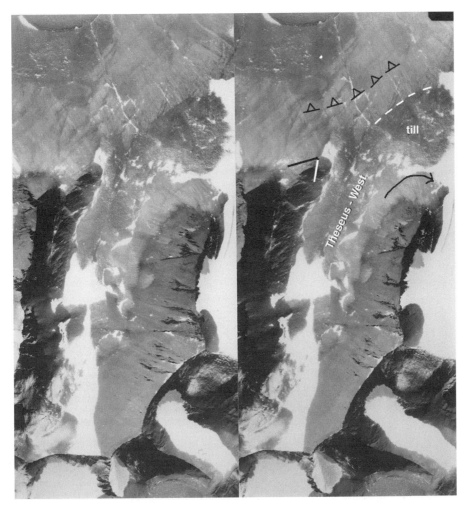

Fig. 15. Stereogram of glacially molded corners on the spurs of the Theseus-west valley in the eastern Olympus Range. The pair of corners is asymmetrically molded to the southeast. The associated alpine-ice flow was steered to the southeast through the Clark Valley into lower Wright Valley. There is also a tongue of unidentified sediment that hooks eastwards out of this tributary. (Source: TMA 2472, # 223, 222.).

northern wall features truncated spurs below Mt. Allen (Figure 3). In the west, the north side is a winding 200–300 meter-high glacial wall. Robertson Ridge is glacially molded up to its top at c. 975 m a.s.l and is cut by two channels with glacially molded walls. The two channels cut into the ridge c. 100 m. Perhaps, the strongest evidence for the glacial carving of Clark Valley is that its corners with Wright Valley are glacially molded (Figure 3). Clark ridge below 750 m a.s.l. is the molded upvalley corner. The eastern downvalley corner of Clark Valley under Mt. Allen is offset inside the Clark ridge. The overdeepening of the northwestern floor of Clark Valley is still further evidence for wet-based glaciation.

On the north side of Victoria Valley, the Vida bench documents wet-based glaciation below 1100 m a.s.l. (Figures 3 and 11). The accordant tributary valleys (labeled B, C, and D in Figure 11) exhibit eastward molded corners. Though the drift is thick, we consider that their transverse profiles tend toward parabolic. The Vida bench slopes from c. 850 m down to a break-of-slope between 750 and 600 m.

Across the trunk-valley floor, the corners at the junction of eastern McKelvey with Victoria Valley are asymmetrically molded eastward (Figure 3). The southeastern corner of the junction is tightly rounded and offset inside the northwestern corner with its linear trend oblique to the trunk valley. Eastern McKelvey

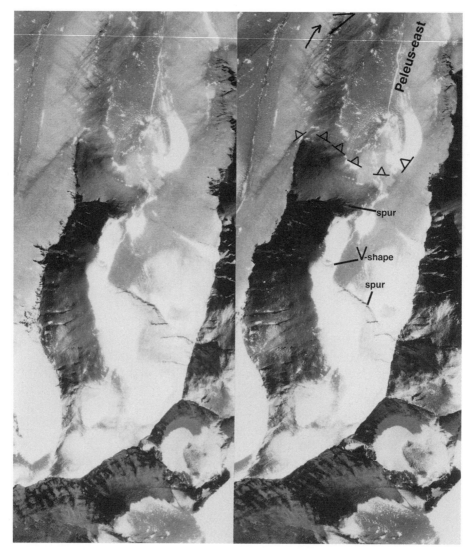

Fig. 16. Stereogram of the Peleus-east valley, eastern Olympus Range, showing fluvial character of the upper basin and glacial character of the lower basin. The upper basin features interlocking, offset spurs and a V-shaped channel. The lower basin exhibits a parabolic profile, eastward molded corners, and a molded right-lateral divide. (Source: TMA 2472, # 223, 224).

also hangs above central Victoria Valley with the break-of-slope at c. 600 m a.s.l. We take this as evidence for significant ice flux out from McKelvey into Victoria Valley with valley floor at an elevation of 600 m.

Trunk valley. There is abundant evidence around central Victoria Valley for glacial incision from 600–750 m a.s.l. to the valley floor at 375–450 m a.s.l. (Figure 3). The valley slope that incises the Vida bench with brink at 650 m a.s.l. has the gentle curve and variable steepness of a glacial slope. The slope below the hanging Clark Valley with brink at 650–700

m a.s.l. looks glacial especially where it sharply truncates Robertson Ridge. Further east, the Packard Valley exhibits eastward molded corners between 500 m and 750 m a.s.l. (Figures 3 and 17). The trunk-valley side of the western corner of the Packard Valley shows classic glacial morphology. Coupled with the parabolic profile in the lower portion of this tributary, the evidence indicates that wet-based alpine ice merged with eastward-flowing trunk-valley ice. In addition to the foregoing, the gross morphologic features such as parabolic transverse profile and 150 m overdeepening indicate glacial carving of the Victoria Valley bottom.

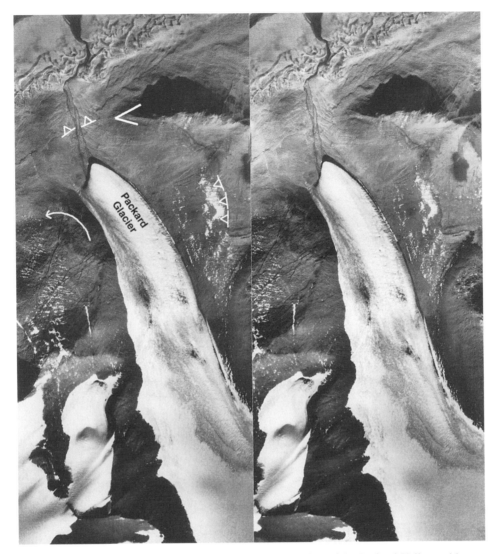

Fig. 17. Stereogram of glacially molded corners at the junction of the Packard Valley and lower Victoria Valley. The pair of corners is asymmetrically molded to the east. The upvalley (left) corner hooks east. The downvalley corner is rounded and offset inside the upvalley corner. The transverse profile of the lower Packard Valley is parabolic. (Source: TMA 2470, # 165, 166).

Glacial landscapes. We tentatively assign the glacial evidence to three different glacial landscapes. The high glacial landscape is represented by the lower highland glacial valleys of the central Olympus Range. The Vida bench as well as the Clark Valley and bordering benches reflect a separate intermediate glacial landscape. It is unlikely that the high glacial landscape and the intermediate glacial landscape were eroded at the same time. We assign the bottom of Victoria Valley below 600–750 m a.s.l. to the low glacial landscape. It is unlikely that the valley-side benches could be eroded coeval with glacial incision to the valley floor.

Central Wright Valley

Lower highlands. There is strong morphologic evidence that the lower highland valleys of the Asgard and Olympus Ranges in central Wright Valley have been modified through glacial erosion. The key evidence is the glacially molded interfluves between many tributary valleys (Figure 3). An example is

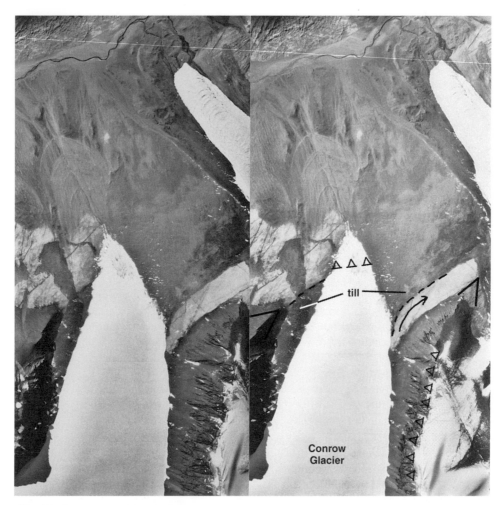

Fig. 18. Stereogram of the glacially molded interfluves of lower Conrow Valley in central Wright Valley. The explanation of symbols is on Figure 3. Location is shown in Figure 4. The inside corners of the interfluve terminations are asymmetrically molded to the east in the downvalley direction (right of the picture). The upvalley corner of the western interfluve is concave toward the Conrow Glacier. The downvalley corner of the eastern interfluve is convex toward the Conrow Glacier. Both basement rock (light-colored) and dolerite are molded. Undifferentiated drift cropping out east of the Conrow Glacier snout can be traced upslope back into the Conrow Valley. (Source TMA 2475, # 60, 61).

provided by the asymmetric interfluves bordering Conrow Valley in the Asgard Range (Figure 18). The molding of the Conrow Valley interfluves is consistent with the presence of a till tongue that swings eastward out of the Conrow Valley (Figure 18). Interfluves bordering Bartley and Meserve Valleys are similarly molded. The corresponding pattern holds for the facing interfluves dividing lower highland valleys in the central Olympus Range across the trunk valley (Figure 3). The interfluves bordering the unnamed valley southeast of Sandy valley are the best example (Figure

13). The asymmetric molding documents the steering of wet-based Asgard and Olympus glaciers eastward as they merged with trunk-valley ice which would also have been wet-based. The glacial morphology of the mouths of the Conrow and Bartley Valleys indicates that the straightened lateral walls in the lower portions of these valleys and parabolic transverse profiles reflect glacial erosion. Upvalley in both Conrow and Bartley Valleys, above c. 1400 m a.s.l., the small tributaries do not exhibit glacial character.

Other evidence for wet-based glaciation in the lower

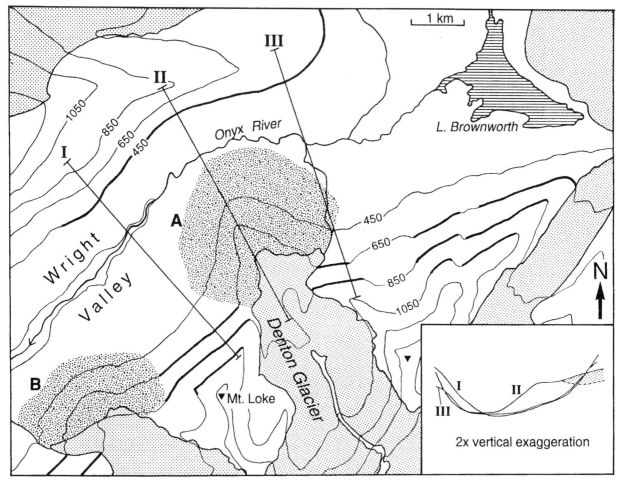

Fig. 19. Sketch map and topographic profiles of the spur-like aprons below Denton and Goodspeed Glaciers in lower Wright Valley. Contours on bare bedrock are thickened.

highlands is glacially scoured benches on the north valley wall. East of Mt. Theseus above the Clark ridge is a glacially scoured bench at an elevation of c. 1300 m (Figure 6). The extensive Olympus Range bench below Mt. Peleus at the corner of Bull Pass and Wright Valley, hereafter the Peleus bench, features scoured basement rock and truncated dolerite spurs at similar elevations. The combined evidence on both sides of the trunk valley documents the former presence of wet-based trunk-valley ice at the level of the lower highlands.

Valley-side benches/tributaries. There are valley-side benches in central Wright Valley that exhibit glacial features. One is the lowest section of the Peleus bench featuring scoured basement (Figure 13). Two others are apparent on the Clark ridge. One at c. 800 m a.s.l. is partially covered by till from wet-based ice

[*Prentice et al.*, 1985; *Hall et al.,* 1993]. This surface is sharply truncated at an elevation of c. 750 m (Figure 6). The other is further east on the molded spur at c. 625 m.

Trunk valley. The hanging valleys of central Wright Valley constitute good evidence for glacial deepening below the valley-side benches. The Conrow, Bartley, and Meserve Valleys hang roughly 700 m above the floor of Wright Valley with thresholds at c. 950 m (Figures 3 and 6). Bull Pass also hangs above Wright Valley. The break-of-slope at the Bull Pass floor threshold is c. 500 m a.s.l. (Figure 12). This threshold and slope below is cut by a V-shaped fluvial canyon that we attribute to glaciofluvial water-flow from retreating ice in Bull Pass.

Critical to the issue of glacial excavation to the floor of Wright Valley is the presence of two spur-like aprons centered below the lips of Goodspeed and Denton Valleys (Figure 6, A and B in Figure 19). The apron below Denton Glacier protrudes much closer towards the valley center than the one below Goodspeed Glacier. These features were interpreted by *Denton et al.* [1993] as remnants of fluvial-valley spurs, indicating that that this part of Wright Valley was fluvially cut down to the present level of its bedrock floor. Previously, *Nichols* [1971] interpreted the aprons as "rock bastions" characteristic of glacial valleys. Critical to the latter interpretation is *Nichols'* [1971] interpretation that they are rock-cored. *Nichols* reported an outcrop of diorite more than 30 m long immediately below Goodspeed Glacier and 15–11 m long outcrops nearby. Below Denton Glacier, *Nichols* described a gneiss outcrop 12 m by 8 m. Considering these rock pieces too big to be boulders, he inferred a rock core. We do not consider the rock below Denton Glacier too big to be a boulder.

Whereas we are unsure whether these features, particularly the Denton apron, are rock-cored, we do think that they are anomalous for a fluvial system. In examination of aerial photos, we have not been able to discern exposed bedrock on the aprons but only on the intervening valley wall. We do not know how close *Nichols'* [1971] rock exposures were to the outer edge of the aprons. What we see is that the Goodspeed and Denton aprons are largely covered by loose deposits, mainly till. Observations from glacial valleys elsewhere are that large obstructions in such valleys almost always show much exposed bedrock. The coexistence of a clean valley wall and a till draped obstruction is anomalous and leads us to think that the aprons are not rock-cored.

We also suggest that the straightened trend of the bedrock wall, where exposed, constrains the interval masked by apron deposits to the same trend. The 600 and 850 m contours are well aligned on bedrock on both sides of Denton Glacier, implying that glacier flow in this section of Wright Valley was once unimpeded by any obstruction at this point. If this interpretation is correct, the apron is a constructional feature.

There is an additional complication with the fluvial spur interpretation. In a normal integrated fluvial system, the spurs constitute the extremities of the interfluves separating tributary valleys. In this sector of Wright Valley, the major interfluves are truncated and flatiron-shaped. The Denton and Goodspeed aprons, on the other hand, are located immediately in front of and below major tributary valleys. This is an anomalous position for a bedrock spur in a fluvial system, but a normal position for a constructional feature built of material supplied from the tributary valley.

Denton et al. [1993] also inferred that a slight protuberance in the north wall of Wright Valley slightly west of and across from the Denton apron was a truncated fluvial spur that was interlocked with the Denton apron. Our examination, however, indicates that this feature is the termination of a broad cuspate indentation in the north wall of the valley that parallels the Denton apron. The valley wall at the axis of this cusp exhibits a curvilinear profile. We interpret the cusp, including the rounded slight protrusion at its western end, as a glacial facet produced as a valley glacier flowed by and was diverted around the Denton apron.

Glacial landscapes. We tentatively assign the glacial features in central Wright Valley to two different glacial landscapes. The high/intermediate landscape covers the glacial benches above an elevation of c. 600 m. This landscape could be multiple. The benches on Clark ridge are consistent with three separate landscapes. The Peleus bench may contain evidence for two landscapes. The low glacial landscape is represented by the lower slopes and floor of the valley.

DISCUSSION

Glacial Erosional Landscapes

Above we assigned glacial bedrock morphologies in the valley sections to different glacial erosional landscapes that are applicable over the study region. The result is regional high, intermediate and low glacial landscapes. These landscapes reflect separate high, intermediate, and low phases of glaciation. Here, we analyze the correlation of the landscapes between valley sections in an effort to evaluate their reliability as well as the characteristics of the implied phases of glaciation.

Intermediate glacial landscape. In our hypothesis, the defining glacial erosional landscape is the intermediate landscape. In the Victoria Valley system, the landscape fits together as follows. Within Barwick Valley we correlated the Fortress bench /Bastion-Haselton valleys at 1300–1000 m a.s.l. to the Vashka bench downvalley at 1200–800 m a.s.l. (Figures 3 and 14). We correlated this landscape to both the Vida bench and the Clark Valley landscape between 1100 and 600 m a.s.l. We assigned most of the slopes of McKelvey and Balham Valleys below

1100 m to the intermediate landscape. Consistent with incorporation of Bastion and Haselton valleys, we correlated the intermediate Balham tributaries as well as western Circe-Rude valley and Rude Spur.

In Wright Valley, we assigned the floor of the intermediate landscape to the surface of Dais at 850 m a.s.l. and the upper bound of the landscape to the break-of-slope truncating Hercules and J. Sykes valleys (Figure 5). In the east, most Bull Pass slopes below the c. 1000 m break-of-slope were assigned to the intermediate landscape. The upper slopes of central and eastern Wright Valley below 1000 m a.s.l. were also classified as intermediate landscape.

An important question is how this intermediate landscape projects west of Dais onto the TAM crest in upper Wright Valley. We speculate that this landscape rises westward onto the easternmost pedestal of the Labyrinth and below most of the upper highland valleys of the Asgard Range which are floored above 1400 m a.s.l. (Figure 3). The intermediate landscape might be represented by the westernmost basins of the Olympus Range floored at c. 1200 m. If ice of this glacial phase flowed into the southern Balham tributary and western Circe-Rude valley, then it flowed into the plateau-bordering western Olympus basins.

High glacial landscape. The high glacial landscape is principally found between c. 1300 m and 1000 m a.s.l. in tributary valleys and on high shoulders of the trunk valleys. In the western Victoria Valley system, we assigned the top surface of the Insel Range, the valley-bench fragments between 1300 and 1150 m a.s.l. on both sides of lower Barwick Valley, and the Aeolus valley to Jason valley bench area of the Olympus Range to this landscape (Figure 3). We assigned the outer channel in the northern Balham tributary with floor sloping from 1900 to 1450 m a.s.l. to this landscape. This yields a reasonable valley-floor profile when projected to 1150 m a.s.l. on the eastern Insel Range. This correlation puts the upper erosional boundary of the high landscape sloping from 2000 m a.s.l. to 1750 m on the south side of the Apocalypse Peaks. In Barwick Valley, we correlated the upper highland valleys of the Apocalypse Peaks to the top of the Insel Range. That correlation fits with our correlation on the south side of the Apocalypse valleys. Because the upper boundary for the high landscape is at the elevation of the main Apocalypse Peaks divide, it is possible that the high north-facing valleys were not cut by alpine glaciers but by outlet glaciers draining the TAM crest.

The lower erosional limit of the high landscape remains c. 1000 m a.s.l. south through Bull Pass and east at the base of the northern tributaries of the central Olympus Range. In Wright Valley, we correlated Olympus and Asgard Range tributary valleys east of Hercules valley to the high landscape. Erosional elevation limits appear well defined between 1000 and 1300 m.

The upper erosional boundary of the high landscape at the transition from TAM crest to trunk valleys is critical for reconstructing high-phase glaciations as well as for recognizing older landscapes. In upper Wright Valley, it is plausible that the upper highland Asgard valleys from Nibelungen Valley west to Sessrumnir Valley are above the high landscape (Figure 3). The floor of Sessrumnir Valley is above 1450 m a.s.l. which is our postulated upper erosional limit for the high landscape at the west end of Balham valley due north.

Low glacial landscape. The low glacial landscape is restricted to the basal portions of Wright, Barwick, and Victoria Valley. In western Wright Valley, we consider the c. 800 m break-of-slope at the edge of Dais as the upper erosional limit for the low landscape. We suggest that the 250 m deep channels in the easternmost Labyrinth were cut during this phase of glaciation and that the top of the channels at 1000 m represents the upper bound of the landscape here. Again, the bottom of these channels at 700 m a.s.l. is well below the 1450 m a.s.l. threshold of the high Asgard valleys directly above indicating that much ice can flow into Wright Valley without impact on the high valleys. In eastern Wright Valley, we correlate the glacial truncation of the Clark ridge below 450 m a.s.l. to this landscape. We date the final overdeepening of Wright Valley and downcutting below the Bull Pass threshold to this glacial phase.

In central Victoria Valley, we take the edge of the Vida bench at c. 650 m a.s.l. as the upper boundary for the low erosional landscape. In lower Barwick Valley, we take the Vashka bench as the equivalent boundary. This elevation is consistent with the upper limit of molded corners fronting the lower Packard valley. Though the latter is an erosional limit, we think that the coeval surface of the Packard Glacier would not have been much higher implying the same for the trunk-glacier surface.

We infer that the upper boundary of the low landscape rises back into upper Victoria Valley through 750 m a.s.l. to the 1000 m break-of-slope in far upper

Victoria Valley. The same pattern is observed in upper Barwick Valley. Given that the floors of Balham and McKelvey Valleys are higher than 600 m a.s.l., we do not think that the low glacial phase accomplished significant erosion there.

Relations to previous hypotheses. Our hypothesis for the glacial landscapes in the northern dry valleys resembles that of *Bull et al.* [1962, 1964]. The principal difference is that they proposed a two-fold zonation of the landscape, carved in their First and Second Glaciations, whereas we propose a three-fold zonation. In their hypothesis, the floors of all valley sections are part of the Second-Glaciation landscape. Alternatively, we put the floors of Wright, Barwick, and Victoria Valley into our low glacial landscape. The difference arises primarily in central Victoria Valley where *Bull et al.* [1962] mapped the entire southern slope of the St. John's Range including the Vida bench and also the entire Clark Valley including the bench ringing it in their First-Glaciation landscape. They also did not use the erosional break between central Victoria Valley and Bull Pass as well as the three erosional surfaces on the southeastern Clark Ridge. We agree with their three erosional landscapes in the Dais region of upper Wright Valley. Because they restricted the threefold landscape division to upper Wright Valley, they inferred that a Third Glaciation was only necessary there. Hence they suggested that the valleys were cut to present level during the Second Glaciation except for the North and South Fork that were subsequently excavated during restricted advance of the upper Wright Glacier in the Third Glaciation. *Calkin et al.* [1970] and *Calkin* [1971] agreed with the concept of multiple elevated glacial surfaces but grouped elevated and valley-floor surfaces into the Insel Glaciation.

Elevated glacial landscapes have been described in the Miller Range in the central Transantarctic Mountains [*Grindley*, 1967]. *Grindley* [1967] mapped three glacial erosion surfaces and inferred progressive glacial incision that he correlated to the Insel Glaciation. Importantly, Miller Range glacial surfaces are not structurally controlled as the bedrock there consists of high-grade metamorphic rocks and granitoids.

Age Constraints

Ash on colluvium and glacial deposits in the upper highland valleys of the Asgard Range indicates that this landscape is older than 15 Ma [*Marchant et al.*, 1993a, 1996]. *Denton et al.* [1993] extended this minimum age to the floor of Wright Valley based on

correlation of inferred similar geomorphic surfaces. Hence the ash dates are very important. The ash primarily occurs within sand wedges developed in the surface of the deposits and so its dates provide minimum ages for the underlying deposits [*Marchant et al.*, 1993a]. Deposits with ash crop out above 1400 m a.s.l. at a minimum of seven locations: one in central Sessrumnir and Nord Valleys, two between central and upper Koenig Valley, and three in Nibelungen Valley. The ashes yield a variety of $^{40}Ar/^{39}Ar$ dates between 10 and 15 Ma.

Though the glacial landscapes proposed here could be younger than 15 Ma, the Asgard ash dates prove that the landscapes could also be middle Miocene or older. An important implication of our vertical subdivision of glacial erosional landscapes is that the age of an elevated glacial landscape provides only a maximum age for a lower landscape. If the Asgard upper valleys are within the high glacial landscape, then the high glacial landscape predates 15 Ma but the lower landscapes need not. The upper highland valleys are also high enough and sufficiently east of the head of upper Wright Valley that they could be above our high glacial landscape in which case our landscapes could all be younger than 15 Ma.

It is difficult to estimate the maximum age of our high glacial landscape. However the oldest age for significant glaciation around Antarctica, c. 34 Ma, represents a meaningful approximation. The oldest glacial sediment drilled in the Ross Sea, from the CIROS 1 core just off the Ferrar Glacier, dates to the early Oligocene [*Hambrey and Barrett*, 1993]. Earliest Oligocene glaciomarine sediment coupled with a significant increase in $\delta^{18}O$ of planktonic and benthic foraminifers in ODP Leg 119 core from the Kerguelen Plateau record an ice sheet of significant proportions [*Barrera and Huber*, 1992; *Zachos et al.*, 1992]. Waterlain tills were recorded in Leg 119 core from the continental shelf of Prydz Bay, East Antarctica, in late Eocene-early Oligocene time [*Hambrey et al.*, 1991]. From the combined evidence, we surmise an earliest Oligocene/latest Eocene age for significant East Antarctic glaciation, c. 34 Ma, using the revised Cenozoic timescale of *Berggren et al.* [1995]. This is our best estimate for the initiation of the high glacial phase.

The oldest sediment on the floor of Wright Valley provides a minimum age for our low glacial landscape. There are two candidate deposits. One is the Jason glaciomarine diamicton from the eastern Lake Vanda basin that indicates that the landscapes are all older than c. 9 ± 1.5 Ma. That date was inferred from the $^{87}Sr/^{86}Sr$ composition of a shell fragment from this deposit [*Prentice et al.*, 1993].

The other candidate deposit is Peleus till that is widespread in central Wright Valley [*Prentice et al.*, 1993] and reflects wet-based East Antarctic Ice Sheet glaciation on a large scale. In our opinion, Peleus till is not constrained in age more closely than between 3.9 and 15.2 Ma and so is less useful than Jason diamicton for constraining the age of the Wright Valley floor. However *Marchant et al.* [1993a] correlated Peleus till to Asgard till in the upper Asgard highland valleys which they inferred dated to 13.6–15.2 Ma. Because Asgard and Peleus till are not continuous but are separated by at least one of our glacial landscapes, we are skeptical of the correlation. Alternatively, Peleus till may be younger than 5.5 ± 0.4 Ma, the maximum age of the Prospect Mesa gravels that directly underlie Peleus till [*Webb*, 1972; *Prentice et al.*, 1993]. Peleus till predates the Hart ash at 378 m a.s.l. in central Wright Valley that has an age of 3.9 ± 0.3 Ma [*Hall et al.*, 1993; *Prentice et al.*, 1993].

Mountain Uplift and Erosion

Recognition of fluvial bedrock morphology can constrain TAM uplift history. We recognize fluvial morphology down to c. 1300 m a.s.l. in the upper Heimdall/Fenrir Valleys of the Asgard Range as well as the Peleus valleys in the Olympus Range (Figures 3 and 16). We suggest that the minimum elevation of the coeval trunk valleys to which these tributaries were graded is 1100 m. We suggest this as the present elevation for the lower limit of definite fluvial morphology. This estimate provides a maximum bound for uplift since initiation of the high glacial phase, c. 34 Ma, because the fluvial morphology formed subaerially. Assuming long-term sea level then was c. 100 m above present [*Haq et al.*, 1988], the northern dry valleys of TAM has not experienced more than 1 km of uplift.

This interpretation differs from that of *Denton et al.* [1993] who inferred the fluvial excavation of the trunk valleys to their present depth. Principal support came from their interpretation of the two spur-like aprons below Goodspeed and Denton Valleys as remnant fluvial bedrock spurs (Figures 6 and 19). We argue on geomorphologic grounds that these spur-like aprons are unlikely to be bedrock spurs. A significant implication of the fluvial-spur interpretation is that uplift has not exceeded the present elevation of the base of these spurs

which is no higher than c. 250 m, the top of the local sediment surface. Another implication of the fluvial-spur interpretation is that the local TAM experienced significant subsidence. Subsidence must be associated with the fluvial-spur interpretation to permit the two marine incursions in Wright Valley dating to 9 ± 1.5 Ma and the early Pliocene.

Based on the abundance of glacial erosional features and questionable evidence for trunk-valley fluvial spurs, our working hypothesis for dry valleys excavation is that erosion of the landscape below c. 1100 m a.s.l., was dominantly by glacial processes. This represents 1100 m of mainly glacial erosion in many trunk valley sections since initiation of the high glacial phase. This magnitude of glacial erosion implies that TAM topography was likely different in the high and intermediate glacial phases than at present. These estimates also afford a maximum estimate for fluvial erosion pre-dating the high glacial phase. We estimate this from relief on the initial high glacial landscape which is 1100 m. The latter estimate derives from the high glacial landscape being c. 1100 m below the average TAM crest elevation, today at 2200 m.

Glaciation Phases and Climates

High phase. During the high phase of glaciation, ice entered the Wright-Victoria Valley system from the upper Victoria, Barwick, Balham/McKelvey, northern Fleming, and southern Fleming accesses. In central Wright Valley east of Bull Pass, the outlet glacier width was c. 8 km whereas, in central Victoria Valley, outlet glacier width was closer to 11 km. Ice exited through both lower Wright and Victoria Valleys, and probably through other valleys. The overall configuration implies an extensive influx of ice from the TAM crest.

High-phase alpine glaciers were pervasive, extensive, and wet-based. Given that these glaciers are cold-based today [*Holdsworth and Bull*, 1970] and that they were closer to sea level during the high glacial phase than at present, they imply a warmer and wetter climate during the high glacial phase than since.

An important question is whether this ice influx was supplied by a TAM ice sheet or requires the presence of an East Antarctic Ice Sheet. We agree with others that linear mountain ice sheets are a likely step in growth of the East Antarctic Ice Sheet [*Bull et al.*, 1962; *Drewry*, 1975; *Mercer*, 1983; *Hambrey and Barrett*, 1993]. One approach to this problem is to estimate how much higher and wider the interior flank of the TAM was during the high glacial phase. The floors of subglacial valleys on the interior flank of the TAM appear to be eroded c. 1700 m below the average TAM crest elevation (Figure 2). This cannot be entirely the result of preglacial fluvial erosion, otherwise the same would have occurred on the seaward flank. Above we inferred that 1100 m is maximum for fluvial erosion on the TAM seaward flank. If this is also a maximum for fluvial erosion of the interior flank, it implies that at least 600 m of interior low-area relief reflects glacial erosion. We speculate that the TAM crest on the interior would have been broader than at present before that amount of glacial erosion occurred. It follows that glacial erosion in upper MacKay Valley to the north and upper Taylor Valley to the south were also significant. Hence the area of TAM crest should have been more extensive meridionally. From this, the probable area of TAM high ground would have been more than sufficient to support a linear mountain ice sheet. Such an ice sheet can be visualized by extending the McMurdo Ice Dome (Figure 1) to the north over interior flank terrain that we suggest was higher during at least the high glacial phase.

Whereas we consider influx from a mountain ice sheet likely during the high glacial phase, we have no expectation regarding a contribution from the East Antarctic Ice Sheet. Further, we cannot distinguish whether high-phase outlet glaciers drained only a local mountain ice sheet or both local ice and the East Antarctic Ice Sheet. Presently, both alternatives are possible.

Intermediate phase. The intermediate glacial landscape indicates ice entry into the Wright-Victoria Valley system from all western access points. We suggest that the Barwick Valley drainage was enlarged relative to today at the expense of the upper Victoria drainage. The tributaries to Barwick, Balham, and McKelvey Valleys were occupied by outlet glaciers draining ice from the TAM crest. Likewise, considerable ice flowed into the Labyrinth area from the Shapeless plateau-Mt. Fleming region and likely from between Mt. Fleming and Tyrol Valley.

As with the high glacial phase, a local TAM ice sheet probably contributed significantly to the intermediate glacial phase. Whether this was the sole ice source or was supplemented by the East Antarctic Ice Sheet is unknown. The evidence for a significant change in ice-flow direction in central Victoria Valley during the intermediate phase is consistent with a transition from a TAM ice sheet regime to the present

regime with a continental ice sheet. The evidence is the cross-cutting bedrock morphology in eastern McKelvey Valley. There, the east-to-west grain of the bedrock shaped by eastward ice-flow out of western McKelvey Valley is crosscut by a north-to-south grain imposed by ice-flow from upper Victoria and Barwick Valleys into eastern McKelvey Valley and Bull Pass. We infer a significant change in ice-surface relief in the Victoria Valley system during the intermediate glacial phase. We speculate that the lowering of the trunk ice-surface slope in western McKelvey and Balham Valleys reflects lowering of a TAM ice sheet through ice-drainage area capture by paleo-Taylor and MacKay Glaciers.

The two-stage morphology of Robertson Ridge at the entrance to Clark Valley is consistent with this proposed ice-dynamics change within the intermediate glacial phase. There, we inferred that channels cut into the ridge top by strong eastward ice flow were crosscut by more restricted flow that eroded below the channel base. The change from stronger to weaker ice-flow into Clark Valley is consistent with the siphoning off of some upper Victoria Valley ice flux into eastern McKelvey Valley and down through Bull Pass.

Wet-based alpine glaciers were also involved in the intermediate glacial phase. Some were located in the St. John's Range and flowed onto the Vida bench. We think that only local snow accumulation contributed to these glaciers. Being wet-based, these small glaciers indicate a warmer-than-present climate during the intermediate glacial phase.

Low phase. The low glaciation phase involved wet-based valley glaciers in Wright, Victoria, and Barwick Valleys. Strong ice flux was likely to the heads of these valleys only. We infer stronger ice flow from the upper Victoria Valley than from upper Barwick Valley because of the greater depth of upper Victoria Valley.

We regard the relatively strong ice flux to the upper Victoria Valley coupled with minimal flux to Balham-McKelvey Valleys as indicating dominance of an East Antarctic Ice Sheet source. A high ice surface in the MacKay Valley to the north is the most plausible way to drive the inferred ice flux to the upper Victoria Valley (Figure 2). Because a high MacKay ice surface is most consistent with the presence of the East Antarctic Ice Sheet, the low glacial phase likely had its primary contribution from the East Antarctic Ice Sheet. The Packard valley provides evidence for minimal wet-based alpine glaciation coeval with East Antarctic Ice Sheet expansion during the low phase. As with the previous phases, the glaciation climate was warmer than at present.

CONCLUSIONS

1) We recognize three glacial erosional landscapes within the northern McMurdo Dry Valleys below an elevation of 1300 m. Strong evidence for glacial erosion within each is the asymmetric molding of paired tributary-valley corners on spurs at intersections with trunk valleys. The glacial features indicate the former confluence of wet-based tributary- and trunk-valley glaciers flowing toward the east to southeast. Hanging benches with glacially molded tributary corners are aligned at two elevations delineating high and intermediate glacial landscapes. Glacially molded corners, hanging valleys, parabolic transverse profiles and overdeepened longitudinal profiles document that deep trunk-valley floors are a separate low glacial landscape.

2) Each glacial landscape appears to be unmodified by subsequent glaciation. This indicates high, intermediate, and low phases of wet-based glaciation that progressively incised the landscape. The continuity of each glacial landscape for over 50 km from Transantarctic Mountain crest toward the ocean indicates significant ice flux from the mountain crest. All of the landscapes predate 9 ± 1.5 Ma, the age of glaciomarine sediment on the floor of central Wright Valley. Because the glacial landscapes decrease in age from high to low elevation, the age of high-landscape deposits may only provide a maximum age for lower landscapes.

3) The present elevation of the lowest definite fluvial landscape, c. 1300 m, constrains dry valleys uplift since fluvial morphology was produced probably prior to 34 Ma. The combined evidence is most consistent with glacial erosion of the stepped glacial landscape below the lower fluvial limit.

4) The TAM crestal area was significantly broader during the high and intermediate glacial phases than at present. It follows that local Transantarctic Mountain ice sheets could account for the high and intermediate glacial phases as well as a lower-than-present East Antarctic Ice Sheet. The East Antarctic Ice Sheet only dominated the low and perhaps later-intermediate glacial phases. The presence of wet-based alpine glaciers during each phase indicates that the coeval climate was significantly warmer and wetter than at present.

Acknowledgments. We thank J. Priscu for inviting us to contribute to and editing this ARS volume. We appreciate the helpful comments of P. Calkin and R. Feldmann that improved the paper. We thank J-C. Thomas (USGS) for the TM scenes. S. Glidden produced Figure 3. S. Williams helped produce the photo figures. M.L.P.'s research was supported by the Office of Polar Programs (OPP-9020975, OPP-9221325, OPP-9627625), National Science Foundation.

REFERENCES

Ackert, R. P., Jr., Surficial Geology and Geomorphology of Njord Valley and Adjacent Areas of the Western Asgard Range, Antarctica: Implications for Late Tertiary Glacial History, M.S. Thesis, University of Maine, 1990.

Aniya, M., and R. Welch, Morphometric analyses of Antarctic cirques from photogrammetric measurements, *Geografiska Annaler*, *63A*, 41–54, 1981.

Augustinus, P. C., and M. J. Selby, Rock slope development in McMurdo oasis, Antarctica, and implications for interpretations of glacial history, *Geografiska Annaler*, *72A* (1), 55–62, 1990.

Barrera, E., and B. T. Huber, Eocene to Oligocene oceanography and temperatures in the Antarctic Indian Ocean, in *The Antarctic Paleoenvironment: a Perspective on Global Change, Part Two*, edited by J. P. Kennett, and D. A. Warnke, pp. 49–66, American Geophysical Union, Washington, D.C., 1993.

Barrett, P. J., The Devonian to Triassic Beacon Supergroup of the Transantarctic Mountains and correlatives in other parts of Antarctica, in *The Geology of Antarctica*, edited by R. J. Tingey, pp. 120–152, Clarendon Press, Oxford, 1991.

Bentley, C. R., and M. B. Giovinetto, Mass balance of Antarctica and sea level change, in *International Conference on the Role of the Polar Regions in Global Change*, edited by G. Weller, C. Wilson, and B. A. B. Severin, pp. 481–488, University of Alaska, Fairbanks, AK, 1991.

Berggren, W. A., D. V. Kent, C. C. Swisher III, and M. P. Aubry, A revised Cenozoic geochronology and chronostratigraphy, in *Geochronology, Time Scales and Global Stratigraphic Correlation*, edited by W.A. Berggren, D. V. Kent, M. P. Aubry, and J. Hardenbol, pp. 129–211, Society for Sedimentary Geology, Tulsa, 1995.

Bull, C., and P. -N. Webb, Some recent developments in the investigation of the glacial history and glaciology of Antarctica, in *Palaeoecology of Africa*, edited by E. M. van Zinderen Bakker, pp. 55–84, A. A. Balkema, Cape Town, Republic of South Africa, 1972.

Bull, C., B. C. McKelvey, and P.-N. Webb, Quaternary glaciations in southern Victoria Land, Antarctica, *Journal of Glaciology*, *4*, 63–78, 1962.

Bull, C., B. C. McKelvey, and P. N. Webb, Glacial benches in south Victoria Land, *Journal of Glaciology*, *5* (37), 131–133, 1964.

Calkin, P. E., Glacial geology of the Mount Gran area,

southern Victoria Land, Antarctica, *Geological Society of America Bulletin*, 75, 1031–1036, 1964.

Calkin, P. E., Glacial geology of the Victoria Valley system, southern Victoria Land, Antarctica, in *Antarctic Snow and Ice Studies II,*, edited by A. P. Crary, pp. 363–412, American Geophysical Union, Washington, D.C., 1971.

Calkin, P. E., Subglacial geomorphology surrounding the ice-free valleys of southern Victoria Land, Antarctica, *Journal of Glaciology*, 13, 415–429, 1974b.

Calkin, P. E., R. E. Behling, and C. Bull, Glacial history of Wright Valley, Southern Victoria Land, Antarctica, *Antarctic Journal of the United States*, 5 (1), 22–27, 1970.

Campbell, I. B., and G. G. C. Claridge, *Antarctica: Soils, Weathering Processes and Environment*, 368 pp., Elsevier, Amsterdam, 1987.

Cotton, C. A., *Climatic Accidents In Landscape Making*, 354 pp., Whitcombe and Tombs, Christchurch, NZ, 1942.

David, T. W. E., and H. E. Priestly, Glaciology, physiography, stratigraphy, and tectonic geology of South Victoria Land, in *British Antarctic Expedition 1907-1909 Reports on the Scientific Investigations, Geology*, pp. 319, Heinemann, London, 1914.

Denton, G. H., M. L. Prentice, and L. H. Burckle, Cenozoic history of the Antarctic ice sheet, in *The Geology of Antarctica*, edited by R. J. Tingey, pp. 365–433, Clarendon Press, Oxford, 1991.

Denton, G. H., M. L. Prentice, D. E. Kellogg, and T. B. Kellogg, Late Tertiary history of the Antarctic ice sheet: Evidence from the Dry Valleys, *Geology*, 12 (5), 263–267, 1984.

Denton, G. H., D. E. Sugden, D. R. Marchant, B. L. Hall, and T. I. Wilch, East Antarctic ice sheet sensitivity to Pliocene climatic change from a dry valleys perspective, *Geografiska Annaler*, 75A (4), 155–204, 1993.

Drewry, D. J., Initiation and growth of the East Antarctic Ice Sheet, *Journal of Geological Society of London*, 131, 255–273, 1975.

Drewry, D. J., Ice flow, bedrock, and geothermal studies from radio-echo sounding inland of McMurdo Sound, Antarctica, in *Antarctic Geoscience*, edited by C. Craddock, pp. 977–983, The University of Wisconsin Press, Madison, 1982.

Drewry, D. J., *Glacial Geologic Processes*, 276 pp., Edward Arnold, London, 1986.

Dyke, A. S., Landscapes of cold-centred Late Wisconsinan ice caps, Arctic Canada, *Progress in Physical Geography*, 17, 223–247, 1993.

Embleton, C., and C. A. M. King, *Glacial and Periglacial Geomorphology*, 608 pp., St. Martin's Press, New York, 1968.

Findlay, R. H., D. N. B. Skinner, and D. Craw, Lithostratigraphy and structure of the Koettlitz Group, McMurdo Sound, Antarctica, *New Zealand Journal of Geology and Geophysics*, 27, 513–536, 1984.

Fitzgerald, P. G., The Transantarctic Mountains of Southern Victoria Land: The application of apatite fission track analysis to a rift shoulder uplift, *Tectonics*, 11, 634–662, 1992.

Flint, R. F., *Glacial and Quaternary Geology*, 892 pp., John Wiley and Sons, New York, 1971.

Glasser, N. F., Modeling the effect of topography on ice sheet erosion, Scotland, *Geografiska Annaler*, 77-A (1–2), 67-82, 1995.

Gleadow, A. J. W., and P. G. Fitzgerald, Uplift history of the Transantarctic Mountains: new evidence from fission track dating of basement apatites in the dry Valleys area, southern Victoria Land, *Earth and Planetary Science Letters*, 82, 1–14, 1987.

Gordon, A. L., B. A. Huber, H. H. Hellmer, and A. Field, Deep and bottom water of the Weddell Sea's western rim, *Science*, 262, 95–97, 1993.

Grindley, G. W., The geomorphology of the Miller Range, Transantarctic Mountains, with notes on the glacial history and neotectonics of East Antarctica, *New Zealand Journal of Geology and Geophysics*, 10, 557–598, 1967.

Grindley, G. W., and G. Warren, Stratigraphic nomenclature and correlation in the western part of the Ross Sea, in *Antarctic Geology*, edited by R. J. Adie, pp. 314–333, North Holland Publishing Co., Amsterdam, 1964.

Gunn, B. M., and G. Warren, Geology of Victoria Land between the Mawson and Mulock Glaciers, Antarctica, *New Zealand Geological Survey Bulletin*, 71, 157, 1962.

Hall, B. L., G. H. Denton, D. R. Lux, and J. G. Bockheim, Late Tertiary Antarctic paleoclimate and ice-sheet dynamics inferred from surficial deposits in Wright Valley, *Geografiska Annaler*, 75A (4), 239–267, 1993.

Hambrey, M. J., and P. J. Barrett, Cenozoic sedimentary and climatic record, Ross Sea Region, Antarctica, in *The Antarctic Paleoenvironment: a Perspective on Global Change, Part Two*, edited by J. P. Kennett, and D. A. Warnke, pp. 91–124, American Geophysical Union, Washington, D.C., 1993.

Hambrey, M. J., W. U. Ehrmann, and B. Larsen, Cenozoic glacial record of the Prydz Bay continental shelf, East Antarctica, in *Proceedings of the Ocean Drilling Program, Scientific Results*, edited by J. Barron, and B. Larson et al., pp. 77–132, Ocean Drilling Program, Texas A&M University, College Station, 1991.

Haq, B. U., J. Hardenbol, and P. R. Vail, Mesozoic and Cenozoic chronostratigraphy and eustatic cycles, in *Sea-Level Changes: An Integrated Approach*, edited by C. S. Wilgus, B. S. Hastings, C. G. St. C. Kendall, H. Posamentier, J. V. Wagoner, and C. A. Ross, pp. 71–108, SEPM, Tulsa, OK, 1988.

Harbor, J. M., Numerical modeling of the development of U-shaped valleys by glacial erosion, *Geological Society of America Bulletin*, 104 (10), 1364–1375, 1992.

Hobbs, W. H., *Characteristics of Existing Glaciers*, 301 pp., Macmillan, New York, 1911.

Holdsworth, G., and C. Bull, The flow law of cold ice; Investigations on Meserve Glacier, Antarctica, pp. 204–216, International Assoc. of Sci. Hydrology and Scientific Committee on Antarctic Research, Cambridge, 1970.

Holtedahl, H., Notes on the formation of fjords and fjord-valleys, *Geografiska Annaler*, 49A, 188–203, 1967.

Hughes, T., Numerical reconstruction of paleo ice sheets, in *The Last Great Ice Sheets*, edited by G. H. Denton, and T. Hughes, pp. 221–261, Wiley-Interscience, New York, 1981.

Kleman, J., The palimpsest glacial landscape in northwestern Sweden - Late Weichselian deglaciation forms and traces of older west-centered ice sheets, *Geografiska Annaler*, *74A* (4), 305–325, 1992.

Kleman, J., and I. Borgström, Glacial land forms indicative of a partly frozen bed, *Journal of Glaciology*, *40*, 255–264, 1994.

Kleman, J., and A. P. Stroeven, Preglacial surface remmants and Quaternary glacial regimes in northwestern Sweden, *Geomorphology*, *19*, 35–54, 1997.

Kyle, P. R., D. H. Elliot, and J. F. Sutter, Jurassic Ferrar Supergroup tholeiites from the Transantarctic Mountains, Antarctica, and their relationship to the initial fragmentation of Gondwana, in *Gondwana Five*, edited by M. M. Creswell, and P. Vella, pp. 283–287, Balkema, Rotterdam, 1981.

Marchant, D. R., G. H. Denton, D. E. Sugden, and C. C. Swisher, Miocene glacial stratigraphy and landscape evolution of the western Asgard Range, Antarctica, *Geografiska Annaler*, *75A* (4), 303–330, 1993a.

Marchant, D. R., G. H. Denton, and C. C. Swisher, Miocene-Pliocene-Pleistocene glacial history of Arena Valley, Quatermain Mountains, Antarctica, *Geografiska Annaler*, *75A* (4), 269–302, 1993b.

Marchant, D. R., G. H. Denton, C. Swisher, and N. Potter, Late Cenozoic Antarctic paleoclimate reconstructed from volcanic ashes in the dry valleys region of southern Victoria Land, *Geological Society of America Bulletin*, *108*, 181–194, 1996.

McKay, C. P., J. A. Nienow, M. A. Meyer, and E. I. Friedman, Continuous nanoclimate data (1985-1988) from the Ross desert (McMurdo Dry Valleys) cryptoendolithic microbial ecosystem, in *Antarctic Meteorology and Climatology: Studies Based on Automatic Weather Stations*, edited by D. H. Bromwich, and C. R. Stearns, pp. 201–207, American Geophysical Union, Washington, 1993.

McKelvey, B. C., and P. -N. Webb, Geological investigations in southern Victoria Land, Antarctica: Part 3 - Geology of Wright Valley, *New Zealand Journal of Geology and Geophysics*, *5* (1), 143–162, 1962.

McKelvey, B. C., P. N. Webb, and B. P. Kohn, Stratigraphy of the Taylor and lower Victoria Groups (Beacon Supergroup) between the Mackay Glacier and Boomerang Range, Antarctica, *New Zealand Journal of Geology and Geophysics*, *20*, 813–863, 1977.

Mercer, J. H., Cold glaciers in the central Transantarctic Mountains, Antarctica: Dry ablation areas and subglacial erosion, *Journal of Glaciology*, *10* (59), 319–321, 1971.

Mercer, J. H., Cenozoic glaciation in the Southern Hem isphere, *Annual Reviews of Earth and Planetary Sciences*, *11*, 99–132, 1983.

Nichols, R. L., Geology of Lake Vanda, Wright Valley, South Victoria Land, Antarctica, in *Antarctica Research: The Matthew Fontaine Maury memorial symposium*, edited by H. Wexler, M. J. Rubin, and J. E. Caskey Jr., pp. 47–52, American Geophysical Union, Washington D.C., 1962.

Nichols, R. L., Glacial geology of the Wright Valley, McMurdo Sound, in *Research in the Antarctic*, edited by L. O. Quam, pp. 293–340, American Association for the Advancement of Science, Washington, D.C., 1971.

Oglesby, R. J., A GCM study of Antarctic glaciation, *Climate Dynamics*, *3*, 135–156, 1989.

Paterson, W. S. B., *The Physics of Glaciers*, Pergamon Press, Tarrytown, 1994.

Prentice, M. L., and R. K. Matthews, Tertiary ice sheet dynamics: the snow gun hypothesis, *Journal of Geophysical Research*, *96* (B4), 6811–6827, 1991.

Prentice, M. L., J. G. Bockheim, S. C. Wilson, G. H. Denton, Geologic evidence for pre-late Quaternary east antarctic glaciation of central and eastern Wright Valley, *Antarctic Journal of the United States*, *20*, 61–62, 1985.

Prentice, M. L., J. G. Bockheim, S. C. Wilson, L. H. Burckle, D. A. Hodell, C. Schlüchter, and D. E. Kellogg, Late Neogene Antarctic glacial history: evidence from Central Wright Valley, in *The Antarctic Paleoenvironment: a Perspective on Global Change, Part Two*, edited by J. P. Kennett, and D. A. Warnke, pp. 207–250, American Geophysical Union, Washington, D.C., 1993.

Pyne, A. R., B. L. Ward, A. J. Macpherson, and P. J. Barrett, McMurdo Sound Bathymetry, in *Miscellaneous series no. 62*, New Zealand Oceanographic Institute Chart, 1985.

Selby, M. J., Slopes and their development in an ice-free, arid area of Antartica, *Geografiska Annaler*, *53A(3–4)*, 235–245, 1971.

Selby, M. J., and A. T. Wilson, Possible Tertiary age for some Antarctic cirques, *Nature*, *229*, 623–624, 1971a.

Selby, M. J., and A. T. Wilson, The origin of the Labyrinth, Wright Valley, Antarctica, *Geological Society of America Bulletin*, *82*, 471–476, 1971b.

Sharp, R. P., *Living Ice: Understanding Glaciers and Glaciation*, 225 pp., Cambridge University Press, Cambridge, 1988.

Shaw, J., and T. R. Healy, The formation of the Labyrinth, Wright Valley, Antarctica, *New Zealand Journal of Geology and Geophysics*, *20*, 933–947, 1977a.

Shaw, J., and T. R. Healy, Rectilinear slope formation in Antarctica, *Annals of the Association of American Geographers*, *67*, 46–55, 1977b.

Smith, H. T. U., Anomalous erosional topography in Victoria Land, Antarctica, *Science*, *148*, 941–942.

Stroeven, A. P., and M. L. Prentice, A case for Sirius Group alpine glaciation at Mount Fleming, South Victoria Land, Antarctica: A case against Pliocene East Antarctic Ice Sheet reduction, *Geological Society of America Bulletin*, in press, 1996.

Stroeven, A. P., H. W. Borns Jr., M. L. Prentice, J. L. Fastook, and R. J. Oglesby, Upper Fleming Sirius till: evidence for local glaciation and warmer climates during the Neogene, in *Landscape Evolution in the Ross Sea Area*, edited by F. M. van der Wateren, A. L. L. M. Verbers, and F. Tessensohn, pp. 117–121, Rijks Geologische Dienst, Haarlem, 1994.

Stroeven, A. P., M. L. Prentice, J. Kleman, On marine microfossil transport and pathways in Antarctica during the late Neogene: Evidence from the Sirius Group at Mount Fleming, *Geology*, 24, 727–730, 1996.

Sugden, D. E., and B. S. John, *Glaciers and Landscape*, 376 pp., John Wiley and Sons, New York, NY, 1976.

Sugden, D. E., G. H. Denton, and D. R. Marchant, Subglacial meltwater channel systems and ice sheet overriding, Asgard Range, Antarctica, *Geografiska Annaler, 73A* (2), 109–121, 1991.

Sugden, D. E., G. H. Denton, and D. R. Marchant, Landscape evolution of the dry Valleys, Transantarctic Mountains: Tectonics implications, *Journal of Geophysical Research, 100*, 9949–9967, 1995.

Taylor, G., Physiography and glacial geology of East Antarctica, *Geographical Journal*, 44: 464–465.

Taylor, G., *The Physiography of the McMurdo Sound and Granite Harbour Region: British Antarctic "Terra Nova" Expedition, 1910-1913*, Harrison & Sons, London, 1922.

Turnbull, I. M., A. H. Allibone, P. J. Forsyth, and D. W. Heron, *Geology of the Bull Pass-St. Johns Range Area, Southern Victoria Land, Antarctica, scale 1:50,000*, 52 pp., Institute of Geological & Nuclear Sciences, Lower Hutt, New Zealand, 1994.

Van Loon, H., and D. J. Shea, A survey of the atmospheric elements at the ocean's surface south of 40°S, in *Antarctic Ocean and Resources Variability*, edited by D. Sahrhage, pp. 3–20, Springer-Verlag, Berlin, 1988.

Webb, P. -N., Geological Investigations in southern Victoria Land, Antarctica. Part 4. Beacon Group of the Wright and Taylor Glacier Region, *New Zealand Journal of Geology and Geophysics, 6* (3), 361–387, 1963.

Webb, P. -N., Wright Fjord, Pliocene marine invasion of an Antarctic dry valley, *Antarctic Journal of the United States, 7* (4), 227–234, 1972.

Wilson, A. T., The great antiquity of some Antarctric landforms-Evidence for an Eocene temperate glaciation in the McMurdo region, in *Palaeoecology of Africa, the Surrounding Islands and Antarctica*, edited by E. M. van Zinderen Bakker Sr., pp. 23–35, Balkema, Cape Town, 1973.

Zachos, J. C., J. R. Breza, and S. W. Wise, Early Oligocene ice-sheet expansion on Antarctica: stable isotope and sedimentological evidence from Kerguelen Plateau, southern Indian Ocean, *Geology, 20* (6), 569-573, 1992.

J. Kleman, Department of Physical Geography, Stockholm University, 10691, Stockholm, Sweden.

M .L. Prentice, Institute for the Study of Earth, Oceans, and Space, University of New Hampshire, Durham, NH 03824.

A. P. Stroeven, Department of Physical Geography, Stockholm University, 10691, Stockholm, Sweden.

(Received March 3, 1997;
accepted May 30, 1997.)

SOLAR RADIATION IN THE MCMURDO DRY VALLEYS, ANTARCTICA

Gayle L. Dana and Robert A. Wharton, Jr.

Biological Sciences Center, Desert Research Institute, University and Community College System of Nevada, Reno, Nevada

Ralph Dubayah

Department of Geography, Laboratory for Global Remote Sensing Studies, and University of Maryland Institute for Advanced Computer Studies, University of Maryland at College Park, College Park, Maryland

Solar radiation is an important driving force for hydrological and biological systems in the dry valleys, influencing sublimation and melting of the glaciers, heating of the soils and air, and providing energy for photosynthesis by the microbial communities in the streams, soils, and perennially ice-covered lakes. We analyzed two years of solar radiation data from eleven meteorological stations positioned on glaciers, lake shores, and lake ice in Taylor, Wright, and Victoria Valleys. Average annual incoming solar radiation ranged from 84 to 117 W m^{-2} during 1994 and 1995. We attribute differences among stations primarily to terrain effects, but coastal cloudiness and orographic effects may also be factors. Average annual net solar radiation was 59 to 76 W m^{-2} at the soil-covered sites, while net solar radiation at glacier and lake-ice sites was lower, 18 to 52 W m^{-2}, due to the high albedo of snow and ice. Terrain obstructions were especially apparent in diurnal time series for Lake Hoare, even in December when the sun is at its highest position. Because of the importance of terrain on solar radiation patterns, we applied a topographic solar radiation model to Taylor Valley, using in situ pyranometer data to drive the model. Considerable topographic variability in solar radiation occurs over the region, even averaged over a monthly time scale, with north facing slopes receiving more energy than south facing slopes. In the valley bottom, differences in incident radiation were discerned among lakes, with Lake Fryxell receiving uniform amounts of energy while Lakes Hoare and Bonney received less energy along their northern shores due to terrain shading. Hourly radiation maps and pyranometer data illustrate that the terminus of the glaciers receive higher levels of solar radiation than their surface, but this intense illumination is of short duration, occurring only when the sun directly strikes the cliff face.

INTRODUCTION

The McMurdo Dry Valleys, located along the western coast of the Ross Sea (76°30′ to 78°30′ S, 160° to 164° E), have been called "oases" because they are among the few ice-free areas on the Antarctic continent. These cold desert regions are dominated by arid soils with a low albedo compared to that of the surrounding snow and ice covered areas, resulting in mean temperatures 7°C warmer in summer and 5° to 7°C cooler in winter than those of adjacent snow-covered coastal areas [*Thompson et al.*, 1971]. In most regions of Antarctica, the large amounts of solar radiation that are reflected by snow and ice, along with the continuous loss of heat by longwave radiation result in a negative annual heat balance [*Rusin*, 1961]. The dry valleys, with their higher absorptivity of solar radiation, are one of the few regions in Antarctica where the annual heat balance is positive [*Thompson et al.*, 1971].

The dry valleys receive all of their solar radiation from August to April with very low light conditions

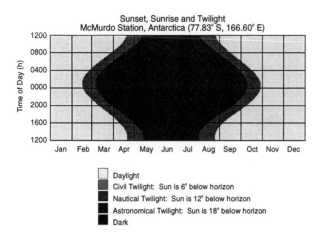

Fig. 1. Sunset, sunrise, and twilight for McMurdo Station (77.83° S, 166.60° E). The dry valleys are about 100 km east and approximately the same latitude as McMurdo Station.

existing the remainder of the year (Figure 1). At the winter solstice, days are typified by darkness or near-darkness. On clear days, the atmosphere over the dry valleys has a very low aerosol optical thickness of 0.04, recorded by sun photometers (349–1022 nm), which results in a high direct beam irradiance from the sun. Complex spatial and temporal patterns of solar radiation are produced in the dry valleys caused by the mountainous terrain and changing solar zenith angle (the angle of the sun's direction with the local upward vertical) and azimuth. In late October, the daily variation of solar zenith angles is from below the horizon (≥90° to 67°, and at the summer solstice it varies between 54.5° and 78.5°. Hence the daily variation in summer heating of the surface is greater in October than in mid summer. Diurnal variations in the solar zenith angle during the summer are sufficient to cause corresponding diurnal changes in temperature, relative humidity, and wind speed [*Clow et al.*, 1988]. Strong daily and seasonal variations in glacier melting and streamflow result from this solar regime [*McKnight et al.*, 1994; *House et al.*, 1995; *Lewis et al.*, 1996; *Conovitz et al.*, this volume; *Fountain et al.*, this volume]. Distribution and productivity of biological communities in the dry valleys are also likely to respond to spatial and temporal patterns in solar radiation [*Lizotte et al.*, 1996].

With the recent formation of the McMurdo Dry Valleys Long-Term Ecological Research (LTER) site, increased emphasis has been placed on understanding ecological processes in this cold desert ecosystem. Physical factors largely control biological processes of the microbial ecosystems in the streams, soils, and perennially ice-covered lakes. Stream ecosystems are controlled by the quantity and timing of glacial meltwater during the austral summer (October through February). Discharge from these streams and thickness of the perennial ice cover (via effects on light availability) are primary factors regulating the lake environment. Soil communities are controlled by moisture availability, temperature, salt concentrations, and allochthonous carbon inputs. These communities are also dependent on the nutrients and gases carried by water from glaciers. Solar radiation exerts a dominant control over almost all the processes listed here: glacial melt, stream discharge, light availability, and soil moisture and temperature. Therefore understanding the solar radiation regime and its spatial and temporal distribution in the dry valleys is central to a greater understanding of hydrological and biological processes in these cold deserts.

Despite its importance to the hydrology and biology of the region, little research has been conducted on radiative fluxes in the dry valleys. *Bull* [1966] was the first to demonstrate that the annual net balance of radiation (short and long wave) in the dry valleys was a gain of 38 W m^{-2}, in contrast to the net loss of 17 W m^{-2} recorded in permanently snow-covered regions of Antarctica. However *Bull's* [1966] values were inferred using cloudiness data from the dry valleys and solar radiation data collected from Scott Base during 1958 to 1962. Only two studies since *Bull* [1966] have assessed annual solar irradiance: one at Lake Vanda in Wright Valley during 1969-1970 [*Thompson et al.*, 1971] and the other at Lake Hoare in Taylor Valley during 1986-1987 [*Clow et al.*, 1988]. *Riordan* [1973] reworked the data of *Thompson et al.* [1971] to calculate the total net radiation for the region. Records of downwelling photosynthetically active radiation (PAR) at the soil surface [*McKay et al.*, 1993] and beneath lake-ice covers [e.g., *Lizotte and Priscu*, 1992; *Palmisano and Simmons*, 1987; *McKay et al.*, 1994], have been presented in conjunction with biological studies for limited time periods. Radiative flux data also have been acquired for shorter time periods during the austral summer to assess thermal forcing of winds in Wright Valley [*Bradley and Martin*, 1995], the heat balance of Lakes Bonney and Vanda [*Ragotzkie and Likens*, 1964], and the contribution of radiative fluxes to melting on the Lower Wright Glacier in Wright Valley [*Chinn*, 1987].

These previous studies provide a basis for evaluating the solar radiation regime in the dry valleys.

However measurements were insufficient in time and space to evaluate spatial and temporal trends in solar radiation. As part of the LTER project in the McMurdo Dry Valleys, a network of 11 meteorological stations was constructed, recording year-round solar radiation, as well as other meteorological data in Taylor, Wright, and Victoria Valleys. The placement of stations was chosen for their proximity to the primary biological and hydrological research sites to aid in interpreting shorter-term, process-level studies and to assess longer-term trends in the hydrology, biology, and climate. However even a well distributed set of point measurements is unlikely to capture the spatial variability in solar radiation caused by topography, especially given the extreme topographic relief present in the dry valleys. Topographic effects on solar radiation can most effectively be analyzed using a physically-based model that incorporates data about the specific terrain of interest. In this paper, we present both in situ measurements of solar radiation along with a topographic modeling approach to assess spatial and temporal trends of solar radiation in the dry valleys.

Modeling Topographic Effects on Solar Radiation

The importance of topographic variability on solar radiation as well as in hydrological and biophysical processes has long been known [e.g., see *Geiger*, 1965; *Dozier*, 1980; *Davis et al.*, 1992]. As a common example, south facing slopes in northern mid-latitudes may receive more direct incident flux than north facing slopes and therefore have different latent, sensible, and soil heat exchanges. Under cloudy skies, sky-obstruction by nearby terrain, such as occurs at the bottom of deep valleys, significantly decreases the diffuse flux reaching the surface. To account for this variability, topographic models have been created that incorporate various topographic effects using digital elevation data [e.g., see *Dozier*, 1980, 1989; *Duguay*, 1993; *Dubayah*, 1994; *Dubayah and Rich*, 1995].

Two important modulators of incoming solar radiation (SW↓), clouds and topography, must be considered in the dry valleys region, where about half of all days during the summer period are cloudy [*Linder et al.*, 1994] and terrain relief ranges from near sea level to 2000 m. The spatial variability in SW↓ is dominated by cloud fields under partly cloudy conditions. Under uniformly clear or cloudy skies, however, the spatial variability in solar radiation at local and regional scales is caused primarily by topography. Variations in slope, aspect, terrain reflectance, shadowing, and sky-obstruction by nearby terrain all affect the amount of insolation incident at any point on the surface. Previous work has shown that, for clear sky cases, the spatial autocorrelation of SW↓ modeled over topographic grids is short, usually less than 300 m, and almost always less than 1000 m for a wide variety of landscapes [*Dubayah et al.*, 1989, 1990; *Dubayah*, 1992, 1994; *Dubayah and van Katwijk*, 1992]. Thus it is unlikely that any set of in situ observations, from pyranometers for example, will capture the spatial variability caused by topography. As a consequence, determining the true energy balance over complex terrain, such as frequently occurs in glacial environments, is impossible without explicit modeling of topographic effects.

The topographic model used in this study is based on *Dozier* [1980], with subsequent versions described in *Dubayah et al.* [1990], *Dubayah* [1992, 1994], *Dubayah and Rich* [1995], and *Dubayah and Loechel* [1997]. The model is improved over other slope radiation models [e.g., *Garnier and Ohmura*, 1968; *Williams et al.*, 1972] in three ways: 1) it uses a physically based calculation of radiation attenuation that accounts for effects caused by varying elevation within the area of interest; 2) it calculates and stores sky view and terrain configuration factors; and 3) it incorporates calculations for reflection from adjacent terrain. This model has been used to accurately capture the spatial variability of solar radiation in other areas of complex and rugged terrain [e.g., *Dozier*, 1980; *Dubayah*, 1992, 1994; *Dubayah and Loechel*, 1997; *Dubayah et al.*, 1990]. The following briefly summarizes the most important aspects of the model.

In the solar spectrum, slopes are illuminated from three sources: 1) direct irradiance, which includes self-shadowing and shadows cast by nearby terrain; 2) diffuse sky irradiance, where a portion of the overlying hemisphere may be obstructed by nearby terrain; and 3) direct and diffuse irradiance reflected by nearby terrain towards the location of interest. Consider a terrain whose topography is represented by a grid of elevation points (a digital elevation model). At each point, the elevation, slope, and aspect may vary; and there may be variable shadowing and reflectance effects. Thus direct and diffuse irradiance incident on a level surface at a particular elevation point must be modified to account for these modulating features of the topography.

The starting points for the topographic model are estimates of the direct and diffuse flux incident on a flat surface at the elevation of the location. Apart from topographic influences that slope orientations cause, points at higher elevation have less atmosphere above them. Under clear skies, the optical depth of the atmosphere decreases with pressure, which in turn decreases with elevation, and therefore the irradiance on a flat surface will increase. For overcast skies, cloud optical thickness generally dominates the clear air column optical thickness either above or below the clouds so that the effects of changes in the elevation of the surface are smaller than for clear conditions. Other factors affecting the fluxes are the particular scattering and absorbing properties of the atmosphere, the variations of these properties with height, the solar zenith angle, and the exoatmospheric flux. Once the direct and diffuse fluxes are known for a flat surface at the given elevation, they are then adjusted for topographic effects.

Modeling Diffuse Irradiance

In general, diffuse irradiance is not isotropic, that is, it varies depending on the direction one looks (e.g., the familiar observation of a brighter sky near the horizon on clear days). However modeling this anisotropy can be complex, especially under partly cloudy conditions. To simplify the problem, we assume that the diffuse radiation coming from the sky is isotropic.

Given the diffuse irradiance on a level surface, a sky view factor V_d is calculated that gives the ratio of diffuse sky irradiance at a point to that on an unobstructed horizontal surface. For example, in a deep valley, much of the overlying sky is blocked and therefore cannot contribute much diffuse irradiance into the valley. The sky view factor accounts for the slope and orientation of a terrain facet and the portion of the overlying hemisphere visible to it, determined by the local horizon in all directions. The horizon can result either from "self-shadowing" by the slope itself or from adjacent topography. The factor V_d must be calculated for every point. However this needs to be done only once for each terrain, as the sky view factor is a property of the terrain (like slope and aspect) and does not change. The diffuse irradiance is then given by

$$\overline{F_x} \downarrow (z) V_d \qquad (1)$$

where $\overline{F_x} \downarrow (z)$ is the average diffuse downwelling at location x which has an elevation z and where V_d varies from 1 (unobstructed) to 0 (completely obstruct-

ed). Under cloudy skies, most of the solar energy reaching the surface is diffuse so that the spatial variability of this diffuse irradiance reflects the variability in sky view factor across the landscape.

Modeling Direct Irradiance

As above, we find an average direct beam irradiance, $\overline{B_x} \downarrow$, around location x and then adjust this flux to the elevation z of x, $\overline{B_x} \downarrow (z)$. The direct irradiance on a slope is then given by

$$\delta \cos\theta_i \ \overline{B_x} \downarrow (z) \qquad (2)$$

where $\cos\theta_i$ is the cosine of the solar illumination angle on the slope. Wherever $\cos\theta_i$ is negative, the point is "self-shadowed," i.e., the sun is below the local horizon caused by the slope itself. This is in contrast to cast shadows which are caused by nearby terrain blocking the sun and which are independent of $\cos\theta_i$. For example, a flat area may be in the shadow of a mountain at low sun angles [Dubayah and Rich, 1995]. Cast shadows are found by calculating local horizon angles from digital elevation data [Dozier and Frew, 1990], and given as δ in Equation 2, a binary shadowing mask set equal to either 0 (shadowed) or 1. In both of these cases, self-shadowing and cast shadows, the direct irradiance is zero.

Modeling Terrain Irradiance

Incoming energy may be reflected from nearby terrain towards the point of interest and can rarely be expected to be isotropic. However because of the complexity of determining the geometric relationships between a particular location and all the surrounding terrain elements, an approximate terrain configuration factor C_t can be calculated [Dozier and Frew, 1990] and used with an average upwelling (terrain reflected) flux, $\overline{F_x} \uparrow (z)$. The counterpart of the sky view factor, the terrain configuration factor, estimates the fraction of the surrounding terrain visible to the point and varies from 0 (only sky visible) to 1 (only terrain visible). The reflected radiation from surrounding terrain is then estimated as

$$C_t \overline{F_x} \uparrow (z) = C_t \overline{R_x} \left[\overline{F_x} \downarrow (z)(1 - V_d) + \cos\theta_o \ \overline{B_x} \downarrow (z) \right] \qquad (3)$$

where $\cos\theta_0$ is the cosine of the solar zenith angle and $\overline{R_x}$ is the average reflectance of the terrain in some local neighborhood centered around x, and which may

TABLE 1. Sites of Solar Radiation Measurements in the McMurdo Dry Valleys

Site	Surface Type	Distance to Coast (km)	Elevation (m)	Sensor Type	Latitude Longitude	Start Date
Taylor Valley						
Commonwealth Glacier	ice/snow	5	290	Eppley PSP	77°34' S 163°18' E	1993/1994
Howard Glacier	ice/snow	16	437	Eppley PSP	77°40' S 163°05' E	1993/1994
Canada Glacier	ice/snow	14	267	LI-COR LI200X	77°37' S 162°58' E	1995/1996
Taylor Glacier	ice/snow	40	327	LI-COR LI200X & Eppley PSP[a]	77°44' S 162°08' E	1994/1995
Lake Fryxell shore	soil	10	20	LI-COR LI200S	77°37' S 163°10' E	1994/1995
Lake Hoare shore	soil	16	72	LI-COR LI200S	77°37' S 162°54' E	1993/1994
Lake Bonney shore	soil	30	60	LI-COR LI200S, LI200X & LI190SB	77°43' S 162°28' E	1993/1994
Lake Bonney ice cover	ice/snow	30	60	LI-COR LI190SB	77°43' S 162°26' E	1992/1993
Wright Valley						
Lake Vanda shore	soil	50	125	LI-COR LI200X	77°31' S 161°41' E	1994/1995
Lake Brownworth shore	soil	25	280	LI-COR LI200S	77°26' S 162°40' E	1994/1995
Victoria Valley						
Lake Vida shore	soil	41	390	LI-COR LI200X	77°23' S 161°50' E	1995/1996

[a]Switched to Eppley November 1995

be dependent on solar zenith and azimuth angles as well as the slope orientation relative to these. The reflected flux term is usually small relative to the direct and diffuse terms, except where the surface albedo is high, such as occurs in snow covered areas.

Modeling Total Irradiance on a Slope

The total irradiance on a slope is the sum of the three components given above

$$SW\downarrow_{slope} = \overline{F_x}\downarrow(z)V_d + \delta\cos\theta_i\overline{B_x}\downarrow(z) + C_t\overline{F_x}\uparrow(z) \quad (4)$$

Thus for a given location, three fluxes are required: average diffuse downwelling, average direct downwelling, and average upwelling. Note that the topographic formulation of Equation 4 is not independent of atmospheric conditions. Different topographic effects will dominate depending on whether the incoming flux is primarily direct (in which case $\cos\theta_i$ dominates), diffuse (where V_d would dominate), or if the reflectance of the surface is high (so that the C_t term is significant).

METHODOLOGY

In Situ Solar Radiation Measurements

In situ measurements of solar radiation were made using a network of meteorological stations initiated in 1993 by the McMurdo Dry Valleys LTER program (Table 1). Six stations were erected in 1993-1994, three in 1994-1995, and two in 1995-1996. *Doran et al.* [1995] provided an overview of the LTER station sensor and sampling. The spatial distribution of LTER stations in the dry valleys is suitable for assessing coarse scale spatial characteristics of solar insolation as well as the influence of surface features on solar fluxes (Figure 2, Table 1). Stations are as near as 5 km to the McMurdo Sound coastline, and as far as 41 km inland. Four of the stations are situated on the ablation zones of the major glaciers; one station is on a lake ice-cover; and the remaining stations are on soils immediately adjacent to ice-covered lakes. Elevations range from 20–390 m for lake shore stations to 250–435 m for glacier sites.

Incoming (SW↓) and outgoing (SW↑) shortwave

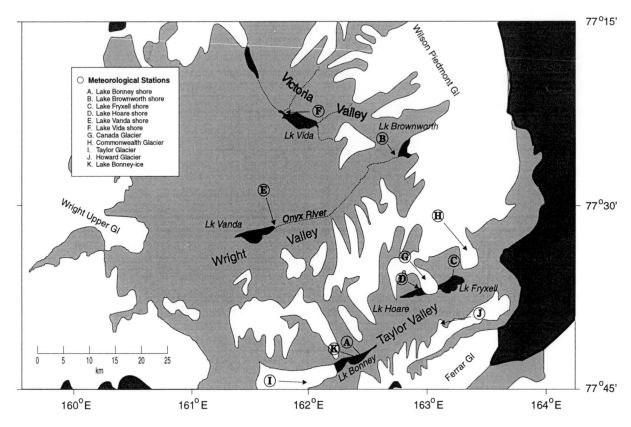

Fig. 2. Locations of LTER meteorological stations in the McMurdo Dry Valleys (adapted from *Doran et al.*, [1995]).

radiation were measured with Eppley model PSP or LI-COR model LI200S and LI200X pyranometers (Table 1). At Lake Brownworth, only SW↓ was measured. The Eppley PSP is a cosine-corrected thermopile receiver with a pair of clear hemispheres (WG295 glass) uniformly transparent to energy between 285 and 2800 nm. Eppley PSPs have a maximum error of ±3% for zenith angles greater than 70°, and ±1% for zenith angles less than 70°. The temperature dependence is ±1% over the temperature range −50° to +40°C. The LI-COR LI200S/X pyranometers contain a cosine-corrected silicon photo-diode with a spectral response which is very low at 400 nm, increases linearly to 950 nm and then decreases nearly linearly to a cutoff near 1200 nm. LI-COR calibrates its LI200S/X against the Eppley PSP to produce output equivalent to the 285–2800 nm range. LI-CORs typically have an absolute error of ±3% which increases to ±5% for zenith angles greater than 80°. Shortwave fluxes at the Lake Bonney ice station were calculated from quantum sensor data since a pyranometer was not installed at this site. The

quantum sensor was a LI-COR model LI190SB, which, like the LI200S/X, is a cosine-corrected silicon diode but measures in the limited range of 400–700 nm (PAR). We used a conversion factor for PAR to solar flux of 0.461 W m⁻² per 1 μmol photons m⁻² s⁻¹ which was derived by averaging the mean hourly ratio of incoming shortwave to PAR measured at the nearby Lake Bonney shore station from October to March, 1993 to 1996. The conversion factor we used in this study is lower than the 0.505 value derived by *Clow et al.* [1988] for Lake Hoare. This difference is likely due to site specific effects of shadowing and cloudiness, as well as the time period over which the ratio was calculated.

Data were collected using Campbell Scientific CR10 (21X at Lake Bonney ice site) data loggers. Sensors were sampled every second and values averaged and stored on a solid state storage module. The averaging frequency ranged between 15 min and 1 h, depending on the station and year, with the exception of winter 1994 when the frequency was 3 h.

Processing of solar radiation data included flagging

erroneous data points that appeared contaminated due to power failures, sensor/datalogger noise, and snow on the sensor (defined here as when SW↑ is greater than SW↓). Only valid ratios of SW↑ to SW↓ were used in computing albedo. Albedos were not reported during the austral winter in periods of very low or no light. Out of range values were flagged and disregarded based on a limits test as defined by the known reflective properties of the surface type. For glacier and lake ice stations, albedos less than 20% and greater than 98% were disregarded; for soil stations, albedos less than 5% and greater than 98% were disregarded.

In calculating hourly and daily averages, we applied certain rules [after *Stone et al.*, 1996]. An hourly average was computed only if all datalogger values stored during that hour were valid. Otherwise, the data point was flagged and disregarded in higher level summaries. Data gaps of up to two hours were filled by linear interpolation to capture temporal variation. A daily average was computed if there were at least 20 hours of valid data for a given day (in the case of 3 h averaging by the datalogger during winter periods, all 8 h values were required for calculating daily means). Monthly averages and totals were computed using valid daily means, and the percentage of valid daily means per month were reported to provide a measure of the reliability of monthly values. Annual averages and totals were reported when possible, but datalogger failures and missing data for months at a time precluded this summary for some stations in some years.

Topographic Solar Radiation Modeling

Driving the topographic model requires the three fluxes given in Equation 4. Various means are available for obtaining these including: 1) in situ atmospheric optical measurements, such as from sun photometer data, in radiative transfer calculations; 2) climatological estimates of the absorbing and scattering properties of the atmosphere, again for use with radiative transfer; 3) satellite estimates of downwelling direct and diffuse flux; and 4) in situ estimates of the fluxes from pyranometers. Although some sun photometer measurements were available from Lake Hoare, the pyranometer record is more complete and was thus chosen to force the topographic model. A detailed methodology for using pyranometer data for such modeling is given in *Dubayah* [1994].

We chose to run the model during the month of December, one of the most biologically and hydrologically active times of the year in the dry valleys, as well as the time when solar zenith angles are the lowest (i.e., the sun is highest in the sky) and the least amount of terrain shadowing occurs. Hourly pyranometer data for December 1994 from the Taylor Glacier station were used to run the model (see Table 1 and Figure 2 for station location and characteristics). This site is relatively free of snow events (which result in unusable data) and is subject to direct shade for only about 2 h each day during December (from 0100–0200 h). The global fluxes are first split into direct/diffuse components using a climatological formulation [*Dubayah*, 1994] which is a function of exoatmospheric flux and sun angle. These down-welling diffuse and direct quantities are the first two fluxes in Equation 4 (i.e., $\overline{F_x}\downarrow(z)$ and $\overline{B_x}\downarrow(z)$). These are then combined with local estimates of surface reflectance derived from Landsat Thematic Mapper data to obtain the average upwelling irradiance $\overline{F_x}\uparrow(z)$, the third flux required in Equation 4. These three fluxes are then adjusted for elevation effects based on pressure [*Dubayah and van Katwijk*, 1992; *Dubayah*, 1994] to obtain the direct and diffuse energy incident on a flat surface at the appropriate elevation z of the grid element x. The remaining topographic effects in Equation 4 are then applied to produce the final topographic radiation map.

Images of $\cos\theta_i$, V_d, C_t, and cast shadows δ, required in Equation 4, are all calculated from digital elevation data. Digitized contour lines were used to produce a digital elevation grid with 30 m grid spacing. The grid which covers Taylor Valley and parts of the Ferrar Glacier is about 30 km by 50 km, covering 1387 km^2. The Landsat Thematic Mapper reflectance image was then registered to this elevation grid. In general, three major sources of errors can arise in the topographic model: 1) errors in the digital elevation model (including misregistration); 2) errors in atmospheric parameters; and 3) errors in the field measurements of radiation fluxes used to drive the model. Another error arises when the model makes flux calculations from the pyranometer data when it is in shadow (in this case for ~2 h a day) and other areas of the grid are in full sun. This results in lower than expected solar radiation for some areas on the grid because the model assumes the whole region is in shadow when the pyranometer at Taylor Glacier is in shadow. Similar errors will occur when the local climate (e.g., clouds) at the pyranometer site used to drive the model is different than other sites in the grid.

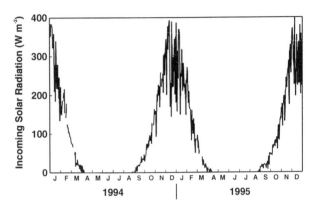

Fig. 3. Mean daily incoming solar radiation at Lake Hoare during 1994 and 1995. Breaks in the line plot are periods of unusable data.

RESULTS AND DISCUSSION

In Situ Solar Radiation Measurements

Incoming solar radiation. The most notable feature of the annual solar irradiance cycle in the dry valleys is the continuous influx of solar radiation during the austral summer months from November to January and its complete absence from May to July. The annual cycle presented in Figure 3 for Lake Hoare is generally representative of the dry valleys region. As the amount of solar radiation increases from spring to summer, so does the heating of the dry valleys. Regional views from satellites clearly show the "oasis" nature of the dry valleys, with soil-covered areas heating significantly from September to January, relative to surrounding ice and snow-covered areas (Plate 1; see also accompanying CDROM). The surface brightness temperatures displayed in these satellite images, which are the product of the thermodynamic temperature and emissivity, are indicators of the radiant energy loading and heat removal processes occurring in the dry valleys. It is this large seasonal accumulation of radiant energy that drives hydrological processes and allows the existence of biological systems in the dry valleys.

The mean annual receipt of shortwave radiation in the dry valleys ranged from 84 to 117 W m^{-2} during 1994 and 1995 (Table 2, Figure 4a). These values are comparable to those previously observed at Lake Vanda, and the yearly averages we report here for Lake Hoare are nearly identical to *Clow et al.'s* [1988], implying similar climate conditions during 1986, 1994, and 1995 (Table 3). To put these values in a continental perspective, other coastal areas of Antarctica and regions on the Polar Plateau report generally higher mean solar irradiance, up to 159 W m^{-2} annually (Table 3). The higher levels of solar radiation at other areas on the continent compared to the McMurdo Dry Valleys region result from: lower latitudinal position (e.g., Oazis station at 66°18' S and Mawson station at 67°34' S), greater transparency and dryness of the air at higher altitudes (e.g., Vostok at 3488 m), and unobstructed view (e.g., South Pole station). Most of the LTER meteorological stations are situated in low areas of extreme topographical relief in which shading and shadowing by nearby ridges and peaks reduce the solar flux.

Topographic variability accounts for much of the variation that we observed in solar irradiance among dry valley stations (Tables 2 and 3, Figure 4a). Stations experiencing significant shading, even during December when the sun is at its highest position in the sky, generally had lower mean annual solar fluxes. For example, Howard Glacier, which is shaded for up to 8 hours a day in December, had a much lower annual solar flux than the Commonwealth Glacier, which is not shaded at all during mid summer. Lake Hoare, which also had low annual solar flux compared to other stations, is shaded by Andrews Ridge over portions of its west shore for up to 5 hours a day in December and by the Asgard Range along portions of its north shore for up to 3 hours daily. By comparison, Lake Fryxell is not shaded in December and receives a higher annual flux. Shading effects become more prominent early and late in the year when solar zenith angles are high. In spring and fall, mean solar irradiance at Lake Fryxell is double that at Lake Hoare; during midsummer, the differences are substantially less (Table 2).

Factors other than topography, such as clouds, are likely to influence the spatial variability of solar irradiance in the dry valleys. Cloudiness may be linked to coastal proximity because low pressure systems and easterly winds pass over McMurdo Sound, transferring moist air over the dry valleys [*Bromley*, 1985]. Precipitation, and presumably cloudiness, decrease westward with distance from the ocean [*Bull*, 1966; *Keys*, 1980]. Given this trend, we would also expect to see a gradient of low to high solar flux from east to west. We do observe that the Commonwealth Glacier, 5 km from the coast, has a lower annual solar flux than the Taylor Glacier, 40 km west (Table 2, Figure 4a), but observations from other stations in our 1994-1995 data set do not favor an east-west solar gradient. Therefore either our assump-

TABLE 2. Mean Monthly and Annual Incoming Shortwave Radiation (W m^{-2}) for the
McMurdo Dry Valleys from December 1993 to January 1996

		Commonwealth Glacier		Howard Glacier		Canada Glacier		Taylor Glacier		L. Bonney ice[a]	
1993	Dec	322	(90)	267	(87)	—		—		304	(97)
1994	Jan	299	(77)	230	(90)	—		—		262	(100)
	Feb	177	(71)	130	(54)	—		—		151	(100)
	Mar	52	(65)	38	(39)	—		—		40	(97)
	Apr	3	(97)	0	(90)	—		—		4	(100)
	May	0	(97)	0	(100)	—		—		0	(100)
	Jun	0	(100)	0	(100)	—		—		0	(100)
	Jul	0	(100)	0	(100)	—		—		0	(100)
	Aug	0	(100)	0	(100)	—		—		1	(100)
	Sep	6	(20)	13	(37)	—		—		—	(0)
	Oct	—	(0)	94	(71)	—		—		—	(0)
	Nov	307	(27)	208	(90)	—		—		247	(87)
	Dec	305	(68)	290	(100)	—		356	(100)	276	(90)
Mean annual		104[b]		84							
Total annual[c]		3002[b]		2639							
1995	Jan	289	(71)	272	(94)	—		335	(100)	273	(100)
	Feb	175	(86)	170	(86)	—		207	(100)	157	(97)
	Mar	63	(100)	56	(39)	—		70	(100)	46	(94)
	Apr	5	(100)	3	(67)	—		4	(100)	4	(100)
	May	0	(100)	0	(100)	—		0	(100)	0	(100)
	Jun	0	(100)	0	(100)	—		0	(100)	0	(100)
	Jul	0	(100)	0	(100)	—		0	(100)	0	(58)
	Aug	1	(100)	1	(90)	—		1	(100)	—	
	Sep	29	(100)	25	(70)	—		29	(100)	—	
	Oct	136	(100)	132	(65)	—		157	(100)	—	
	Nov	282	(87)	242	(93)	—		297	(100)	—	
	Dec	344	(77)	288	(94)	324	(100)	298	(100)	—	
Mean annual		110		99				117			
Total annual[c]		3474		3115				3663			
1996	Jan	291	(74)	244	(74)	268	(77)	275	(77)	—	

Values in parentheses are the percentage of days used in calculating the monthly average
[a]calculated from PAR; see text
[b]annual statistics based on incomplete year
[c]Units are in MJ m^{-2} yr^{-1}

tion about an east-west cloudiness gradient is incorrect, or the complicating influence of terrain obstruction ameliorates the effects of cloudiness.

Finally orographic effects may contribute to spatial variability of solar irradiance in the dry valleys. We already mentioned the contribution of topography to the low solar irradiance observed on the Howard Glacier (Table 2, Figure 4a). Situated 16 km from the coast, coastal cloudiness may also influence this site. In addition to these factors, at 437 m, the Howard Glacier meteorological station is the highest in our network and might be expected to have higher solar fluxes due to thinner atmosphere. However its eleva-

tion and position among the Kukri Hills may be great enough to induce some orographic cloudiness. Observations of the Howard Glacier from the Lake Hoare camp support this idea, as the glacier is often obscured by clouds while surrounding areas in the valley are in full sunshine. The three influences on solar radiation we have discussed here (terrain, proximity to ocean, and orographic effects) have been noted for other regions of the planet as well [e.g., *Aguado*, 1986].

We point out that some differences in solar irradiance among stations in the dry valleys may be due to measurement error. It was noted earlier that

TABLE 2. Mean Monthly and Annual Incoming Shortwave Radiation (W m^{-2}) for the
McMurdo Dry Valleys from December 1993 to January 1996 (Continued)

		L. Fryxell shore		L. Hoare shore		L. Bonney shore		L. Brownworth shore		L. Vanda shore		L. Vida shore	
1993	Dec	—		325	(39)	309	(100)	—		—		—	
1994	Jan	275	(74)	285	(100)	270	(97)	—		—		—	
	Feb	189	(93)	146	(86)	167	(100)	—		—		—	
	Mar	83	(71)	40	(94)	44	(100)	—		—		—	
	Apr	10	(87)	2	(97)	2	(100)	—		—		—	
	May	0	(100)	0	(100)	0	(100)	—		—		—	
	Jun	0	(100)	0	(100)	0	(100)	—		—		—	
	Jul	0	(100)	0	(100)	0	(100)	—		—		—	
	Aug	3	(90)	0	(100)	0	(100)	—		—		—	
	Sep	44	(90)	22	(100)	14	(100)	—		—		—	
	Oct	136	(90)	89	(94)	108	(100)	—		—		—	
	Nov	264	(97)	217	(97)	237	(100)	—		—		—	
	Dec	295	(100)	290	(100)	292	(90)	—		293	(100)	—	
Mean annual		108		91		94							
Total annual^c		3400		2865		2971							
1995	Jan	279	(100)	257	(100)	288	(81)	224	(100)	277	(100)	—	
	Feb	167	(100)	142	(93)	—		133	(100)	166	(96)	—	
	Mar	59	(65)	37	(77)	60	(48)	48	(100)	30	(48)	—	
	Apr	7	(97)	3	(100)	2	(30)	2	(100)	2	(70)	—	
	May	0	(100)	0	(100)	0	(100)	0	(100)	0	(100)	—	
	Jun	0	(100)	0	(100)	0	(100)	0	(100)	0	(100)	—	
	Jul	0	(100)	0	(100)	0	(94)	0	(100)	0	(100)	—	
	Aug	2	(100)	1	(100)	—		0	(26)	1	(90)	—	
	Sep	30	(83)	23	(100)	—		—	(0)	19	(77)	—	
	Oct	121	(100)	87	(100)	—		—	(0)	92	(97)	—	
	Nov	255	(100)	224	(100)	304	(30)	249	(33)	265	(100)	—	
	Dec	333	(100)	299	(100)	304	(100)	275	(100)	269	(100)	308	(100)
Mean annual		104		90						93			
Total annual^c		3285		2819						2935			
1996	Jan	282	(74)	241	(74)	268	(81)	240	(71)	248	(71)	263	(73)

LI-COR sensors have an increasing cosine-collection error at solar zenith angles greater than 80°. In midsummer, solar zenith angles in the dry valleys are mostly less than 80°, but earlier and later in the year, they are greater than 80° for a significant portion of the diurnal cycle. Therefore small differences in solar irradiance between stations configured with LI-COR sensors may be due to measurement error rather than topography or climate. Reconfiguration of all pyranometers in the LTER meteorological network to the Eppley sensors would reduce the uncertainty of our measurements and allow for a more standardized comparison among stations.

Net shortwave radiation. Net shortwave radiation, that amount of the incoming energy absorbed by the surface, is biologically and hydrologically important because it is the energy available to drive processes such as photosynthesis, melting and sublimation of the glaciers, and air and soil heating. Large differences in annual net solar radiation were observed among sites in the dry valleys during 1994-1995 (Table 4, Figure 4b). These differences are largely determined by variation in reflectance properties among the soils, ice, and snow-covered regions in the dry valleys (Table 5, Figure 4c). Soil sites with their low albedo had the highest net annual solar radiation ranging from 59 to 76 W m^{-2}. In contrast, the high albedo of the snow-covered Commonwealth and Howard Glaciers resulted in the comparatively low net solar radiation of 18 to 52 W m^{-2}. The higher incidence of snow cover on the Commonwealth and Howard Glaciers has been attributed to the precipitation gradient noted earlier

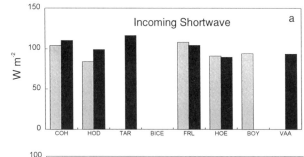

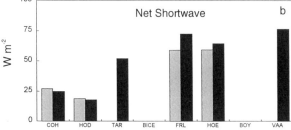

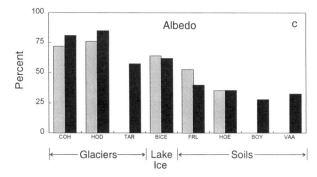

Fig. 4. Mean annual incoming shortwave radiation (a), net shortwave radiation (b), and albedo (c) for eight meteorological stations in the dry valleys. 1994 is represented by the striped bars and 1995 by the solid bars. Absence of a bar indicates either no or insufficient data were collected at that station. Meteorological stations are as follows: COH, Commonwealth Glacier; HOD, Howard Glacier; TAR, Taylor Glacier; BICE, Lake Bonney ice; FRL, Lake Fryxell shore; HOE, Lake Hoare shore; BOY, Lake Bonney shore; VAA, Lake Vanda shore.

[see also *Fountain et al.,* this volume]. The mostly ice-covered Taylor Glacier and Lake Bonney ice sites were intermediate in albedo and annual net solar radiation. Similarities in albedo between lake ice and glacier ice create difficulties when trying to delineate lake and glacier boundaries from satellite data based on reflectance properties.

Net radiation and albedo varied among soil sites (Figures 4b and 4c), most likely due to differences in soil type and color. Low albedo soils are dark or aggregated, containing relatively large irregular particles that decrease reflection by trapping radiation

due to multiple reflection between adjacent faces [*Monteith and Unsworth*, 1990]. Light-colored soils, as well as homogenous and highly polished soil surfaces such as sandstone, have a higher albedo. We have insufficient soils data for the dry valleys region to relate the measured albedos to soil characteristics at each site.

We caution that albedos measured by LI-COR pyranometers may not, in some cases, be directly comparable to those measured by Eppleys. LI-COR pyranometers measure solar radiation only in the silicon detection range (400–1200 nm), despite being calibrated to the broader range Eppley sensors (285–2800 nm). For materials such as snow and ice, whose spectral reflectance changes greatly from the visible near infrared to the short wave infrared regions, LI-COR measured albedos will be higher than the corresponding albedos measured with Eppley sensors.

Inter-annual differences in solar radiation. There were differences in solar radiation between 1994 and 1995. Incoming solar radiation increased at glacier stations but decreased at lake shore stations from 1994 to 1995 (Figure 4a). The trend was reversed for net solar radiation which decreased from 1994 to 1995 at glacier stations but increased at lake shore stations (Figure 4b), due largely to albedo effects (Figure 4c). *Dana et al.* [1996a] observed that increased albedo on the glaciers in 1995 was due to greater precipitation and snow cover at those sites. However the lower albedo observed at Lake Fryxell implies less snow cover in 1995. To reconcile this apparent paradox, it may be that climate patterns such as increased cloudiness lead to increased precipitation at the higher elevations on the glaciers, and also a reduction in radiant cooling, thereby enhancing snow melt on the valley floor. Future research should explore these ideas since the mean annual temperature in both Lake Hoare and Lake Fryxell basins increased by 1.3°C between 1994 and 1995. The influence of snow cover in reducing the amount of solar radiation absorbed by the soils, lakes, and glaciers is an important consideration when modeling spatial and temporal patterns of heat fluxes in the dry valleys.

Diurnal patterns of solar radiation. The diurnal solar irradiance cycle exerts considerable influence on daily patterns in the magnitude and timing of the hydrology in the dry valleys [e.g., *Conovitz et al*, this volume; *Lewis et al.*, 1996]. Typical diurnal solar irradiance curves for December 1994 are shown in Figure 5a for each dry valley station and demonstrate further the influences of terrain obstructions noted

TABLE 3. Mean Annual Solar Irradiance at Selected Sites in Antarctica

Station	Latitude	Elevation (m)	Solar Flux (W m^{-2})	Year
McMurdo Dry Valleys, present study				
Commonwealth Glacier	77°34'	290	104, 110	1994, 1995
Howard Glacier	77°40'	437	84, 99	1994, 1995
Taylor Glacier	77°44'	327	117	1995
Lake Fryxell	77°37'	20	108, 104	1994, 1995
Lake Hoare	77°37'	72	91, 90	1994, 1995
Lake Bonney	77°43'	60	94	1994
Lake Vanda	77°31'	125	93	1995
McMurdo Dry Valleys, previous studies				
Lake Hoare[a]	77°37'	72	92	1986
Lake Vanda[b]	77°31'	125	104	1970
Other Coastal Stations[c]				
Oazis	66°18'	28	118	1956-1958
Mirny	68°33'	35	131	1960-1961
Port Martin	66°49'	14	104	1951
Mawson	67°34'	8	132	1961-1963
Roi Baudouin	70°26'	37	129	1965-1966
Norway	70°30'	56	130	1958-1959
Novolazarevskaya	70°46'	87	123	1963-1965
Maudheim	77°03'	38	126	1949-1952
Scott Base	77°51'	15	122	1957-1958
Little America V	78°11'	40	99	1957-1958
Inland Stations[c]				
Charcot	69°22'	2400	152	1957-1958
Pionerskaya	69°44'	2740	143	1956-1958
Vostok	78°24'	3488	159	1959
Plateau	79°15'	3625	143	1966-1967
South Pole	90°00'	2800	137	1957-1966

[a]*Clow et al.* [1988]
[b]*Thompson et al.* [1971]
[c]from Table 1 in *Kuhn et al.* [1977]

earlier. Locations without terrain obstructions, such as the Commonwealth Glacier, have a relatively smooth, Gaussian-type radiation curve reaching a peak at solar noon and tapering off late in the day. Locations like Lake Hoare that are strongly influenced by shading from nearby topography show departures from a Gaussian-shaped solar irradiance curve. We explore in detail below how differences in location and aspect influence diurnal variability of solar irradiation.

The interaction of a site's daily solar irradiance curve with changes in reflectance determines its diurnal net solar radiation pattern. For example, peak incident solar radiation on the Commonwealth Glacier occurs at 1300 h, while its peak net solar radiation is maintained from 1300 to 1600 h due to albedo effects

(Figures 5a, b, and c). The daily albedo cycle on the Howard Glacier creates an unusual diurnal pattern of net solar radiation. Between 0600 and 2100 h, albedo on the Howard Glacier increases and then decreases at a rate similar to the solar irradiance creating a "flat" net radiation curve (Figure 5b). This is in contrast to other stations where solar irradiance increases faster than albedo changes such that net radiation curves tend to follow the same trend as solar irradiance.

Albedo is affected by changing sun angle and azimuth, clouds, and the slope and roughness of the surface. For example, the albedo of snow increases with both solar zenith angle and cloud cover [*Warren*, 1982], and the apparent albedo of a sloping snow surface will change more throughout the day than a

TABLE 4. Mean Monthly and Annual Net Shortwave Radiation (W m^{-2}) for the McMurdo Dry Valleys from December 1993 to January 1996

		Commonwealth Glacier		Howard Glacier		Canada Glacier		Taylor Glacier		L. Bonney ice[a]	
1993	Dec	144	(90)	87	(90)	—		—		108	(197)
1994	Jan	106	(77)	82	(90)	—		—		99	(100)
	Feb	41	(71)	19	(50)	—		—		44	(100)
	Mar	14	(65)	9	(39)	—		—		11	(97)
	Apr	1	(97)	0	(90)	—		—		2	(97)
	May	0	(97)	0	(100)	—		—		0	(100)
	Jun	0	(100)	0	(100)	—		—		0	(100)
	Jul	0	(100)	0	(100)	—		—		0	(100)
	Aug	0	(100)	0	(100)	—		—		0	(100)
	Sep	2	(20)	2	(37)	—		—		—	(0)
	Oct	—	(0)	19	(71)	—		—		—	(0)
	Nov	73	(27)	40	(90)	—		—		137	(87)
	Dec	64	(71)	55	(100)	—		149	(100)	119	(90)
Mean annual		27[b]		19							
Total annual[c]		787[b]		597							
1995	Jan	68	(71)	45	(94)	—		135	(100)	124	(100)
	Feb	46	(86)	30	(86)	—		98	(100)	85	(100)
	Mar	14	(100)	7	(39)	—		21	(100)	13	(94)
	Apr	1	(100)	0	(67)	—		2	(100)	1	(100)
	May	0	(100)	0	(100)	—		0	(100)	0	(100)
	Jun	0	(100)	0	(100)	—		0	(100)	0	(100)
	Jul	0	(100)	0	(100)	—		0	(100)	0	(58)
	Aug	0	(100)	0	(90)	—		1	(100)	—	
	Sep	7	(100)	3	(70)	—		14	(100)	—	
	Oct	34	(100)	15	(65)	—		77	(100)	—	
	Nov	65	(87)	44	(93)	—		155	(100)	—	
	Dec	63	(100)	71	(97)	123	(100)	121	(100)	—	
Mean annual		25		18				52			
Total annual[c]		780		565				1628			
1996	Jan	59	(74)	40	(74)	80	(77)	100	(77)	—	

Values in parentheses are the percentage of days used in calculating the monthly average
[a]calculated from PAR data; see text
[b]annual statistics based on incomplete months
[c]Units are in MJ m^{-2} yr^{-1}

flat surface due to the change in the sun angle relative to the slope [Grenfell et al., 1994]. Soil albedo is also influenced by the sun angle, as well as the soil's physical properties, such as moisture and ice formation, which can change during the course of a day.

Typical diurnal curves for December 1994 demonstrate that albedo does change throughout the day for most stations in the dry valleys (Figure 5c). Diurnal changes in albedo are the greatest for the Howard Glacier, which has a moderate surface roughness and the greatest slope of all the meteorological sites in the dry valleys. The effect of aspect on the diurnal cycle of albedo is apparent from the inverse albedo curves of the Commonwealth Glacier, which faces southeast, and the Howard Glacier, which faces north. Diurnal patterns for soil sites were of less magnitude and more irregular than glacier and lake-ice sites.

The effect of terrain obstructions on daily patterns of solar radiation becomes more dramatic earlier and later in the season because overall solar zenith angles are larger and diurnal changes in sun angle are greater. For example, the terrain effects on the diurnal solar radiation cycle previously noted for Lake Hoare during December 1994 (Figure 5a) becomes more marked in October 1995 (Figure 6a). The departure of the solar irradiance curve from a Gaussian shape becomes more

TABLE 4. Mean Monthly and Annual Net Shortwave Radiation (W m^{-2}) for the McMurdo
Dry Valleys from December 1993 to January 1996 (Continued)

		L. Fryxell shore		L. Hoare shore		L. Bonney shore		L. Brownworth shore	L. Vanda shore		L. Vida shore	
1993	Dec	—		237	(39)	—		—	—		—	
1994	Jan	224	(74)	210	(100)	—		—	—		—	
	Feb	74	(93)	72	(86)	—		—	—		—	
	Mar	20	(71)	24	(94)	—		—	—		—	
	Apr	3	(80)	2	(100)	—		—	—		—	
	May	0	(100)	0	(100)	—		—	—		—	
	Jun	0	(100)	0	(100)	—		—	—		—	
	Jul	0	(100)	0	(100)	—		—	—		—	
	Aug	1	(90)	0	(100)	—		—	—		—	
	Sep	16	(90)	15	(100)	—		—	—		—	
	Oct	40	(90)	45	(94)	—		—	—		—	
	Nov	126	(97)	138	(97)	—		—	—		—	
	Dec	204	(100)	202	(100)	241	(71)	—	—		—	
Mean annual		59		59								
Total annual[c]		1862		1866								
1995	Jan	207	(100)	184	(100)	241	(84)	—	236	(100)	—	
	Feb	118	(100)	102	(93)	—		—	135	(96)	—	
	Mar	24	(65)	19	(77)	34	(42)	—	16	(48)	—	
	Apr	3	(97)	2	(100)	1	(30)	—	2	(70)	—	
	May	0	(100)	0	(100)	0	(100)	—	0	(100)	—	
	Jun	0	(100)	0	(100)	0	(100)	—	0	(100)	—	
	Jul	0	(100)	0	(100)	0	(94)	—	0	(100)	—	
	Aug	1	(100)	1	(100)	—		—	1	(90)	—	
	Sep	21	(83)	16	(100)	—		—	18	(77)	—	
	Oct	74	(100)	63	(100)	—		—	71	(97)	—	
	Nov	162	(100)	149	(100)	261	(30)	—	219	(100)	—	
	Dec	261	(100)	236	(100)	253	(100)	—	216	(100)	245	(100)
Mean annual		73		64					76			
Total annual[c]		2284		2028					2395			
1996	Jan	226	(74)	192	(74)	226	(77)	—	203	(71)	216	(73)

significant for Lake Hoare and also for Lake Vanda at this time of year. Also evident during October are earlier sunrises for some stations (e.g., Lake Vanda) and later sunsets for others (e.g., Taylor Glacier). For stations without significant terrain obstructions, the peak irradiance occurs about 3 hours later in the day compared to December. Differences in the timing of peak net shortwave radiation among stations become greater in October, with peaks occurring at 1100 h (Lake Vanda), 1400 h (Lake Hoare), and 1600 h (Lake Fryxell and remaining stations in Figure 6b). Daily cycles of albedo in October (Figure 6c) are less apparent than in December. Such differences in timing of peak net solar radiation may bear on the daily production cycles of biological communities in the dry valley lakes.

From our in situ observations of solar radiation in the dry valleys during 1994 and 1995, we see that topography greatly influences spatial and temporal distribution of incoming solar radiation in the McMurdo Dry Valleys. Many of the differences in both diurnal cycles and in mean annual solar radiation between meteorological stations in the LTER network can be explained by the degree to which terrain shading occurs at individual sites. The solar radiation regime at some sites may additionally be modified by cloudiness due to coastal proximity or orographic effects. Intersite differences in net radiation are primarily determined by differences in albedo between soil, ice, and snow covered surfaces. As indicated by comparisons between 1994 and 1995, year-to-year changes in snow cover on the soil and ice surfaces in

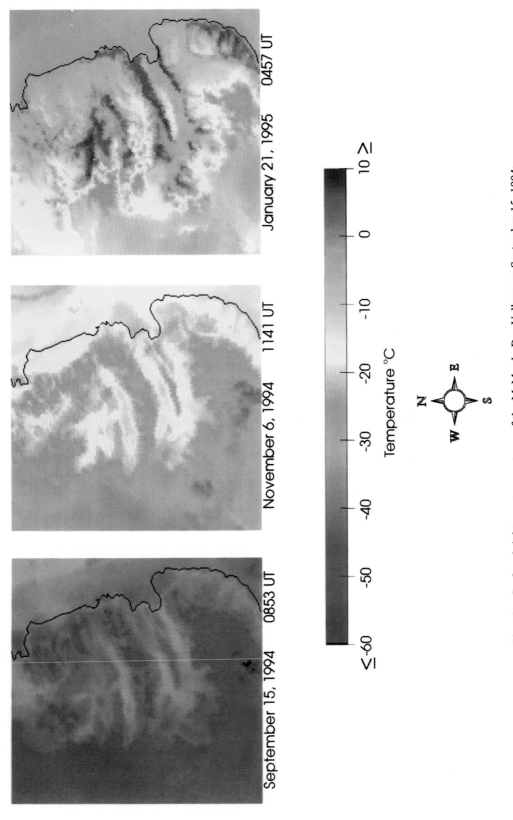

Temperature °C

Plate 1. Surface brightness temperatures of the McMurdo Dry Valleys on September 15, 1994; November 16, 1994; and January 21, 1995. The dry valleys, located to the left of the coastline (black line), clearly show up as soils warm to temperatures higher than the surrounding snow and ice covered regions. McMurdo Sound, to the right of the coastline, also warms significantly during the austral summer. The Polar Plateau is seen as the cooler expanse to the left of the dry valleys. Bounding coordinates are: 77°00'–78°15' S; 159°00'–164°27' E. Images were derived from AVHRR satellite data. See *Dana et al.*, [1996b] for processing methodology.

TABLE 5. Mean Monthly and Annual Albedo (%) for the McMurdo Dry Valleys from December 1993 to January 1996

		Commonwealth Glacier		Howard Glacier		Canada Glacier		Taylor Glacier		L. Bonney ice	
1993	Dec	56	(100)	60	(94)	—		—		63	(100)
1994	Jan	65	(97)	61	(97)	—		—		60	(100)
	Feb	79	(96)	76	(96)	—		—		69	(100)
	Mar	72	(100)	75	(58)	—		—		74	(100)
	Apr	63	(37)	84	(7)	—		—		63	(90)
	May	—		—		—		—		—	
	Jun	—		—		—		—		—	
	Jul	—		—		—		—		—	
	Aug	—		—		—		—		75	(94)
	Sep	66		84	(33)	—		—		65	(17)
	Oct	—		74	(87)	—		—		—	
	Nov	79	(47)	74	(97)	—		—		49	(97)
	Dec	80	(100)	80	(100)	—		58	(100)	56	(100)
Mean annual		72		76						64	
1995	Jan	79	(97)	82	(94)	—		59	(100)	55	(100)
	Feb	79	(100)	83	(96)	—		54	(100)	46	(97)
	Mar	81	(100)	90	(77)	—		68	(100)	71	(100)
	Apr	88	(67)	89	(40)	—		66	(100)	76	(90)
	May	—		—		—		—		—	
	Jun	—		—		—		—		—	
	Jul	—		—		—		—		—	
	Aug	89	(55)	93	(29)	—		54	(95)	—	
	Sep	80	(100)	87	(93)	—		54	(100)	—	
	Oct	78	(100)	86	(100)	—		51	(100)	—	
	Nov	79	(100)	82	(100)	—		49	(100)	—	
	Dec	79	(81)	74	(100)	62	(100)	62	(100)	—	
Mean annual		81		85				58		62	
1996	Jan	82	(74)	82	(74)	70	(77)	66	(81)	—	

Values in parentheses are the percentage of days used in calculating the monthly average

the dry valleys can cause corresponding changes in annual net solar radiation. In the next section, we present the results of the topographic modeling of solar radiation in the McMurdo Dry Valleys to further explore some of the terrain effects observed from the in situ measurements.

Topographic Solar Radiation Modeling

Spatial distribution of solar radiation. Figure 7 shows the monthly solar radiation map of Taylor Valley derived from the topographic model (see also accompanying CDROM). Average values for December 1994 range from 103 to 449 W m^{-2}. Considerable topographic variability in irradiance can be seen in Figure 7, even averaged over a monthly time scale, with north facing slopes receiving more energy than south facing slopes. The broad bottom of Taylor Valley has a fairly uniform radiation regime, with little effects of shading and shadowing visible. The glaciers do show variability in radiation incident on their terminus cliffs versus their tops. For example, the Commonwealth Glacier, seen as the prominent lobe in the northeast portion of the map, has average monthly energy fluxes less than 320 W m^{-2} along its eastern and southern sides, respectively, but fluxes of about 340 W m^{-2} on its top. The lighter tones on the western edge of the Commonwealth Glacier, representing fluxes greater than 350 W m^{-2}, indicate that this cliff surface receives on average slightly more radiation than the top. High solar flux values were also present on the north-facing terminus of the Howard Glacier, located just north of the meteorological station marker on the south side of Taylor Valley in Figure 7.

TABLE 5. Mean Monthly and Annual Albedo (%) for the McMurdo Dry Valleys from December 1993 to January 1996 (Continued)

		L. Fryxell shore		L. Hoare shore		L. Bonney shore		L. Brownworth shore		L. Vanda shore		L. Vida shore	
1993	Dec			33	(42)	—		—		—		—	
1994	Jan	20	(84)	29	(100)	—		—		—		—	
	Feb	57	(100)	55	(100)	—		—		—		—	
	Mar	74	(100)	42	(94)	—		—		—		—	
	Apr	64	(87)	44	(20)	—		—		—		—	
	May	—		—		—		—		—		—	
	Jun	—		—		—		—		—		—	
	Jul	—		—		—		—		—		—	
	Aug	55	(68)	8	(10)	—		—		—		—	
	Sep	55	(90)	28	(80)	—		—		—		—	
	Oct	67	(100)	43	(100)	—		—		—		—	
	Nov	49	(100)	39	(100)	—		—		—		—	
	Dec	32	(100)	30	(100)	16	(87)	—		—		—	
Mean annual		53		35									
1995	Jan	28	(100)	28	(100)	18	(100)	—		18	(100)	—	
	Feb	36	(100)	36	(96)	21	(27)	—		27	(96)	—	
	Mar	64	(97)	55	(90)	42	(100)	—		44	(90)	—	
	Apr	51	(63)	30	(40)	40	(27)	—		54	(27)	—	
	May	—		—		—		—		—		—	
	Jun	—		—		—		—		—		—	
	Jul	—		—		—		—		—		—	
	Aug	36	(55)	40	(13)	—		—		46	(10)	—	
	Sep	46	(80)	44	(93)	—		—		35	(27)	—	
	Oct	39	(100)	30	(100)	—		—		30	(71)	—	
	Nov	39	(100)	37	(100)	17	(30)	—		21	(97)	—	
	Dec	22	(100)	22	(100)	16	(100)	—		20	(100)	21	(100)
Mean annual		40		36		26				33			
1996	Jan	20	(77)	21	(77)	16	(81)	—		18	(71)	18	(73)

We note that the monthly averaged values on the surface of the Commonwealth Glacier are ~35 W m^{-2} higher than those observed from pyranometer data on the surface of the glacier for the month of December 1994 (Table 2). However these values are not directly comparable as the model output is corrected for the slope of the glacier while the pyranometer values are for a level, horizontal surface. A thorough comparison of model-generated solar radiation to in situ measurements is merited but beyond the scope of this study.

Diurnal variability of solar radiation. The diurnal variability of solar radiation that occurs in Taylor Valley can be evaluated on different slopes by examining averaged radiation fields for steep north-facing slopes, steep south-facing slopes, and level slopes (Figure 8). The north-facing curve shows a Gaussian shape typical for radiation under normal day and night conditions (at more northerly latitudes), with the peak at solar noon and tapering off towards sunset. The curve does not have a perfect Gaussian shape because of cloud effects, preferential slope orientation relative to the sun, and topographic effects that are not dependent directly on slope, such as shadows cast by nearby terrain or sky obstruction of the diffuse field (sky view factor effects). Because the sun does not set in December at this latitude, there is always some diffuse irradiance reaching these slopes so that the tails of the curve never reach zero. In contrast to north-facing slopes, the level slope curve is quite atypical from the Gaussian shape that would be found at lower latitudes. While there is still a peak at solar noon, the radiation does not fall off precipitously on either side because these areas are under constant illumination. The unique interaction of solar irradiance with this landscape is seen in the south-facing diurnal curve of Figure 8. There are peaks in irradiance at

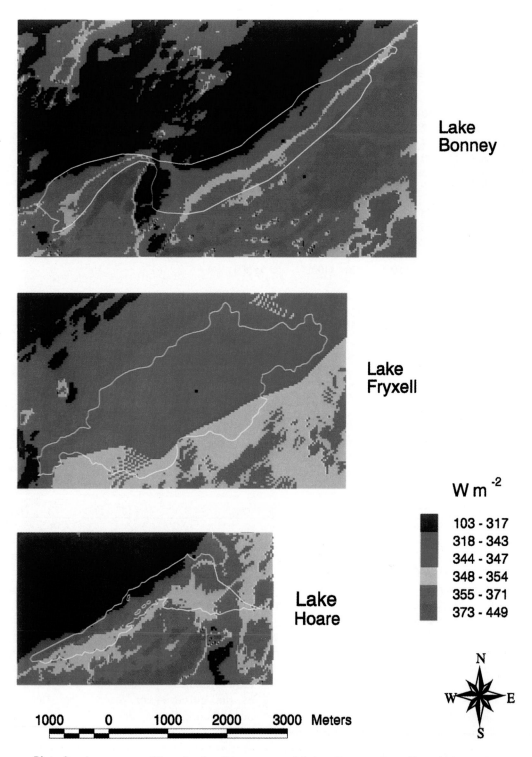

Plate 2. Average monthly solar irradiance maps of Lakes Bonney, Fryxell, and Hoare for December 1994, derived from the topographic radiation model. Images were subset and enlarged from Figure 7. To bring out differences in solar flux, color enhancement was accomplished by equal-area which results in unequal ranges in each of the six light levels. Solar irradiance ranges from 103 (dark blue) to 449 W m-2 (dark red). Small black squares indicate the locations of meteorological stations.

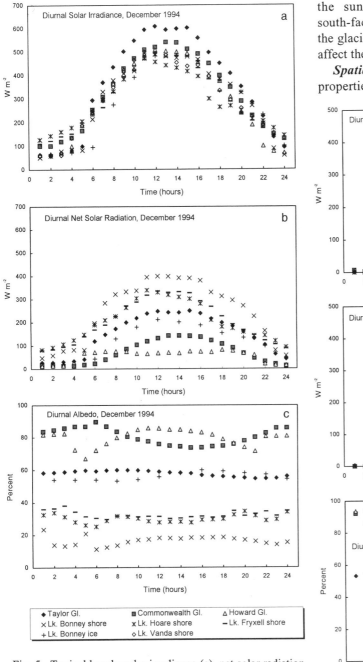

Fig. 5. Typical hourly solar irradiance (a), net solar radiation (b), and albedo (c) at eight meteorological stations during the austral summer in the dry valleys. Hourly values shown were averaged over all days during December 1994. Only solar irradiance is shown for Lake Vanda because reflected shortwave radiation was not measured at that station during this period.

the sun again swings to the south. For steep south-facing slopes, such as those found on some of the glaciers, these bursts of radiation may significantly affect the local energy balance.

Spatial autocorrelation. Spatial autocorrelation properties of the monthly radiation field for Taylor

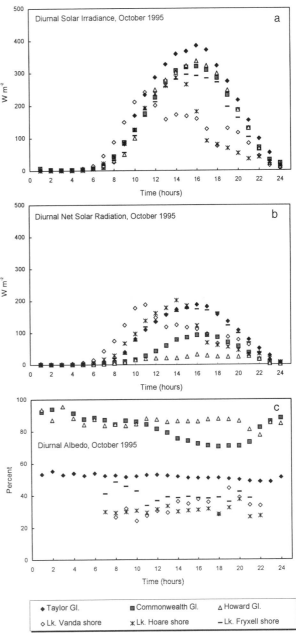

Fig. 6. Typical hourly solar irradiance (a), net solar radiation (b), and albedo (c) for six meteorological stations during the austral spring in the dry valleys. Hourly values shown were averaged over all days during October 1995. During low light conditions between 2200 and 0700 h, albedos at some stations could not be computed.

about 0600 h and 2200 h, when the sun is in the southern half of the sky. As the sun moves towards solar noon, the irradiance drops as these slopes go into shadow and rises again only much later in the day, as

Taylor Valley
Average Incoming Solar Radiation for December 1994

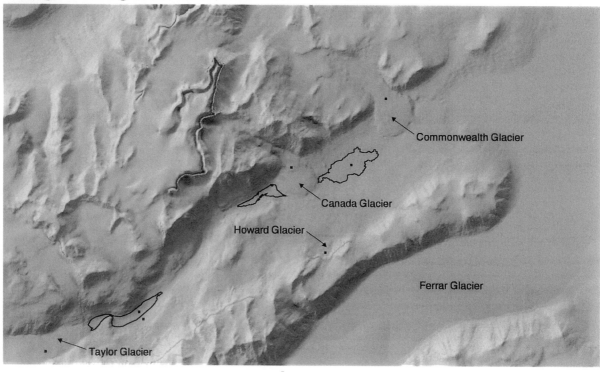

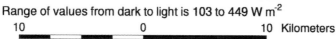

Range of values from dark to light is 103 to 449 W m⁻²

Fig. 7. Average incoming solar radiation in Taylor Valley for December 1994, derived from the topographic radiation model. Grey scale ranges 103 to 449 W m-2, from dark to light. Overlaid onto the image are outlines of the three largest lakes in Taylor Valley, from west to east: Bonney, Hoare, and Fryxell. Meteorological stations are depicted by black squares. Pyranometer data from the Taylor Glacier meteorological station (far lower left corner) were used to drive the model.

Valley are illustrated in the semi-variogram (Figure 9). Semi-variograms are defined as one-half of the squared difference between pairwise combinations of samples over a specific distance. Semi-variograms describe the distance over which solar radiation values are correlated. They are useful for determining over what distances in situ solar radiation measurements can be interpolated and for defining the scale at which solar radiation should be modeled in the dry valleys. The semi-variogram for the Taylor Valley radiation field shows a rapid increase in variability in the first 2000 m; it is within this distance that sites and their associated solar radiation values are more likely to be spatially dependent. Beyond 2000 m, the semi-variogram slowly rises (no sill is ever reached), reflecting non-stationarity in the mean and variance of the field.

The relatively long spatial autocorrelation distance

(i.e., the distance over which solar radiation values are spatially correlated) shown in the semi-variogram is attributable to the long period of integration. As we aggregate over longer time scales, shading, shadowing, and other topographic effects caused by changes in illumination angle become smaller so that the solar radiation field is more uniform than at an instantaneous or hourly time signal. The long spatial autocorrelation distance can also be attributed to the large areas of relatively uniform solar radiation, such as in the bottom of the valleys and the Ferrar Glacier south of the valley (located in the lower right hand corner of Figure 7). In essence, two different radiation regimes can be invoked: one for the large flat areas and another for topographic areas. Because these areas have very different variability and tend to be aggregated into specific geographic regions, the variance of the field is

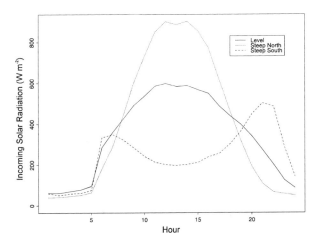

Fig. 8. Typical diurnal irradiance for December 1994 for steep north slopes, steep south slopes, and flat terrain derived from the topographic radiation model. The hourly values are the average of the 31 radiation maps for a particular hour. The magnitude of a steep slope was set at 30° or greater, and a level slope as 2° or smaller. All terrain elements falling into one of the three classes were then used to compute the diurnal curves. These curves thus represent both a temporal average for the month and a spatial average across the image for the particular slope class. Because these are hourly maps, the incoming irradiance is much higher than shown on the monthly map in Figure 7.

not stationary and is different depending upon where it is measured (in the valleys or in the areas of high relief). This results in the continued increase in semi-variance with distance.

Solar radiation distribution across lakes. The monthly radiation map in Figure 7 shows that incident energy was fairly uniform over the valley bottoms. However by enhancing differences not readily visible on the map, we can show that there are differences in the monthly amount and distribution of radiation received by the three lakes in Taylor Valley. Lake Fryxell, which lies in a more open basin than the other two lakes, received an average of 345 W m-2 almost uniformly during December 1994 (Plate 2). In contrast, Lakes Bonney and Hoare, which lie in more confined basins, receive less energy along their northern shores and more energy along their southern shores. These distribution patterns are caused by shadowing due to terrain obstructions that occur during certain times of the day. An example is shown in Figure 10. At 0300 h (and for 5 h prior), the western portion of Lake Hoare is in the shadow of Andrews Ridge lying to the south of the lake. Later in the day, at 1500 h (and for 2 h prior), the northern shore of Lake Hoare is in the shadow of the Asgard Range. In

contrast, Lake Fryxell remains shadow-free during the entire day. The higher intensities of light along the southern shores of Lakes Hoare and Bonney (Plate 2) could be of importance to the distribution and productivity of the under-ice algal mats and phytoplankton communities that inhabit these lakes, as well as microbial communities present in the soils [for a discussion on algal mats, see *Simmons et al.*, 1993; for phytoplankton communities, see *Lizotte and Priscu*, this volume, and *Neale and Priscu*, this volume; for soil communities, see *Campbell et al.*, this volume, and *Freckman and Virginia*, this volume].

Diurnal patterns of solar radiation on glaciers. The time series of hourly radiation maps in Figure 10 (see also accompanying CDROM) also demonstrate a prominent diurnal radiation feature of the glaciers. In all of these maps, the tops of the glaciers receive more solar energy than the terminus sides, except on cliff faces that are receiving direct sunlight. This is illustrated best in the 0900 h map where the east side of the Canada Glacier receives more energy than the surface and at 2100 h when illumination along west and south cliffs of the glacier is greater than the surface. At 1500 h, there is also a peak of radiation on the western side of the glacier on the edge of the shadow. The diurnal nature of illumination on the glacier cliffs is shown in the in situ pyranometer measurements presented in Figure 11. During most of the day, solar flux on the western cliff of the Canada Glacier is much lower than what is received on the surface. However at 1500 h radiation on the cliff equals that of the surface; at 2100 h, cliff irradiance is triple that on the surface. The time between the two peaks corresponds to a

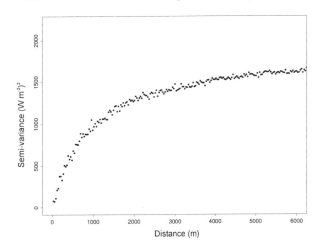

Fig. 9. Semi-variogram generated from the topographic radiation model output illustrating the autocorrelation properties of the monthly map shown in Figure 7.

Canada Glacier Region
Average Hourly Incoming Solar Radiation

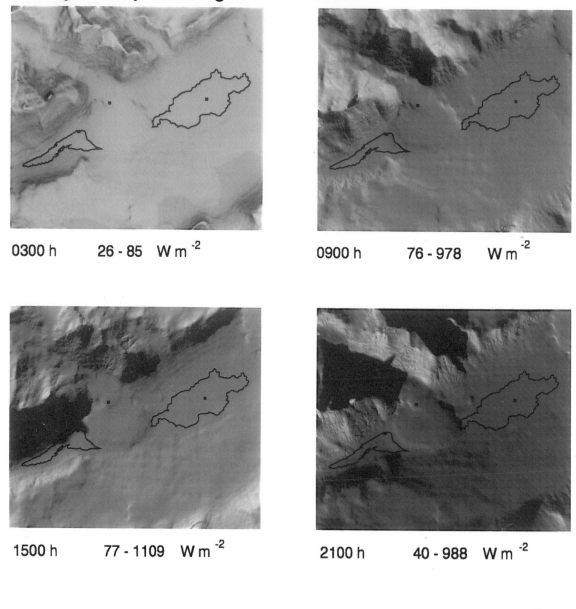

Fig. 10. Typical hourly irradiance of the Canada Glacier region during the austral summer at four times during the day. Each map shown is the average of the hourly values from the topographic radiation model over all days during December 1994. Values presented under each map range from dark to light. Each image has been histogram equalized separately to highlight detail so that direct comparison among grey levels is misleading. Lake Hoare's outline is to the left and Lake Fryxell's to the right of the Canada Glacier. Locations of meteorological stations are shown as black squares.

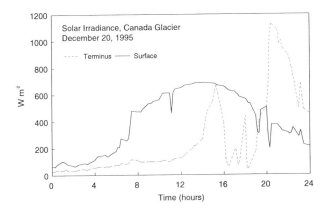

Fig. 11. Solar irradiance on the surface (solid line) and at the terminus (dotted line) of the Canada Glacier for December 20, 1995. Values are 15 min averages. See Figure 2 for the location of the meteorological station on the surface of the Canada Glacier; location of the terminus station was on the west side of the glacier just a few hundred meters from the Lake Hoare meteorological station.

period of terrain shadowing. Although the data presented in Figures 10 and 11 are from two different years, the patterns illustrated should be similar to those observed during any year. However, since the hourly values presented in Figure 10 are averaged over the entire month, their magnitude will be dampened compared to those presented in Figure 11. Because the peaks of solar flux on the terminus are infrequent and of short duration, the average daily solar irradiances on the terminus are lower than on the surface (Figure 12). It has been shown recently that these strong "pulses" of solar radiation on the vertical faces of the glacier are the dominant control on daily streamflow variation [*Chinn*, 1987; *Lewis et al.*, 1996; *Fountain et al.*, this volume].

CONCLUSIONS

We have characterized the spatial and temporal distribution of solar radiation in the McMurdo Dry Valleys using both a network of in situ point measurements and a topographic modeling approach. The topographic model demonstrated the differential receipt of solar radiation which results from topographic variation in the dry valleys. Steep north-facing slopes received more energy than steep south-facing slopes, and the amount received by level areas depended on their sky view factor. In the monthly map derived from the model, values of solar radiation were spatially correlated at lengths of 2000 m or less. From the in situ measurements, we also found that terrain variabil-

ity figured heavily in determining the patterns of solar radiation in the dry valleys at all time scales. The annual average incoming solar radiation to the dry valleys is less than other areas of Antarctica, in part due to obstruction by the high relief of the mountains. Topography could also account for differences in monthly averages and in diurnal patterns of solar radiation among sites. A good example is Lake Hoare, in which solar fluxes are lowered early in the day from shadowing and shading by Andrews Ridge and later in the day due to shading by the Asgard Range. These diurnal patterns of shading lead to low monthly and annual averages of solar radiation compared to sites such as Lake Fryxell and Commonwealth Glacier that have relatively unobstructed views. Mean monthly solar flux at Lake Hoare in the spring and fall when the terrain effects become more severe. The topographic model demonstrated that even when averaged over the month of December, when the sun is at its highest, gradients in incident radiation occurred across Lake Hoare as well as Lake Bonney, with lower solar fluxes present along their northern shores and higher fluxes along their southern shores. Gradients were not present for Lake Fryxell, which had a uniform distribution of solar radiation across the surface.

While terrain features exert a large influence on solar radiation patterns in the dry valleys, other factors such as cloudiness due to coastal proximity and orographic effects may be present. Evidence for greater cloudiness in down valley areas near the coast is suggested by the lower amounts of solar flux and the

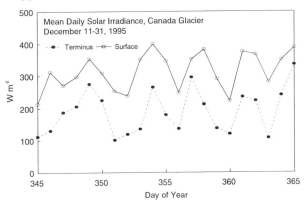

Fig. 12. Mean daily solar irradiance on the surface (solid line, open circles) and at the terminus (dotted line, solid circles) of the Canada Glacier for December 11–31, 1995. See Figure 2 for the location of the meteorological station on the surface of the Canada Glacier. The location of the terminus station was on the west side of the glacier just a few hundred meters from the Lake Hoare meteorological station.

higher amounts of precipitation received by the Commonwealth Glacier (5 km from the coast) compared to the Taylor Glacier (40 km from the coast), which receives a higher solar flux and very low amounts of precipitation. However not all the in situ solar flux measurements demonstrated a valley gradient. It may be that terrain obstruction masks the effects of cloudiness or that the gradient does not exist. We have often observed clouds over the Howard Glacier when other nearby areas are cloud-free. This observation, in conjunction with the low solar fluxes observed at the Howard Glacier, is evidence that orographically-induced clouds influence solar radiation patterns in the dry valleys. A climate model of the dry valleys at the appropriate scale could help characterize patterns of cloudiness and aid in our interpretation of solar radiation data.

The local climate-induced differences in solar irradiance described above can lead to biased output by our topographic solar radiation model which is driven by pyranometer data from only one station. The model distributes solar irradiance across the landscape considering topography and solar angle only and does not account for variations caused by local climate. For the model to accurately capture the distribution of solar irradiance, pyranometer input from multiple stations would be required.

The distribution of net solar radiation is ecologically important since it is that energy which is absorbed. Soil sites in the dry valleys have a higher net radiation than glacier and lake-ice sites because of their lower albedo. However snow cover on any of these surfaces will result in a reduction in net radiation. The decrease in net radiation from 1994 to 1995 that we observed on both the Howard and Commonwealth Glaciers is an example of such an effect. Lake Fryxell provides another example in which a decrease in mean albedo from 1994 to 1995 resulted in higher net radiation. Less snow cover in 1995 is the most plausible explanation for the albedo decrease. Thus seasonal and year-to-year differences in snow cover will dramatically alter the distribution of net solar radiation and will be a critical parameter in models of net radiation in the dry valleys. Topographic models such as the one used in the present study can also be used to map the spatial distribution of net solar radiation [e.g., *Dubayah*, 1992] but require adequate characterization of the reflectance distribution over the landscape. The retrieval of reflectance from satellite data may offer one of the few avenues for effectively characterizing the spatial distribution of albedo that changes temporally

in the dry valleys due to snow cover variations.

Results of the solar topographic model reveal that the cliff faces of the glaciers receive higher amounts of solar radiation than their surface but only during the time of day when the sun's rays directly strike the face. Pyranometer data collected simultaneously from the terminus and the surface of the Canada Glacier show that while peak values of solar radiation on the terminus can be as much as triple that on the surface, its average daily radiation is only about half. While this may be true for the Canada Glacier, the monthly-averaged solar map indicated slightly higher flux levels along the western terminus face on the Commonwealth Glacier and along the northern face of the Howard Glacier, compared to the fluxes found on the surfaces of these two glaciers. This could prove to be an important point when evaluating spatial differences in streamflow, which is strongly controlled by the daily solar cycle.

The spatial variability of solar radiation demonstrated here may bear on distribution and productivity patterns of the biotic communities in the soils, streams, and lakes of the dry valleys. The dominant control that solar radiation exerts on the timing and magnitude of glacial melt and streamflow, which is of critical importance to dry valley ecosystems, has already been discussed. Ice-cover thickness of the lakes, which determines how much light reaches the lake communities, is dependent in part on the amount of sunlight absorbed [*McKay et al.*, 1985; *Adams et al.*, this volume; *Fritsen et al.*, this volume]. Variability in solar radiation could also influence patterns of primary production within the lakes. From diurnal curves of solar irradiance in the spring (October), we see that lakes with more terrain shading, such as Hoare and Vanda, have a shorter period of direct illumination and therefore have a more limited window of time for daily primary production compared to lakes such as Fryxell with fewer terrain obstructions. The gradients in illumination across Lakes Hoare and Bonney demonstrated by the topographic model may result in differential distribution and/or production of algal mats which inhabit the benthic regions of these lakes. While we did not evaluate patterns of solar radiation over stream or soil areas in this study, such pursuits may reveal relationships between solar radiation levels, which influence moisture, heating, light available for photosynthesis, and the distribution and biological processes that occur in these areas.

In closing, we submit that our observations and

analyses of solar radiation in the dry valleys are pertinent to climate change issues in this region of Antarctica. The changing thickness of ice-covers and rising lake levels of some dry valley lakes since the 1970s [*Wharton et al.*, 1993] are indications that the climate of the dry valleys may be changing. As ice-cover thins, *Wharton et al.* [1993] predict that lake productivity will increase due to the increased light allowed to enter the water column. Because solar radiation exerts a primary influence on both ice cover dynamics [*Fritsen et al.*, this volume] and glacial meltwater entering the lakes [*Fountain et al.*, this volume], continued monitoring of solar radiation in the dry valleys and additional modeling of its spatial and temporal distribution will aid in understanding how climate change will influence the overall ecology of the McMurdo Dry Valley ecosystem.

Acknowledgments. This research was supported by National Science Foundation grants OPP-9211773 to R. Wharton and OPP-9117907 and OPP-9419423 to J. Priscu, grants from NASA's Topography and Surface Change Program to R. Dubayah (NAGW-2928) and NASA's Exobiology Program to R. Wharton (NAGW-1947). Additional support came from a DRI Nevada Medal Research Fellowship and a NASA Global Change Fellowship to G. Dana. We thank Peter Doran, Karen Lewis, Paul Langevin, and Paul Sullivan for assistance in the field, and Robert Stone and Robert Davis for their invaluable advice on solar radiation measurements in cold regions. Karen Lewis graciously provided the solar radiation data from the Canada Glacier terminus. The contributions by David Shirey to the solar modeling efforts and by Tim Wade to the GIS data manipulations were greatly appreciated. We also thank the International Center for Antarctic Information Research, Christchurch, New Zealand, for providing the DEM. Finally, we appreciate the comments of two anonymous reviewers, which greatly improved the manuscript.

REFERENCES

Adams, E. E., J. C. Priscu, C. H. Fritsen, S. R. Smith, and S. L. Brackman, Permanent ice covers of the McMurdo Dry Valley Lakes, Antarctica: bubble formation and metamorphism, this volume.

Aguado, E., Local-scale variability of daily solar radiation-San Diego County, California, *J. Climate Appl. Meterol.*, 25, 672-678, 1986.

Bradley, S. G. and P. D. Martin, Intense thermal flows in an Antarctic dry valley, *American Meteorol. Soc., 7th Conference on Mountain Meteorology*, Breckenridge, CO, pp. 57-60, July 17-21, 1995.

Bromley, A. M., Weather observations in Wright Valley, Antarctica, *Infor. Publ. 11*, 37 pp., New Zealand Meteorological Service, Wellington, 1985.

Bull, C., Climatological observations in ice-free areas of southern Victoria Land, Antarctica, in *Studies in Antarctic Meteorology, Antarct. Res. Ser., vol. 9*, edited by M. J. Rubin, pp. 177-194, AGU, Washington, DC, 1966.

Campbell, L. B., G. G. C. Claridge, D. I. Campbell, and M. R. Balks, The soil environment of the McMurdo Dry Valleys, Antarctica, this volume.

Chinn, T. J., Accelerated ablation at a glacier ice-cliff margin, dry valleys, Antarctica. *Arctic and Alpine Res.*, 19, 71-80, 1987.

Clow, G. D., C. P. McKay, G. M. Simmons, Jr., and R. A. Wharton, Jr., Climatological observations and predicted sublimation rates at Lake Hoare, Antarctica. *J. Climate*, 1(7), 715-728, 1988.

Conovitz, P. A., D. M. McKnight, L. M. McDonald, A. Fountain, and H. R. House, Hydrologic processes influencing streamflow variation in Fryxell Basin, Antarctica, this volume.

Dana, G. L., A. G. Fountain, and R. A. Wharton, Jr., McMurdo Dry Valleys LTER: Solar radiation on glaciers in Taylor Valley, Antarctica, *Antarct. J. of the U.S.*, in press, 1996a.

Dana, G. L., M. A. Wetzel, and R. A. Wharton, Jr., Satellite-derived surface temperatures in the McMurdo Dry Valleys, Antarctica, in *Proc. Internat. Radiation Symposium*, in press, 1996b.

Davis, F., S. Schimel, M. A. Friedl, T. G. F. Kittel, R. Dubayah, and J. Dozier, Covariance of biophysical data with digital topographic and land use maps over the FIFE site, *J. Geophys. Res., 97*, 19009-19002, 1992.

Doran, P. T., G. L. Dana, J. T. Hastings, and R. A. Wharton, Jr., The McMurdo LTER Automatic Weather Network (LAWN). *Antarct. J. of the U.S.*, in press, 1995.

Dozier, J., A clear-sky spectral radiation model for snow-covered mountainous terrain, *Water Resources Res., 16*, 709-718, 1980.

Dozier, J., Spectral signature of alpine snow cover from the Landsat Thematic Mapper, *Remote Sensing Environ., 28*, 9-22, 1989.

Dozier, J. and J. Frew, Rapid calculation of terrain parameters for radiation modeling from digital elevation data, *IEEE Transactions Geosci. Remote Sensing, 28*, 963-969, 1990.

Dubayah, R., Estimating net solar radiation using Landsat Thematic Mapper and digital elevation data, *Water Resources Res., 28*, 2469-2484, 1992.

Dubayah, R., A solar radiation topoclimatology for the Rio Grande River Basin, *J. Vegetation Sci., 5*, 627-640, 1994.

Dubayah, R. and S. Loechel, Modeling topographic solar radiation using GOES data, *J. Applied Meteorol., 36(2)*, 141-154, 1997.

Dubayah, R., and P. M. Rich, Topographic solar radiation models for GIS, *Internat. J. GIS, 9*, 405-419, 1995.

Dubayah, R. and V. van Katwijk, The topographic distribution of annual incoming solar radiation in the Rio Grand River basin, *Geophys. Res. Letters, 19*, 2231-2234, 1992.

Dubayah, R., J. Dozier, and F. W. Davis, The distribution of clear-sky radiation over varying terrain, *Proc. IGARSS '89, Vancouver, IEEE No. 89 ch2768-0*, pp. 885-888, 1989.

Dubayah, R., J. Dozier, and F. W. Davis, Topographic distribution of clear-sky radiation over the Konza Prairie, Kansas, *Water Resources Res., 26*, 679-690, 1990.

Duguay, C. R., Radiation modeling in mountainous terrain review and status, *Mountain Res. and Develop., 13*, 339-357, 1993.

Fountain, A. G., G. L. Dana, K. J. Lewis, B. H. Vaughn, and D. M. McKnight, Glaciers of the McMurdo Dry Valleys, Southern Victoria Land, Antarctica, this volume.

Freckman, D. W. and R. A. Virginia, Ecological interactions between soil organisms and their environment, this volume.

Fritsen, C. H., E. E. Adams, C. M. McKay, and J. C. Priscu,

Permanent ice covers of the McMurdo Dry Valleys, Antarctica: liquid water content, this volume.

Garnier, B. J., and A. Ohmura, A method of calculating the direct shortwave radiation income of slopes, *J. Appl. Meteorol., 7,* 796–800, 1968.

Geiger, R. J., *The Climate Near the Ground,* Harvard University Press, Cambridge, 1965.

Grenfell, T. C., S. G. Warren, and P. C. Mullen, Reflection of solar radiation by the Antarctic snow surface at ultraviolet, visible, and near-infrared wavelengths, *J. Geophys. Res., 99, No. 9,* 18,669–18,684, 1994.

House, H.R., D. M. McKnight, and P. von Guerard, The influence of stream channel characteristics on streamflow and annual water budgets for lakes in Taylor Valley, *Antarct. J. of the U.S.,* in press, 1995.

Keys, J. R., Air temperature, wind, precipitation and atmospheric humidity in the McMurdo region, *Publ.17,* 57 pp., Geology Department, Victoria University, Wellington, NZ, 1980.

Kuhn, M., L. S. Kundla, and L. A. Stroschein, The radiation budget at Plateau Station, Antarctica, 1966-1967, Paper 5 in *Meteorological Studies at Plateau Station, Antarctica, Antarct. Res. Ser., vol. 25,* edited by J. A. Businger, pp. 41–73, AGU, Washington DC, 1977.

Lewis, K., A. Fountain, and P. Langevin, The role of terminus cliff melt in the hydrological cycle, Taylor Valley, McMurdo Dry Valleys, *Antarct. J. of the U.S.,* in press, 1996.

Linder, B. L., C. P. McKay, and G. D. Clow, Links between lake ice and climate variations, in *6th Climate Variations, Am. Meteorol. Soc.,* Boston, Massachusetts, pp. 377–378, 1994.

Lizotte, M. P. and J. C. Priscu, Spectral irradiance and bio-optical properties of perennially ice-covered lakes of the dry valleys (McMurdo Sound, Antarctica), in *Contributions to Antarctic Research, Antarct. Res. Ser., vol. 57,* pp. 1–14, 1992.

Lizotte, M. P. and J. C. Priscu, Pigment analysis of the distribution, succession, and fate of phytoplankton of the McMurdo Dry Valley Lakes of Antarctica, this volume.

Lizotte, M. P., T. R. Sharp, and J. C. Priscu, Phytoplankton dynamics in the stratified water column of Lake Bonney, Antarctica. I. Biomass and productivity during the winter-spring transition, *Polar Biology, 16:*155–162, 1996.

McKay, C. P., G. D. Clow, R. A. Wharton, Jr., and S. W. Squyres, Thickness of ice on perennially frozen lakes, *Nature, 313,* 561–562, 1985.

McKay, C. P., J. A. Nienow, M. A. Meyer, and E. I. Friedmann, Continuous nanoclimate data (1985-1988) from the Ross desert (McMurdo Dry Valleys) cryptoendolithic microbial ecosystem, in *Antarctic Meteorology and Climatology: Studies based on automatic weather stations, Antarct. Res. Ser., vol. 61,* pp. 201–207, 1993.

McKay, C. P., G. D. Clow, D. T. Andersen, and R. A. Wharton Jr., Light transmission and reflection in perennially ice-covered Lake Hoare, Antarctica, *Jour. Geoph. Res., vol. 99, C10,* 20,427–20,444, 1994.

McKnight, D. M., H. House, and P. von Guerard, McMurdo LTER: Streamflow measurements in Taylor Valley, *Antarct. J. of the U.S.,* in press, 1994.

Monteith, J. L. and M. Unsworth, *Principals of Environmental Physics,* 291 pp., Routledge, Chapman and Hall, New York, 1990.

Neale, P. J. and J. C. Priscu, Fluorescence quenching in phytoplankton of the McMurdo Dry Valleys lakes (Antarctica): implications for the structure and function of the photosynthetic apparatus, this volume.

Palmisano, A. C. and G. M. Simmons, Spectral downwelling irradiance in an Antarctic lake, *Polar Biol. 7,* 145–151, 1987.

Ragotzkie, R. A. and G. E. Likens, The heat balance of two Antarctic lakes, *Limnol. Oceanogr., 9,* 412–425, 1964.

Riordan, A. J., The climate of Vanda Station, Antarctica, in *Climate of the Arctic,* edited by G. Weller and S.A. Bowling, pp. 268–275, 1973.

Rusin, N. P. Meteorological and radiational regime of Antarctica, Leningrad, (Translated by the *Israel Program for Scientific Translations,* Jerusalem, 1964) 355 pp. 1961.

Simmons, G. M., Jr., J. R. Vestal, and R. A. Wharton, Jr., Environmental regulators of microbial activity in continental Antarctic lakes, in *Physical and Biogeochemical Processes in Antarctic Lakes, Antarct. Res. Ser., vol. 59,* edited by W. J. Green and E. I. Friedmann, pp. 165–195, 1993.

Stone, R., T. Mefford, E. Dutton, D, Longnecker, B. Halter, and D. Endres, 1996. Barrow, Alaska, surface radiation and meteorological measurements: January 1992 to December 1994. *NOAA Data Report ERL CMDL-11.* 81 pp., 1996.

Thompson, D.C., R. M. F. Craig, and A.M. Bromley. Climate and heat balance in an Antarctic dry valley. *New Zealand J. Sci., 14,* 245–251, 1971.

Warren, S. G., Optical properties of snow, *Rev. Geophys., 20,* 67–89, 1982.

Wharton, R. A., Jr., C. P. McKay, G. D. Clow, and D. T. Andersen, Perennial ice covers and their influence on Antarctica lake ecosystems, in *Physical and Biogeochemical Processes in Antarctic Lakes, Antarct. Res. Ser., vol. 59,* , edited by W. J. Green and E. I. Friedmann, pp. 53–70, 1993.

Williams, L. D., R. G. Barry, and J. T. Andrews, Application of computed global radiation to areas of high relief. *J. Appl. Meteorol., 11,* 526–533, 1972.

Gayle L. Dana, Biological Sciences Center, Desert Research Institute, P.O. Box 60220, Reno, NV 89506.

Ralph Dubayah, Department of Geography, Lefrak Hall, University of Maryland at College Park, College Park, MD 20742.

Robert A. Wharton, Jr., Biological Sciences Center, Desert Research Institute, P.O. Box 60220, Reno, NV 89506.

(Received November 19, 1996;
accepted 8 March, 1997)

GLACIERS OF THE MCMURDO DRY VALLEYS, SOUTHERN VICTORIA LAND, ANTARCTICA

Andrew G. Fountain[1], Gayle L. Dana[2], Karen J. Lewis[3], Bruce H. Vaughn[4], and Diane McKnight[5]

The glaciers of the McMurdo Dry Valleys are fundamental to the hydrology and biology of the valleys because they are the only significant source of water. Understanding the controls on the glacial extent and meltwater runoff is fundamental to a process-oriented approach to studying the dry valleys ecosystem. The elevation of the equilibrium-line of the alpine glaciers changes dramatically in the dry valleys, probably a result of large gradients in precipitation. Temporally, they have been relatively constant since the Pliocene (3.5 million years), the furthest extent is not more than a few hundred meters from their present positions. Ablation (all forms of mass loss) from the glaciers is dominated by sublimation, which accounts for more than 70% of the total mass loss. However the magnitude of sublimation is about the same as that of temperate glaciers. The salient difference from temperate glaciers is the relatively small fraction of ablation due to melt and results from a combination of very cold ice and insufficient sensible heat for much melting. The cliff faces are crucial for initiating and maintaining stream flow because they are the first part of the glacier to start melting and the last to stop. During melt periods the distribution of glacier area with altitude can control the response of stream flow to temperature variations.

INTRODUCTION

The McMurdo Dry Valleys region, adjacent to McMurdo Sound at 76°30' to 78° 30'S, 160° to 164°E (Figure 1), is the largest of the ice-free regions in Antarctica, which accounts for 2.4% of the continental area [Drewry et al., 1982]. The physical appearance of the valleys includes sandy-gravel floors with large expanses of exposed bedrock, perennially ice-covered lakes, and glaciers descending from the surrounding mountains. Ephemeral streams flow from the glaciers to the lakes. The lack of extensive ice cover in the Mc-Murdo Dry Valleys is due to two factors. First the

Transantarctic Mountains block much of the flow of the Eastern Antarctic Ice Sheet toward McMurdo Sound [Chinn, 1990]. Second, the little snow that falls in the valley bottom is typically lost to sublimation and does not accumulate. Thus at the level of the valley floor ablation exceeds accumulation during all seasons.

The climate of the dry valleys is that of a cold desert. Average air temperature over the valley floor is about –20°C and the warmest months in the summer average about 1.4°C [Keys, 1980]. In 1995 several meteorological stations in Taylor Valley measured mean annual air temperatures from –16 to –21°C and at one station at Lake Vanda in Wright Valley of –19°C. Precipitation falls as snow, although small amounts of rain occur every several years during the summer [Keys, 1980]. Precipitation is quite low in the valleys, similar to much of Antarctica. For three years of record at Lake Vanda, the average annual snowfall was 0.6 cm water equivalent with an annual maximum of 10 cm and a minimum of 0.6 cm [Bromley, 1985]. From observations in Wright and Taylor Valleys, precipitation decreases westward as the distance from

[1] U.S. Geological Survey, Denver, CO. Currently on leave to Department of Geology, Portland State University, Portland,OR
[2] Biological Sciences Center, Desert Research Institute, Reno, NV
[3] INSTARR, University of Colorado, Boulder, CO and U.S. Geological Survey, Denver, CO
[4] INSTARR, University of Colorado, Boulder, CO
[5] U.S. Geological Survey, Boulder, CO, now at INSTARR, University of Colorado, Boulder CO

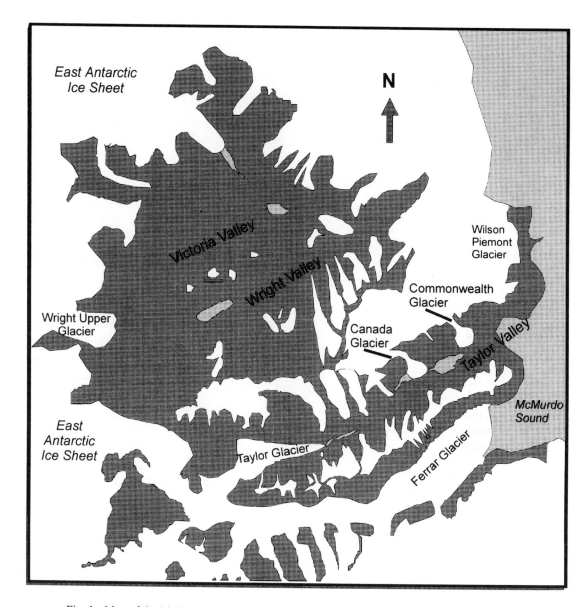

Fig. 1. Map of the McMurdo Dry Valleys, Southern Victoria Land, Antarctica. The Kukri Hills are located between the Ferrar Glacier and Taylor Valley, and the Asgard Range separates Taylor and Wright Valleys. Lake Fryxell is located between the Canada and Commonwealth Glaciers, and Lake Hoare is located on the west side of Canada Glacier.

the ocean increases [*Bull*, 1966; *Keys*, 1980]. Precipitation comes with easterly winds as low pressure systems pass over open water and transfer moist air over the dry valleys [*Bromley*, 1985]. Precipitation is greater in Taylor Valley than in Wright Valley. At the eastern end of Wright Valley, the Wilson Piedmont Glacier is 300–400 m high and forms a precipitation shadow in the valley by blocking the moisture-bearing storms. Winds are typically high with monthly average wind speeds ranging from 2–9 m s^{-1} for Wright Valley

[*Bromley*, 1985] and 2–4 m s^{-1} in Taylor Valley [*Clow et al.*, 1988]. In 1995, the range of mean annual wind speeds we recorded at different meteorological stations was 2.7–3.9 m s^{-1} in Taylor Valley, whereas at Lake Vanda in Wright Valley the mean wind speed was higher, 4.3 m s^{-1}. Similarly, peak daily average winds in Taylor Valley ranged from 11- 15 m s^{-1} and at Lake Vanda the peak was 20 m s^{-1}. The common occurrence of ventifacts in both valleys is a testament to the windy environment.

During the summer months from (late November and to mid-February) glacial melt provides the only source of water for the streams [*Conovitz et al.*, this volume]. Snowfall in the valleys does not contribute to the streams or to the general hydrology because it usually sublimates before melting [*Chinn*, 1981]. One exception to this generalization is the accumulated snow piled against the glacier termini by winds that sweep across the valley floor or by snow drifting off the glaciers. It usually disappears, by melting and/or sublimating, early in the summer season. In Taylor, Wright, and Victoria Valleys, streams transport glacial meltwater to lakes that have no outlets. These enclosed lakes lose water only through sublimation and evaporation. In other valleys, such as Miers Valley (not shown in Figure 1), lakes may have an outlet to the ocean. Life inhabits the waters of the dry valleys; mosses and algae grow in the streams [*Howard-Williams and Vincent*, 1989; *McKnight and Tate*, in press] and benthic algal mats and planktonic microorganisms grow in the lakes [*Simmons et al.*, 1993; *Priscu*, 1995; *Lizotte et al.*, 1996]. These ecosystems depend on glacial meltwater for their environment and for transporting the vital nutrients. Thus glaciers are fundamental to the biology and the hydrology of the dry valleys.

This report characterizes the glaciers of the McMurdo Dry Valleys with emphasis on the Taylor and Wright Valleys where much of the glaciological work to date has been focused. The glacial processes described are those relevant to understanding the mass balance as it relates to the hydrology of the valleys. The resulting conclusions about the glacier fluctuations, processes controlling mass changes, and influence on hydrologic systems will apply generally to glaciers in other valleys.

GLACIER CHARACTERISTICS

Most of the glaciers in the dry valleys are small alpine glaciers averaging a few km^2 in area (Figure 1). These glaciers flow from the mountain ranges that divide the region. In two locations, lobes of the East Antarctic Ice Sheet penetrate the Transantarctic Mountains and terminate in the valleys: Taylor Glacier in Taylor Valley and Wright Upper Glacier in the Wright Valley. In other locations, lobes of the ice sheet divide the dry valleys and flow directly into the ocean (e.g., Ferrar Glacier, MacKay Glacier). Piedmont glaciers extend along much of the coast line and fill the seaward extension of the Wright and Victoria Valleys with ice. They flow towards the ocean and into the

valleys. The topography under the piedmont glaciers rises several hundred meters such that sufficient snow can accumulate to form glaciers [*Calkin*, 1974]. Taylor Valley has an ice-free, low elevation outlet to the ocean.

The pattern of glacierization in the dry valleys is complex and results from an interplay of local topography and climate. For example, glaciers in the Kukri Hills are smaller than those in the Asgard Range, although net annual snow accumulation is about the same [*Fountain et al.*, in press], because the Kukri Hills have a smaller area suitable for snow accumulation. Fewer and smaller glaciers exist in the Olympus Range, despite large areas at elevation, presumably because they reside in the precipitation shadow of several mountain ranges. In Taylor Valley, we observe decreasing snow accumulation on the glaciers at constant altitude with increasing distance from the ocean [*Fountain et al.*, in press]. Also, solar radiation increases with distance from the ocean [*Dana et al.*, in press], which warms the ice and increases ablation. Together these observations suggest a decrease in cloudiness with distance from the ocean and supports anecdotal field observations. This climatic gradient is reflected in the altitude of the equilibrium line of the glaciers, which separates the accumulation zone (annual net snow accumulation), and the ablation zone (snow and ice ablation exceed accumulation). In the dry valleys, the equilibrium-line altitude rises about 1100 m over the 40 km distance from the ocean (Figure 2).

The dry valley glaciers are polar glaciers; their interior and basal temperature is well below freezing, closely matching the mean annual air temperature of −20°C [*Holdsworth and Bull*, 1968]. Inspection of the base of glacier margins in the dry valleys, further indicates that the glaciers, like most glaciers in Antarctica, are frozen to the ground underneath. Tunnels dug into the glacier margin indicate that the interior ice is also frozen to the base [*Holdsworth and Bull*, 1968]; (Conway, personal communication; Fitzsimmons, personal communication). The combined effect of geothermal heat flux and frictional heating produced by the glacier movement is insufficient to warm the basal ice to melting temperature. This may not be so for Taylor Glacier, a tributary of the East Antarctic Ice Sheet, which may produce sufficient frictional heat to melt the basal ice under parts of the lower glacier [*Robinson*, 1984].

The temperature of a glacier is important for two reasons. First, a glacier frozen to the bed moves

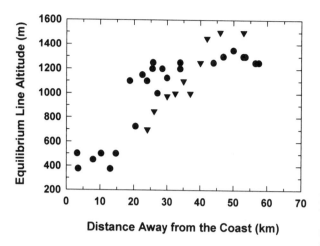

Fig. 2. Equilibrium line altitude for glaciers in Taylor Valley (circles) and in the Wright Valley (triangles).

entirely from the internal deformation of the ice and more slowly than a temperate glacier that also slides along its base. The ice streams, which drain inland ice to the Ross Ice Shelf, best exemplify the contrast between these two situations. The streams are lubricated at the base and flow at speeds of about 500 m yr[-1], much faster than the non-streaming ice that flows at only about 5 m yr[-1] [*Whillans et al.*, 1987]. Comparing individual alpine glaciers is more difficult because of differences in slope and width. However the flow speed of several glaciers in the Wright Valley ranges from 0.1 to 3 m yr[-1] [*Bull and Carnein*, 1970; *Chinn*, 1981]. For comparison, South Cascade Glacier, a relatively small temperate glacier in the North Cascade Range of Washington, USA, exhibits flow speeds of 2–20 m yr[-1] [*Meier and Tangborn*, 1965].

Second, ice temperature influences the movement of meltwater. On temperate glaciers, water flows from the surface, via crevasses, into interior and basal passages. These passages converge and appear at the base of the glacier margin as discrete streams flowing from the glacier [*Paterson*, 1994]. Subsurface water passages can only develop when the heat carried by surface meltwater exceeds the heat conducted into the ice. In the dry valleys, the conditions of cold ice and low meltwater flux results in the water freezing in crevasses and eliminates the development of an internal drainage system. Thus runoff is restricted to surface, or near-surface, streams that cascade off the edge of the glacier.

Another characteristic feature of most dry valley glaciers is the steep terminus, which usually forms a near-vertical cliff. Many glaciers in Taylor Valley have ice cliffs that range from 70°–90° from horizontal.

The specific cause of these cliffs is not known, but they probably result from a number of factors. For glaciers terminating on land, steep marginal cliffs are associated with advancing glaciers because ice advection to the terminus exceeds ablation and the glacier terminus steepens and advances. Also, glaciers move faster at the top than at the base because of friction against the base. That temperate glaciers do not form 20 m high near-vertical cliffs like those of polar glaciers probably results from the difference in ablation rates (see the section on Glacier Mass Balance) and angle of the sun. For glaciers of the dry valleys, the terminus near the base ablates 3 to 5 times more that the sub-horizontal surface at the top of the terminus. This is due, in part, to the difference in air temperature between the air close to the ground and a few tens of meters above. Summer air temperatures in the dry valleys are typically around freezing and when the temperature is just below freezing, solar radiation can heat the ice to the melting point. The solar angle is lower at polar latitudes and the solar intensity on a vertical ice cliff is 3 to 5 times greater than on the sub-horizontal surface [*Lewis et al.*, in press]. Thus the terminus can be warmed to melting temperatures while the sub-horizontal surface remains below freezing. These factors reinforce the vertical slope of the ice cliff. By comparison, the solar angle in temperate latitudes is much greater and most termini face north or east, which reduces ablation on a vertical terminus cliff compared to the sub-horizontal surface on top of the terminus. These conditions work towards reducing the height of the terminus cliff.

GLACIAL HISTORY

Fluctuations in the extent of the Ross Ice Shelf and outlet glaciers of the East Antarctic Ice Sheet (e.g., Taylor Glacier) dominate the glacial history of the dry valleys. Taylor Glacier advanced to the western end of Lake Fryxell about 70,000–100,000 yr B.P. [*Hendy et al.*, 1979]. A lobe of the Ross Sea Ice Sheet then entered Taylor Valley from McMurdo Sound about 23,800 yr B.P. and also eventually reached the western end of Lake Fryxell [*Denton et al.*, 1989]. The ice lobe retreat was underway about 13,000 C[14] yr B.P. and retreated from the valley nearly 6,000 C[14] yr B.P. During the same time the ice lobe was entering Taylor Valley, the Wilson Piedmont Glacier thickened and expanded toward the coast [*Hall and Denton*, 1995] and into the Wright and Victoria Valleys. The expansions of the ice lobe and Wilson Piedmont Glacier coincide with the presence of large lakes that filled much of the Taylor, Wright, and Victoria Valleys [*Hall et al.*, in press]. These lakes were up to several hundred meters deep and the present-day lakes are only small remnants [*Hall et al.*, in press]. For a detailed review of the paleolimnology see *Doran et al.* [1994]. We speculate that if the glaciers were close to the positions observed today that high lake levels flooded the lower elevations of several glaciers and floated some. This would have dramatically affected the glaciers. Land-based glaciers near floating experience large mass losses through calving [*Brown et al.*, 1982] and unless snowfall during that time increased to offset the loss, the glaciers would have certainly thinned and retreated to near the lake margin. Once the lakes reduced in size and the glaciers would be thicken and readvance.

Geologic evidence suggests that the alpine glaciers in the dry valleys have not significantly advanced from their present position since the Pliocene (~3.5 million years). Such evidence includes dating of alpine drift in Wright Valley [*Hall et al.*, 1993] and on the lack of glacial reworking of cinder cones next to alpine glaciers in Taylor Valley [*Wilch et al.*, 1993]. Currently, the alpine glaciers are in their most extended position relative to the past 12,000 to 14,000 years [*Denton et al.*, 1989]. Historic photography of the glaciers in the dry valleys started in 1911 with Scott's expedition, which sent a party into the dry valleys [*Taylor*, 1922]. The party came down the Taylor Glacier and photographed Taylor, Rhone, and Canada Glaciers in Taylor Valley. In 1957, the glaciers were rephotographed and a comparison showed that no detectable change occurred during the 46 intervening years [*Péwé and Church*, 1962]. Subsequently *Chinn and Cumming* [1983] initiated a photographic survey in the Wright and Taylor Valleys and based on a comparison of photos taken at a 5–6 yr interval in the late 1970s concluded that most glaciers were advancing.

GLACIER MASS BALANCE PROCESSES

Glaciers exist where snow accumulation exceeds loss and sufficient mass accumulates to form ice that moves downhill [*Paterson*, 1994]. Their size depends on a balance between mass accumulation and ablation. Like most glaciers, those of the dry valleys exhibit a gradient of mass balance with elevation (Figure 3). This gradient changes from year to year depending on snow accumulation patterns, which, in turn, are effected by the winds. We have observed snow accumulations patterns to spatially vary from storm to storm. In

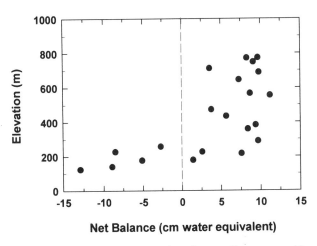

Fig. 3. Mass balance versus elevation on Commonwealth Glacier, November 1994 to November 1995.

temperate zones, changing patterns of snow accumulation may not significantly affect the glacier balance, but in the dry valleys with low ablation rates, snow accumulation on an ice surface may persist for years and protect that ice surface from ablation.

Net annual snow accumulation is small in the dry valleys, on the order of 0.1 m water equivalent, averaged over the accumulation area. This value is typical of Antarctica but contrasts sharply with glaciers in temperate regions that have accumulations of order 1–10 m. Correspondingly, the annual mass loss in the ablation zone is on the order of 0.1 m. The estimated time scale response of the glaciers in the dry valleys, based on the mass loss at the terminus and average thickness of the glaciers (order 100 m), is of order 1000 years [*Johannesson et al.*, 1989]. Therefore the glacier advances observed today results from conditions roughly 1000 years ago.

Glaciers in the dry valleys accumulate snow any time of the year [*Chinn*, 1985]. During the 1970s in the Wright Valley, net accumulation was largest during the summer season, which coincides with the season of greatest ablation. This situation is similar to the conditions on the Tibetan Plateau [*Ageta*, 1994]. In the early 1990s, however, glaciers in Taylor Valley gained the most snow during the winter [*Fountain et al.*, in press]. Comparing observations between valleys is difficult because of the time interval and because a single snowstorm can alter significantly the seasonal mass balance. Average seasonal snow accumulation on the glaciers in the Wright Valley ranged up to 4 cm of water equivalent [*Chinn*, 1980]. In Taylor Valley, average snow accumulation ranged up to 8 cm [*Fountain et al.*, in press]. The difference in these values is consistent with other observations that suggest more precipitation in Taylor Valley.

Not all glaciers in the valleys form from snow accumulation. Some glaciers, like the lower Calkin in Taylor Valley, exists by the accumulation of calved ice at the base of a cliff on the top of which a glacier terminates.

The most important ablation process in the dry valleys is sublimation, which accounts for 70–90% of the total ablation. Melting accounts for about 15–30% of the ablation and calving accounts for 1–3% [*Bull and Carnein*, 1970; *Lewis et al.*, 1995]. The large fraction of sublimation is common for Antarctica [*Clow et al.*, 1988; *Stearns and Weidner*, 1993; *Bintanja and van den Broeke*, 1995]. Thus for typical annual ablation values of roughly 10 cm water equivalent [*Chinn*, 1980; *Fountain et al.*, in press], about 7–9 cm are lost to sublimation These values of sublimation are similar to that measured on Hintereisferner, a temperate glacier in Austria, and probably typical of temperate glaciers overall [*Kaser*, 1982]. Unlike the dry valleys, the most important component of ablation at Hintereisferner is melting, which accounts for 97% of the total, while sublimation represents only 3% [*Greuell and Oerlemans*, 1989].

Calving of ice from the cliff margins largely occurs where the glacier is expanding into the valley and the ice cliffs are steepest. Comparatively little calving occurs where the cliffs are forming as the glacier flows out of the accumulation basin and cliff heights are 10 m or less. We observe no seasonal trend, between winter and summer, to the calving, suggesting that calving results from the stress and strain of glacier flow rather than fatigue caused by daily thermal expansion and contraction due to solar heating [*Fountain et al.*, in press]. Calving from dry valley glaciers is an exfoliation of a thin skin of ice, about 0.5 m thick, over the cliff. In summer, the ice talus is more rapidly ablated as it lies on the warm and dark soil compared to ice calved during the winter.

Cliff ablation by sublimation and melt is typically 5 to 10 times that of the lateral glacier surface on top of the cliffs with annual values ranging from 20 to 50 cm, water equivalent [*Fountain et al.*, in press]. There are several reasons for the enhanced ablation. The vertical cliffs present a surface more perpendicular to the sun's rays compared with the top surface of the glacier. At the summer solstice, on December 21, the solar incident angle of about 55° from the normal to a horizontal surface (zenith), and 35° from the normal of a vertical surface. Thus vertical surfaces receive more intense solar radiation than horizontal surfaces because the flux density of incident radiation is proportional to the cosine of the angle between the sun and the normal to the surface [*Dana et al.*, this volume]. For a short period each day the difference in measured intensity is large. Simultaneous measurement that horizontal ice surface rarely exceeds 300 W m^{-2}, whereas radiation on cliff faces often exceeds 500 W m^{-2} [*Lewis et al.*, in press]. Although daily average radiation on the cliff faces is much less than on top of the glaciers (50 and 150 W m^{-2}, respectively), peak intensity on the cliffs may be just sufficient to melt the ice. The terminus cliffs are also at the lowest elevation and close to the soil covered valley floor and therefore are warmer than the top surface of the glacier. Furthermore, unlike the horizontal surfaces, the ice cliffs do not become covered with snow and maintain comparatively lower albedos all season. The enhanced ablation makes a relatively small contribution to the total mass balance of the glacier because the area of the cliff face is small compared to that of the lateral surface on top of the glacier.

MELT PROCESSES AND FLOW HYDRAULICS

Melting of ice is the only significant source of water to the streams and lakes in the dry valleys [*Chinn*, 1981], therefore to understand variations in stream flow and lake level changes knowledge of glacier melt is required. Snow and ice melt occur below 1,500 m elevation [*Chinn*, 1981], however the elevation limit for runoff must be much lower. For example, we commonly observed on Commonwealth Glacier at an elevation of 400 m, ice lenses in the snowpack with dry snow above and below. This indicates that surface melt refroze in the snowpack with no runoff from that elevation. On the ice surface in the ablation zone, the elevation limit to melt contributing to runoff is difficult to detect because of evaporation. For the glaciers descending to low altitudes, large melt water channels are incised in the lower ablation zone. However we have rarely seen water flowing in the channels during our field observations from December 1992 to January 1996, except very close to the glacier margin. Our observations may be restricted to unusually cold years with meltwater produced only close to the glacier margin and on the ice cliffs. Certainly warmer summers have occurred in the past to create the observed channels.

The ice cliffs are particularly important sources for

TABLE 1. Comparison of Andersen Creek FlowVolume and Adjacent Cliff Melt

Summer (Nov-Jan)	Stream Volume (m^3)	Cliff Ablation (m)	Total Melt (m^3)	Fraction of Stream Volume
1993-1994	53252	0.34	13600	0.26
1994-1995	15988	0.16	6400	0.40
1995-1996	39121	0.16	6400	0.16

stream flow because they are the first part of the glacier to melt in the spring, continue to melt during otherwise cool summer periods, and the last to melt in the autumn. These characteristics result from enhanced radiation on the cliff face and warmer air temperatures due to the proximity to the warm ground, as previously described. In certain topographic situations, radiation from a nearby rock valley wall can accelerate ablation of these cliffs [Chinn, 1987]. The aspect of the cliff face also determines the timing of peak stream flow near the glacier [Lewis et al., in press]. Evidence for the importance of cliffs on stream flow can be seen in two ways. First, the ablation of the cliff face is 5 to 10 times greater than on the adjacent top surface, and evaporation is reduced because the length of the flow path to the streams is much shorter than for the flow on top of the glacier [Fountain et al., in press]. Second, a comparison of seasonal stream volume and wall ablation indicates that the melt from the walls account for a large fraction of the runoff. Andersen Creek drains the west side of Canada Glacier and is part of the stream measurement program in Taylor Valley [House et al., in press]. We estimate that the ice cliff bordering the stream is about 20 m tall and 2000 m long, yielding an area of 4×10^4 m^2. By assuming that the measured values of cliff ablation from Canada Glacier along Andersen Creek are equal to the average melt of the cliff area draining to the creek we can estimate the contribution of cliff melt to stream flow (Table 1). For the 3 years of observations to date, the ice cliff has contributed 16–40% of the total stream flow in Andersen Creek. The potential drainage area of the top of the glacier (excluding the ice cliffs) to the creek is roughly 1.13×10^6 m^2 and the total area is 1.17×10^6 m^2. Thus about 3% of the total drainage area contributes 5–13 times as much melt as the remaining drainage area.

As previously mentioned the flow of glacial meltwater is restricted to the surface of the glaciers. However the hydraulic system of these polar glaciers may not be quite that simple. Water may also flow in a shallow zone of near-surface ice, presumably along the crystal boundaries of the ice crystals, much like near-surface inter-flow in soils. Intergranular veins can form when the ice approaches the melting temperature [Nye and Frank, 1973]. Solar radiation can enhance this process. This hypothesis is based on two observations: 1) shallow cylindrical holes drilled into the ice filled with meltwater and refroze during a period when we observed no meltwater on the ice surface, and 2) we have observed a seepage face on the ice surface such that water percolated from the ice into a surficial channel. Water may form at the surface of the glacier and percolate along the crystal boundaries [Nye and Frank, 1973] into the near surface zone of the ice and/or melt water may also be generated inside the ice. Subsurface melting results from a greenhouse effect, whereby the heat flux to the interior ice from solar radiation exceeds conductive heat losses through the surface [Paige, 1968]. In the absence of phase changes, the heat budget in the ice can be expressed as,

$$\rho c \frac{\partial T}{\partial t} = \frac{\partial}{\partial z}(K \frac{\partial T}{\partial z}) + I(z) \qquad (1)$$

where ρ is the ice density and c specific heat capacity, respectively, T is the temperature, K is the thermal conductivity, I is the solar radiation, and t and z are the coordinates of time and depth in the ice, respectively. The term on the left represents the change in ice temperature with time at a depth z. The first term on the right is the heat conduction at a depth z and the second term is the rate of heat gain at depth z from solar radiation. Thus when the heating due to solar radiation exceeds the heat conducted away to the colder ice at the surface and at depth, the temperature of the ice will warm with time. The heat gained from solar radiation can be expressed as,

$$I(z) = \mu I_o (1-\alpha) e^{-\mu z} \qquad (2)$$

where, I_o is the solar radiation incident on the ice surface, α is the albedo of the ice, and, μ is the extinction coefficient. *Brandt and Warren* [1993] showed, theoretically, that the penetration of radiation into clear ice can increase its temperature by 14°C. It is an interesting coincidence that surface melt has been observed at air temperatures as low as −15°C [*Chinn*, 1990]. In the extreme case subsurface melting creates subsurface pools. They only occur in regions where the ice is sufficiently clear and free of bubbles [*Endo*, 1970] and have been observed in sea ice [*Endo*, 1970; *Ishikawa and Kobayashi*, 1985] and in glacier ice [*Paige*, 1968]. These pools are common on the lower part of Canada and Taylor Glaciers in Taylor Valley and range in size up to about 10 m in diameter and depths up to a meter have been observed. Lake ice covers in the dry valleys have also been shown to melt internally [*Fritsen et al.*, this volume]. The processes that create subsurface water are not limited to the greenhouse heating effect. We have observed thin ice covers, formed during a cold period, covering flowing streams. Also, apparently frozen water falls were observed to be hollow with water falling inside.

The importance of subsurface melt and subsurface water flow to stream flow from the glaciers is currently unknown. The magnitude of subsurface melt is uncertain at this time, moreover, it is uncertain that subsurface melt can reach a stream channel if part of the path is in shadow. However it has two potentially significant implications. First subsurface melt is a component of ablation that is not measured directly from surface measurements of ablation stake heights. Therefore ablation is underestimated and so is the calculated melt, if based on the difference between calculated sublimation and stake height. Second subsurface flow is not subject to evaporation that affects surface transport; thus the melt flux created should reach its destination.

INFLUENCE ON STREAM FLOW AND LAKE LEVELS

The response of stream flow, and hence lake levels, to glacier melt is expected to be different depending on the distribution of glacier area with elevation in the watershed [*Chinn*, 1993]. Figure 4 illustrates the basic concept. As the freezing level in the atmosphere rises, more area of Canada Glacier will be exposed to melting temperatures more quickly, compared with the Suess Glacier. Therefore we expect that streams flowing from the Canada Glacier to have a more flashy response to increasing air temperatures than the Suess Glacier. However the *rate* of increasing stream flow will peak when the freezing level reaches 200 m, after

which the *rate* of increase will slow. The 200 m elevation represents the maximum glacier area per elevation contour in the ablation zone. A simple calculation can be made by applying a glacial melt model that depends on temperatures alone. The increase in summer lake level is directly related to air temperature [*Wharton, et. al.*, 1992]. For our purposes, we consider the following model,

$$M = \begin{matrix} 0.1T \, for \, T > 0 \\ 0 \, for \, T \leq 0 \end{matrix} \qquad (3)$$

Where M is the daily melt in cm of water and T is the average daily air temperature in °C. The temperature change with elevation is assumed to be 7 °C per 1000 m [*Keys, 1980*],

$$T = -0.007 \, (z\text{-}z_0) + T_0 \qquad (4)$$

where z is the elevation above datum and T_0 is the temperature at the datum elevation, z_0. The melt from a glacier as a function of area above freezing is,

$$G = \int_{z_o}^{z} M(z)W(z) \, dz \qquad (5)$$

Equation 5 is solved using Equations 3 and 4 and knowing the width-elevation distribution of the glaciers (Figure 4). Figure 5 shows the results of Equation 5; meltwater flux from the glacier increases as the surface temperature (freezing level) increases. The more rapid increase of meltwater from Canada Glacier versus Suess Glacier is a consequence of the different distributions of area with elevations. The area of Canada Glacier rapidly increases with elevation up to 200 m, while the Suess Glacier maintains a relatively narrow and unchanging profile (Figure 4). The model is certainly not quantitatively accurate because it does not include the other meteorological variables, solar radiation and sublimation, that affect glacier mass balance and water runoff. The intent is to provide a sense of the effect of glacier area distribution with elevation on stream flow and lake level rise. This is an important consideration when attempting to compare stream flow variations and summer lake level rises.

CONCLUSIONS

The climatic environment of the McMurdo Dry Valleys is a polar desert, with low annual precipitation in the bottom of the dry valleys of about 0.6 cm water equivalent, mean annual air temperatures of about –20°C, and almost constant mean annual winds of 2–9 m s^{-1}. The spatial distribution and spatial characteristics of dry valley glaciers are quite variable and dramatically affects the ephemeral streams and perennially ice-covered lakes. The glaciers are large and numerous in the Asgard Range, but they are small and few in number only 20 km away in the Olympus Range. Furthermore the equilibrium line altitude of the glaciers dramatically rises about 1000 m across a distance of about 40 km in a direction away from the ocean. We believe that these characteristics result from strong gradients of precipitation and solar radiation, such that precipitation decreases and solar radiation increases with distance from the ocean. Although the climate is spatially variable, the temporal variation over large time scales is small. Geologic evidence indicates that the alpine glaciers have not changed more than a few hundred meters in the past 3.5 million years. In contrast many alpine glaciers in the northern hemisphere have retreated several hundred meters since the little ice age a few hundred years ago. Currently glaciers in the dry valleys are in their most extended position for the past 12,000 to 14,000 years, and in historic times they seem to be advancing.

The main component of glacier ablation is sublimation, which accounts for more than 70% of total ablation. Melting accounts for no more than 25% of the ablation and calving of ice makes up 5% or less.

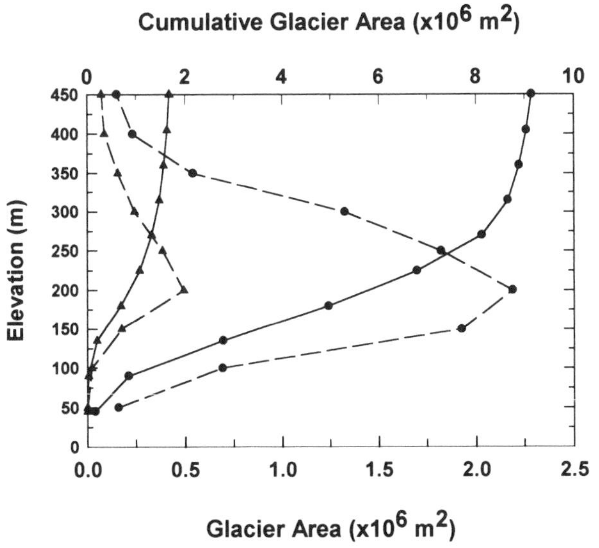

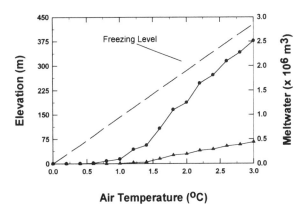

Fig. 5. Potential daily meltwater flux as a function of daily freezing level. Circles and solid line indicate Canada Glacier and triangles and dashed line indicate Suess Glacier.

The relatively large component of sublimation results from the windy, dry, and cold environment. The magnitude of sublimation is about 9 cm of water equivalent, similar to temperate glaciers. An important distinguishing difference for the ablation of temperate and dry valley polar glaciers is sensible heat. The relative lack of sufficient sensible heat in the dry valleys requires that much of the solar radiation is used for warming the ice to melting temperatures, while in temperate glaciers a greater proportion of the solar radiation goes directly into melting the ice. During the summer, when air temperatures warm to near freezing, glacier melt initiates and water begins to flow. The flow is restricted to the surface of the glacier because the body of the ice is close to the mean annual temperature of –20°C preventing an internal hydraulic system from developing. Some water may flow in the near-surface ice where heat gained by solar radiation exceeds the rate conducted away. The interior ice can warm to melting temperatures although the surface skin may be below freezing. Anecdotal evidence suggests that water may seep through this warm layer of ice. Water flow from the ice cliffs is particularly important to stream flow. Because cliffs are vertical, they receive more intense solar radiation than the lateral upper surface. They are also the lowest part of the glacier and nearest the warm ground. For these reasons the cliffs initiate the melt in the spring and maintain the stream flow during cool periods during the summer and extend runoff later into the season. Furthermore the aspect of the cliffs control the timing of peak stream flow during cool periods.

The distribution of glacier area with elevation is an important factor in determining stream runoff and lake

Cumulative Glacier Area (x10^6 m^2)

Fig. 4. Area-elevation curves for Canada Glacier (circles) and Suess Glacier (triangles). The solid lines are cumulative area with elevation and dashed lines are glacier areas for 50 meter elevation intervals.

level rise. Glaciers that fan out on the valley floor have relatively large areas at low elevations and create highly variable stream flows. A non-linear response of stream flow to changing freezing level results from the non-linear distribution of glacier area with elevation in the ablation zone. Under the same conditions, a glacier that has parallel sides will increase stream flow in a linear fashion. The glaciers of the dry valleys vary significantly in shape and altitude. Accordingly the rate of change in stream discharge will differ from glacier to glacier. Summer lake levels will also vary individually, depending on the distribution and hypsometry of glaciers in each watershed. However the rise in level is also the aggregate of the contribution of all the glaciers in each watershed, which will smooth the unique responses of each.

Acknowledgments. Funding for our research was provided by the National Science Foundation's Office of Polar Programs grant OPP-9211773. We acknowledge the assistance of many field team members who have helped with the collection of the data and made this report possible. We are particularly grateful to Harry House, Paul Langevin, and the members of the Berg Field Center mountaineering field guides including Steve Dunbar, Bill McCormick, Brooks Montgomery, and Kim Reynolds.

REFERENCES

Ageta, Y., Characteristics of mass balance of summer accumulation type glaciers in the Himalayas and Tibetan Plateau, in Glacier mass balances: measurements and reconstructions, *Univ. Innsbruck, Inst. of Meteorology and Geophysics*, 1994.

Bintanja, R., and M.R. van den Broeke, The surface energy balance of Antarctic snow and blue ice. *J. App. Met.,* 34, 902–926, 1995.

Brandt, R.E., and S.G. Warren, Solar-heating rates and temperature profiles in Antarctic snow and ice, *J. Glaciol.*, 39, 9910., 1993.

Bromely, A.M., Weather observations Wright Valley, Antarctica, *Infor. Publ.* 11, 37 pp, New Zealand Meteorological Service, Wellington, 1985.

Brown, C.S., M.F. Meier, and A. Post, Calving speed of Alaska tidewater glaciers, with applications to Columbia Glacier, U.S. Geological Survey Professional Paper 1258–C, 13 pp., 1982.

Bull, C., Climatological observations in ice-free areas of Southern Victoria Land, Antarctica, in *Studies in Antarctic Meteorology, Ant. Res. Ser.,* vol. 9, edited by M.J. Rubin, pp.177–194, AGU, Washington, DC, 1966.

Bull, C., and C.R. Carnein, The mass balance of a cold glacier: Meserve Glacier, south Victoria Land, Antarctica, in *International Symposium on Antarctic Glaciological Exploration, Int. Assoc. Hydrol. Sci.*, Publ. 86, edited by A.J. Gow and others, pp. 429–446, 1970.

Calkin, P.E., Subglacial geomorphology surrounding the ice-free valleys of Victoria Valley system, southern Victoria Land, Antarctica, *J. Glaciol.*, 69, 415–429, 1974.

Chinn, T.J., Glacier balances in the dry valleys area, Victoria Land, Antarctica, Int. Assoc. Hydrol. Sci., publ. no. 126, 237–247, 1980.

Chinn, T.J., Hydrology and climate in the Ross Sea area, *Jour. Royal Soc. N. Z.*, 11, 373–386, 1981.

Chinn, T.J., Structure and equilibrium of the dry valley glaciers, *N.Z. Ant. Rec.*, 6, 73– 88, 1985

Chinn, T.J., Accelerated ablation at a glacier ice-cliff margin, dry valleys, Antarctica, *Alpine and Arctic Res.*, 19, 71–80, 1987.

Chinn, T.J., The dry valleys in Antarctica: the Ross Sea Region, 137-153, *Department of Scientific and Industrial Research, Wellington, NZ*, 1990.

Chinn, T.J., Physical limnology of the dry valley lakes, in *Physical and biogeochemical processes in Antarctic lakes, Antarct. Res. Ser.,* vol. 59, edited by W.J.Green and E.I. Friedmann, pp. 1–51, AGU Washington, D.C., 1993

Chinn, T.J., and R.J. Cumming, Hydrology and glaciology, Dry Valleys, Antarctica annual report for 1978-79, *Rep WS 810*, N.Z. Min. of Works Develop., Christchurch, 1983.

Clow, G.D., C.P. McKay, G.M. Simmons, Jr., and R.A. Wharton, Jr, Climatological observations and predicted sublimation rates at Lake Hoare, Antarctica, *J. Climate*, 1, 7, 715–728, 1988.

Conovitz, P.A., D.M. McKnight, L.M. McDonald, A.G. Fountain, and H.R. House, Hydrologic processes influencing streamflow variations in Fryxell Basin, Antarctica, this volume.

Dana, G.L., A.G. Fountain, and R.A. Wharton, Jr., Solar radiation on glaciers in Taylor Valley, Antarctica, *Antarctic J. of the U.S.*, in press.

Dana, G.L., R.A. Wharton Jr., and R. Dubayah, Solar Radiation in the McMurdo Dry Valleys, Antarctica, this volume

Denton, G.H., J.G. Brockheim, S.C. Wilson, and M. Stuiver, Late Wisconsin and early Holocene Glacial History, Inner Ross Embayment, Antarctica, *Quat. Res.*, 31, 151–182, 1989.

Doran, P.T., R.A. Wharton, Jr., and W.B. Lyons, Paleolimnology of the McMurdo Dry Valleys, Antarctica, *J. Paleolimnol.*, 10, 85–114, 1994.

Drewry, D.J., S.R. Jordan, and E. Jankowski, Measured properties of the Antarctic ice sheet: surface configuration, ice thickness, volume and bedrock characteristics, *Ann. Glaciol.*, 3, 83–91, 1982.

Endo, Y., Puddles observed on sea ice from the time of their appearance to that of their disappearance in Antarctica, *Low Temp. Sci., Ser. A*, 28, 203–213, 1970.

Fountain, A.G., K.J. Lewis, G.L. Dana, Spatial variation of glacier mass balance in Taylor Valley, Antarctica, *Antarctic J. U.S.*, in press.

Fritsen, C.H., E.E. Adams, C.M. McKay, and J.C. Priscu, Permanent ice covers of the McMurdo Dry Valley lakes, Antarctica: Liquid water content, this volume.

Greuell, W., and J. Oerlemans, Energy balance calculations on and near Hinteriesferner (Austria) and an estimate of the effect of greenhouse warming on ablation, in *Glacier fluctuations and climatic change*, edited by J. Oerlemans, pp. 325–343, Kluwer Academic Publishers, Boston, 1989.

Hall, B.L., and G.H. Denton, Late Wisconsin/Holocene history of the Wilson Piedmont Glacier, *Antarctic J. U.S.*, 24, 5, 20–22, 1995.

Hall, B.L, and G.H. Denton, Late Quaternary lake levels in the dry valleys, Antarctica, *Antarctic J. U.S.*, in press.

Hall, B.L., G.H. Denton, D.R. Lux, and J.G. Bockheim, Late Tertiary Antarctic paleoclimate and ice-sheet dynamics inferred from surficial deposits in Wright Valley, *Geog. Ann*, 75A, 239–268, 1993.

Hendy, C.H., T.R. Healy, E.M. Rayner, J. Shaw, and A.T. Wilson, Late Pleistocene glacial chronology of the Taylor Valley, Antarctica, and the global climate, *Quat. Res.*, 11, 172–184, 1979.

Holdsworth, G., and C. Bull, The flow law of cold ice; investigations on Meserve Glacier, Antarctica, *Contrib. 125*, Institute of Polar Studies, Ohio State University, Columbus, Ohio, 1968.

House, H.R., D.M. McKnight, and P. von Guerard, The influence of stream channel characteristics on stream flow and annual water budgets for lakes in Taylor Valley, *Antarctic J. U.S.*, 30, 284–287, 1995.

Howard-Williams, C., and W.F. Vincent, Microbial communities in southern Victoria Land streams (Antarctica) I. Photosynthesis, *Hydrobiologia*, 172, 27–38, 1989.

Ishikawa, N., and S. Kobayashi, On the internal melting phenomenon (puddle formation) in fast sea ice, East Antarctica, *Ann. Glaciol.*, 6, 138–141, 1985.

Johannesson, T., E.D. Waddington, and C.F. Raymond, A simple method for determining the response time of glaciers, in *Glacier fluctuations and climatic change*, edited by J. Oerlemans, pp. 343–352, Kluwer Academic Publishers, Boston, 1989.

Kaser, G., Measurement of evaporation from snow, *Arch. Met. Biokl. Ser. B*, 30, 333–340, 1982.

Keys, J.R., Air temperature, wind, precipitation and atmospheric humidity in the McMurdo Region, *Publ.* 17, 57 pp, Geology Department, Victoria University, Wellington, NZ, 1980.

Lewis, K., G. Dana, A. Fountain, and S. Tyler, The surface energy balance of the Canada Glacier, Taylor Valley, *Antarctic J. U.S.*, 30, 280–282, 1995.

Lewis, K., A. Fountain, and P. Langevin, The role of terminus cliff melt in the hydrological cycle, Taylor Valley, McMurdo Dry Valleys, *Antarctic J. U.S.*, in press.

Lizotte, M.P., T.R. Sharp and J.C. Priscu, Phytoplankton dynamics in the stratified water column of Lake Bonney, Antarctica I. Biomass and productivity during the winter-spring transition. *Polar,* 16, 155–162, 1996.

Priscu J.C., Phytoplankton nutrient deficiency in lakes of the McMurdo Dry Valleys, Antarctica, *Freshwater Biology*, 34, 215–227, 1995.

McKnight, D.M., and C.M. Tate, Algal mat distribution in glacial meltwater streams in Taylor Valley, Southern Victoria Land, Antarctica, *Antarctic J. of the U.S.*, 287–289, 1995.

Meier, M.F., and W.V. Tangborn, Net budget and flow of South Cascade Glacier, Washington, *J. Glaciol.*, 5, 547–566, 1965.

Nye, J.F. and F.C. Frank, Hydrology of the intergranular veins in a temperate glacier, *Hydrology of Glaciers*, number 95, 157–161, Intl. Assoc. of Hydrol. Sci., Wallingford, England, 1973.

Paige, R.A., Sub-surface melt pools in the McMurdo Ice Shelf, Antarctica, *J. Glaciol.*, 7, 511–516, 1968.

Paterson, W.S.B., *The Physics of Glaciers, 3rd Ed.*, Pergamon Press, Tarrytown, 1994.

Péwé, T., and F. Church, Glacier Regimen in Antarctica as reflected by glacier-margin fluctuation in historic time with special reference to McMurdo Sound, *IAHS Publ, 58*, 295–305, 1962.

Robinson, P.H., Ice dynamics and thermal regime of Taylor Glacier, south Victoria Land, Antarctica, *J. Glaciol.*, 30, 153–160, 1984.

Simmons, G.M., Jr, J.R. Vestal, and R.A. Wharton, Environmental regulators of microbial activity in continental Antarctic lakes, in *Physical and Biogeochemical Processes in Antarctic Lakes, Antarctic Res. Ser.*, vol. 59, edited by W.J. Green and E.I. Friedmann, pp. 165–195, AGU, Washington, D.C., 1993.

Stearns, C.R., and G.A. Weidner, Sensible and latent heat flux estimates in Antarctica, in *Antarctic Meteorology and Climatology, Ant. Res. Ser.*, vol 61, edited by D.H. Bromwich and C.R. Strearns, pp. 109–138, 1993.

Taylor, G., *The physiography of the McMurdo Sound and Granite Harbour Region: British Antarctic "Terra Nova" Exped., 1910-13*, Harrison & Sons, London, 1922.

Whillans, I.M., J. Bolzan, and S. Shabtaie, Velocity of Ice Stream B and C, Antarctica, *J. Geophys. Res.*, 92, 8895–8902, 1987.

Wharton, Jr., R.A., C.P. McKay, G.D. Clow, D.T. Andersen, G.M. Simmons, Jr., and F.G. Love, Changes in ice cover thickness and lake level of Lake Hoare, Antarctica: Implications for local climate change, *J. Geophys. Res.*, 97, 3503–3513, 1992.

Wilch, T.R., G.H. Denton, D.R. Lux, and W.C. McIntosh, Limited Pliocene glacier extent and surface uplift in Middle Taylor Valley, Antarctica, Geog. Ann., 75A, 331–351, 1993.

G. Dana, Biological Sciences Center, Desert Research Institute, PO Box 60220, Reno, NV 89506.

A. Fountain , U.S. Geological Survey, PO Box 25046, MS-412, Denver, CO 80225, currently on leave to the Department of Geology, Portland State University, Portland, OR 97207-0751.

K. Lewis, INSTAAR, Campus Box 450, University of Colorado, Boulder, CO 80309 and the U.S. Geological Survey, P.O. Box 25046, MS-412, Denver, CO 80225.

D. McKnight, now at INSTAAR, Campus Box 450, University of Colorado, Boulder, CO 80309.

B. Vaughn, INSTAAR, Campus Box 450, University of Colorado, Boulder, CO 80309.

(Received October 4, 1996;
accepted March 28, 1997.)

GEOCHEMICAL LINKAGES AMONG GLACIERS, STREAMS AND LAKES WITHIN THE TAYLOR VALLEY, ANTARCTICA

W. Berry Lyons, Kathy A. Welch, Klaus Neumann, Jeffrey K. Toxey,
Robyn McArthur, Changela Williams

Department of Geology, University of Alabama, Tuscaloosa, Alabama

Diane M. McKnight

USGS-WRD, Boulder, Colorado

Daryl Moorhead

Department of Biology, Texas Tech University, Lubbock, Texas

Aquatic systems in the Taylor Valley, Antarctica have characteristics of other desert aquatic systems in that the amount of water is limited and highly variable during the year. As part of the McMurdo Dry Valleys LTER, we have examined the major element chemistry of the three largest lake basins in order to investigate the geochemical "continuum" and the geochemical processes occurring within the Taylor Valley. During the summer, meltwater is generated from the glaciers and flows through streams to perennially ice-covered lakes in the valley bottom. As water moves through the system, solute concentrations increase by orders of magnitude. The glacier data suggest that some amount of salt is recycled from the soils and blown by winds onto the glaciers. Spatial differences in glacier chemistry have been observed and these, along with characteristics of the glacial meltwater streams, result in differences in stream chemistry within the valley. Dissolution of evaporite salts within the stream channels, as well as the weathering of Si minerals appear to be significant geochemical processes especially in the longer streams. The differences in modern day stream chemistry would lead to different chemical evolutionary pathways for the different lakes. High interannual variability of stream flow has also been observed which leads to differences in the amount of fresh water and solutes entering into the lakes each season. In addition, seasonal chemical changes occur within the lakes due to the inflow of fresh water and biological activity. For example, changes in calcite saturation in the lakes have been observed through the austral summer period. Based on our work, it appears that long-term systematic monitoring of stream and lake hydrology and chemistry is needed in order to quantitatively evaluate water and solute balances for the lakes, as well as to understand lake dynamics.

INTRODUCTION

The geochemistry of the aquatic systems, especially the lakes, in the Taylor Valley, Southern Victoria Land, Antarctica, has been investigated since the International Geophysical Year in the late 1950s. Yet, even after nearly 40 years of study, there is still great controversy over the origin and evolution of solutes in these lake systems [e.g., *Wilson, 1979; Heywood, 1984; Green et al., 1988; Lyons and Mayewski, 1993*]. In addition, with the exception of the seminal work of *Green et al.* [1988] there has been little attempt to investigate the system as a "watershed," that is, to relate in a systematic manner, the geochemistry and hydrology of the glaciers and streams to the lakes themselves. Since the austral summer of 1993-1994,

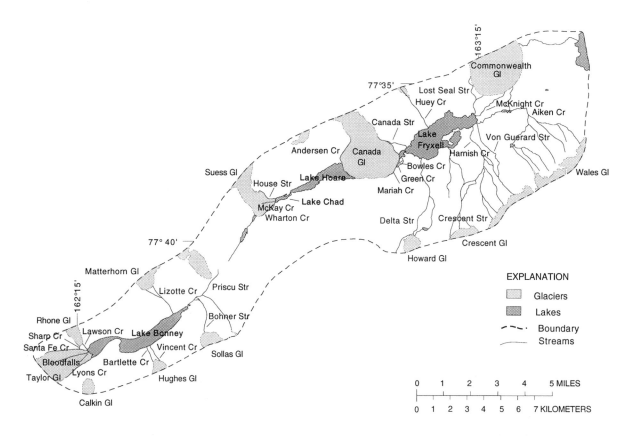

Fig. 1. The Taylor Valley, Antarctica with names of glaciers, streams, and lakes discussed in the text.

members of the McMurdo Dry Valleys (MDV) Long Term Ecological Research (LTER) site scientific teams have attempted to do this. This effort builds on studies of the instream processes and short-term dynamics of streams begun in 1990 in the Lake Fryxell basin [*McKnight and Andrews*, 1993; *von Guerard et al.*, 1995]. The results presented within are the first attempt to describe the geochemical dynamics of the entire Taylor Valley system.

In reality, the Taylor Valley "watershed" is four watersheds. These include three major basins which consist of the glaciers and streams entering Lake Bonney, the Lake Chad/Lake Hoare complex, and Lake Fryxell (Figure 1). These lakes are closed-basin lakes having no outflow. In these lakes, water is gained through surface inflow and lost through sublimation of their perennial ice-covers [*Clow et al.*, 1988]. The fourth "watershed" is a small one in the easternmost portion of the valley (Figure 1) consisting of flow from Commonwealth Glacier into Commonwealth Stream which enters McMurdo Sound. Although we have been monitoring Commonwealth Stream, the data will not be discussed here. Instead we will focus our attention

on the three internally drained watersheds in the valley.

This paper should be viewed as a preliminary assessment in that, although we now have detailed seasonal geochemical profiles of the lakes over the 1993-1994, 1994-1995, and 1995-1996 field seasons, chemical monitoring of the glaciers has not been accomplished in a completely systematic manner. Chemical monitoring of the streams has been accomplished with 3 to 5 samples collected from each of the streams each year during this period; however these data have not yet been analyzed using the continuous conductivity record. In addition, gauging of the major streams in the three internal watersheds was undertaken (due to logistic considerations) at differing times over this three-year period. Our most detailed and lengthy gauging records come from the Lake Fryxell basin due to the fact that previous, non-LTER, activities established the gauges in many of the streams there [*von Guerard et al.*, 1995].

The goal of this paper is two-fold: first, to describe the geochemical continuum that exists between glaciers, streams, and lakes in these watersheds, addressing the important geochemical processes occurring; and second, to add to the previously detailed knowledge on the major element dynamics within the lakes themselves. As mentioned above, *Green et al.* [1988] have developed the initial approach that we use here and applied it to Lakes Fryxell and Hoare. We acknowledge greatly their pioneering work and their "watershed" view of the McMurdo Dry Valleys. We have added to it, however, by including the Lake Bonney basin, thereby evaluating all three internally drained watersheds in Taylor Valley. We have also begun to address the geochemical consequences for the lakes of the substantial interstream and interannual hydrologic variations in the McMurdo Dry Valleys. This analysis provides more than just a one season view of the hydrological and geochemical dynamics within the valley. One of the important conclusions of this work is that only through long-term monitoring can a realistic understanding of the system be gained.

STUDY AREA

The McMurdo Dry Valleys are among a small group of ice-free areas that lie along the coastal regions of the Antarctic continent [*Green and Friedmann*, 1993]. The chemical nature of the lakes in Taylor Valley (Figure 1) is reasonably well known [e.g., *Green et al.*, 1988; *McKnight et al.*, 1991, 1993; *Spigel and Priscu, 1996*] and the general watershed dynamics are generally understood [*Chinn, 1993*]. The bedrock in the area consists primarily of undifferentiated granitic and metamorphic rocks and to the south injected sills and sheets of dolerite. The valley floors are covered with glacial drift, moraines, and related surficial materials. The annual mean temperature is between $-20°$ and $-25°C$ and precipitation is thought to be approximately 10 cm water equivalent per year [*Heywood,* 1984]. Detailed explanations of recent climatological and hydrological studies can be found elsewhere in this volume [*Dana et al.*, this volume; *Fountain et al.*, this volume; *Conovitz et al.*, this volume].

ANALYTICAL METHODS

Samples for geochemical analysis were obtained with great care in order to minimize contamination. Major cations and anions were analyzed by ion chromatography. Details of both sample collection and the ion chromatographic measurements are found in *Welch et al.* [1996] and will not be repeated here. Carbonate speciation in the lake waters was calculated from ΣCO_2 data (measured via infrared gas analyzer in

the field at Lake Hoare camp) and pH (measured in the field, immediately upon sample collection), while carbonate speciation for the stream samples was obtained through alkalinity titration and pH. The relative standard deviation of the alkalinity measurements is less than ±3%. The mean value for ionic balances of the streams was 3.4 ± 2.5%, while those for the lakes ranged between 3.0 ± 2.0% for Lake Fryxell and 1.4 ± 1.0% for the west lobe of Lake Bonney. Bicarbonate (i.e., "alkalinity") was estimated on the glacier samples by subtracting the sum of the Cl^-, NO_3^-, and SO_4^{2-} equivalents from the total cation equivalents in order to yield a value of zero or an equal ionic balance. Reactive silicate was determined using the colorimetric method of *Mullin and Riley* [1955]. The relative standard deviation of these measurements is less than ± 3%. For the more saline lake water samples, the method was modified in order to minimize any salt effect [*Toxey et al.*, in review].

GEOCHEMICAL CONTINUUM

Figures 2 through 5 show all the snow/ice/stream /lake surface water data on the aquatic geochemical continuum from Taylor Valley. The lake data are from just below the ice-cover down to 6 m in Lake Fryxell and 8 m in Lakes Hoare and Bonney. These figures demonstrate the major chemical constituents, Na^+, Ca^{2+}, Mg^{2+}, K^+, SO_4^{2-}, alkalinity (HCO_3^-), and H_4SiO_4 plotted versus Cl^- for the glaciers, streams, and lake surface waters. There are two obvious points to make here. One is that many of the major chemical constituents of interest have concentration ranges of six orders of magnitude. For example, Cl^- concentrations are as low as 400 nM and as high as 300 mM. The second is that, although a few of the plots are very close to "straight" lines with little variation or noise of the constituent of interest versus Cl^- (i.e., Na^+, Mg^{2+}), some show considerable variability (such as H_4SiO_4, K^+, Ca^{2+} and alkalinity), while others fall somewhere between these two extremes (SO_4^{2-}).

Although Cl^- salts in the soils around the McMurdo Dry Valleys do decrease away from the coast, NaCl is one of the ten most abundant salts observed throughout the McMurdo Dry Valley regions [*Keys and Williams*, 1981; *Campbell et al.*, this volume]. The dissolution of salts, including Cl^- salts such as NaCl is a major process within the dry valleys [*Campbell and Claridge*, 1987]. In these plots, Cl^- is used to monitor the extent of salt dissolution and transport through this desert system. Although Cl^- rich salts are probably blown

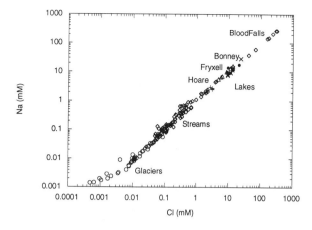

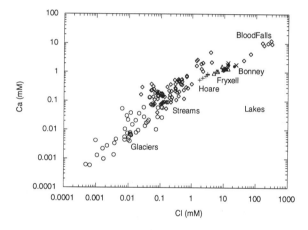

Fig. 2. Na⁺ and Ca²⁺ versus Cl⁻ concentration for Taylor Valley glaciers, streams, and surface lake water samples.

into the stream beds during the year and may be precipitated when stream water evaporates at the end of the austral summer, Cl⁻ is conservative (i.e., not precipitating) in the system during the flow season, as all the streams are grossly undersaturated with respect to halite. (The exception to this may be Blood Falls, the saline discharge from underneath the Taylor Glacier that flows into the west lobe of Lake Bonney.)

In all these plots (Figures 2 to 5), the samples with the Cl⁻ concentrations that are higher than the surficial lake waters are from Blood Falls. This iron-rich saline discharge emanating from beneath the Taylor Glacier has been estimated to contribute 2000 ± 500 m³ of water per year into Lake Bonney [Keys, 1979]. At an average Cl⁻ concentration of 218 mM (from our 1993-1996 measurements), the Cl⁻ flux from this source would be 4.4 x 10⁵ moles yr⁻¹. The samples enriched in Ca²⁺ and SO₄²⁻ at approximately 2 mM Cl⁻ (Figures 2 and 4) are those from Taylor Glacier

streams, excluding Blood Falls. These same samples also are slightly depleted in K⁺ and H₄SiO₄, relative to their overall trends versus Cl⁻ (Figures 3 and 5). The input of solutes from these sources lead to the higher Cl⁻, Na⁺, Ca²⁺ and SO₄²⁻ concentration in Lake Bonney than in the surface waters of Lake Fryxell and Lake Hoare. In general, Lake Bonney surface water is depleted in K⁺ and H₄SiO₄ relative to Lake Fryxell because of the Blood Falls input, as well.

Glaciers

The glacier ice and snow show the lowest values for all the chemical constituents. These data represent surface snow values with only one sample (from the Canada Glacier) being deeper than 15 cm. The average values for the Commonwealth, Canada, Howard, and Taylor Glaciers are given in Table 1. These four glaciers are where the majority of our samples have been obtained. For comparison to other glaciers in the

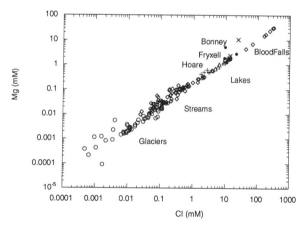

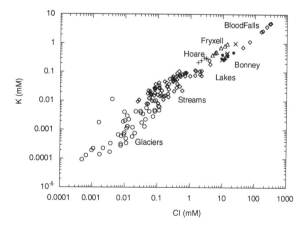

Fig. 3. Mg²⁺ and K⁺ versus Cl⁻ concentration for Taylor Valley glaciers, streams, and surface lake water samples.

region, average values from two snow pits on the Newell Glacier in the Asgard Range north of Taylor Valley at elevations of 1600 and 1700 m, and the mean values from two snow pits on Taylor Dome at 2450 m, west of Taylor Valley are also shown (Table 1). Note that the alpine glaciers flowing into Taylor Valley and snow from the snout of Taylor Glacier have much higher concentrations of cations and anions than the glaciers at elevation. Numerous processes can affect the chemical composition of surficial glacier snow such as dry deposition onto the surface, the input of terrestrially derived wind-blown debris, the sublimation of snow, and hence, concentration of salts due to mass loss, as well as the chemical signature of the original precipitation.

The influence of marine aerosol on the glaciochemistry can be clearly seen, as the mean molar Na:Cl ratios of the surface on Commonwealth, Canada, Howard, and Taylor Glaciers are 0.85, 0.86, 0.83, and

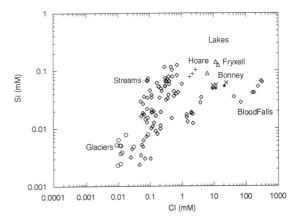

Fig. 5. Reactive silicate (H_4SiO_4) versus Cl^- concentration for Taylor Valley glaciers, streams, and surface lake water samples.

0.815, respectively. The seawater ratio is 0.86. As one moves inland the amount of sea salt influence decreases [*Keys and Williams*, 1981], and an aerosol influenced by stratospheric transport begins to influence the snow chemistry [*Mayewski et al.*, 1995]. The elevated Ca^{2+} concentrations in the snows of the alpine glaciers flowing into Taylor Valley also suggest a large input of terrestrial dust (i.e., high $CaCO_3$), especially compared to the Taylor Dome site at elevation, tens of kilometers away from the ice-free regions. Probably much of the Ca^{2+}, Mg^{2+}, alkalinity, and a portion of the K^+ and SO_4^{2-} leaving the glaciers as they melt and entering the streams may have originated from the valley floor as wind-blown salt. Even very small amounts of H_4SiO_4 are present in water flowing off the glacier surfaces with values of 4 to 8 μM observed on the Canada and Taylor Glaciers, respectively, suggesting that dissolution of silicate dusts is occurring as glacier snow melts. The origin of this dust must be the valley floor.

The lowest chemical values are those from older Taylor Glacier ice. The Na^+, Cl^-, and NO_3^- values are as low, or lower than present-day snow accumulating on Taylor Dome (Table 1). The Mg^{2+}, Ca^{2+}, and SO_4^{2-} are higher than at Taylor Dome, possibly suggesting that this older ice has been affected by the input of aeolian transported materials. With the exception of Ca^{2+} and alkalinity (aeolian transported $CaCO_3$), the Taylor Glacier snow has higher concentrations of salts than does the Canada Glacier. This suggests that the higher sustained wind speeds up the valley toward the Taylor Glacier terminus may negate the influence of distance from the coast (marine aerosol source), at least at these lower elevations. Chemical

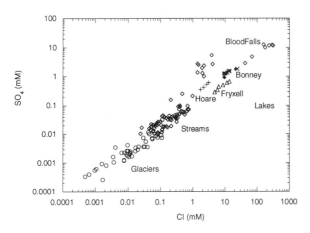

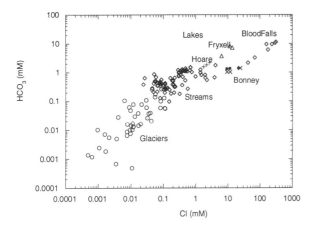

Fig. 4. SO_4^{2-} and HCO_3^- versus Cl^- concentration for Taylor Valley glaciers, streams, and surface lake water samples.

TABLE 1. Glaciochemistry of Taylor Valley (Values in µM).

Location	Cl⁻	NO₃⁻	SO₄²⁻	Na⁺	K⁺	Mg²⁺	Ca²⁺	Alk[a]	Na:Cl
Commonwealth	35.6	3.1	3.8	30.3	1.9	6.7	13.9	27.1	0.85
Canada	21.4	2.3	3.7	18.4	4.3	5.7	36.9	76.9	0.86
Howard	55.4	3.5	7.2	46.1	4.3	10.2	71.2	139.9	0.83
Taylor (snow)	42.2	8.3	11.6	34.4	2.3	6.9	10.4	20.4	0.815
Taylor (ice)	1.6	0.6	0.92	2.9	1.8	1.2	2.2	7.3	1.82
Newell [b]	5.15	1.16	1.5	3.02	0.11	0.43	0.64	–	0.59
Taylor Dome[b]	1.2- 2.0	1.05- 1.12	0.47- 0.54	0.5- 1.0	–	0.07- 0.10	0.05- 0.08	–	–

[a] Determined by difference (see text).
[b] From: *Welch* [1993]

loading to these glaciers, and hence to the streams and lakes, appear to be both a function of distance from the coast [*Keys and Williams*, 1981] and the influence of windblown deposition onto the glaciers themselves. In this way, some of the salt that originates in the valley is "recycled" by the winds onto the glaciers where it eventually is redissolved during ablation and melt during the austral summer. This might help lead to variations in the bulk stream chemistry from glacier to glacier. For example, melt waters from the Canada Glacier should have higher K:Cl and higher total Ca²⁺ concentrations than streams originating from the Commonwealth or Howard Glaciers (Table 1). This, in turn, could have an important influence on the chemical evolution of the lakes. It is unknown if these surficial samples chemically represent the bulk composition of the entire glacier. Obviously an ice coring program is needed in order to evaluate this problem.

Streams

There is a reasonable amount of controversy over the mechanisms of solute incorporation to the McMurdo Dry Valley streams. Although all authors agree that the dissolution of previously deposited salts is a major component to the stream solute loading, there is disagreement as to whether chemical weathering of silicate minerals is a major contributor. There are essentially two schools of thought on the matter: one argues that chemical weathering is a relatively minor circumstance [*Wilson*, 1979; *Campbell and Claridge*, 1987]; the other more recent position argues that chemical weathering is a major process within the stream reaches and flood plains, where liquid water

exists during the austral summer [*Green et al.*, 1988; *Blum et al.*, 1996; *Lyons et al.*, in review]. (The idea of weathering being a major contributor to solute loading is not altogether new, as it was originally proposed for Lake Vanda in the Wright Valley by *Jones and Faure* [1978]. In addition to surface water initiation of chemical weathering, arguments for subsurface weathering via groundwaters (at least in Wright Valley) also have been made [*Lyons and Mayewski*, 1993]. In general these two different schools have based their conclusions regarding the importance of chemical weathering in the McMurdo Dry Valleys on two different sets of evidence: the lack of solid weathering residues (i.e., clay minerals) within the soil zones (soil scientists) versus the concentrations of dissolved species, especially HCO₃⁻ and H₄SiO₄ (aquatic geochemists) within the streams themselves. The disagreement may actually be more semantic than real in nature, as the aquatic geochemistry argument for chemical weathering emphasizes that weathering reactions can only occur where liquid water is present, and therefore, are minimal to nonexistent in the true soil zones away from the streambeds and flood plains.

The soil-based argument against chemical weathering is based on the notion that, due to lower temperature and low moisture content at high latitudes, weathering is dominated by physical and mechanical weathering processes [*Matsuoka*, 1995]. Strong arguments have been made, however, that even physical weathering processes are minimal in the polar regions, especially Antarctica, because of the lack of moisture [*Campbell and Claridge*, 1987]. The effectiveness of water-based weathering processes in the McMurdo Dry Valleys will not be debated here. What will be demon-

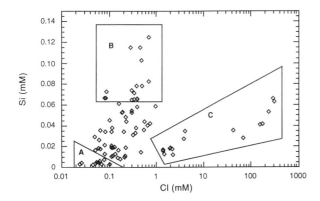

Fig. 6. Reactive silicate (H$_4$SiO$_4$) versus chloride concentration for all streams in Taylor Valley. (Group A = Suess Glacier streams; Group B = southern Lake Fryxell streams; Group C = west lobe of Lake Bonney streams).

strated, however, is that chemical weathering is a significant process within the streams of Taylor Valley. Recent work in the Antarctic [*Blum et al.*, 1996; *Lyons et al.*, in review], as well as the northern polar regions (J. Edmond, personal communication) has shown unequivocally that where running water exists, chemical weathering rates can be as rapid as in temperate or tropical regions.

Evidence for silicate weathering. Figure 6 is a plot of reactive silicate versus chloride for all the streams during both the 1994-1995 and 1995-1996 seasons. It is very clear that the vast majority of these streams have reactive silicate concentrations greater than 10 μM. In fact, only group A has concentrations at or below this level. These samples also have low Cl⁻ concentrations. These data are from very short streams which flow from the Suess Glacier into Lake Chad (Figure 1). Data from group C (Figure 6) have intermediate reactive silicate values and high Cl⁻ concentrations. These data are from streams (Santa Fe, Lyons, and the Red River) that flow from Taylor Glacier. Group B has relatively low Cl⁻ and the highest reactive silicate concentrations. These samples were all collected from the streams flowing from the south and east of Lake Fryxell and include McKnight Creek, Aiken Creek, Von Guerard Stream, Harnish Creek, Crescent Stream, and Delta Stream. These streams represent some of the longest streams in Taylor Valley (Figure 1). The differences in stream chemistry reflect the differences in the source water flowing from the glaciers, the chemistry of the rocks and soils, and also the length of time the water is in contact with the regolith. As mentioned above, although there are low

levels of reactive silicate in the melt waters from the glaciers, it is certain that concentrations of H$_4$SiO$_4$ above 10 μM must originate from silicate mineral weathering within the stream reaches. The higher concentrations of reactive silicate in group B rival those found in more humid, lower latitude river systems [*Edmond et al.*, 1996].

Figure 7 is a dissolved alkali/alkaline earth fractionation diagram [*Shiller and Frilot*, 1996]. The space to the right of the lines in the diagram represents most rock compositions, and the upper continental crustal average [*Wedepohl*, 1995] is represented by a single point. As recently stated by *Shiller and Frilot* [1996], because Na⁺ and Ca²⁺ are more easily solubilized from continental rocks than K⁺ and Mg²⁺ [*Nesbitt et al.*, 1980], this diagram can be used to distinguish between transport-limited and weathering-limited erosional regimes [*Stallard and Edmond*, 1983; *Stallard*, 1985]. Transport-limited regions should plot closer to the rock compositions, while weathering-limited regimes will plot to the left and below that of the rock compositions [*Shiller and Frilot*, 1996].

Not surprisingly, both weathering and transport-limited streams appear to exist in Taylor Valley. This is supported by Figure 8 which indicates that as the Si:Na+K–Cl ratio increases, the amount of Ca²⁺ being derived from other sources besides CaCO$_3$ dissolution also increases. Note in both figures the negative values along the abscissa indicating a source of Cl⁻ that cannot be accounted for by Na⁺ and K⁺ association alone. The majority of these data are from the aforementioned Suess Glacier streams that flow into Lake

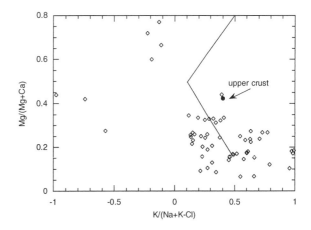

Fig. 7. Alkali/alkaline earth fractionation diagram for Taylor Valley streams. The lines delineate the field of typical rock compositions. The solid point represents an average upper continental crust composition.

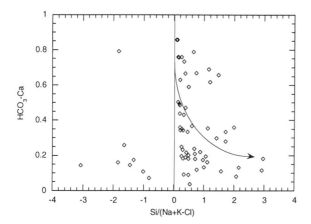

Fig. 8. HCO$_3$-Ca versus reactive silicate (H$_4$SiO$_4$)/Na$^+$+K$^+$–Cl$^-$ plot of Taylor Valley streams. Negative values on the x-axis represent Cl not balanced by Na plus K. The arrow represents increasing Si mineral weathering.

Chad and the streams with high total dissolved solids flowing from the Taylor Glacier into the west lobe of Lake Bonney (Figure 1). The streams with extremely high Mg/(Mg+Ca) ratios are from Blood Falls, the saline discharge into the west lobe of Lake Bonney. As chemical weathering intensifies, the Si:Na+K–Cl molar ratio should increase to values of five or greater [*Stallard and Edmond*, 1987; *Palmer and Edmond*, 1992; *Stiller and Frilot*, 1996]. The Taylor Valley streams never approach this ratio (Figure 8), suggesting that incongruent dissolution of silicates is the major silicate weathering feature in the stream reaches.

This is supported by other evidence as well. Figure 9 is a stability field diagram taken from *Stallard and Edmond* [1987]. The majority of the stream data from Taylor Valley plot in the right-most portion of the graph which is "below" the Na-feldspar plus muscovite to kaolinite line. This suggests that the production of 2:1 clays such as smectites, illites, and vermiculites could be occurring in the stream reaches as incongruent dissolution occurs [*Stallard and Edmond*, 1987]. Although these weathering residues have been observed in small amounts in dry valley soils [*Campbell and Claridge*, 1987], to our knowledge there has been no attempt to delineate them in the stream floodplains themselves. Where moisture is present in soils, smectite has been observed in appreciable amounts [*Claridge*, 1965; *Campbell and Claridge*, 1982]. Actual mineral identification in the floodplain areas would help better resolve this controversy as to the significance of silicate weathering.

Stream chemistry and lake geochemical evolution.
The surface waters of all the lakes in Taylor Valley are,

in general, enriched relative to seawater in major cations and anions relative to Cl$^-$ (Table 2). The lone exception is the Na:Cl ratio of Lake Chad (Table 2). Lake Hoare and Lake Chad have almost an order of magnitude enrichment in K:Cl relative to seawater, and all the lakes are enriched in Mg:Cl by at least 50% to that of seawater. (Lake Chad is enormously enriched.) Lake Hoare has the greatest Ca:Cl, SO$_4$:Cl and HCO$_3$ (alkalinity):Cl ratios as well. The reason for this is not clear. The influence of silicate weathering in the stream reaches, with the exception of Anderson Creek, is minimal in the Lake Hoare and Lake Chad system due to the short stream lengths and direct water input through contact with the Canada Glacier (Figure 6). Certainly the higher Mg:Cl ratios in the surface water of Lake Hoare may be due, in part, to the inflow of the very high Mg:Cl ratio waters from Lake Chad, but why the ratios are high in the Lake Chad inflow waters is unclear.

Hardie and Eugster [1970] were the first to point out the quantitative significance of the ionic ratios of inflow waters to the subsequent geochemical evolution of closed-basin lakes. If inflow waters have HCO$_3^-$ to Ca^{2+} molar ratios greater than 2, calcium deficient, bicarbonate-rich lake waters will develop. This important geochemical concept was first applied to the Taylor Valley lakes by *Green et al.* [1988]. The majority of the Taylor Valley streams lie above the chemical "divide" (Figure 10), indicating that the lakes fed by these streams should evolve to a Na$^+$-HCO$_3^-$ rich water, depleted in Ca^{2+}. The exceptions to this trend are the very short Suess Glacier streams flowing into Lake Chad whose ratios fall on the line and the

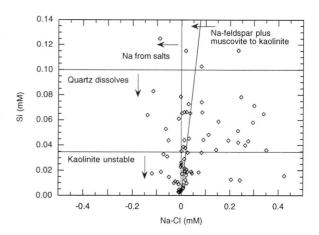

Fig. 9. Stability field diagram from *Stallard and Edmond* [1987] showing Taylor Valley stream chemistry in relationship to silicate mineral stabilities.

TABLE 2. Molar Ratios of Major Ions in the Lakes of Taylor Valley. The Data Represent the Average of Five Different Values Except for Lake Chad Where Only One Measurement at 4m Depth is Presented.

Water	Na:Cl	K:Cl	Mg:Cl	Ca:Cl	SO_4:Cl	HCO_3:Cl
W. Bonney	0.90	0.027	0.143	0.128	0.123	0.133
E. Bonney	0.88	0.030	0.148	0.135	0.117	0.100
Hoare	0.97	0.109	0.190	0.285	0.202	0.874
Fryxell	1.18	0.077	0.147	0.155	0.057	0.605
Chad	0.67	0.108	0.359	0.614	0.214	–
Seawater [a]	0.86	0.019	0.97	0.019	0.052	0.004

[a] From: *Bruland* [1983].

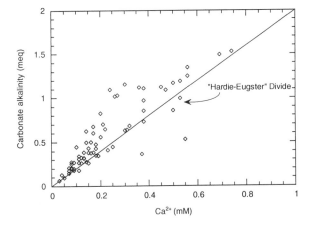

Fig. 10. Carbonate alkalinity versus calcium plot of all Taylor Valley streams. The line represents the *Hardie and Eugster* (1970) chemical divide.

Taylor Glacier derived streams having high total dissolved solids whose ratios fall below the line. These data indicate that Lake Fryxell and Lake Hoare should be evolving to HCO_3^- rich waters. *Green et al.* [1988] observed a similar situation for Lake Fryxell in 1982-1983, but a slightly different situation for Lake Hoare, when the HCO_3^-:Ca^{2+} ratios were less than 2. Certainly Lakes Fryxell and Hoare are evolving toward Ca^{2+} depleted waters, as acknowledged by *Green et al.* [1988] and our earlier work [*Welch and Lyons*, in press]. However if one goes through the evolutionary scheme to the next chemical divide as outlined by *Hardie and Eugster* [1970], Lake Fryxell and Lake Hoare diverge in their geochemical behavior. In the Lake Fryxell surface waters (6 to 7 m depth), the Alk>2 Mg^{2+} even after Ca^{2+} removal. Based on the *Hardie and Eugster* [1970] classification, these waters, through evapoconcentration, would evolve to true Na^+-HCO_3^-, CO_3^{2-} rich waters similar to those of Lake Magadi in the East African Rift Valley, Kenya. In the case of Lake Hoare, $2Mg^{2+}$>Alk after Ca^{2+} removal, and these waters would evolve to a Na^+, Mg^{2+}, Cl^-, SO_4^{2-} water similar to the Great Salt Lake or the Dead Sea.

The surface waters of Lake Bonney, on the other hand, have $2Ca^{2+}$>Alk and SO_4^{2-}>Ca^{2+} (after initial Ca^{2+} removal via $CaCO_3$) and so would, theoretically, also evolve to a Na^+, Mg^{2+}, SO_4^{2-}, Cl^- rich water, like Lake Hoare, but from a very different path [*Hardie and Eugster*, 1970]. The very interesting fact is that the hypersaline brine at depth in the east lobe of Lake Bonney has evolved to this composition, with the exception of its very low SO_4^{2-} content [*Spigel and Priscu*, 1996]. The west lobe has lower Mg^{2+} and higher Ca^{2+} concentrations suggesting very different histories as originally suggested by *Hendy et al.* [1977]. There is little doubt, however, that the small variations in the geochemistry of the streams flowing into the lakes do exert a major influence on the surface water chemistry of the lakes, and undoubtedly, during their evolution, as lake levels rise and fall due to variation in hydrologic parameters derived by climate change.

Taylor Valley Lakes

Major element geochemistry. As mentioned above, the chemistry of the Taylor Valley lakes has been studied since the early 1960s. Recent advances in analytical techniques have allowed for better, more precise measurements of the major ions in these lakes. This can easily be seen in Figure 11 where the Na^+, Ca^{2+}, K^+, Mg^{2+}, HCO_3^- and SO_4^{2-} to Cl^- ratios are plotted versus depth for Lake Fryxell over the years. These data include that of *Angino et al.* [1962] from 1962, *Green et al.* [1988] from 1983-1984 and our data from 1994-1995. It is probably the case that this "variation" through time of these ratios versus depth is

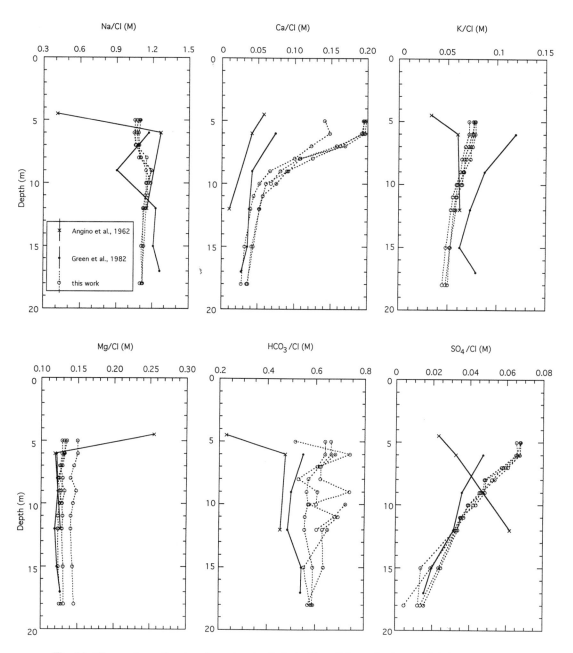

Fig. 11. Comparison of major element ratios in Lake Fryxell from *Angino et al.* [1962], *Green et al.* [1988] and this work.

not due solely to geochemical or biochemical processes as previously thought [e.g., see discussion, p. 304 in *Heywood*, 1984], but rather to the improvement of analytical instrumentation. The exception to this is in the uppermost waters where the inflow from glacier melt and the freezing of water to produce ice can change the geochemistry of these waters through the year.

Chloride "ages" of lakes. Numerous authors have utilized the Cl⁻ fluxes into and Cl⁻ inventories of closed

basin lakes to determine lake ages. Simply done, the total Cl⁻ mass in the lake (i.e., the inventory) is divided by the annual Cl⁻ flux via stream/river input and an age is obtained. Because Cl⁻ behaves conservatively (i.e., it is not precipitating out of solution), the residence time of Cl⁻ in the water body is thought to be the "age" of the water body. One of the limitations of this method is that the Cl⁻ age determination will underestimate the lake age due to diffusion of Cl⁻ into the lake sediments, thereby requiring a longer period of Cl^- accumulation to give the observed watercolumn concentrations. This method also assumes that there is no significant input of solutes via groundwater. This technique has been extensively used for the Rift Valley lakes of East Africa [*Gaudet and Melack*, 1981; *Yuretich and Cerling*, 1983; *Barton et al.*, 1987; *Ojiambo and Lyons*, 1996]. *Green et al.* [1988] used a similar approach to determine the ages of Lake Hoare and Lake Fryxell. These authors, however, acknowledged the possibility that the ages obtained were determined on only one melt season, and that substantial "noise" or variation exists from season to season. They calculated that this variation could be as large as 35% [*Green et al.*, 1988]. *Green et al.* [1988] determined Cl^- ages of 2241 years and 2997 years for Lakes Hoare and Fryxell, respectively.

We have used the flow data from the LTER database to calculate Cl^- ages for the melt seasons 1993-1994, 1994-1995 and 1995-1996 for Lakes Fryxell and Hoare. The ages for each lake in each melt season are shown in Table 3. Needless to say there is great variation in these "ages" from year to year. Using our data and those of *Green et al.* [1988] from the 1982-1983 season, mean values can be determined. First of all, our values are considerably greater than those of *Green et al.* [1988]. Secondly, our values indicate a much larger error (close to ± 100% for Lake Fryxell) in using this technique to calculate lake ages. These values are 10,020 ± 9744 and 7412 ± 5350, respectively for Lake Fryxell and Lake Hoare. This large variation is primarily due to variations in discharge. For example, lower discharges lead to lower annual Cl^- inputs, which in turn, produce larger "ages." Currently we have no way to evaluate whether the lower discharge volumes observed by us in the mid-1990s or the higher values obtained by *Green et al.* [1988] in the early 1980s and *von Guerard et al.* [1995] in the early 1990s are more typical of these watersheds over a longer (thousands of years) period of time. Until more discharge data are available, we recommend that the use of Cl^- ages as true ages of these lake systems be discontinued. It is apparent, as *Green et al.* [1988] first pointed out, the hydrological variation in these systems is quite large and simplistic approaches such as those using residence time or short-term mass balances yield wildly differing numbers depending on the year. It is also very apparent that only through long-term monitoring, such as that undertaken by the LTER, can this matter be better resolved.

CaCO₃ dynamics. We have used the computer

TABLE 3. Chloride Ages for Lakes Fryxell and Hoare.

	Water Year	Age in Years
Lake Fryxell	82–83 [a]	2,997
	93–94	2,770
	94–95	23,524
	95–96	10,787
Lake Hoare	82–83 [a]	2,241
	93–94	4,242
	94–95	14,257
	95–96	8,908

[a] From: *Green et al.* [1988].

code, PHREEQE (for Lakes Fryxell and Hoare) and PHRQPITZ (for Lake Bonney) to calculate the saturation indices for the $CaCO_3$ minerals, aragonite and calcite, in the Taylor Valley lakes. Previous work by *Green et al.* [1988] had shown that Lake Fryxell was saturated with respect to calcite from top to bottom, while Lake Hoare was saturated with respect to calcite in the upper 15 m, and close to equilibrium to slightly undersaturated below this depth. In addition, mass balance calculations by these authors indicated that large amounts of Ca^{2+} and HCO_3^- had been lost from the watercolumn, suggesting $CaCO_3$ precipitation to be an important process in both lakes. They hypothesized that this process was probably more important in the shallow portions of the lake, implying that photosynthesis may play an indirect role in controlling $CaCO_3$ formation, in part, through CO_2 uptake and pH increase [*Green et al.*, 1988].

The calculated saturations are shown in Figures 12a and 12b. Data are for both November and January in the austral summer of 1994-1995. It is clear that strong seasonal changes occur, especially in the surface waters. In all the basins, the surface waters become more strongly undersaturated late in the season, although in Lake Hoare the topmost water sample only approaches calcite saturation late in the season (Figure 12a). In the case of the west lobe of Lake Bonney and Lake Fryxell, the water above the chemocline becomes undersaturated in January. In addition, the deep water of Lake Hoare becomes more undersaturated later in the summer. These calculations suggest that, if the most important locus of $CaCO_3$ formation in the lakes is the shallow surface water, it can act as a source only early in the austral summer, as later in the summer condi-

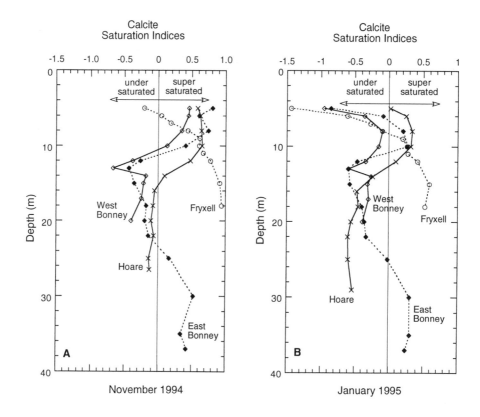

Fig. 12. CaCO₃ saturation indices, log (IAP/Ksp), for the Taylor Valley lakes through the 1994-1995 austral summer.

tions change from supersaturated to undersaturated with respect to CaCO₃. The shapes of both the Lake Bonney profiles indicate a constant depth zone of undersaturation at the top of the chemocline, where any CaCO₃ produced above in the early parts of the summer may be dissolved as it falls toward the underlying supersaturated bottom waters (Figures 12a, 12b). The amount of dissolution that occurs in this region is unknown, but it would be determined by the degree of undersaturation and other kinetic factors [*Morse*, 1978]. Current research by our group is directed at determining this process. There is evidence that in Lake Hoare a portion of the CaCO₃ is produced within the benthic mats at the sediment-water interface [*Wharton et al.*, 1982; *Wharton*, 1994]. There is also carbon isotopic evidence that the dissolution of CaCO₃ occurs in the bottom waters of Lake Hoare [*Wharton et al.*, 1993]. Our saturation profiles certainly support this notion and indicate that the preservation of CaCO₃ is only certain in the east lobe of Lake Bonney and Lake Fryxell where supersaturation is maintained throughout the austral summer. In addition late season inflow of calcite undersaturated stream water and its interlayering

underneath the ice, as well as calcite undersaturated meltwater from the ice cover increase the solubility in the uppermost meters of the watercolumn. Nonetheless these lakes appear to be much more dynamic in regard to CaCO₃ geochemistry than previously thought, with large changes in CaCO₃ saturations observable through the spring-summer period.

Silica saturation. Figure 13a shows the concentration of H₄SiO₄, reactive silicate, with depth in all of the Taylor Valley lake basins. Our values for Lakes Fryxell and Hoare are similar to what have been reported previously [*Green et al.*, 1988]. We have also plotted the H₄SiO₄ to Cl⁻ ratio versus depth for each lake (Figure 13b). The low ratio values for Lake Bonney reflect the high solute loadings of the streams entering Lake Bonney even though the reactive silicate values in the streams are relatively high (Figure 5). As discussed above, Lake Fryxell streams carry the most H₄SiO₄, yet Lake Hoare has the highest ratio. The reason for this is probably two-fold: first, although the Fryxell streams have the highest H₄SiO₄ values, they also have high Cl⁻ loadings (relative to Lake Hoare); and second, the Si might be removed through uptake by

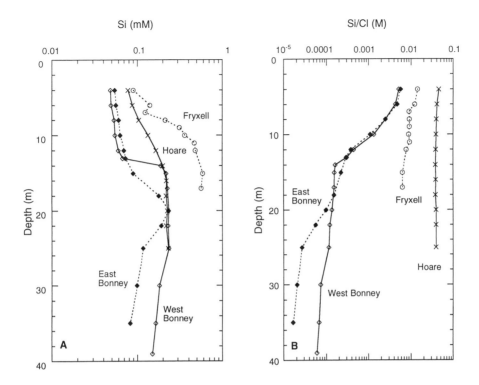

Fig. 13. Profiles of reactive silicate (H_4SiO_4) versus depth (A) and reactive silicate (H_4SiO_4) to Cl ratio versus depth (B) in the Taylor Valley lakes.

diatoms in benthic algal mats which are abundant in the shallow reaches of Lake Fryxell [*Spaulding et al.*, in press]. Saturation indices for both amorphous SiO_2 (*AS*) and quartz were calculated using the computer codes, PHREEQE and PHRQPITZ. Only below about 20 m in the east lobe of Lake Bonney is *AS* supersaturated (Figure 14a). In all the lacustrine waters of Taylor Valley, quartz is saturated (Figure 14b). Even though abundant diatomaceous materials have been observed in both the benthic algal mats [*Wharton et al.*, 1983] and lake sediments [*Spaulding et al.*, in press], only in the east lobe of Lake Bonney should they be permanently preserved. Because the diatom population of Lake Hoare appears to be dominated by benthic taxa [*Spaulding et al.*, in press], saturation conditions could exist at or within the sediment-water interface, and thereby explain the preservation of diatom debris within the sediment column. The dissolution of biogenically produced *AS* is a complex phenomenon affected by numerous parameters, such as temperature, surface area, and surface coating, as well as the degree of mineral saturation [*Hurd*, 1972]. The circa-neutral pH's of these lake waters may also hinder dissolution [*Barker et al.*, 1994]. Rapid burial rates, especially in the algal mats, could also lead to slower dissolution rates.

The relatively high concentrations of H_4SiO_4 in these lakes, especially Lake Fryxell, support the contention that chemical weathering is an active agent within the stream channels in the valley [*Green et al.*, 1988]. Values observed in the surface waters of the Taylor Valley lakes are similar to, or higher than, other more temperate lacustrine systems both in humid and arid conditions where chemical weathering is occurring (e.g., Aral Sea with surface values 42 to 105 μM [*Lyons et al.*, in press] and approximately 200 to 400 μM in Lawrence Lake [*Wetzel*, 1975]).

CONCLUSIONS

The Taylor Valley aquatic systems exist within a polar desert environment, and thus they have close similarities to other lower latitude desert aquatic systems in that water is of limited availability and its abundance varies dramatically on a seasonal basis. In addition, closed basin saline lakes represent the "sinks" for water and solutes in the system. Because of this, solute concentrations increase by orders of magnitude

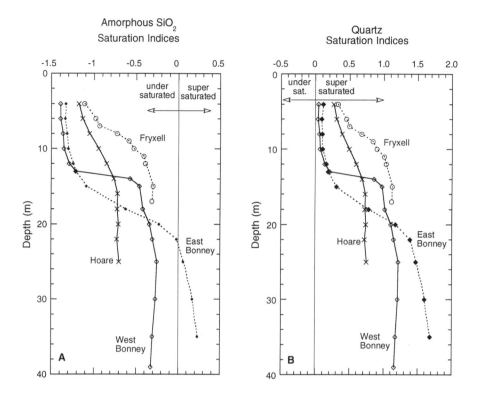

Fig. 14. Amorphous Si (A) and quartz (B) saturation indices, log (IAP/Ksp), for the Taylor Valley lakes.

as one moves downstream through the system from the glaciers to the streams and finally to the lakes. The glacier chemistry data suggest that some amount of salt is recycled from the valley floor soils and blown by the strong katabatic winds up onto the glacier surfaces.

Major element versus Cl^- plots in the Taylor Valley aquatic system fall close to a straight line and represent evaporative concentration and/or dissolution of evaporite salts within the flood plains in this extremely arid environment. For some of the chemical constituents, including Si, Ca^{2+}, HCO_3^-, and K^+, that do not fall on a straight line, there are additional geochemical processes, besides evapoconcentration and salt dissolution, occurring in order to explain their distribution. These processes include silicate mineral weathering within the stream floodplains. It is important to note that chemical weathering appears to be a significant contributor to stream and, hence, lake solute input even though liquid water exists for only a few months of the year. This is demonstrated by the fact that H_4SiO_4 concentration in many of the Lake Fryxell basin streams are comparable to values in streams from lower latitudes.

The year-to-year variability of stream flow is

extremely large due to what we think to be subtle changes in local climate [*Clow et al.*, 1988]. Therefore long-term systematic monitoring of both stream hydrology and chemistry is needed in order to quantitatively evaluate water and solute mass balances to the lakes. For example Cl^- "ages" determined for the lakes vary by an order to magnitude from year-to-year over the three years of this study.

A more systematic evaluation of the stream geochemical input into the lakes in Taylor Valley will allow for a better understanding of the future geochemical evolution of these lakes. Based on our current information, even though these lakes are situated in close proximity to one another in similar geologic terrain, they should evolve to different lake compositions over time. The subtle differences in stream chemistries leading to these different evolutionary paths have not been previously acknowledged.

Acknowledgments. This work was supported by NSF grant-OPP-9211773. Thanks are given to Antarctic Support Associates and VXE-6 for technical and logistical support. We greatly appreciate the help and dedication of our LTER colleagues who helped collect these samples. They include R. Edwards, J. Thomas, A. Butt, P. Feher, K. Lewis, C. Takas,

P. Langevin, G. Dana, H. House, J. Putscher, A. Legg, J. Schindler, and P. Soper. S. Wilder helped with the chemical analyses of the streams. Discussions with A. Fountain, J. Priscu, R. Wharton, Jr., P. Doran, W. Green, C. Howard-Williams, and C. Bowser and their insights into the mysteries of the McMurdo Dry Valley geochemistry and hydrology that were provided are gratefully acknowledged. We thank S. Tyler and W. Green for their thoughtful reviews of the manuscript. We thank P. Smith for typing the manuscript and E. Graham for her help with the graphics. JKT was supported by a REU supplement to the original NSF grant, RM was supported, in part, through an ACS-SEED grant to WBL and CW was supported by the NIH's NCRR Minority Initiative Program.

REFERENCES

Angino, E. E., K. B. Armitage, and J. C. Tash, Chemical stratification in Lake Fryxell, Victoria Land, Antarctica, *Science*, *138*, 34–36, 1962.

Barker, P., J.-C. Fontes, F. Gasse, and J.-C. Druart, Experimental dissolution of diatom silica in concentrated salt solutions and implications for paleoenvironmental reconstruction, *Limnol. Oceanogr.*, *39*, 99–110, 1994.

Barton, C. E., D. K. Solomon, J. R. Brown, T. E. Cerling, and M. D. Seyer, Chloride budgets in transient lakes: Lake Baringo, Naivasha and Turkana, *Limnol. Oceanogr.*, *32*, 745–751, 1987.

Blum, A. E., D. M. McKnight, and W. B. Lyons, Silicate weathering rates along a stream channel draining into Lake Fryxell, Taylor Valley, Antarctic (abstract), *Abstracts with Programs*, Annual GSA Meeting, Denver, CO., 1996.

Campbell, I, G. G. C. Claridge, D. I. Campbell, and M. R. Balks, The soil environment of the McMurdo Dry Valleys, Antarctica, this volume.

Campbell, I. B., and G. G. C. Claridge, The influence of moisture on the development of soils of the cold deserts of Antarctica, *Geoderma*, *28*, 221–238, 1982.

Campbell, I. B., and G. G. C. Claridge, *Antarctica: Soils Weathering Processes and Environment*, 368 pp., Elsevier, Amsterdam, 1987.

Claridge, G. G. C., The clay mineralogy and chemistry of some soils from the Ross Dependency, Antarctica, *N.Z. Jour. Geol. Geophys.*, *8*, 186–220, 1965.

Clow, G. D., C. P. McKay, G. M. Simmons, Jr., and R. A. Wharton, Jr., Climatological observations and predicted sublimation rates at Lake Hoare, Antarctica, *J. Clim.*, *7*, 715–728, 1988.

Conovitz, P. A., D. M. McKnight, L. M. McDonald, A. Fountain, and H. R. House, Hydrologic processes influencing streamflow variation in Fryxell Basin, Antarctica, this volume.

Dana, G. L., R. A. Wharton, and R. Dubayah, Solar radiation in the McMurdo Dry Valleys, Antarctica, this volume.

Edmond, J. M., M. R. Palmer, C. I. Measures, E. T. Brown, and Y. Huh, Fluvial geochemistry of the eastern slope of the northeastern Andes and its foredeep in the drainage of the Orinoco in Columbia and Venezuela, *Geochim. Cosmochim. Acta*, *60*, 2949–2976, 1996.

Fountain, A. G., K. J. Lewis, and G. L. Dana, Glaciers of the McMurdo Dry Valleys, southern Victoria Land, Antarctica, this volume.

Gaudet, J. J. and J. M. Melack, Major ion chemistry in a tropical African lake basin, *Freshwater Biol.*, *11*, 309–333, 1981.

Green, W. J., and E. I. Friedmann (Eds.), *Physical and Biogeochemical Processes in Antarctic Lakes*, Antarctic Res. Series, Vol. 59, 216 pp. AGU, Washington, D.C., 1993.

Green, W. J., M. P. Angle, and K. E. Chave, The geochemistry of Antarctic streams and their role in the evolution of four lakes in the McMurdo Dry Valleys, *Geochim. Cosmochim. Acta*, *52*, 1265–1274, 1988.

Hardie, L. A., and H. P. Eugster, The evolution of closed-basin brines, *Mineral. Soc. Amer. Spec. Publ.* 3, 273–290, 1970.

Hendy, C. H., A. T. Wilson, K. B. Popplewell, and D. A. House, Dating of geochemical events in Lake Bonney, Antarctica, and their relation to glacial and climate changes, *N.Z. Jour. Geol. Geophys.*, *20*, 1103–1122, 1977.

Heywood, R. B., Inland waters, in *Antarctic Ecology*, edited by R. M. Laws, pp. 279–343, Academic Press, London, 1984.

Hurd, D. C., Factors affecting solution rate of biogenic opal in seawater, *Earth Planet. Sci. Lettr.*, *15*, 411–417, 1972.

Jones, L. M., and G. Faure, A study of strontium isotopes in lakes and surficial sediments of the ice-free valleys, southern Victoria Land, Antarctica, *Chem. Geol.*, *22*, 107–120, 1978.

Keys, J. R., Saline discharge at the terminus of Taylor Glacier, *Antarctic J. U.S.*, *14*, 82–85, 1979.

Keys, J. R., and K. Williams, Origin of crystalline, cold desert salts in the McMurdo region, Antarctica, *Geochim. Cosmochim. Acta*, *45*, 2299–2309, 1981.

Lyons, W. B., and P. A. Mayewski, The geochemical evolution of terrestrial waters in the Antarctic: The role of rock-water interactions, in *Physical and Biogeochemical Processes in Antarctic Lakes*, edited by W. J. Green and E. I. Friedman, pp. 135–143, AGU, Washington, D.C., 1993.

Lyons, W. B., J. K. Toxey, V. Zhamoida, and N. Kurinny, Nutrient geochemistry of the Aral Sea: Influence of declining water levels, in *Proceedings of Third USA/CIS Joint Conf. on Environ. Hydrology and Hydrogeol., Water: Sustaining a Critical Resources*, A.I.H. Publ., in press.

Lyons, W. B., K. A. Welch, D. M. McKnight, K. Crick, J. K. Toxey, C. A. Nezat, and J. A. Mastrine, Chemical weathering rates and reactions in the Lake Fryxell Basin, Taylor Valley: Comparison to temperate river basins, in *Proceedings of Inter. Workshop on Antarctic Desert Ecosystems*, edited by C. Howard-Williams, W. B. Lyons and I. Hawes. Balkema, Rotterdam, in press.

Matsuoka, N., Rock weathering processes and landform development in the Sør Rondane Mountains, Antarctica, *Geomorphol.*, *12*, 323–339, 1995.

Mayewski, P. A., W. B. Lyons, G. Zielinski, M. Twickler, S. Whitlow, J. Dibb, P. Grootes, K. Taylor, P.-Y. Whung, L. Fosberry, C. Wake, and K. Welch, An ice-core-based, late Holocene history for the Transantarctic Mountains, Antarctica, in *Antarctic Res. Series*, Vol. 67, Contribution to Antarctic Res. IV, AGU, Washington, D.C., pp. 33–45, 1995.

McKnight, D. M., G. R. Aiken, and R. L. Smith, Aquatic fulvic acids in microbially-based ecosystems: Results from two Antarctic desert lakes, *Limnol. Oceanogr., 36*, 998–1006, 1991.

McKnight, D. M., G. R. Aiken, E. D. Andrews, E. C. Bowles, and R. A. Harnish, Dissolved organic material in dry valley lakes: A comparison of Lake Fryxell, Lake Hoare and Lake Vanda in *Physical and Biogeochemical Processes in Antarctic Lakes*, edited by W. J. Green and E. I. Friedmann, pp. 119–133, AGU, Washington, D.C., 1993.

McKnight, D. M., and E. D. Andrews, Hydrologic and geochemical processes at the stream-lake interface in a permanently ice-covered lake in the McMurdo Dry Valleys, Antarctica, *Verh. Internat. Verein. Limnol., 25*, 957–959, 1993.

Morse, J. W., Dissolution kinetics of calcium carbonate in sea water: VI. The near equilibrium dissolution kinetics of calcium carbonate-rich deep sea sediments, *Am. Jour. Sci., 278*, 344–353, 1978.

Mullin, J. B., and J. P. Riley, The colorimetric determination of silicate with special reference to sea and natural waters, *Anal. Chim. Acta, 12*, 162–176, 1955.

Nesbitt, H. W., G. Markovics, and R. C. Price, Chemical processes affecting alkalis and alkaline earths during continental weathering, *Geochim. Cosmochim. Acta, 44*, 1659–1666, 1980.

Ojiambo, B. S., and W. B. Lyons, Residence times and solute fluxes of major ions in Lake Naivasha, Kenya and their relationship to lake hydrology, in *The Limnology, Climatology and Paleoclimatology of the East African Lakes*, edited by T. C. Johnson and E. O. Odada, pp. 267–278, Gordon and Breach, Amsterdam, 1996.

Palmer, M. R., and J. M. Edmond, Controls over the strontium isotope composition of river water, *Geochim. Cosmochim. Acta, 56*, 2099–2111, 1992.

Shiller, A. M., and D. M. Frilot, The geochemistry of gallium relative to aluminum in Californian streams, *Geochim. Cosmochim. Acta, 60*, 1323–1328, 1996.

Spaulding, S. A., D. M. McKnight, E. F. Stoermer, and P. T. Doran, Diatoms in sediments of Lake Hoare, Antarctica, *J. Paleolimnol.*, in press.

Spigel, R. H., and J. C. Priscu, Evolution of temperature and salt structure of Lake Bonney, a chemically stratified Antarctic lake, *Hydrobiol., 321*, 177–190, 1996.

Stallard, R. F., River chemistry, geology, geomorphology and soils in the Amazon and Orinoco basins, in *The Chemistry of Weathering*, edited by J. T. Drever, pp. 293–316, Reidel, Dordrect, 1985.

Stallard, R. F., and J. M. Edmond, Geochemistry of the Amazon 2. The influence of geology and weathering environment on the dissolved load, *J. Geophys. Res., 88*, 9671–9688, 1983.

Stallard, R. F., and J. M. Edmond, Geochemistry of the Amazon 3. Weathering chemistry and limits to dissolved inputs, *J. Geophys. Res., 92*, 8293–8302, 1987.

Toxey, J. K., D. A. Meese, K. A. Welch, and W. B. Lyons,

The measurement of reactive silicate in saline-hypersaline lakes: Examples of the problem, *Inter. J. Salt Lake Res.*, in press.

von Guerard, P., D. M. McKnight, R. A. Harnish, J. W. Gartner, and E. D. Andrews, Streamflow, water-temperature, and specific-conductance data for selected streams draining into Lake Fryxell, Lower Taylor Valley, Victoria Land, Antarctica, 1990-92, USGS/OFR 94-545, 65 pp., Denver, CO, 1995.

Wedepohl, K. H., The composition of continental crust, *Geochim. Cosmochim. Acta, 59*, 1217–1232, 1995.

Welch, K. A., Glaciochemical investigations of the Newell Glacier, Southern Victoria Land, Antarctica, MSc. Thesis, University of New Hampshire, 92 pp. 1993.

Welch, K. A., and W. B. Lyons, Comparative limnology of the Taylor Valley lakes: the major solutes, *Antarctic J. US*, in press.

Welch, K. A., W. B. Lyons, E. Graham, K. Neumann, J. M. Thomas, and D. Mikesell, Determination of major element chemistry in terrestrial waters from Antarctica by ion chromatography, *Jour. Chromatogr. A, 739*, 257–263, 1996.

Wetzel, R. G., *Limnology*, Saunders, Philadelphia, 743 pp., 1975.

Wharton, Jr., R. A., Stromatolitic mats in Antarctic lakes in *Phanerozoic Stromatolites II* edited by J. Bertrand-Sarfati and C. Monty, pp. 53–70, Kluwer, The Netherlands, 1994.

Wharton, Jr., R. A., B. C. Parker, G. M. Simmons, Jr., K. G. Seaburg, and F. G. Love, Biogenic calcite structures forming in Lake Fryxell, Antarctica, *Nature, 295*, 403–405, 1982.

Wharton, Jr., R. A., B. C. Parker, and G. M. Simmons, Jr., Distribution, species composition and morphology of algal mats in Antarctic dry valley lakes, *Phycologia, 22*, 355–365, 1983.

Wharton, Jr., R. A., W. B. Lyons, and D. J. DesMarais, Stable isotopic biogeochemistry of carbon and nitrogen in a perennially ice-covered Antarctic lake, *Chem. Geol., 107*, 159–172, 1993.

Wilson, A. T., Geochemical problems of the Antarctic dry areas, *Nature, 280*, 205–208, 1979.

Yuretich, R. F., and T. E. Cerling, Hydrogeochemistry of Lake Turkana, Kenya: Mass balance and mineral reactions in an alkaline lake, *Geochim. Cosmochim. Acta, 47*, 1099–1109, 1983.

W. B. Lyons, R. McArthur, K. Neumann, J. K. Toxey, K. A. Welch, and C. Williams, Department of Geology, Box 870338, University of Alabama, Tuscaloosa, AL, 35487-0338.

D. McKnight, INSTAAR, 1560 30th Street, Campus Box 450, Boulder, CO 80309-0450.

D. Moorhead, Department of Biology, Texas Tech University, Lubbock, TX 79409-3131.

(Received September 10, 1996;
accepted April 25, 1997)

HYDROLOGIC PROCESSES INFLUENCING STREAMFLOW VARIATION IN FRYXELL BASIN, ANTARCTICA

Peter A. Conovitz[1], Diane M. McKnight[2], Lee H. MacDonald[1], Andrew G. Fountain[3,5],

and Harold R. House[4]

In the McMurdo Dry Valleys, glacial meltwater streams are a critical linkage between the glaciers and the lakes in the valley bottoms. This paper analyzes the physiographic characteristics and six years of discharge data from five streams in order to better characterize the dynamic inputs into Lake Fryxell, a closed basin in Taylor Valley. These feeder streams typically flow only for six to eight weeks during the summer, and streamflow is highly variable on an interannual as well as daily basis. During low flow years, the shorter streams contributed a higher proportion of the total annual inflow into the lake; this pattern may reflect the greater losses to wetting the hyporheic zone. Comparisons of the period of direct sun on the glacier faces with the time of peak flow suggested that solar position and melt from the glacier faces are the dominant controls on the diurnal fluctuations in streamflow. An analysis of streamflow recession showed considerable variability between streams and in some cases, over time. For example, recession coefficients for Canada Stream, a short stream with an incised channel, were fairly invariant with streamflow. In contrast, the recession coefficients for Lost Seal Stream, an unconfined, low gradient stream, increased significantly with increasing discharge. These observations lead to hypotheses for the control of streamflow dynamics in the McMurdo Dry Valleys by climate, solar position, and geomorphic factors.

INTRODUCTION

The McMurdo Dry Valleys region of southern Victoria Land, Antarctica is a large polar desert located along the west coast of the Ross Sea. Climate is extremely cold and dry; air temperatures range from a mean daily low of about –45°C during the winter to a mean daily high of about 7°C during the short summer [*Keys*, 1980]. Annual precipitation is less than 10 cm yr-1[*Bromley*, 1985], and much of this is quickly lost to sublimation. Solar radiation, which is the driving force of the melt cycle, is subject to considerable topographic

variability. Average incoming solar radiation on the glaciers ranged from 18 to 52 W m-2 during the 1994-1995 season [*Dana et al.*, this volume.]

The McMurdo Dry Valleys contain a number of lakes fed by meltwater streams from the surrounding glaciers. These lakes have a perennial ice cover and most are closed hydrologic systems, as few streams cross through the coastal ridges to the sea [*Chinn*, 1993]. When precipitation and inflows to the lakes are balanced by evaporation and sublimation losses, lake levels remain constant. During the past century, lake levels have been rising because inflows have exceeded sublimation and evaporation [*Chinn*, 1993].

Streamflow is highly variable on an interannual, seasonal, and daily basis [*House et al.*, 1996; see streamflow data in accompanying CDROM]. Glacial melt and resultant streamflow begins in late November to mid-December and ends in mid-January to early February. Between the glaciers and the lake, the streams flow through unconsolidated alluvium. This

---

[1] Department of Earth Resources, Colorado State University, Fort Collins, Colorado

[2] INSTAAR, University of Colorado, Boulder, Colorado

[3] U.S. Geological Survey, Denver, Colorado

[4] U.S. Geological Survey, Madison, Wisconsin

[5] Currently on leave to the Department of Geology, Portland State University, Portland, Oregon

alluvium is mostly sand-sized particles interbedded with large cobbles and boulders, predominately composed of gneiss, diorite and schist [*Bockheim*, 1997]. The unconsolidated alluvium and absence of vegetation results in unstable streambanks and large sediment loads at high flows. In some streams the stream bed is stabilized through the formation of stone pavements by a combination of periglacial and fluvial processes, while in other streams bed surfaces remain irregular. A deep, impenetrable permafrost layer lies about 0.5 m below the surface of the alluvium. Substantial interactions between the streams and the permafrost have yet to be documented.

Because the streams are both dynamic aquatic ecosystems and the primary means of delivering water and materials to the lakes, a better knowledge of streamflow processes and dynamics is critical to understanding the entire dry valley ecosystem. Streams with a stone pavement have abundant algal mats within the main stream channel, while those streams with less stable substrates have algal mats on the edge of the adjacent parafluvial zone [*McKnight and Tate*, 1996 a, 1996 b; *Alger et al.*, 1997]. These algal mats are important sources of primary production and nutrient transformations. The combination of hydrologic, geochemical and ecological processes within the streams and hyporheic zone largely determine the inputs of water, nutrients, and organic matter to the lakes [*Simmons et al.*, 1993; *Spaulding et al.*, 1994]. Temporal variations in streamflow can serve as an index of short- and long-term climatic fluctuations and glacial melt rates. Hence the purpose of this paper is to evaluate the temporal and spatial variability of streamflow and then to relate these patterns to predicted glacial melt and the physical characteristics of the individual streams. This information will strengthen our understanding of the physiographic controls on streamflow and the fluctuating inputs of water and materials to the lakes. The analysis presented here will also help guide future, more process-based research on the hydrology of the dry valley streams.

The first objective was simply to assess the interannual variability in streamflow among the gauged streams flowing into Lake Fryxell (Figure 1). If the respective contributions are consistent between years or can be related to some of the physical characteristics of the different streams, this would help predict the relative contributions of the ungauged streams flowing into Lake Fryxell.

The dry valley streams exhibit a large diel variation in flow [*von Guerard et al.*, 1994], and this is due to both "source" and "instream" processes. Source processes control the generation of meltwater and are driven largly by climatic factors such as cloud cover, temperature regime, and solar energy inputs to the glacier faces and glacier surfaces. Discharge increases rapidly in response to warmer temperatures and high fluxes of solar radiation. Conversely cold and/or cloudy periods during the summer months will immediately reduce melt, and streamflow can decline by an order of magnitude within a few hours (Figure 2). Much of the melt is believed to come from direct radiation on the nearly vertical glacier faces rather than melt on the more horizontal glacier surfaces [*Fountain et al.*, this volume; *Dana et al.*, this volume]. Thus the second objective of this study was to evaluate the relationship between sun angle on the different glacier faces and the observed daily patterns of flow in five of the dry valley streams.

After the meltwater drains from the glaciers, a different set of instream processes are hypothesized to control stream discharge patterns. Physical characteristics such as the gradient, length, bed morphology, and characteristics of the alluvial material may regulate flow velocity, control the storage of water within the hyporheic zone, attenuate the meltwater hydrograph, and influence sediment transport. It was estimated that 1.3×10^5 m^3 of water was stored in the hyporheic zone along the Onyx River (C. Howard-Williams, personal communication), while a tracer experiment has shown a very rapid exchange between the hyporheic zone and the main channel [*McKnight and Andrews*, 1993; *Runkel et. al.*, in press]. Thus the third objective of this study was to analyze streamflow recession rates as a means of assessing the storage of water in the hyporheic zone and the dynamics of drainage when meltwater generation ceases.

SITE DESCRIPTION

Fryxell Basin

Lake Fryxell Basin is the easternmost basin in the Taylor Valley (Figure 1). The lake is large, shallow and permanently ice-covered. Inflows come from thirteen streams that drain the various glaciers surrounding the basin. The two most prominent glaciers are the Canada and Commonwealth Glaciers, and these are the source of seven of the thirteen streams. The landscape through which the streams flow is a relatively uniform unconsolidated alluvium. The porosity or hydraulic conductivity of this material presumably does not vary between the different streams.

Fryxell Basin Reference Map

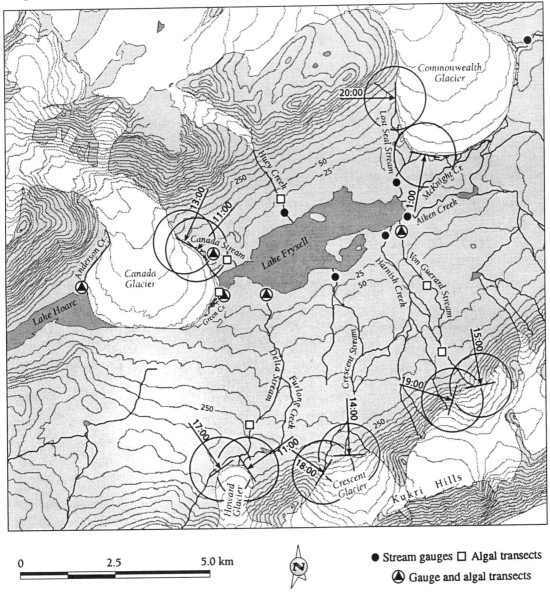

Fig 1. Map of the lower Taylor Valley, showing Lake Fryxell and inflowing streams. The aspects of the glacial meltwater source areas on the alpine glaciers draining into five of the streams are indicated, as is the time of day when the sun is directly facing these aspects. Elevations are in meters.

Streamflow, water temperature and specific conductance have been measured since the 1990-1991 field seasons except for the austral summer of 1992-1993. Of the thirteen streams draining into Lake Fryxell, streamflow data have been collected at 15-minute intervals at eight streams and periodically at five streams [*von Guerard et al.*, 1994]. Periodic measurements are of limited value due to the high short-term variability of streamflow.

Five streams were selected for analysis in this paper: Canada, Lost Seal, von Guerard, Crescent, and Delta Streams (Figure 1). These streams were selected because

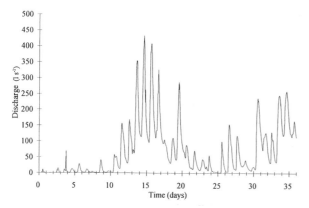

Fig 2. Hydrograph for Canada Stream from November 26 to December 31, 1991. Time zero is midnight on 26 November 1991.

of the quality of the discharge records and because they are representative of the streams in Fryxell Basin with regard to length, gradient, and aspect, as well as duration and magnitude of flow [*von Guerard et al.*, 1994]. The length, gradient and other characteristics of these streams are summarized in Table 1 and described below.

Canada Stream

Canada Stream drains the north-facing section of the Canada Glacier and is one of the largest sources of inflow to Lake Fryxell. Canada Stream is usually the first stream to begin flowing into the lake, as it reaches the lake in late November to early December. It is the shortest of the five streams and has a relatively low gradient of 0.03 m m^{-1} (Table 1). Meltwater enters the channel from a large section along the face of the Canada Glacier. Water draining from the northern section of the face enters a well-defined channel, while water from the more southern part of the source area drains into a pond. Outflow from this pond then enters the channel and continues to the gauge located about 1 km upstream of Lake Fryxell. Between the gauge and the lake, the stream flows through an incised channel with boulders in the streambed. In the last 0.5 km reach the flow spreads out through a large alluvial fan.

Lost Seal Stream

Lost Seal Stream drains a large source area along the western and southern faces of the Commonwealth Glacier. It is slightly longer and considerably flatter than Canada Stream (Table 1). The channel itself is wide and shallow and is gauged about 0.3 km from the

lake. At high flow ponds may form near the base of the glacier.

Von Guerard Stream

Von Guerard Stream drains the northern face of a glacier in the Kukri Hills and flows into the southeastern end of Lake Fryxell. It is intermediate in length and has the steepest average gradient. The steeper upper reaches of the stream have a bed of large cobbles and boulders, and this section may remain snow-covered into the summer due to the accumulation of wind-blown snow. Below this steep upper reach, the stream enters a wide, flat area and then flows through a long reach of moderate gradient before the channel widens at the outlet. The stream gauge is located approximately 0.25 km upstream of the inlet to Lake Fryxell.

Crescent Stream

Crescent Stream also drains from the glaciers of the Kukri Hills into the south-central region of Lake Fryxell. Most of the meltwater is derived from three main source areas along the northern face of the glaciers. These three tributaries converge into one channel below the very steep upper reaches located just below the face. The channel is considered to be stable, with no observable scour, fill, or sand transport. The stream gauge is located approximately 0.25 km upstream of the inlet.

Delta Stream

Delta Stream originates from Howard Glacier in the Kukri Hills and flows into the southernmost section of Lake Fryxell. It is the longest of the five streams, and in contrast to von Guerard and Crescent Streams, it has a relatively uniform gradient. The bed consists of coarse particles arranged in a flat pavement and is also considered relatively stable.

METHODS

Inflows to Lake Fryxell

Existing data were used to calculate the correlations in daily flows between streams and the interannual variability in discharge for each stream. To evaluate the relative contribution of each stream to the annual water budget for Lake Fryxell, flows were estimated for the intermittently gauged streams and direct glacier

TABLE 1. Physiographic Characteristics an Estimated Accuracy of the Gauging Record for the Streams Used in This study. The Pond at the Base of the Source Glacier for Lost Seal Stream exists at High Flows Only.

Stream	Length (km)	Gradient (m m⁻¹)	Length/ gradient	Pond at base of glacier face	Elevation at base of source glacier (m)	Type of gauge	Estimated error in discharge (%)
Canada	1.5	0.03	50	yes	100	rectangular weir	<10
Lost Seal	2.2	0.02	110	yes	125	6 in. Parshall flume	10–15
Von Guerard	4.9	0.08	61	no	375	channel control	10
Upper reach	0.5	0.55					
Lower reach	4.4	0.05					
Crescent	5.6	0.07	80	no	350–450	channel control	10–15
Upper reach	0.7	0.28					
Lower reach	4.9	0.05					
Delta	11.2	0.03	370	no	300	channel control	10–15

inflows. The basis for these estimates is discussed in detail by *House et al.* [1996]. Flows in the intermittently gauged streams (McKnight Creek, Harnish Creek, Bowles Creek, Mariah Creek, and Andrews Creek) were derived largely from numerous direct measurements during the 1993-1994 field season. Visual observations indicated that the sum of the direct glacier inflows were approximately equal to the average of the annual flows for Canada Stream and Green Creek. To evaluate the interannual and daily variability, correlation matrices were constructed using total annual and mean daily flows for the 1990-1991 through 1995-1996 seasons

To further characterize the variability in the daily flow pattern among the five streams, the estimated proportion of melt from the glacier surfaces and glacier faces were inferred from daily hydrographs for eight different days in the 1990-1991 and 1991-1992 field seasons. To separate these two sources of melt it was assumed that glacial surface melt provided the baseflow component of the hydrograph, while the peak of the daily hydrograph occurred in response to melt from the face of the source glacier. The separation of baseflow from peakflow was made using the straight line method [*McCuen*, 1989], and the ratio of peak height to baseflow was determined for each of the eight days for the five streams.

Solar Position and Lag Times

To determine the time lag between peak solar intensity and time of peak daily discharge, it was first necessary to determine the period of peak insolation on the faces of the glaciers that feed the five streams. Aspects of the source areas were identified from the 1:50,000 map of the Lake Fryxell quadrangle (Figure 1). The maximum period of time that the glacier faces were exposed to direct radiation was determined for each stream from the mapped aspects and the known path of the sun during the melt season. Peak solar intensity is reached when the solar azimuth is normal to the face. In the Lake Fryxell Basin, the sun revolves counterclockwise about the basin while maintaining a continuous low angle to the horizon and is approximately due north at 1400 h.

The average time of daily peak streamflow was calculated for each stream for each year of record. Periods at the start of the season, when the streambed was becoming saturated, and at the end of the season, when melt had slowed, generally exhibited more variability with regard to the timing of the daily peak flow and were excluded from this analysis. Days during the main flow period with little of no melt were also excluded from this analysis. The mean time of peak flow and corresponding standard deviation were compared with the range of time that each source area receives direct solar radiation. Thus two lags were determined, with the first lag being the time between the first direct solar radiation on the glacier face and the average time of peak flow. The second lag was the time between when the source area last receives direct solar radiation and the average time of peak flow. The calculated mean lags were then related to the length and gradient of each of the five streams.

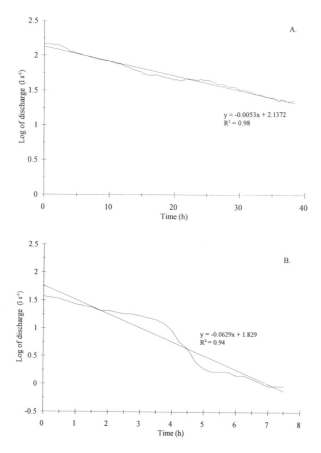

Fig 3. Log of discharge versus time for A. Canada Stream from December 18 to December 20, 1990, and B. Delta Stream from January 14 to January 15, 1991. Time zero is the beginning of the recession period.

Recession Analysis

An analysis of recession curves was conducted in order to assess the storage of water within the hyporheic zone and the drainage characteristics of each stream. We selected periods from the existing discharge records when there was no detectable increase in discharge due to the daily melt cycle. Hydrographs were plotted at 15-minute time intervals to confirm the identified recession period (Figure 2). For those hydrograph periods that showed no diurnal melt patterns, the log of the discharge was plotted against time [*Eagleson*, 1970]. The slope of this line was defined as the recession coefficient (Figures 3a and 3b).

X-Y scatterplots were constructed to assess the relationships between recession coefficients and magnitude of flow. Since the recession coefficients were expected to vary with the extent of hyporheic zone

saturation, we only used recession coefficients obtained after the seasonal maximum flow. Given the high interannual variability in streamflow, we also tested the relationship between these coefficients and seasonal maximum flow.

RESULTS

Interannual and Daily Variability in Discharge

Average annual streamflows in the Lake Fryxell basin show tremendous interannual variability (Table 2). For each stream except Lost Seal Stream, the total discharge in 1990-1991 was 15–30 times the discharge in 1994-1995. These large annual differences imply a high interannual variability of both source and instream processes, and a need to consider the volume of annual streamflow when evaluating other measurements.

The results in Table 2 also show considerable interannual variations in relative flow. Green Creek, for example, contributed less than seven percent of the estimated total inflow to Lake Fryxell in 1993-1994 but 18% in 1994-1995. Streams with similar locations within Lake Fryxell Basin do not necessarily follow similar patterns from year to year. Total discharge in 1991-1992 for Von Guerard, Crescent, and Delta Streams, all on the south side of Lake Fryxell, was respectively 88%, 38%, and 64% of their respective discharges in 1990-1991. These differences between streams and between years may stem from the variation in the pattern of snowfall on the glaciers. Increased albedo on snow-covered ice reduces melt, particularly in these conditions where solar radiation is so important. Annual discharge of streams in the Lake Bonney basin exhibits less interannual variability. This is attributed to the lack of snow on the lower parts of the glaciers and lower overall precipitation as compared to Fryxell Basin (Fountain, unpublished data). Differences between streams may also stem from the different aspects of the source glaciers, the interannual variations in solar position during warm periods when most of the flow is generated, and physical characteristics of the individual streams.

In general, the total annual discharge was significantly correlated among the five different streams (Table 3). The strongest relationships were between Delta and Canada Stream ($R^2 = 0.99$) and Von Guerard and Lost Seal Stream ($R^2 = 0.99$). These two pairs of streams are at the eastern and western ends of Lake Fryxell, respectively, and they have at least one

TABLE 2. Annual Streamflow in the Lake Fryxell Basin and Estimated Percent Contribution to Lake Fryxell.

Stream site	Length km	1990-1991 x10³ m³	%	1991-1992 x10³ m³	%	1993-1994 x10³ m³	%	1994-1995 x10³ m³	%
Canada Stream	1.5	510	15	300	13	130	14	32	20
Huey Creek	2.1	200	6	72	3	51	5.5	0	0
Lost Seal Stream	2.2	530	15	470	20	190	21	19	12
§Aiken Creek	1.3	250	7	280	12	57	6.2	8.7	5
Von Guerard Stream	4.9	160	5	140	6	68	7.4	7.2	5
Crescent Stream	5.6	210	6	80	3	39	4.2	4.8	3
Delta Stream	11.2	220	6	140	6	63	6.8	3.1	2
Green Creek	1.2	330	10	220	9	61	6.6	28	18
McKnight Creek	2.0	*	*	*	*	41	4.5	*	*
Harnish Creek	5.1	*	*	*	*	53	5.8	*	*
Bowles Creek	0.9	*	*	*	*	22	2.4	*	*
Mariah Creek	0.9	*	*	*	*	26	2.8	*	*
Andrews Creek	1.5	*	*	*	*	20	2.2	*	*
Intermittent streams	*	610	18	430	17	NA	NA	26	16
Direct glacier inflow	*	420	12	260	11	99	11	31	19
Total		3440	100	2390	100	920	100	160	100

*Indicates data not available. NA = not applicable. §The stream length for Aiken Creek is only for the reach between Many Glaciers Pond and Lake Fryxell.

TABLE 3. Coefficients of Determination (R²) for Total Annual Flow and for Mean Daily Flow for 1990-1991 through 1995-1996. In Each Case the Number in Parentheses below the Coefficient is the p-Value.

Stream	Total Annual Flow					Mean Daily Flow				
	Canada	Crescent	Delta	Lost Seal	Von Guerard	Canada	Crescent	Delta	Lost Seal	Von Guerard
Canada	—	0.95 (0.022)	0.99 (0.004)	0.89 (0.054)	0.89 (0.057)	—	0.58 (0.0001)	0.69 (0.0001)	0.56 (0.0001)	0.68 (0.0001)
Crescent	—	—	0.92 (0.041)	0.74 (0.14)	0.74 (0.14)	—	—	0.69 (0.0001)	0.68 (0.0001)	0.69 (0.0001)
Delta	—	—	—	0.94 (0.032)	0.94 (0.031)	—	—	—	0.71 (0.0001)	0.75 (0.0001)
Lost Seal	—	—	—	—	0.99 (0.002)	—	—	—	—	0.74 (0.0001)
Von Guerard	—	—	—	—	—	—	—	—	—	—

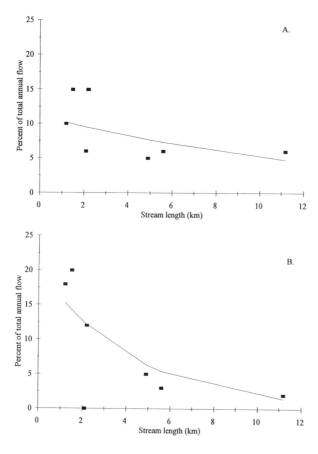

Fig 4. Relationship between percent of total inflow to Lake Fryxell and stream length for A. summer of 1990-1991 and B. summer of 1994-1995.

Fryxell. Intermittent measurements made in the other three years were consistent with this estimate. The 20-fold variation in total inflows to Lake Fryxell mean that the contributions of the streams to the basin-scale water budget change quite substantially, and one would also expect considerable interannual variation in the amounts of sediment and nutrients delivered to the lake.

In addition to variation in source processes, the interannual variability in flow may be related to differences in physiographic characteristics. Figure 4 shows stream length versus percent of total flow for the the high flow year of 1990-1991 and the low flow year of 1994-1995. The exponential regression lines indicate that in the low flow year, a greater portion of flow was contributed by streams with short stream lengths, but the relationship with stream length was not as strong in the high flow year. The greater contribution of shorter streams in a dry year may be related to the amount of meltwater storage in the hyporheic zone, as the meltwater initially encounters a dry streambed and must saturate the channel before advancing. During a low flow year this recharge and storage will be a greater percentage of the total meltwater runoff in the longer streams. This stored water is believed to be largely lost to evaporation rather than carried over to the next melt season (Howard-Williams, personal communication). Further investigation into the interactions between the stream, the hyporheic zone, and the permafrost is needed to better understand the storage and loss of meltwater along the channel.

Hydrograph Separation

The estimated proportions of melt from the glacier surface (baseflow) and glacier faces (peakflow) as determined from the daily hydrographs showed considerable variation between days and between streams (Table 4). A few very high ratios caused the mean values for some streams to be much larger than the median, and for this reason the interbasin comparisons are based primarily on the medians. Three of the streams (Canada, Lost Seal, and Delta) had median peakflow/baseflow ratios of less than five, while the median ratios for Von Guerard and Crescent Streams were greater than ten. Both Canada and Lost Seal Streams are draining relatively large and flat glacier surfaces, and the high baseflow discharge (generally greater than 70 l s[-1]) in these two streams indicates that there is a more even input of energy to these glacier surfaces than to the steeper glaciers of the Kukri Hills.

source area with a similar aspect. However annual discharge in Crescent Stream was not well correlated with the annual discharge in either Lost Seal or Von Guerard Streams, even though the latter also drain northwards from the Kukri Hills. Nevertheless the generally high correlations suggest that regional climate is the dominant control on streamflow. In a warm versus a cold year, all streams will have higher annual flows despite the fact that their relative contributions may vary. Correlations were not as strong when comparing mean daily flows for the five streams, although all comparisons were significant ($p<0.0001$) (Table 3). The poorer correlations for shorter time periods suggests that each stream is subject to considerable short-term variability in source and possibly instream processes.

The more intensive measurements during the 1993-1994 summer indicated that the ungauged streams contributed about 20% of the total inflow to Lake

TABLE 4. Peakflow, Baseflow, and Peakflow/Baseflow Ratios for Randomly Selected Dates During the 1990-1991 and 1991-1992 Seasons. Discharge Values are in Liters per Second.

	18 Dec 1990	22 Dec 1990	27 Dec 1990	11 Dec 1991	30 Dec 1991	1 Jan 1991	12 Jan 1991	7 Jan 1992
Canada Stream								
Peakflow	150	300	650	410	260	480	340	220
Baseflow	70	100	100	90	110	125	95	20
Peakflow / Baseflow	2.1	3.0	6.5	4.6	2.4	3.8	3.6	11
Mean ratio = 4.6	Median ratio = 3.7							
Lost Seal Stream								
Peakflow	410	220	350	400	1000	750	450	750
Baseflow	125	70	70	200	175	195	100	20
Peakflow / Baseflow	3.3	3.1	5.0	2.0	5.7	3.8	4.5	38
Mean ratio = 8.1	Median ratio = 4.2							
Von Guerard Stream								
Peakflow	120	125	305	425	250	210	260	125
Baseflow	10	10	20	40	45	20	25	5
Peakflow / Baseflow	12	13	15	11	5.6	11	10	25
Mean ratio = 13	Median ratio = 11.5							
Crescent Stream								
Peakflow	195	240	315	170	245	300	375	110
Baseflow	15	15	25	25	25	50	25	10
Peakflow / Baseflow	13	16	13	6.8	9.8	6.0	15	11
Mean ratio = 11	Median ratio = 12.0							
Delta Stream								
Peakflow	135	130	300	340	210	210	130	140
Baseflow	25	30	50	80	60	70	20	7
Peakflow / Baseflow	5.4	4.3	6.0	4.3	3.5	3.0	6.5	20
Mean ratio = 7	Median ratio = 4.9							

Delta Stream does not have such a large glacial surface and baseflow is less than Canada and Lost Seal Streams. Conversely the face of Howard Glacier has aspects that range from northwest to northeast (Figure 1), which may sustain the higher baseflow and account for the lower peakflow/baseflow ratio. Crescent and Von Guerard Streams both drain from the steeper glaciers of the Kukuri Hills, and the range of aspects for these faces is not as great as for Howard Glacier. Thus the discharge in Crescent and Von Guerard Streams exhibits greater diel variability and appears to be more dependent on solar position.

Solar Position and Lag Times

The time periods that the glacier faces were receiving direct solar radiation ranged from a minimum duration of two hours for Canada stream to a maximum duration of six hours for Delta Stream. The faces contributing to Canada, Crescent, Delta, and Von Guerard Streams receive direct sunlight from late morning (1100 h) to the evening (1900 h). The glacier faces contributing to Lost Seal Stream receive direct sun from the evening (2000 h) until early morning (0100 h).

With the exception of Lost Seal Stream, average times of peak flow (Table 5) were generally consistent from year to year. Canada Stream showed the least variation in time of average peak flow within and between years, while the interannual variability in time of peak flow was 2–3 hours for Von Guerard, Crescent and Delta Streams. As might be expected, there was considerable variation in the timing of daily peak flows within each year, and this presumably reflects hourly-scale variations in temperature and cloud cover.

The eight-hour variation in the time of peak flow for Lost Seal Stream may stem from its topographic situation and the range of contributing aspects of the source glacier. Streamflow records show a general trend of two separate peak flows, one in the late evening and another in the early morning. The times of peak flow for Lost Seal Stream are an average of these peaks, and this averaging adds another source of variability to the calculation of mean lags. The source area for Lost Seal Stream is in close proximity to the mountain that lies to the west of Commonwealth Glacier. Thus the dual peak in daily flow is most likely caused by the mountain shadow. More precise calculations based on the true period of illumination should reduce this variability. Intermittent ponding at the base of the glacier during higher flows and the flatter gradient of Lost Seal Stream may also reduce streamflow velocities and allow for more storage; these both might contribute to an increased variability in the recorded times for peak flow.

In addition to the aspect and range of source areas, the time of peak flow and the calculated lags will depend on the time necessary for the melt water to reach the gauging station, and this will be a function of both width and gradient. The mean lag between the time of first sun on the glacier faces and time of peak flow was shortest (4.6 h) for Canada Stream. Conversely, the longest average lag was eleven hours for Delta Stream, and this is also the longest stream. Thus lags increased

with stream length, with the exception of Lost Seal Stream. The rate at which the peak discharge wave moves downstream will also be affected by the gradient. There was a stronger correlation with the length-to-gradient index in Table 1 than either length or gradient alone (Table 6).

Recession Analysis

The recession coefficients for each stream were calculated from periods of no melt which lasted up to a few days. High recession coefficients (i.e., a steeper decline in discharge over time) are believed to indicate more rapid drainage, while low recession coefficients imply slower drainage. Because the slope of the recession curves indicate the relative speed at which water moves out of the adjacent saturated areas, variations in coefficients may be related to the physiographic characteristics of the different streams and the volumes of water stored in the hyporheic zone.

Of the five streams analyzed, Canada Stream had the lowest average recession coefficient (Table 7) even though it is also the shortest stream. The average slope of the recession limb was −0.011, and individual values ranged from a minimum of −0.005 to a maximum of −0.021. Canada Stream is the only stream with a pond at the base of the glacier face that functions at all flow levels, and this pond may store sufficient water to maintain relatively consistent flows even after melt has ceased.

Von Guerard Stream has the steepest average recession slope and the steepest gradient. The average slope of recession was −0.04, with individual values ranging from a minimum of −0.012 (the average slope for Canada Stream) to a maximum of −0.099. These values are consistent with the stream's steep channel and moderate length (Table 1).

The other three streams had similar mean recession coefficients (Table 7). Because Delta Stream had the highest length-to-gradient value and Lost Seal Stream the lowest gradient, it is likely that the recession coefficients are controlled by other variables in addition to stream length and gradient.

DISCUSSION

Solar Position and Lag Times

The timing of peak streamflow appears to be strongly related to the period of peak solar radiation on the source glaciers. Because the glacier surfaces share

TABLE 5. Periods of Peak Solar Insolation, Times of Peak Flows, and Resultant Lag Times in Hours and Minutes. In Three Cases Less Than 40 Days Were Used to Determine Average Peak Flows and in These Cases the Number of Days is Shown in Parentheses.

Stream	Period with Direct Radiation (h)	Year					Mean lag For Period Of Record
		1990-1991	1991-1992	1993-1994	1994-1995	1995-1996	
Canada Stream	1100–1300						
Average time of peak		14:42	15:37	15:32	15:59	16:07	
Standard deviation		1:23	1:24	1:10	1:01	1:29	
Mean lag from earliest direct radiation		3:42	4:37	4:32	4:59	5:07	4:35
Mean lag from latest direct radiation		1:42	2:37	2:32	2:59	3:07	2:35
Lost Seal Stream	2000–0100						
Average time of peak		22:32	18:05	23:44	2:06 (14)	23:40 (30)	
Standard deviation		3:28	3:17	2:47	6:45	1:06	
Mean lag from earliest direct radiation		2:32	22:05	3:44	7:06	3:40	7:51
Mean lag from latest direct radiation		21:32	17:05	22:44	25:06	22:40	21:49
Von Guerard Stream	1430–1900						
Average time of peak		20:06	21:41	23:01	22:24	21:33	
Standard deviation		2:12	2:48	1:20	2:15	3:50	
Mean lag from earliest direct radiation		5:36	7:11	8:31	7:54	7:03	7:15
Mean lag from latest direct radiation		1:06	2:41	4:01	3:24	2:33	2:45
Crescent Stream	1400–1800						
Average time of peak		21:41	22:09	20:53	19:18	*	
Standard deviation		3:07	1:56	3:10	2:50	*	
Mean lag from earliest direct radiation		7:41	8:09	6:53	5:18	*	7:00
Mean lag from latest direct radiation		3:41	4:09	2:53	1:18	*	3:00
Delta Stream	1100–1700						
Average time of peak		22:05	22:07	22:58 (22)	20:55	*	
Standard deviation		1:43	1:56	6:00	2:16	*	
Mean lag from earliest direct radiation		11:05	11:07	11:58	9:55	*	11:01
Mean lag from latest direct radiation		5:05	5:07	5:58	3:55	*	5:01

* Indicates data not available.

TABLE 6. Comparison of Lag Times to Length to Gradient Index

Stream	Length (km)	Gradient (m m⁻¹)	Length/Gradient	Mean Annual Lag (h)	
				Lag from Earliest Direct Radiation	Lag From Latest Direct Radiation
Canada	1.5	0.03	50	4:35	2:35
Lost Seal	2.2	0.02	110	7:51	21:49
Von Guerard	4.9	0.08	61	7:15	2:45
Crescent	5.6	0.07	80	7:00	3:00
Delta	11.2	0.03	370	11:01	5:01

similar aspects with the faces, it is difficult to separate meltwater contributions from these two source areas. The nearly vertical glacier faces do receive nearly twice the intensity of solar radiation than the glacier surfaces during the short period of direct insolation [*Lewis et al.*, in press]. Because the temperature of the ice is well below freezing, the high intensity of solar radiation on the glacier faces may be necessary to both warm the ice to 0°C and generate melt [*Fountain et al.*, this volume]. The faces are also at lower elevations than the surfaces, closer to the dark-colored valley floors, and maintain lower albedos throughout the season; all these factors favor a larger contribution from the glacier faces relative to the surfaces [*Fountain et al.*, this volume]. Water flowing in channels along the glacier surface is rarely observed, indicating that surface melt at higher elevations can refreeze before running off [*Fountain et al.*, this volume]. The relatively steep diurnal peaks and the high peakflow/baseflow ratios also suggest a shorter and more concentrated input of meltwater than would be the case if the glacier surface was the primary source of melt.

Velocity data were used to estimate the travel time of the peak discharge wave from the base of the glacier to the stream gauge. Average velocities at each gauging site were obtained by dividing discharge by the cross-sectional area of the channel. Average velocities at the gauging station for Canada and Lost Seal Streams were 0.51 m s⁻¹ and 0.48 m s⁻¹, respectively (USGS, unpublished data). These values are considered representative for these streams because the Canada Stream gauge is located midway between the glacier and Lake Fryxell, and the low gradient at the Lost Seal Stream gauge is representative of the entire stream. For Delta and Von Guerard Streams, we used velocities of 0.27 m s⁻¹ and 0.32 m s⁻¹, respectively, as these had been measured during moderate flows at algal transects 2–4 km above the gauge [*Alger et al.*, 1996]. Velocities

from the gauging stations were not believed to be representative and the measured velocity in Von Guerard Stream also may be low because the algal transect was located in a low gradient reach. We would expect flow velocities in Delta Stream to be greater than Lost Seal Stream due to the differences in gradient. However the velocity recorded for Lost Seal Stream is nearly twice the velocity of Delta Stream, suggesting that the former was taken at higher flows. Velocity data were not available for Crescent Stream.

From these measurements, the estimated travel times for Canada, Delta, Lost Seal, and Von Guerard Streams were 0.8, 12, 1.3, and 4.3 h, respectively. It must be noted that these times are rough estimates, because the velocity measurements were taken at only one cross-section. For Canada, Delta, and Von Guerard Streams the estimated travel times were within or somewhat less than the range of lags between the period of direct sun on the glacier face and the average time of peak flow. The consistency between lag times and estimated travel times for Lost Seal Stream is less clear due to the ambiguity in determining average time of peak flow. As one would expect, lags and estimated travel times increased with increasing stream length.

Previous study has shown that there is a delay between the time of melt generation and the influx of meltwater to the stream channel [*Lewis*, 1996]. This lag is attributed to the time between the initial absorption of shortwave radiation and the delivery of melt water to the stream channel, as well as the size and extent of the source area. In Andersen Creek, where there is only a 100 m reach between the glacier terminus and the stream gauge, a two hour lag was observed between peak radiation input and peak discharge [*Lewis*, 1996]. However, the source area extends two km upstream from Andersen Creek, suggesting that part of the lag is due to the travel time within the source area.

TABLE 7. Recession Periods and Corresponding
Coefficients. SD is Standard Deviation.

Canada Stream	Coefficient	Crescent Stream	Coefficient
18-20 Dec 90	0.005	9-10 Jan 91	0.047
23-25 Dec 90	0.006	14-15 Jan 91	0.074
5-6 Feb 91	0.015	2-4 Jan 92	0.007
6-7 Feb 91	0.017	8-9 Jan 92	0.015
13-14 Dec 91	0.007	11 Jan 92	0.022
20 Dec 91	0.007	19-20 Jan 92	0.012
11-12 Jan 92	0.017	21-22 Jan 94	0.003
14-15 Jan 92	0.015	24 Jan 94	0.003
19-21 Dec 93	0.013	12-13 Jan 95	0.045
20-22 Dec 94	0.021	Mean	0.025
31 Dec 95-1 Jan96	0.010	SD	0.025
11-14 Jan 96	0.008		
Mean	0.011		
SD	0.005		

Lost Seal Stream	Coefficient	Von Guerard Stream	Coefficient
17 Jan 91	0.035	7 Dec 90	0.099
18 Jan 91	0.033	22-23 Dec 91	0.018
10-11 Jan 92	0.028	8-9 Jan 92	0.012
17 Jan 92	0.052	16 Jan 92	0.033
28-29 Dec 93	0.021	17 Jan 92	0.044
11 Jan 95	0.015	17-18 Jan 92	0.013
29 Dec 95	0.024	17-18 Jan 92	0.059
12 Jan 96	0.023	13 Jan 94	0.039
Mean	0.029	Mean	0.040
SD	0.011	SD	0.029

Delta Stream	Coefficient
9-10 Jan 91	0.018
14-15 Jan 91	0.063
16 Jan 91	0.045
19-20 Jan 91	0.019
19 Dec 92	0.014
8-9 Dec 93	0.011
17 Jan 94	0.025
23-24 Jan 55	0.048
Mean	0.030
SD	0.019

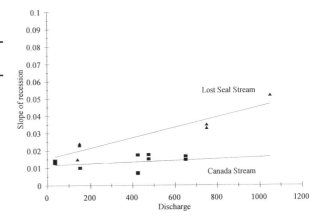

Fig 5. Discharge versus slope of the recession limb for Canada and Lost Seal Streams.

Recession Analysis

Interpretation of the recession curves for the streams in Lake Fryxell Basin is a departure from typical studies of subsurface flow within a catchment. Baseflow can be defined as the component of flow that is derived from groundwater storage or other delayed sources [*Hall*, 1968], but in Fryxell Basin subsurface flow is probably limited to the areas immediately adjacent to the stream. Because groundwater flow has not been previously observed [*Chinn*, 1993], this was not been considered in our interpretation of the recession coefficients. Hence we believe that the recession coefficients should depend primarily on the available storage in the hyporheic zone rather than baseflow in the traditional sense.

In the case of Canada Stream, which had the lowest average recession coefficients, flows may be regulated by the pond at the base of Canada Glacier. The volume of this pond is sufficiently large relative to daily flows to regulate meltwater inputs into the stream. Canada Stream has higher baseflows, the lowest peakflow/baseflow ratio, and the most consistent timing of peakflows. These characteristics all support the concept that the pond serves to attenuate high flows and supplement low flows during periods with little or no melt.

For Canada Stream the recession coefficients were relatively independent of discharge within a season and between years (Figure 5). It is the shortest of the five streams and has a relatively narrow, incised channel. The area adjacent to the channel that can be saturated is therefore limited and this could limit the effect of higher flows on the slope of the recession limb. In both low flow and high flow years the extent of the saturated hyporheic zone would be similar.

In contrast to Canada Stream, recession coefficients for Lost Seal Stream increase with an increase in seasonal peak flow (Figure 5). The recession coefficient for the period after the seasonal peak in 1994-1995 was 0.015, while the corresponding value in 1990-1991 was 0.052, nearly four times larger. In a wide, low-gradient channel, a small increase in water level could create a relatively large increase in saturated area. Lost Seal Stream is the flattest of the five streams, and therefore should have a potentially larger saturated area adjacent to the channel. At lower flows the saturated volume adjacent to the stream will be less, and a drop in stream stage will not create a very strong hydraulic gradient back to the stream channel. At high flows there would be a larger saturated volume and a stronger gradient as stage decreases. Hence the effect of discharge on the recession coefficients should be more pronounced in the wider, lower-gradient streams that have a potentially larger hyporheic zone.

A conceptual model for the different patterns in the extent of the hyporheic zone in Canada and Lost Seal Streams is shown in Figure 6. This model can be used to develop more quantitative hypotheses to explain how differences between Canada Stream's steep, incised channel and Lost Seal Stream's wide and flat channel could be influencing drainage. Because the alluvium is generally uniform throughout the basin, we hypothesize that these geomorphic differences are significant in explaining differences between these streams.

Relationships for the other three streams are less clear. The recession coefficients for Crescent Stream increase with higher streamflow except for the lowest melt season (1994-1995). Recession coefficients for Delta and Von Guerard Streams display no consistent pattern within a season, between seasons, or in relation to magnitude of streamflow. Von Guerard and Crescent Streams are both longer and steeper on average than Lost Seal and Canada Streams and have more variation in topography along their length. Both Von Guerard and Crescent Streams drain steep upper reaches before flattening out. Thus their recession curves may be a composite of different drainage patterns from the different reaches, and this may obscure any relationship with total length, average gradient, or seasonal peak discharge.

A final problem with interpreting the recession coefficients is the sensitive temperature balance between melt generation and freezing. When temperatures drop and glacial melt shuts off, some of the water in the hyporheic zone may freeze in situ and cannot drain back into the channel. The amount of water frozen in the saturated zone will vary with the temperature regime during each recession period, and this may greatly affect the observed rates of recession. The extreme sensitivity of streamflow to small changes in temperature and other climatic factors means that we have to expect considerable variation in any streamflow characteristic.

CONCLUSIONS

Streams in the McMurdo Dry Valleys exhibit tremendous variability in flow on daily, seasonal, and interannual time scales. The proportion of flow between streams is not necessarily consistent from year to year. The relative consistency of lags between the period of sun on the glacier face and average time of peak flow between years and between streams demonstrates the importance of solar position and glacial melt in controlling the daily flow regime. A more detailed analysis of how solar position, solar radiation, and temperature control runoff would help in identifying the source processes and how they might control stream and lake responses to climatic fluctuations. Additional years of streamflow and meteorological data will aid in the understanding of watershed processes and the stream ecosystems within the dry valleys.

Although we have put forth a variety of hypotheses to explain the dynamics of streamflow and evaluate the role of the hyporheic zone, there are more controls on the system than we can account for in these analyses. Of the five streams, only Canada and Lost Seal display some consistency in their relationship between recession coefficient and discharge. Recession analysis of Delta, Crescent, and Von Guerard Streams illustrate that other methods are necessary to evaluate how stream channel geomorphology governs instream channel processes. Additional tracer experiments and hydrometric measurements are needed to quantify the hyporheic zone and evaluate its significance for the dry valley streams and lakes, particularly since drainage controls the duration of the summer growth period.

Acknowledgments. We acknowledge K. Lewis, S. Spaulding, E. Andrews, and C. Howard-Williams for their helpful discussions and suggestions. We also thank K. Lewis for letting us review a draft copy of her thesis.

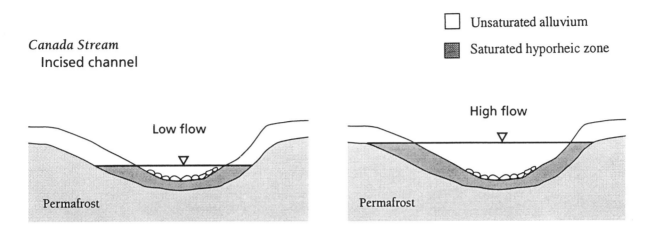

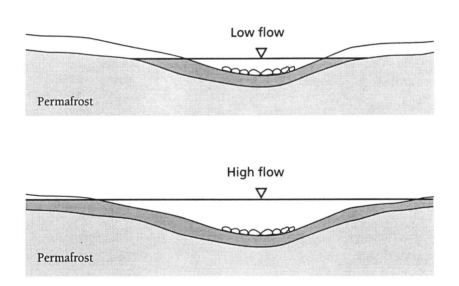

Fig 6. Schematic showing variation in the extent of the hyporheic zone under low and high flow conditions in Canada and Lost Seal Stream.

REFERENCES

Alger, A.S., D.M. McKnight, S.A. Spaulding, G.H. Shupe, A. Welch, R. Edwards, E.D. Andrews, and H.R. House. Ecological processes in a cold desert ecosystem: the abundance and species distribution of algal mats in glacial meltwater streams in Taylor Valley, Antarctica. Occasional paper No. 51. Institute of Arctic and Alpine Research, 1997.

Bockheim, J.G., *Properties and classification of cold desert soils from Antarctica*. Soil Sci. Soc. Am. J. 61:224–23. 1997.

Bromley, A.M., Weather observations, Wright Valley,

Antarctica, Infor. Publ. 11, 37 pp., New Zealand Meteorological Service, Wellington, 1985.

Chinn, T.J., Physical hydrology of the dry valley lakes. In *Physical and Biogeochemical Processes in Antarctic Lakes, Antarctic Research Series, V. 59*, W.J. Green and E.I. Friedman, eds. Washington, D.C. 1–51. 1993.

Dana, G.L., R.A. Wharton, R. Dubayah. Solar radiation in the McMurdo Dry Valleys, Antarctica. This volume.

Eagleson, P.M., Dynamic hydrology. McGraw-Hill Book Co. New York. 1970.

Fountain, A.G., G.L. Dana, K.J. Lewis, B.H. Vaughn, and D.M. McKnight. Glaciers of the McMurdo dry valleys, Southern Victoria Land, Antarctica. This volume

Hall, F.R., *Base-flow recessions--a review.* Water Resources Research 4: 973–983. 1968

House H.R., D.M. McKnight, and P. von Guerard. *The influence of stream channel characteristics on streamflow and annual water budgets for lakes in Taylor Valley.* Antarctic Journal of the United States. (in press). 1996.

Keys, J.R., Air temperature, wind, precipitation and atmospheric humidity in the McMurdo Region. Publ. 17, 57 pp. Geology Department, Victoria University,Wellington, NZ, 1980

Lewis, K. Processes controlling glacial melt in Taylor Valley, M.S. Thesis, University of Colorado.1996.

McKnight, D.M., and E.D. Andrews. *Hydrologic and geochemical processes at the stream-lake interface in a permanently ice- covered lake in the McMurdo Dry Valleys, Antarctica.* Veh. Internat. Verein. Limnol. 25:957–959, 1993.

McKnight, D.M. and C.M. Tate. *Algal mat distribution in glacial meltwater streams in the Taylor Valley, southern Victoria Land, Antarctica.* Antarctic Journal of the United States, 1996.

McKnight, D.M. and C.M. Tate. *Canada Stream: a glacial meltwater stream in Taylor Valley, South Victoria Land, Antarctica.* Journal of North American Benthological Society, 1996.

McKnight, D.M., H.R. House, and P. von Guerard. *Streamflow measurements in Taylor Valley.* Antarctic Journal of the United States,1995.

McCuen, R.H. *Hydrologic analysis and design.* Ch. 8, pp 355–360, Prentice Hall, New Jersey. 1989

Runkel, R.L., D.M., McKnight, and E.D. Andrews. Analysis of transient storage subject to unsteady flow: diel flow variation in an Antarctic stream. In preparation.

Simmons, G.M., Jr., J.R. Vestal, and R.A. Wharton, Jr. Environmental regulators of microbial activity in continental lakes. In: Physical and Biogeochemical Processes in Antarctic Lakes,Antarctic Research Series V.59, W.J. Green and E.I. Friedman, eds. Washington D.C. 165–195. 1993

Spaulding, S.A., D.M. McKnight, R.L. Smith and R. Dufford. *Phytoplankton population dynamics in perennially ice-covered Lake Fryxell, Antarctica.* J. Plank. Res. 16:527–541.1994

von Guerard, P., D.M. McKnight, R.A. Harnish, J.W. Gartner, and E.D. Andrews. Streamflow, water temperature, and specific conductance data for selected streams draining into Lake Fryxell, Lower Taylor Valley, Victoria Land, Antarctica. U.S. Geological Survey. Open-file Report 94–545. 65 pp. 1994.

P. Conovitz, Department of Earth Resources, Colorado State University, Fort Collins, CO 80523

A. Fountain, Department of Geology, Portland State University, Portland, OR 97207-0751

H. House, U.S. Geological Survey, 6417 Normandy Ln., Madison, WI 53719

L. MacDonald., Department of Earth Resources, Colorado State University, Fort Collins, CO 80523

D. McKnight, INSTAAR, Campus Box 450, University of Colorado, Boulder, CO 80303

(Received September 6, 1996:
accepted April 30, 1997)

LONGITUDINAL PATTERNS IN ALGAL ABUNDANCE AND SPECIES DISTRIBUTION IN MELTWATER STREAMS IN TAYLOR VALLEY, SOUTHERN VICTORIA LAND, ANTARCTICA

Diane M. McKnight[1], Alex Alger[2], Cathy M. Tate[3], Gordon Shupe[4], and Sarah Spaulding[5]

The abundance and distribution of algal mats were studied in three streams flowing into Lake Fryxell, located in lower Taylor Valley. Algal mats were most abundant at sites which have moderate gradients and streambeds composed of large cobbles arranged in a flat stone pavement through periglacial processes. Algal abundance was less at high gradient and deltaic sites. Most of the length of the three streams can be characterized as "large cobble" and the total chlorophyll-a in each stream was estimated from measurements made at representative sites. Four algal mat types were used to characterize the stream biota. Black-, orange-, and green-colored algal mat types occurred at most sites, but red-colored mats occurred in only one of the streams. At all sites, black-colored mats were found near the channel margins and green-colored mats were found on the underside of rocks in the main channel. Orange- and red-colored mats occurred in flowing water habitats, either in the main channel or in rivulets draining the hyporheic zone at the stream margins. Thus similarities in physical characteristics of the stream habitat appeared to determine the occurrence of the different algal mats rather than differences in water quality. The species composition of the different mat types was consistent among sites. The black-colored algal mats were dominated by *Nostoc* sp., with a low average evenness of 0.13 ± 0.07. The green-colored algal mats were also essentially unialgal, composed chiefly of *Prasiola calophylla* or *P. crispa*, and having a low average evenness of 0.17 ± 0.10. The orange- and red-colored algal mats were composed of species of *Oscillatoria* and *Phormidium* and were much more diverse in composition, with average evenness values of 0.55 ± 0.16 and 0.48 ± 0.21, respectively. The orange- and red-colored mats also had a high degree of intrasite heterogeneity.

INTRODUCTION

Taylor Valley is one of the many valleys comprising the McMurdo Dry Valleys in south Victoria Land, Antarctica. Glacial meltwater streams are common

features of this area and are fed by alpine, piedmont, and terminal glaciers. These streams flow for four to eight weeks during the austral summer. Algal mats and mosses that persist from summer to summer are found in most of these streams. Algal photosynthesis begins soon after the mats are wetted by the first streamflow [*Vincent and Howard-Williams*,1986], and the mats are in a "freeze- dried" state during the winter.

The initial descriptions of the algal flora of Victoria Land were presented by *West and West* [1911] and *Fritsch* [1912, 1917], and subsequent descriptions from a broad range of habitats have been done by *Seaburg et al.* [1979]. Recently, *Vincent et al.* [1993] discussed the characteristics of algal communities occurring in the

[1]Institute of Arctic and Alpine Research, University of Colorado, Boulder, Colorado
[2]Water Resources Division, U.S. Geological Survey, Boulder, Colorado
[3]Water Resources Division, U.S. Geological Survey, Denver, Colorado
[4]National Mapping Division, U.S. Geological Survey, Reston, Virginia
[5]Department of Invertebrate Zoology and Geology; California Academy of Sciences, San Francisco, California

full range of flowing water habitats in Antarctica. *Broady* [1982] conducted a floristic study which focused on Canada Stream, a meltwater stream that flows into Lake Fryxell in Taylor Valley. Canada Stream, which was formerly called Fryxell Stream, has abundant algal mats throughout the 1.5 km stream length. *Broady* [1982] showed that species of filamentous cyanobacteria were predominant and examined longitudinal changes in stream microhabitat and species distribution. He found that in the upper, steeper reach characteristic algal species grew under and between large rocks in the streambed. In the middle and lower stream reaches, characteristic algal species were abundant in the mats in the streambed, and in epilithic crusts on rarely-wetted rocks. *Broady* [1982] stated that "in view of the considerable number of summer meltwater streams in the ice-free areas, most supporting rich algal growths, it is suggested that their biology deserves more detailed investigation."

During January 1994, we conducted a floristic survey of twelve streams in Taylor Valley (Figure 1). The survey included streams where algal mats were abundant, such as Canada Stream, and streams with no apparent algal mats [*Alger et al.*, 1997]. Several general patterns in the location of algal mat types were observed. In the main channels, and in parafluvial seeps along the stream margin, mats are orange in color and dominated by several species of *Phormidium* and *Oscillatoria*. The other common and abundant mat type is black-colored, occurs in damp areas at the stream margin, and is composed primarily of one of two *Nostoc* species. A common but less abundant algal mat type is green-colored, which often occurs on the underside of rocks on the main channel, and is dominated by *Prasiola* sp., as described by *Broady* [1982]. A red-colored algal mat was observed in a few streams and is composed chiefly of *Phormidium* species.

In this chapter, we examine the longitudinal changes in algal mats for Delta and Von Guerard Streams, two streams in Taylor Valley which are several times longer than Canada Stream. Delta and Von Guerard Streams have reaches with characteristics not found in Canada Stream, such as steep, incised channels with extensive snowbanks at upstream reaches, and sandy deltaic reaches near the outlet to the lake. We also reexamine longitudinal changes in Canada Stream. The seven sites on the three streams compared here are a subset of the sixteen sites in Taylor Valley described by *Alger et al.* [1997].

In addition to containing more varied habitats, longer streams exhibit a progressive downstream increase in solute concentrations. Stream length influences water chemistry because dissolution of marine aerosols and calcite, as well as primary weathering, contribute solutes in the hyporheic zone. Hyporheic zone water continually exchanges with water in the channel. At the outflow to Lake Fryxell, Canada Stream is very dilute, with a specific conductance of 27 μS cm^{-1}, whereas Delta and Von Guerard Streams have greater solute concentrations, with specific conductance values of about 132 and 144 μS cm^{-1}. One purpose of this study was to qualitatively assess the influence of habitat and water chemistry on the species composition of the algal mats by comparing algal mat from similar habitats in different streams.

SITE DESCRIPTION

Taylor Valley (78°S, 164°E) is bordered by the Asgard Range to the north and the Kukri Hills to the south. The Canada and Commonwealth Glaciers are sources for the streams entering Lake Fryxell on the western and northern side, and several small alpine glaciers in the Kukri Hills are sources for the streams on the south side. Fryxell Basin is broad and open, and the streams range in length from 1.5 to 11 km (Figure 1). The strand lines on the valley walls are evidence of the higher lake stands which occurred 10,000 to 20,000 years ago [*Doran et al.*, 1994]. The streams on the south side cut through perched deltas in which buried paleo-algal material is abundant [*Doran et al.*, 1994]. It is estimated that about 1,000 to 3,000 years ago Lake Fryxell became dry, and the lake has been refilling since then. Since the early 1900's when Taylor Valley was first explored, the levels of lakes throughout the McMurdo Dry Valleys, including Lake Fryxell, have been rising due to warmer climate and associated increased streamflow [*Chinn*, 1993].

Glacial meltwater is the only source of water to the streams; there is no vegetation and the land surface is an unconsolidated alluvium with large (5 m) hexagonal patterned ground features caused by periglacial processes. Permafrost begins about 0.5 m below the land surface. A network of stream gauges was established in Fryxell Basin in November 1990 [*von Guerard et al.*, 1994]. Streamflow varies considerably during the summer, depending upon insolation and air temperature [*Conovitz et al.*, this volume]. A moist area bordering the stream develops as the summer progresses; in this hyporheic zone there is flow in the downstream direction through the alluvium. Water in the hyporheic zone exchanges with the water in the main channel, bringing solutes into the stream.

Fryxell Basin Reference Map

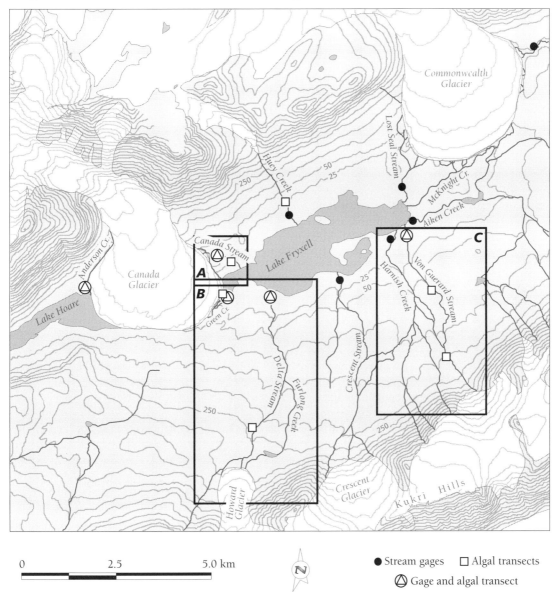

| 0 | 2.5 | 5.0 km |

● Stream gages   ☐ Algal transects

⬙ Gage and algal transect

Fig. 1. Map showing the basin of Lake Fryxell and the inflowing streams. The areas covered by the maps of the three study streams are shown. The contours are in meters.

Because the bank vegetation is limited to small patches of moss, the stream banks are steep and unstable. Undercutting during high flow periods can result in 1–2 cm thick deposits of sediment in lower reaches.

To learn more about the stream ecosystems in Taylor Valley, detailed sample collections were made during January 1994 at stream sites containing permanent transects. Transects varied in length with a typical transect being approximately 40 m across. Table 1 lists the locations of the transect endpoints for the three streams in Fryxell Basin that are examined in this report. In the detailed descriptions of the three streams presented below, moderate gradient refers to a

TABLE 1. Latitude (South) and Longitude (East) Measurements of Taylor Valley Stream Transects

SITE		LATITUDE	LONGITUDE
Canada Stream	(start)	77 36 47.669	163 03 9.675
near gauge	(end)	77 36 48.237	163 03 2.883
Canada Stream	(start)	77 36 52.543	163 04 11.498
at delta	(end)	77 36 53.345	163 04 12.421
Von Guerard Stream	(start)	77 38 04.424	163 18 19.031
at upper site	(end)	77 38 03.163	163 18 26.391
Von Guerard Stream	(start)	77 36 33.213	163 15 19.076
at gauge	(end)	77 36 34.111	163 15 10.822
Von Guerard Stream	(start)	77 37 09.483	163 17 18.526
at lower site	(end)	77 37 10.770	163 17 14.857
Delta Stream	(start)	77 39 11.564	163 05 58.855
at upper site	(end)	77 39 10.270	163 05 51.629
Delta Stream	(start)	77 37 31.796	163 06 38.887
at gauge	(end)	77 37 33.157	163 06 30.297

gradient between 0.1 m m^{-1} and 0.5 m m^{-1}, and a wide streambed refers to an active channel 8 or more meters across.

Delta Stream

Delta Stream, which drains Howard Glacier in the Kukri Hills, is located on the south side of the valley and is the longest stream (11.2 km) in Fryxell Basin (Figure 2). The stream cuts through a series of perched deltas deposited during previous high lake stands; hence its name. For most of the stream (above 1 km from the lakeshore), there is a moderate gradient and a wide streambed composed of large rocks embedded in sediment to form a flat stone pavement. Algal mats are abundant in the long reach delimited on the map as being between "a" and "b" (Figure 2). The stream gauge is located about 100 m above the outlet of the stream to the lake. The reach between "a" and the lakeshore includes the gauge and has a shallow

gradient, with a steep bank on the west side of the channel. Two transects were established on Delta Stream, one representative of reach a–b with abundant algal mats and one representative of the deltaic reach (Figure 2).

Von Guerard Stream

Von Guerard Stream is similar to Delta Stream in most characteristics, also draining an alpine glacier in the Kukri Hills. Three transects (upper, lower, and gauge) were established on Von Guerard Stream (Figure 3). The upper transect site is representative of a steep gradient reach located between "c" and the glacier face (Figure 3). In this transect, the streambed has incised the glacial till, leaving steep-sided banks. The streambed is composed of large cobbles and boulders embedded with sand in an irregular manner. The stream channel contains large snow banks from the accumulation of wind-blown snow, and the streambed

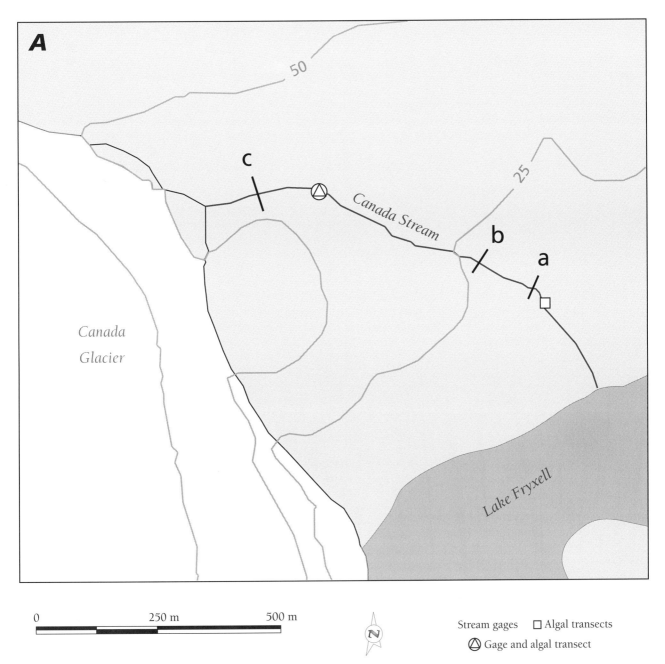

Fig. 2. Map of Delta Stream (area B on Figure 1). The lines "a" and "b" across the stream channel delimit reaches for which the transects are representative.

may be snow covered for some time into the summer. Below the steep upper reach, the gradient becomes shallow between points "c" and "b" on the map. Most of the flow (about 75%) is directed down a primary channel on the east. The remaining flow goes to the west and drains out into a broad, flat area. During low flow periods a large area of alluvium is saturated, but shallow pools form during high flow. No transect was located in this reach.

The lower transect site on Von Guerard Stream is located below point "b" in a reach with moderate gradient and wide streambed. Also in this reach, large, rounded rocks form a flat stone pavement embedded in sediment. Algal mats were visibly abundant in this

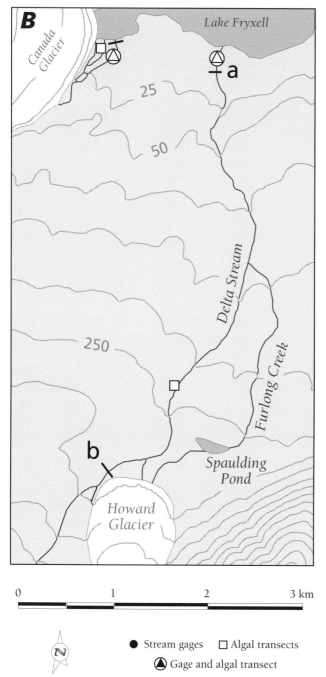

Fig. 3. Map of Von Guerard Stream (area C on Figure 1). The lines "a," "b," and "c" across the stream channel delimit reaches for which the transects are representative.

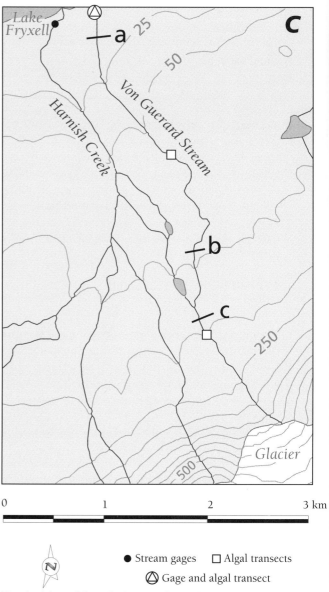

Fig. 4. Map of Canada Stream, formerly known as Fryxell Stream, (area A on Figure 1). The lines "a," "b," and "c" across the stream channel delimit reaches for which the transects are representative.

site is representative of the reach between point "a" and the lakeshore.

Canada Stream

Canada Stream is one of the major sources of inflow to Lake Fryxell, and is typically the first stream in the basin to begin flowing in the austral summer. Canada Stream drains the east side of the Canada Glacier (Figure 4) and is the location of the Site of

reach. This lower transect is representative of conditions extending down to point "a" within 0.5 km of the outlet to the lake.

The transect site at the gauge is located in a sandy deltaic area where the stream channel is braided. This

Special Scientific Interest in Fryxell Basin. The stream gauge is located at the site of a previous rock weir about 1.0 km above the lakeshore. At the gauge, flow draining the upper glacier has joined with flows from a large pond at the base of the glacier. The pond contains abundant moss and algae. A transect site was established just above the gauge and is representative of the reach between "c" and "b." In this reach, the rocks in the streambed are rounded and wedged together forming a stone pavement and algal mats are abundant. Below point "a" the stream spreads into a wide alluvial fan, which is about 100 m across at the lakeshore. The transect in this reach was located on the east side and did not extend completely across the stream. Algal mats were also very abundant in this reach.

METHODS OF SAMPLE COLLECTION AND ANALYSIS

Within each transect, algal mats were visually identified as either orange-colored, red-colored, black-colored, or green-colored, although not all types of algal mats were found within every site. The exact location of sample collection was mapped on small-scale stream contour maps. A maximum of five algal samples of each color type was collected from each transect using a #13 cork borer (diameter = 17 mm, area = 227 mm^2). Algal samples were preserved in 10% formalin for laboratory analysis. Additional algal samples were collected for chlorophyll determination. Water samples were collected in polyethylene bottles for nutrient and major ion analyses.

Streamflow was measured using a pygmy current meter; as a matter of practicality, measurements were made at the narrowest point in the stream within the mapped area. Therefore, the velocities represent the maximum velocities occurring in the reach for the discharge at the time of mapping. Pebble counts were made by a random walk approach [Wolman, 1954]. The intermediate axis (i.e., b-axis) of the bed particles was measured. The smallest rocks sampled were 1.4 cm. At sites with abundant algal mats, pebble counts were done in reaches immediately above and below the mapped reach, in order to minimize disturbance. In reaches where the rocks were embedded in a stone pavement, the rocks were not picked up, but measured in place.

Preserved algal samples were examined in the laboratory to determine species composition and relative abundance using a Nikon Diaphot phase contrast microscope. Samples were well mixed, and then subsamples of 2 ml were withdrawn. Each subsample was placed in a 2 ml settling chamber with a 26 mm diameter [Utermöhl, 1958] and examined at x400 magnification. To obtain a representative count of relative species abundance, seven random fields (a total area of 0.44 mm^2) were examined per subsample. Within each field, the algae were identified to genus and species where possible. Diatoms were counted as a group but were not identified more specifically because the taxonomy of the diatoms in the dry valley streams is lacking at this time. The descriptions of the morphotypes of filamentous cyanobacteria and the other algal species are presented in Table 2. Percent cover was determined in each field by measuring the length and width of all specimens using an ocular micrometer. Cell "depth" was also measured in order to estimate biovolume. All seven fields were then tallied, and the percentage of the total algal biomass was determined for each taxa. A total of nineteen species of algae were identified in the four algal mat types. In order to present the variability in species distribution, evenness values were calculated for each mat sample [Zar, 1996] and an average was calculated for the orange-, black-, red-, and green-colored algal mats at each site.

Samples for chlorophyll were filtered through GF/C glass fiber filters. Chlorophyll was extracted in buffered acetone and analyzed spectrophotometrically using the trichromatic method [Strickland and Parsons, 1972]. Nutrient samples (PO_4^{-3}, and NO_3^-) were analyzed at McMurdo Station, using standard colorimetric methods.

RESULTS AND DISCUSSION

Table 3 presents summary characteristics of the transect sites on the three streams. Alger et al. [1997], categorized stream sites in the Taylor Valley as "high gradient," "large cobble," or "deltaic" and found that there were similarities in the algal mat type and abundance within categories. The assignment of the sites in this study to these categories are indicated in Table 3. Large cobble sites have a streambed composed of large rocks wedged together with their flat side upwards, forming a flat pavement. The steep upper Von Guerard Stream site is typical of a high gradient site where the rocks in the streambed are larger and are uneven, or jumbled. At deltaic sites the rocks are smaller and movement of sand-sized particles occurs during flow periods, and the sites at the outflows of

TABLE 2. Descriptions of Algal Species found in Taylor Valley.

Family Oscillatoriaceae

Morphotype A / Oscillatoria subproboscidea W. & G.S. West. Trichomes between 7.5–10 μm in width (including sheath in this and following taxa). Terminal cell rounded and sometimes swollen with calyptra. Trichomes often found within a sheath. Granules present along transverse walls.

Morphotype B / Oscillatoria subproboscidea W. & G.S. West. Trichomes between 9–12 μm in width. Similar to Morphotype A but wider and with more numerous granules. These granules may be gas vacuoles, providing protection from the high levels of ultraviolet light encountered in Antarctica.

Morphotype C / Oscillatoria irrugua Kützing. Trichomes between 7–9 μm in width and often surrounded by a sheath. Many cells have transverse walls with a convex shape (perhaps separation disks). Terminal cell often attenuated.

Morphotype D / Oscillatoria irrugua Kützing. Trichomes between 7.5–9.5 μm in width, sometimes surrounded by a sheath. Dark, refractory surface obscures the cell interior. Cell walls thin. Terminal cell often with a calyptra.

Morphotype E / Phormidium crouani Gomont. Trichomes between 8–10 μm in width. Dark, refractory surface obscures cell interior. Granules scattered throughout the trichome. Terminal cell slightly attenuated.

Morphotype F / Oscillatoria koettlitzi Fritsch. Trichomes between 7–9.5 μm in width. Terminal cell swollen and often with a calyptra. Cells narrow with a distinct light-dark pattern. Necridia common (1–5) within each trichome.

Morphotype G / Oscillatoria koettlitzi Fritsch. Trichomes between 7–9.5 μm in width. Similar to Morphotype E but with much shorter cell length. Terminal cell swollen.

Morphotype H / Oscillatoria koettlitzi Fritsch. Trichomes between 8–10 μm in width. Dense cells with a dark, refractory surface and no apparent sheath. Terminal cells often capitate with no visible calyptra. Distinct constrictions at the cell walls. Necridia rare.

Morphotype I / Phormidium autumnale (Agardh) Gomont. Trichomes between 4–6 μm in width and often with a slight terminal hook. Variable position and quantity of granules. Terminal cell with a calyptra. Could be independent trichomes of *Microcoleus vaginatus* (see discussion in *Broady*, 1991, p.44).

Morphotype J / Phormidium autumnale (Agardh) Gomont. Trichomes between 2.5–3.5 μm in width. Generally, cell walls clearly visible; granules rare. Terminal cell with a calyptra.

Morphotype K / Phormidium autumnale (Agardh) Gomont. Trichomes between 2–3.5 μm in width. Nearly identical to Morphotype J except that trichomes arose from a common "clump." Trichomes also appeared to taper slightly at the terminal end. Rare occurrence.

Morphotype L / Phormidium frigidum Gomont. & *O. deflexa* W. & G.S. West. Trichomes all less than 2.5 μm in width. Very thin trichomes—probably more than one species. Usually no cell structures were visible, but occasionally cells were box-shaped with distinct constrictions at the transverse walls.

Morphotype M / Microcoleus vaginatus (Vaucher) Gomont. Trichomes between 4–6 μm in width. Nearly identical to Morphotype I except that all trichomes were contained within a common sheath. A few trichomes would protrude from the apex of the sheath. Terminal cell with a distinct calyptra. Could be colony of *Phormidium autumnale* (see discussion in *Drouet*, 1962 and *Broady*, 1991).

Family Rivulariaceae

Calothrix cf. *intricata* Fritsch. Distinct, thick (1–8 μm), yellow sheaths surround these trichomes. Cell width varied from 6–9 μm at base to 4–8 μm at apex. Cells 2–3 μm long. Basal heterocyst often present within sheath. Granules scattered throughout trichome. Gradual attenuation of cells from basal to terminal end. Akinetes rare. Terminal cell slightly pointed. Immature trichomes with thinner, greenish colored sheaths. Occasionally, multiple trichomes intertwined together in thin mats but majority as independent trichomes. Similar to *Calothrix* sp. described by *Broady* [1982].

TABLE 2. Descriptions of Algal Species found in Taylor Valley. (Continued)

Family Prasiolaceae

Prasiola calophylla (Carmichael) Meneghini. Young thalli consisting of uniseriate filaments 8–15 μm wide with cells 5–10 μm long. Most common growth in narrow, foliose ribbons ranging from 5–50+ mm in length. Growth processes form uniseriate to biseriate filaments with cell division in two planes. Ribbons generally taper to a uniseriate ending often with a holdfast. Cells with stellate chloroplast and central pyrenoid. Aplanospores rare.

Prasiola crispa (Lightfoot) Meneghini. This species most closely matched the description given by *Fritsch* [1917]. Although *Broady* [1979] and others commonly found *Prasiola crispa* near penguin rookeries, we found thalli (matching the description by Fritsch) in the hyporehic seeps of Taylor Valley streams (non-rookery locations). Most common as young thalli with uniseriate filaments 10–15 μm wide and cells 4–12 μm long. Axial chloroplast and central pyrenoid generally visible. Biseriate and multiseriate filaments produced after cell division in two planes. Cells often remaining in clusters of four. "Phormidium" form was common with its oval-shaped, opaque cells that appeared to be in a moribund state.

Family Nostocaceae

Nostoc spp. Ranged in growth from individual trichomes to dense, irregularly shaped colonies. Trichomes yellowish, 3–5 μm wide, and closely packed within colonies. Heterocysts present at both ends and within the trichomes. Akinetes not observed. Thin sheath surrounded the trichomes while a thick, yellow, mucilaginous sheath surrounded the colonies. Size of colonies ranged from 10 mm in diameter to over 10 cm². Juvenile colonies with intercalary heterocysts also found. Difficult to identify to species level without cultures (see *Mollenhaur*, 1988 for discussion and detailed life cycles).

Nodularia cf. *harveyana* Thuret. Trichomes 4–6 m wide with no visible sheath. Cells 3–5 m long and terminal cell slightly conical. Heterocysts large (5–8 μm long) and circular.

Family Chroococcaceae

Chroococcus cf. *minutus* (Kützing) Nägeli. Spherical cells, 5–8 μm in diameter. Cells surrounded by a thin sheath. Cells have larger, colorless mucilage envelopes (4–10 μm) immediately surrounding the sheath. One-, two-, and four-cell clusters were found but most common in colonies of two cells.

cf. *Gloeocapsa kuetzingiana* Nägeli. Generally, a dark orangish brown colony of individual cells, 2–5 μm in diameter. Cells arranged irregularly in a mucilaginous sheath. Colonies ranged from 6–78 μm in width and contained 3–100+ cells.

Gloeocapsa sp. Cells ovoid-ellipsoidal, 1–3 μm in width, and contained within a sheath. Colonies spherical with 1–5 cells, often found growing adjacent to other colonies. Rare.

Family Mesotaeniaceae

Actinotaenium cf. *cucurbita* (Brébisson) Teil. Cells 25–33 μm by 13–20 μm. Rounded hemi-cell with a slight median constriction. One chromatophore per half cell with four ribs and a central, spherical pyrenoid.

Family Palmellaceae

Asterococcus sp. Spherical cells, 3 μm in diameter, found in a common gelatinous matrix. Each cell with an asteroidal chloroplast with arms radiating from a central pyrenoid. Rare.

Family Ulotrichaceae

cf. *Binuclearia tectorum* (Kützing) Beger. Long, uniseriate filaments with cells 6–11 μm in width and 8–13 μm long. Filaments surrounded by a mucilaginous sheath. Cells with a parietal chloroplast and a distinct pyrenoid.

Family Chlorellaceae

Chlorella sp. Globular, mucilaginous colonies, 8–10 μm in width, consisting of several (20) green, spherical cells, 1–2 μm in diameter. Rare.

TABLE 3. Characteristics of Stream Sites Located on Three Streams in Fryxell Basin

	Gradient	Most abundant rock size	Streambed Character	Maximum velocity	Category	Conductivity	PO_4^{3-}	NO_3^-
	$(m\ m^{-1})$	(cm)		$(m\ s^{-1})$		$(\mu S\ cm^{-1})$	(μM)	(μM)
Delta Stream								
at upper site	0.1	11	pavement	0.46	large cobble	89	0.05	0.74
at gauge	0.05	4	delta	0.37	deltaic	144	0.09	0.81
Von Guerard Stream								
at upper site	0.2	11	uneven	0.68	high gradient	93	0.44	1.76
at lower site	0.06	8	pavement	0.86	large cobble	112	1.08	0.94
at gauge	0.05	ND	delta	1.11	deltaic	132	0.42	0.89
at gauge seep	--	--	--	--	--	154	1.35	3.40
Canada Stream								
at gauge	0.05	11	pavement	0.52	large cobble	24	0.13	1.12
at alluvial fan	0.03	8	pavement	0.77	large cobble	27	0.25	0.74

* ND indicates not determined).

Delta and Von Guerard Streams had both these features. Black-colored algal mats were found at all sites; orange- and green-colored algal mats were found at all but two sites; red-colored algal mats were found only in Canada Stream.

For all streams, the conductivity increased downstream, and the increases in Delta and Von Guerard Streams were greater because of their greater length. The nitrate concentrations were similar at all sites, with the exception of the seep at the Von Guerard gauge. The phosphate concentrations were higher at the Von Guerard sites than at the other sites. These findings are consistent with the nutrient sources in the dry valleys being primarily weathering of apatite and dissolution of nitrate in atmospheric deposition. These sources are distributed such that they would be continuous sources of nutrients through hyporheic exchange processes.

Algal Mat Type and Abundance

The small-scale topographic maps of the large cobble and deltaic sites in Delta Stream (Figures 5 a and b) are illustrative for these categories of stream sites in Fryxell Basin. At large cobble sites, orange-colored algal mats were very abundant, typically covering the streambed for the width of the wetted zone, and black-colored mats were plentiful along the stream margins. In contrast, deltaic sites had much lower algal abundance. As shown in Figure 5b, algal mats were not found in the main active channel where sediment can be resuspended and deposited during high flow. Black-colored mats were found at the margins in areas that were wetted intermittently at high flow. Orange-colored mats were found in parafluvial seeps, small rivulets at the channel margins which drain the hyporheic zone. These rivulets continue flowing after high flow has receded and thus represent a stable and sustained flowing water habitat. The nutrient data for the gauge site on Von Guerard Stream (Table 3) show that nutrient concentrations are higher in the seep water than in the water from the main channel, further enhancing this habitat.

The steep upper reach of Von Guerard Stream also contains only sparse algal mats compared to large cobble reaches and is similar to other high gradient reaches described by *Alger et al.*[1997]. In these sites algal mats occur near the stream margins and not as a covering in the main channel. An occasional green-colored algal mat can be found under rocks in the main channel. Again, the scour and high sediment transport and deposition during high flow may prevent orange-colored mats from becoming established in the main channel.

Table 4 lists estimates of representative widths of streambed covered by a mat type based upon the map of the site and field notes. This estimate was then used

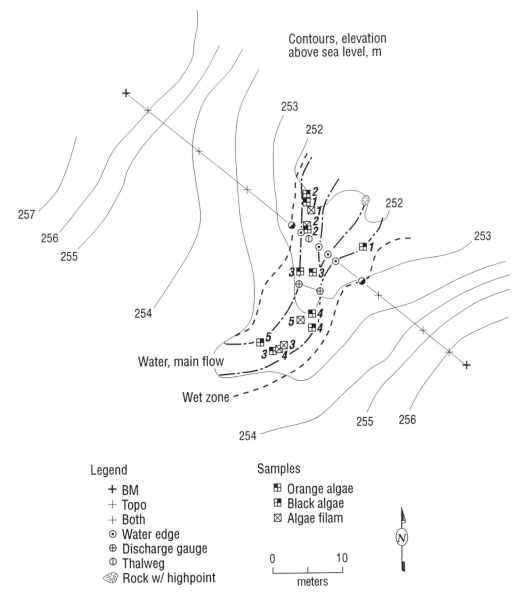

Fig. 5. Maps of stream sites showing location of significant features of the stream and the location of algal mat samples: a) upper Delta Stream; b) Delta Stream at the gauge.

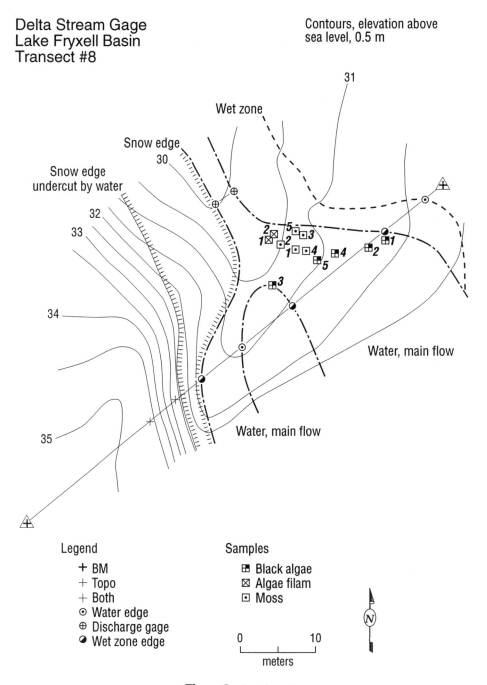

Figure 5. (continued)

TABLE 4. Chlorophyll-*a* Content of Algal Mat Types in Three Streams (NF indicates not found)

| Site | Category | Length (km) | Orange Mat | | | Black Mat | | | Green Mat | Red Mat |
			Chl-*a* (μg cm-2)	width (m)	Total Chl-*a* (kg)	Chl-*a* (μg cm-2)	width (m)	Total Chl-*a* (kg)		
Delta Stream										
at upper site	large cobble	4	12	3	1.4	41	2	3.3	present	NF
at gauge	deltaic	0.5	NF	--	--	28	2	0.3	present	NF
Von Guerard Stream										
at upper site	high gradient	1	4.7	0.5	0.02	NS	--	--	present	NF
at lower site	large cobble	3.5	12	3	1.3	31	1	1.1	NF	NF
at gauge	deltaic	0.5	3.1	0.5	0.001	9	1	0.05	NF	NF
Canada Stream										
at gauge	large cobble	0.5	6.4	3	0.1	63	1	0.3	present	present
at alluvial fan	large cobble	0.5	NF	--	--	64	3	0.1	present	present

NF indicates not found, and NS indicates not sampled.

to further estimate the quantity of chlorophyll-*a* present in that mat type in the total reach of stream for which the site was representative. Clearly this estimate of total chlorophyll-*a* may be subject to large error. This estimate is largely determined by the measured algal abundance at the large cobble site and the length of the stream. The high gradient and deltaic sites account for only a short reach of stream and contribute little to the total chlorophyll-*a* in each stream.

Vincent et al. [1993] note that there is little evidence for nutrient limitation of algal populations in Antarctic streams. The apparent primary control of habitat on the abundance of algal mat types found in this study, and the minimal evidence for a nutrient control is consistent with what has generally been observed for Antarctic streams. However the water chemistry and the availability of nutrients could also have an influence at a secondary or subtle level. For example, red-colored algal mats were only found in Canada Stream, which is the most dilute.

Algal Species Distribution

Filamentous cyanobacteria were abundant in all mats except the green-colored algal mats. Of the total of 29 species found at these three streams, 10 were found at all 7 sites in one mat type or another, and 17 were found at every stream. The species composition of the different mat types was generally consistent among

the sites. The black-colored algal mats were dominated by one of two species of *Nostoc*. The green-colored algal mats were also essentially unialgal, composed chiefly of *Prasiola calophylla* or *P. crispa*. The orange- and red-colored algal mats were composed of a number of species of *Oscillatoria* and *Phormidium*.

Given the apparent similarity of mat types in the stream habitats, the uniformity of mat types at the species level becomes an important question for understanding these stream ecosystems. Figure 6 (a and b) presents the variation in species composition for orange-colored and black-colored algal mat samples from the three sites on Von Guerard Stream. Table 5 summarizes the evenness values for the different algal mats at the sites studied. The low evenness values for black-colored and green-colored algal mats show that these mat types are essentially unialgal. For these mat types there is little variation in species composition among samples from the same site or between sites in the same stream. In contrast, the species distribution was much more even for the orange- and red-colored algal mats, with evenness values two to three times greater than those for the black- and green-colored mats.

Numerous algal species comprise the orange- and red-colored algal mat samples from the same site (Figure 6a). The average number of species in a single mat sample was 10, which is lower than the total number of algal species (19) found for all 5 orange mat

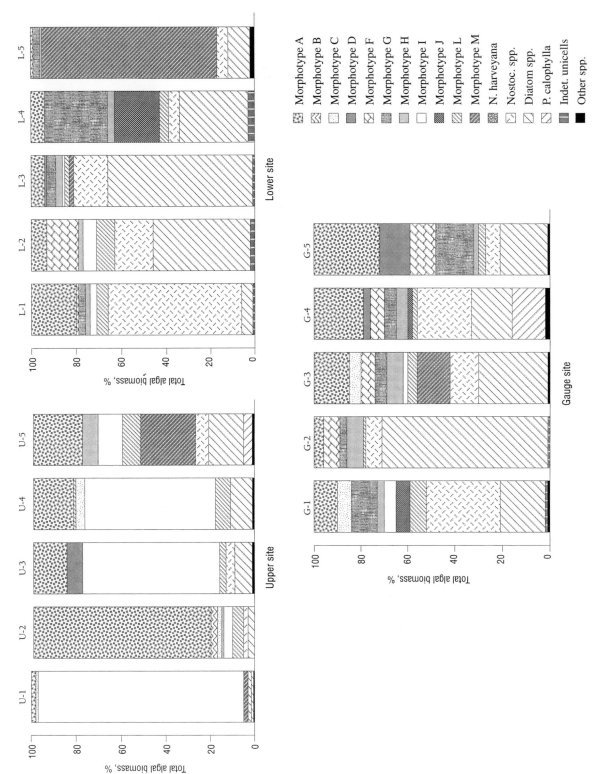

Fig. 6. Species distribution in algal mat samples from three sites in Von Guerard Stream: a) Orange-colored algal mats; the species are described in Table 2; b) black-colored algal mats.

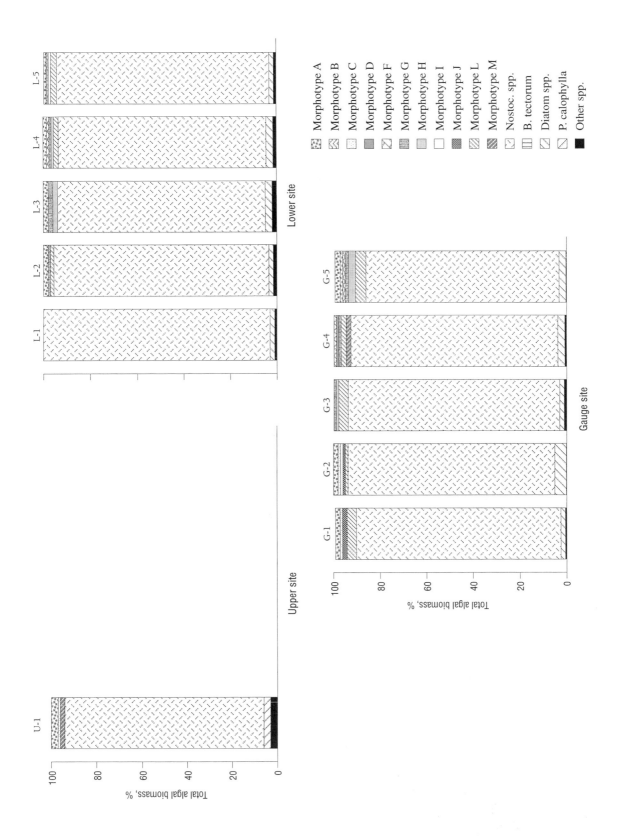

TABLE 5. Eveness Values for Different Algal Mat Types in Streams in Fryxell Basin

Site	Category	Orange mat	Black mat	Green Mat	Red Mat
Delta Stream					
at upper site	large cobble	0.68	0.08	0.22	NF
at gauge	deltaic	NF	0.18	0.23	NF
Von Guerard Stream					
at upper site	high gradient	0.39	0.21	0.27	NF
at lower site	large cobble	0.46	0.15	NF	NF
at gauge	deltaic	0.62	0.19	NF	NF
Canada Stream					
at gauge	large cobble	0.61	0.07	0.05	0.37
at alluvial fan	large cobble	NF	0.10	0.13	0.58
Mean and Standard Deviation		0.55±0.16	0.13±0.07	0.18±0.10	0.48±0.21

NF indicates not found.

samples from each of the 3 sites (upper, lower, and gauge, respectively). There was also a high degree of heterogeneity in the relative abundance of algal species in the different samples of orange-colored mat from the same site.

Table 6 summarizes and compares the occurrence of algal species in the black-, orange-, green-, and red-colored mats from all three streams. The species list from these three streams includes all but three of the species identified in the study of 16 different sites in Taylor Valley, illustrating the similarity in species composition among sites of low and high algal abundance in Taylor Valley. This result is consistent with the conclusion of *Vincent and James* [1996] that environmental extremes plus biogeographical isolation control the biodiversity of algal communities in the Ross Sea Sector of Antarctica.

However the unialgal character of the black- and green-colored mats provide evidence of specialization with respect to habitat. Furthermore of the 32 algal species or types identified in the overall Taylor Valley study, only four were found in all mat types at the three streams. These were *Oscillatoria subproboscidea* morphotype A, *P. frigidum & O. deflexa* morphotype L, *Nostoc* spp. and diatoms. The ubiquity of diatoms in mats suggests that species level identification of the diatoms may be useful in refining our understanding of the streams and in developing tools for paleolimnological studies of stream channels.

Although the black-colored algal mats are dominated by *Nostoc* spp., the total number of species in black mats is comparable to the other mat types because of

the trace species, ranging from 10 to 18 other species of filamentous cyanobacteria. Based upon the results of *Alger et al.* [1997], this conclusion can be extended to other streams in the Taylor Valley.

Only two of the species observed in this sampling from three Fryxell Basin streams were always trace, *Chroococcus minutus* and *Chlorella* sp. The two *Nostoc* species which comprised the black-colored algal mats were a common trace or abundant species in the other mat types; whereas the two species of *Prasiola* were either dominant or abundant, primarily in the green-colored mats and never occurred as trace species.

Hyporheic Zone Processes

Our results suggest that geomorphic factors primarily control the overall abundance of algal mats and the spatial position of the mat types along a transect across the stream. Abundant algal mats occurred in moderate gradient reaches with a stone pavement providing a stable habitat. We hypothesize that these stone pavements form as a result of periglacial processes occuring in the saturated hyporheic zone. These stone pavements are similar to stone pavements formed around the shores of alpine lakes. Over numerous cycles of freeze and thaw of the hyporheic zone, large rocks would be worked toward the surface and become wedged together. This flat surface could armor the streambed, thus limiting erosion, and influencing the flow regime at the stream-water interface. Both these effects would promote growth of algal mats.

TABLE 6. List of Algal Taxa from Taylor Valley Streams Showing Occurrence in Different Algal Mat Types

Taxa	Delta Stream		Von Guerard Stream			Canada Stream	
	Upper	Gauge	Upper	Lower	Gauge	Gauge	Fan
CYANOPHYTA							
Oscillatoria subproboscidea W. & G.S. West							
morph. A	O,B*,G*	B*G+	O+,B,G+	O,B*	O,B*	O,B*,G*,R*	B*,G*,R
morph. B	O*,B*,G*	B*	O+,B*,G*	O*,B*	B*	O*,B*,G*,R*	B*,G*,R
Oscillatoria irrigua Kützing							
morph. C	O,B*,G*	G	O,B*,G*	O*,B*	O,B*	O,B*	B*,R
morph. D	O*,B*,G*	B*	B*	B*	O,B*	O*,B*,G*,R	B*,G*,R
Phormidium crouani Gomont							
morph. E			G*	B*	O*	O,B*,R*	R
Oscillatoria koettlitzi Fritsch							
morph. F	O,B*,G*	B*,G*		O,B*	O,B*	O,B*,G*,R*	B*,G*,R
morph. G	O,B*,G*	B*	B*	OB*	O,B*	O,B*,G*,R*	B*,R
morph. H	O,B*,G	B*,G*		O,B*	O,B*	O,B*,G*,R*	B*,G*,R
Phormidium autumnale (Agardh) Gomont							
morph. I	O,B*	B	O+,B*	O,B*	O	O*,B*,G*,R*	B*,R
morph. J	B*			O,B*	O	O,B*,G*,R+	B*,R
morph. K							
P.frigidum Gomont & *O.deflexa* W. & G.S.West.							
morph. L	O,B*,G	B*,G*	O,B*,G	O,B*	O,B	O,B*,G*,R	B*,G*,R
Microcoleus vaginatus (Vaucher) Gomont							
morph. M	O*,B*,G*		O,B	O	O	O,R*	B*
Calothrix intricata Fritsch	O*,B*		G*			R	G*
Chroococcus minutus (Kützing) Nägeli	O*,G*				B*	B*	B*
Gleocapsa kuetzingiana Nägeli	B*,G*					O,B*,G*,R*	R*
Gleocapsa sp.							
Nodularia harveyana Thuret	B*					B*,G*,R*	B*,R
Nostoc spp.	O,B+,G	B+,G	O,B+,G*	O,B+	O,B+	B+,G*,R	B+,G,R

(Categories are abbreviated as follows: O, orange; B, black; G, green; R, red. * Trace (> 1%), + > 50%)

TABLE 6. List of Algal Taxa from Taylor Valley Streams Showing Occurrence in Different Algal Mat Types (Continued).

Taxa	Delta Stream		Von Guerard Stream			Canada Stream	
	Upper	Gauge	Upper	Lower	Gauge	Gauge	Fan
CHLOROPHYTA							
Actinotaenium cucurbit (Brébisson) Teil						O,B*,G*,R*	B*,G*,R*
Asterococcus sp.							
Binuclearia tectorum (Kützing) Beger	O	G*	O*,B*,G*		B*	O,R*	
Chlorella sp.	O*,G*			O*	O*		R*
Desmid sp.						B*	
Indeterminate unicells	O*,B*,G*	B*,G*	O*,B*,G*		O*	O*,B*,G*,R*	B+,G*,R*
Indeterminate branched filament				O*		O*	
Indeterminate colony	B*						
Prasiola calophylla (Carmichael) Meneghini			B,G+				
Prasiola crispa (Lightfoot) Meneghini	G+				O	G+	G+
BACILLARIOPHYTA							
Diatom spp.	O,B*,B*	B*,G*	O,B*,G	O+,B	O,B	O,B*,G*,R	B*,G*,R
Marine diatom fragment					O*		
CHRYSOPHYTA							
Chrysophyte cysts	O*					B*,R	R*
EUGLENOPHYTA							
Euglena sp.						G	

(Categories are abbreviated as follows: O, orange; B, black; G, green; R, red. * Trace (> 1%), + > 50%)

The stream gradient may control the extent to which the streambed is influenced by periglacial processes. A reach may be steep enough that the hyporheic zone drains rapidly before becoming frozen as discussed by *Conovitz et al.* [this volume]. In which case, periglacial processes may have little influence, and sediment scour and deposition may control the nature of the streambed.

The differences in species distribution in samples of orange- and red-colored algal mats from the same site do not appear to be related in an obvious way to small differences in position relative to features in the streambed, such as the thalweg. However the species distribution may be related to spatial heterogeneity in the exchange of water between the main channel and the hyporheic zone. Different patches could be zones in which streamwater recharges the hyporheic zone or zones of hyporheic discharge into the stream. The mats could therefore be subjected to different small scale hydrologic and chemical regimes that would be difficult to observe visually. Because of the importance of the hyporheic zone as a source of solutes, including nutrients, it may be that small scale spatial variations in supply of solutes account for the variations in species distribution of orange- and red-colored algal mat within a stream site.

Results of this study highlight the need for quantitative understanding of hydrologic and geomorphic processes in the streams in order to understand linkages between glacier, stream, soil and lake components of dry valley watersheds. Specific processes can be quantified through experiments and observation. For example, studies of hyporheic zone interactions could test the hypothesis that spatially heterogeneous hyporheic interactions control species distribution in the orange-colored algal mats.

Acknowledgements. Support for our research was provided by the National Science Foundation Office of Polar Programs grant OPP-9211773. We acknowledge field assistance provided by P. Doran and A. Butt, and assistance from K. Bourke on the ecological analysis. Helpful comments on the manuscript were provided by W. Dodds, W. Vincent, and K. Lohman.

REFERENCES

Alger, A.S., McKnight, D.M., Spaulding, S.A., Tate, C.M., Shupe, G.H., Welch, K.A., Edwards, R., Andrews, E.D., and House, H.R., 1997, Ecological processes in a cold desert ecosystem: The abundance and species distribution of algal mats in glacial meltwater streams in Taylor Valley, Antarctica, 102 pp., *Occasional Paper No. 51*, Institute of Arctic and Alpine Research, Boulder, Colorado, 1997.

Broady, P.A., Taxonomy and ecology of algae in freshwater stream in Taylor Valley, Victoria Land Antarctica, *Archivs fur Hydrobiologie, 32, supplement 63.3 (Algolocial studies)*, 331–349, 1982.

Chinn, T.H., Physical hydrology of the Dry Valley Lakes, in *Physical and biogeochemical processes in Antarctic lakes*, edited by W.J. Green and E.I. Friedmann, pp. 1–52, Antarctic Research Series, 59, American Geophysical Union, Washington, D. C., 1993.

Doran, P.T., Wharton, R.A., Jr., and Lyons, B.W., Paleolimnology of the McMurdo Dry Valleys, Antarctica, *J. Paleolimnology, 10*, 85–114, 1994.

Fritsch, F.E., *Freshwater algae: National Antarctic Expedition, Natural History*, v. 6, pp. 1–66. British Museum (Natural History), 1912.

Fritsch, F.E., *Freshwater algae: British Antarctic (Terra Nova) Expedition 1910-13: Botany*, v. 1, pp. 1–16. British Museum (Natural History), 1917.

McKnight, D.M., and Tate, C.M., Canada Stream: Glacial Meltwater Stream in Taylor Valley, South Victoria Land, Antarctica, *J. North Amer. Bentholog. Soc, 16*, 14–17, 1997.

Seaburg, K.G., Parker, B.C., Prescott, G.W., and Whitford, L.A., The algae of southern Victoria Land, *Bibliothecia Phycologica*, 46, 169, 1979.

Strickland, J.D.H., and Parsons, T.R., A practical handbook of seawater analysis (2d ed.), Fisheries Research Board of Canada, Bulletin 167, 310 pp., Ottawa, Ontario, 1972,

Utermöhl, H., Zur Vervollkommnung der quantitativen phytoplankton-methodik: *Mitt. Int. Ver. Limnology*, 9, 1–38, 1958.

Vincent, W.F., and James, M.R., Biodiversity in extreme aquatic environments: lakes, ponds, and streams of the Ross Sea sector, Antarctica, *Biodiversity and Conservation, 5*, in press, 1996.Vincent, W.F., Howard-Williams, C., and Broady, P.A., Microbial Communities and Processes in Antarctic Flowing Waters, *Antarctic Microbiology*, pp. 543–569, Wiley-Liss, 1993.

Vincent, W.F., Howard-Williams, C., Antarctic stream ecosystems: physiological ecology of a blue-green algal epilithon, *Freshwater Biol.*, 16, 219–233, 1986.

von Guerard, P., McKnight D.M., Harnish, R.A., Gartner, J.W., and Andrews, E.D., Streamflow, water-temperature, and specific-conductance data for selected streams draining into Lake Fryxell, Lower Taylor Valley, Victoria Land, Antarctica, 1990-92. U.S. Geological Survey Open-File Report 94-545, 65 pp., 1994.

West, W., and West, G.S., Freshwater algae, in Biology: Reports on the scientific investigations of the British Antarctic Expedition, 1907-1909, vol. 1, pp. 263–298 edited by J. Murray, 1911.

Wolman, M.G., A method for sampling coarse river-bed material, 35, 951–956, 1954.

Zar, Jerrold H., *Biostatistical Analysis*, Prentice Hall, New Jersey, 1996.

Alex Alger, Water Resources Division, MS 415, U.S. Geological Survey, Denver Federal Center, Boulder CO 80303

Diane M. McKnight, Institute of Arctic and Alpine Research, 1560 30th Street, Boulder, CO 80309

Gordon Shupe, Mapping Division, 12201 Sunrise Valley Drive, U.S. Geological Survey, Reston, VA 22092

Sarah Spaulding, Department of Invertebrate Zoology and Geology, California Academy of Sciences, Golden Gate Park, San Francisco, CA 94118

Cathy M. Tate, Water Resources Division, MS 415, U.S. Geological Survey, Denver Federal Center, Denver, CO 80225

(Received September 12, 1996; accepted April 30, 1997)

PRIMARY PRODUCTION PROCESSES IN STREAMS OF THE MCMURDO DRY VALLEYS, ANTARCTICA

Ian Hawes and Clive Howard-Williams

National Institute for Water and Atmospheric Research, Christchurch, New Zealand

New and published information on production of microbial communities in streams of the McMurdo Dry Valleys is reviewed. The dominant community in many of these streams is a thick, cohesive cyanobacterial mat. Light-photosynthesis relationships of microbial mat communities from a range of streams tended to show a surprising degree of convergence. Gross rates of photosynthesis typically approached an upper limit of 4 µg C cm^{-2} h^{-1} at ambient temperature (0–8°C), and community light saturation intensities were almost always below incident irradiance during the period when streams were flowing. Net and gross photosynthesis increased with increasing temperature, and our analysis supports previous views that temperature is the prime determinant of the rate of net production in these communities. There were generally higher respiration rates in thicker mat communities, resulting in these mats tending toward a zero net gas exchange, i.e., where gross photosynthesis approximately equalled respiration. Accumulation of new material on exposed surfaces was slow, and most communities were clearly at least 3 to 4 years old. We argue that the development of high biomass communities, which are balanced or near-balanced with respect to gas exchange, is possible due to the lack of disturbance within areas of these streams, which lack macroscopic grazers or flood disturbance, a high rate of overwinter survival, and the constancy of growth conditions during the flow period.

INTRODUCTION

Running water ecosystems are common in the McMurdo Dry Valleys, where they typically arise from the melting of glaciers or permanent ice sheets. Snow accumulations and summer precipitation are not common in this environment and hence snow-fed streams are absent. Where groundwater is present, it is as a highly concentrated brine forming a thin layer on the surface of the permafrost [*Wilson,* 1979]. When this groundwater reaches the surface it forms salt evaporites rather than spring-fed streams. A characteristic feature of McMurdo Dry Valley streams is therefore that flow is dependent on melting of ice, and discharge is determined by temperature and, to a lesser extent, insolation [*Chinn,* 1993; *Conovitz et al.,* this volume].

The climate of the region ensures that periods of ice melt, hence flow, are short, usually confined to parts of December and January.

This short period of ice melt represents the entire growing season for the stream flora. It is interrupted by the Antarctic winter when flow stops, and water in stream channels drains or freezes. This ice then typically ablates, leaving freeze-dried organisms exposed to temperatures which may descend to –50°C. Despite the extreme winter conditions, and the short growing season, microbial communities develop in these water bodies, often accumulating high biomass [*Howard-Williams and Vincent,* 1986; *Vincent et al.,* 1993; *McKnight et al.,* this volume]. While temperatures are low, 24 h daylight, adequate nutrient supply and an absence of macroscopic herbivorous organisms or large flood flows provide a benign environment for growth and accumulation [*Howard-Williams et al.*, 1986].

In catchments essentially devoid of vegetation, organic material in these streams is entirely autochthonous. It is dependent on photosynthesis-driven transformations of dissolved carbon dioxide and nutrients into biomass. The significance of this nutrient transformation to downstream ecosystems is discussed by *Moorhead et al.* [this volume]. The biomass which can accumulate and provide organic carbon to other components of the stream-lake ecosystems is, as in all such systems, dependent on the balance between the rate of production during the short growing season and the loss processes, both during summer flow and winter freezing. In Antarctic stream communities, photosynthesis has been relatively well studied, respiration somewhat less so, and other loss processes hardly at all. In this contribution, we review existing information on primary production in dry valley streams and present new data on the dependence of photosynthesis and respiration on irradiance and temperature in a variety of communities. We argue that temperature is the most important determinant of production in these systems.

PRIMARY PRODUCERS

A variety of algae and cyanobacteria have been identified as growing in McMurdo Dry Valley streams. Three main growth forms can be identified: encrusting communities, trailing filamentous forms, and cohesive mats [*Broady*, 1982; *Vincent et al.*, 1993; *McKnight et al.*, this volume; *Niyogi et al.*, in press].

Encrusting Communities

These communities are widespread and typically comprise dark brown films dominated by *Gloeocapsa* spp. and, in some places, *Schizothrix*. In parts of the Alph River, in southern Victoria Land, the dark crusts also contain *Calothrix*, and here they have been shown to attain a biomass of 11.7 µg chlorophyll-*a* cm^{-2} [*Howard-Williams and Vincent*, 1989]. These encrusting communities are confined to the upper surfaces of substrate particles large enough to remain stable during flowing conditions. For example, in the Onyx River system of the Wright Valley, they are common in areas of large cobbles and boulders. However in reaches of the river which flow through sand-dominated substrates, they occur only on the larger particles associated with relatively highly armored banks on the outsides of meanders.

The universally dark coloration of these organisms

is due to the presence of an extracellular sheath pigment, scytonemin, which absorbs strongly in the UV-A region of the spectrum [*Garcia-Pichel and Castenholz*, 1991]. This pigment is common in cyanobacteria exposed to high ambient irradiance and has been shown to protect them from photoinhibition [*Garcia-Pichel et al.*, 1992]. Unprotected cyanobacteria have been shown to be particularly prone to prolonged photoinhibitory damage by high light [*Demmig-Adams et al.*, 1990].

Though encrusting communities are common, little is known of their ecological significance. In this paper we present the first data on photosynthesis for an encrusting community, in this case a *Gloeocapsa* crust collected from the Onyx River.

Trailing Filamentous Forms

Though common in streams in the maritime Antarctic [*Hawes*, 1989; *Hawes and Brazier*, 1991], trailing filamentous forms are sparsely distributed in McMurdo Dry Valley streams. Where they do occur, they comprise streamers of the chlorophytes *Binuclearia tectorum* and *Prasiola calophylla*, and the xanthophyte *Tribonema elegans*. These communities occur both on the surfaces of stones and in the highly shaded environment under stones [*Broady*, 1982, 1989]. The filamentous forms are reported to have a high over-winter survival rate [*Vincent and Howard-Williams*, 1986a] and may show biomass accumulation over the first few weeks of flow [*Vincent et al.*, 1993]. Biomass of up to 30 µg chlorophyll-*a* cm^{-2} has been reported [*Vincent and Howard-Williams*, 1986a, *Howard-Williams and Vincent*, 1989], which is very high even by temperate stream standards.

Prasiola from beneath rocks has been shown to be highly shade adapted, with a light saturated rate of photosynthesis (P_{max}) of approximately 0.09 µg C fixed (µg chlorophyll-*a*)$^{-1}$ h^{-1}, and a light saturation parameter (E_k), the irradiance at which the slope of the photosynthesis-light curve at low irradiance intercepts P_{max}, of 20 µmol photons m^{-2} s^{-1} [*Howard-Williams and Vincent*, 1989]. As might be expected, *Binuclearia*, which has a less shaded habitat, is less shade adapted, with an E_k of 70 (µmol photons m^{-2} s^{-1} and a P_{max} of approximately 0.17 µg C fixed (µg chlorophyll-*a*)$^{-1}$ h^{-1} [*Howard-Williams and Vincent*, 1989].

Cohesive Mat Communities

These are the most widespread and, in terms of biomass, most important communities in the McMurdo

Dry Valley streams. They comprise a matrix of filamentous cyanobacteria, mostly species of *Phormidium* and *Oscillatoria*. In addition, mats and individual colonies dominated by *Nostoc commune* are also found, particularly on the margins of streams and in areas which are alternately wetted and dried during the summer period. These mats can be epilithic, occur on loose substrata, or in gaps between larger particles. They occur in full sunlight, and *Nostoc* mats show the dark pigmentation observed in the *Gloeocapsa* crusts described earlier. Interestingly, lower layers of *Nostoc* mats are olive-green with no brown pigments, suggesting some degree of plasticity in pigment synthesis. Oscillatoriacean mats are usually confined to areas where water occurs' most of the time (when flowing), and typically have a reddish-brown appearance, with no evidence of scytonemin. In these communities, carotenoids appear to play a major role in protecting cells from excessively high light [*Demmig-Adams et al.*, 1990; *Vincent et al.*, 1994]. Of this pigment group, the xanthophylls canthaxanthin and myxoxanthophyll are particularly abundant in the McMurdo Dry Valley stream mats (I. Hawes and C. Howard-Williams, unpublished data).

Oscillatoriacean mats can accumulate to high biomass, with up to 40 mg chlorophyll-*a* cm^{-2} having been recorded [*Vincent et al.*, 1993; *Hawes*, 1993]. As well as accumulating biological material, they frequently entrap inorganic particles within the mat matrix giving an ash content of up to 95%, and can be up to 10 mm thick (unpublished data). The surface layers consist of stacked sheets of trichomes embedded in mucilage, with a lower layer of trichomes binding sediment grains together [*Vincent and Howard-Williams*, 1986b]. Thick mats do not develop within a single growing season but rather reflect the high success of overwintering of this growth form [*Vincent and Howard-Williams*, 1986a]. *Hawes* [1993] has shown that where a cyanobacterial mat does develop annually, as in some maritime Antarctic locations with long growing seasons, it has markedly different characteristics to the perennial films characteristic of the McMurdo Dry Valley streams. In particular, mat thickness and sediment content is much lower, while chlorophyll-*a* and photosynthesis per unit areas are similar to older mats.

There have been a number of estimates of photosynthetic production of these mat communities. *Howard-Williams and Vincent* [1989] reported light saturated areal rates of net photosynthesis of around 2-2.5 µg C cm^{-2} h^{-1} for both *Nostoc* and Oscillatoriacean mats, with similar E_k of 150–200 µmol photons m^{-2} s^{-1}. These authors had previously reported slightly lower rates of net photosynthesis for these communities 0.4–2.2 µg C cm^{-2} h^{-1} [*Vincent and Howard-Williams*, 1986], though these are of similar magnitude. These measurements of photosynthesis have been made by determining the rate of change of CO_2 in the gas phase overlying moist mats. In this paper we report the first measurements of photosynthesis and respiration made using changes in oxygen concentration in the aqueous phase and compare these with previous measurements.

Other Communities

Two other communities that occur in the streams of the McMurdo Dry Valleys are the mosses and their epiphytes which may occur on the margins of streams, and a diatom assemblage found on mobile sediment surfaces. The diatom assemblage is dominated by motile forms, notably species of *Navicula*, *Hantzschia*, and *Stauroneis*. To date there is no information on production processes in this growth form.

METHODS

Study Sites

Material collected from two study sites during the 1995-1996 summer, the Onyx River and Cripple Creek, was used in the experiments described here. The Onyx River has been fully described by *Howard-Williams et al.* [1986]. It is the longest known river in Antarctica, flowing for over 30 km through the Wright Valley to discharge into Lake Vanda. Material used in this study was mainly collected from the "boulder pavement," an area where the river is highly braided and the substrate comprises large, stable rocks, and cobbles. It is particularly rich in microbial mat and encrusting communities, though trailing forms are absent [*Howard-Williams et al.*, 1986]. Two morphotypes were sampled: 1) a red-brown colored *Phormidium*-based mat that formed a cohesive film some 3–4 mm thick between boulders, and 2) a dark brown encrusting *Gloeocapsa* community which occurred on large rocks close to the median water level. These two morphotypes were the only two encountered in the study area.

The second site, unofficially known as Cripple Creek, was a small stream flowing on the surface of the McMurdo Ice shelf. The McMurdo Ice Shelf is a sediment covered, partially grounded part of the Ross Ice Shelf, which borders the coast of the McMurdo Dry Valleys. The features of the area have been described in *Howard-Williams et al.* [1989a]. Cripple Creek flows for approximately 150 m between two ponds, has an average width of 1.5 m and is 1–5 cm deep. Substrates vary from sand to small cobbles; the location where samples were taken comprised a flat area of gravel and sand. It was completely covered by a cohesive, *Phormidium*-based mat 1–2 mm thick. Similar dominance by *Phormidium*-based mat was observed throughout this stream, and others close by, though occasional tufts of *Binuclearia* were observed.

In each case where a mat community was sampled, collections were biased towards "mature" communities. These thick, leathery mats could be easily removed from their substrata with no damage to mat structure. Thin, poorly developed mats or those heavily broken up were not easily sampled. All measurements made will therefore represent those of older communities and may not reflect rate processes in developing ones or those subject to higher levels of disturbance. Older, thicker mats are visibly dominant in most stream channels, particularly those with low or moderate velocities. However sparse Oscillatoriacean communities, which formed discrete tufts rather than cohesive mats, were also seen growing epilithic on cobbles and boulders, particularly in fast-flowing areas and these communities are likely to have different photosynthetic and respiratory characteristics to the mature ones. In particular, the ratio of heterotrophs to autotrophs is likely to increase with increasing mat thickness.

Measurements of Photosynthesis and Respiration

Two methods of estimation of rates of photosynthesis and respiration were employed. The first involved measurements of rate of release of oxygen by intact samples of mat communities at a range of light intensities (including dark). The second was used to measure photosynthesis in encrusting algae and involved measuring uptake of ^{14}C-bicarbonate by communities as flakes scraped from rocks. Measurements were made at a range of light intensities and temperatures. In order to compare rates of photosynthesis between techniques and with other published data, oxygen-derived photosynthesis and respiration were converted to carbon, assuming a 1:1 molar equivalency for both processes.

Oxygen flux experiments were conducted on Onyx River and Cripple Creek mat communities by sealing known areas of mat (approximately 2 cm^2) in 24–ml vials of filtered river water. In the Onyx River experiments, vials were incubated at a range of light intensi-

ties by positioning them different distances from a quartz-halogen light source. Temperature was maintained at 4°C ± 1°C in an insulated water bath. In Cripple Creek experiments, the light gradient was generated from incident daylight and increasing numbers of layers of neutral density filters and vials were incubated in the stream. Temperature in this experiment varied between 2-3°C. Irradiance was measured for each vial using a LI-Cor Li 190 PAR (photosynthetically available radiation) sensor.

In a second series of experiments, designed to investigate the effects of temperature on light-saturated photosynthesis and dark respiration, incubations of Cripple Creek material were carried out in dark and at 600 μmol photons m^{-2} s^{-1} at 1.5, 5, 10, and 20°C. This irradiance had been shown by preliminary experiments to be above saturation for photosynthesis with no evidence of photoinhibition. Temperature was maintained to ±1°C in insulated water baths by addition of warm water or ice.

To avoid oxygen dissolution at high rates of photosynthesis in all experiments, the oxygen content of the water was reduced (prior to filling the incubation vials to approximately 80%) by blending with partially de-gassed river water. Water was de-gassed by heating under vacuum, then cooled in sealed bottles. Preliminary experiments had shown that this had no effect on the rates of photosynthesis or respiration.

Incubations were kept short (0.3 to 1 h depending on irradiance) to minimize effects enclosure and the establishment of diffusion gradients in the static boundary layer overlying the mat. Net gas exchange through the mat-water interface was determined as the change in oxygen concentration in the overlying water over the period of the incubation relative to a control vial containing no mat. Oxygen concentration in sub-samples withdrawn from experimental and control vials at the end of incubations was determined by couloximetry [*Hawes and Schwarz*, 1996; in press]. Evolution or consumption of oxygen, subsequently normalized to mat area and time, was calculated as the difference in oxygen content between experimental vials and control vials.

Biomass of these communities was determined as chlorophyll-*a* and ash-free dry mass (AFDM), on an areal basis. AFDM was determined as the weight change of dried samples on combustion at 500°C, while chlorophyll-*a* was estimated by spectrophotometric analysis of 90% acetone extracts with correction for phaeophytin by acidification [*Marker et al.,* 1980].

Photosynthetic rates of encrusting communities were

too low to be measured using the oxygen change technique so a ^{14}C-bicarbonate method was developed. A *Gloeocapsa* suspension was prepared by gently scraping material off of 60 cm^2 rocks and diluting to 2.5 l with river water. A portion (23 ml) of this suspension was then introduced to each four replicate 24 ml vials, allowed to settle, and inoculated with 1 μCi (3.7 x 10^4 Bq) of ^{14}C bicarbonate. *Gloeocapsa* flakes rapidly settled to form a thin film loosely attached to the vial surface. Incubations were for two hours in insulated water baths at each of 11 irradiances ranging from 0 to 100% ambient. The light gradient was generated from incident daylight using increasing numbers of layers of neutral density shade cloth. The experiment was carried out at 5 and 15°C. Vials were pre-incubated in the dark for 30 minutes to allow for temperature equilibration.

Incubations were terminated by shaking the vial to re-suspend the *Gloeocapsa,* then rapid filtration onto a 25 mm GF/F filter (Whatman). Filters were frozen, then returned to New Zealand for ^{14}C uptake determination by liquid scintillation counting. Biomass of *Gloeocapsa* used in the experiments was estimated as chlorophyll-*a* after filtration of three 10-ml sub-samples of the suspension onto Whatman GF/F filters. Chlorophyll-*a* analysis was by fluorometry of 90% acetone extracts.

Curves were fitted to the irradiance-photosynthesis data using the Jassby-Platt hyperbolic tangent equation [*Jassby and Platt*, 1976], modified to include respiration. This relates photosynthetic rate (P) to light-saturated rate of photosynthesis (P_{max}), irradiance (E), respiration rate (R), and the slope of the P versus E curve as E approaches zero (α) thus

$$P = P_{max} \tanh(E\ \alpha/P_{max})\text{-}R \qquad (1)$$

Inclusion of the R term means that P_{max} calculated with this equation will be gross photosynthesis and that R will be dark respiration. For the *Gloeocapsa* experiments using ^{14}C, the R term could not be included.

In order to relate measured relationships between photosynthesis, irradiance and temperature to real conditions, incident irradiance and water temperature in the Onyx River were measured throughout January 1995, using a LI-Cor LI 190 PAR sensor and a Campbell Instruments 107 temperature probe respectively. These were connected to a Campbell CR10 data logger, which interrogated sensors every 60 s and recorded a mean value every 15 min.

Biomass Accumulation on New Surfaces

Rate of colonization and accumulation of biomass on newly exposed surfaces was estimated using artificial substrates. These comprised a 500 x 250 mm sheet of 45 molded plastic hemispheres, each 38 mm in diameter, secured to a high density polyethylene baseplate. The surface of the hemispheres was slightly rough. At each of five sites in the Onyx River, five substrates were secured in early January, 1995. The sites were located immediately above (site 5), within (sites 3 and 4) and immediately below (site 2) the microbially rich boulder pavement area. Site 1 was within a well defined channel approximately 1 km downstream of the boulder pavement.

In September 1995, while the stream bed was still dry, and again in early January 1996 when the river was flowing, three replicate hemispheres were removed from each substrate and analyzed for chlorophyll-*a*. The September sampling was taken to represent accumulation during the second half of the 1994-1995 summer (i.e., mid-January to mid-February) and the January 1996 samples to reflect further accumulation during the first half of the 1995-1996 season (mid-December to mid-January). Chlorophyll-*a* was extracted into 90% ethanol, by heating to 78°C for 2 minutes, followed by fluorometric analysis [*Hawes and Schwarz*, 1996]. Chlorophyll-*a* was expressed per unit hemisphere surface area. Accumulation of chlorophyll-*a* will reflect colonization and subsequent growth on these substrates.

RESULTS

Irradiance and Temperature

Irradiance incident to the Onyx River during the period of flow in 1994-1995 varied between 50 and 1200 μmol photons m^{-2} s^{-1} (Figure 1a). Water temperature followed a similar diel cycle, ranging from zero to almost 9°C (Figure 1b). Daily variation in temperature could cover this entire range. The minimum diel range was 4°C.

Photosynthesis and Respiration of Mat Communities

Photosynthesis-light curves for the two Oscillatoriacean mat communities were similar (Figure 2). The mats showed similar light saturated rates of gross photosynthesis (4.3 and 4.0 μg C cm^{-2} h^{-1}, Onyx River, Cripple Creek) though irradiance at which saturation

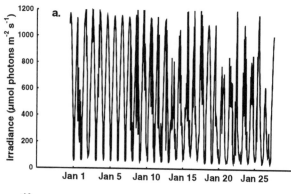

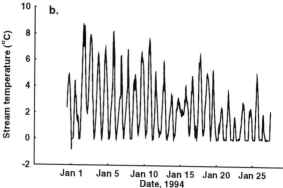

Fig. 1. Incident irradiance, as photosynthetically active radiation (a) and water temperature at the Onyx River site during the flow period of 1994 (b).

began in the Onyx River mats was slightly lower than that of the Cripple Creek community (approximately 120 to 200 μmol m^{-2} s^{-1}) (Table 1). We found no evidence of photoinhibition at the highest irradiance used. However maximum ambient irradiance (Figure 1) was 50% higher than the highest experimental irradiance. Respiration rate was higher in the Onyx River than in Cripple Creek (Table 1), which tended to offset the light-net photosynthesis curve to a lower position (Figure 2), resulting in a lower net rate of photosynthesis at light saturation. Light compensation points, the irradiance at which net photosynthesis was zero, were estimated from the fitted curves as 40 and 60 μmol m^{-2} s^{-1} for Onyx River and Cripple Creek mats respectively.

The differences between the Onyx River and Cripple Creek mats may reflect their degree of development. Cripple Creek mats contained more chlorophyll-a per unit area than Onyx River, 20.8 ± 4.9 to 12.8 ± 3.5 μg cm^{-2} (mean ± s.d. n = 10), but less AFDM, 4.4 ± 1.8 to 10.4 ± 3.2 mg cm^{-2} (mean ± s.d. n = 10). Rates of photosynthesis and respiration normalized to these measures of biomass are shown in Table 2. There was

a close agreement between the sites for respiration when normalized to AFDM. Photosynthesis normalized to either chlorophyll-a or AFDM differed between sites much more than when normalized to surface area.

All of these parameters were estimated at ambient temperature (25°C). The sensitivity of the parameters in the Cripple Creek mat to temperature, within and outside the typical ambient range, are shown in Figure 3. These data provided a close fit to an exponential function, which indicated a Q_{10}, the proportional rise in activity for a 10°C rise in temperature, of 1.7 for dark respiration and 1.6 for gross photosynthesis over the range 2 to 12°C.

The *Gloeocapsa* dominated encrusting community showed similar photosynthesis-light-temperature relationships to the mats (Figure 4). At 5°C, photosynthesis saturated at less than 100 μmol m^{-2} s^{-1}. Increasing temperature caused an increase in saturated rate of photosynthesis, but little change in the light- photosynthesis relationship below saturation. Alpha therefore changed little with temperature, while P_{max} increased, with a Q_{10} in this case of 2.0. As ^{14}C methodology was used for this community, no estimate of respiration was obtained. In order to obtain a comparative value of P_{max} to use in Table 1, it was necessary to estimate areal photosynthesis. This was done by normalizing photosynthesis per unit chlorophyll-a to chlorophyll-a per unit area. Since a suspension technique was used with this material, such a calculation was at best an estimate of areal activity. Scraping will have altered both the physical structure of the community and its light attenuating properties.

Accumulation of Biomass on New Surfaces

Colonization of replicated hemispheres deployed in the Onyx River in mid-January 1995, as measured by accumulation of chlorophyll-a showed little difference between the five sites in September 1995 (Figure 5). Median values were 0.001 to 0.003 μg cm^{-2}, with some replicates below detection levels, and distributions within individual sites, and in all sites pooled, showed a log-normal pattern. Such a distribution pattern might be expected where the lower boundary is constrained by zero. When the substrates were re-sampled one year after deployment median values were again similar between the five sites, but there were more extreme values than in September. This applied particularly to sites three and four, which were located within the microbially rich boulder pavement area. Values of up to 0.35 μg cm^{-2} were recorded here, while median

TABLE 1. Parametes of the Photosynthesis-Irradiance Relationship for Some Anarctic Stream Communities

Community	P_{max} (µg C cm^{-2} h^{-1})	E_k (µmol m^{-2} s^{-1})	α (P_{max}/E_k)	R (µg C cm^{-2} h^{-1})	Source	Comments
Oscillatoriaceae	3.0-3.6	300	0.01	NR	*Hawes,* 1995	Maritime communities
	1.8	105	0.02	NR	*Howard-Williams and Vincent, 1989*	Canada Stream, Taylor Valley
	0.4-1.41	NR	NR	0.07-1.23	*Vincent and Howard-Williams, 1986*	Various locations
	4.3	75	0.01	2.5	This study	Onyx River
	4.0	190	0.02	1.2	This study	Cripple Creek
	6.5	120	0.06	5.7	Unpublished, Duff and Tate	Green Creek
Nostoc	0-2.15	NR	NR	0.24-1.22	*Vincent and Howard-Williams, 1986*	Alph River
	2.6	150	0.02	NR	*Howard-Williams and Vincent, 1989*	Canada Stream, Taylor Valley
	8.0	60	0.15	3.2	Unpublished, Duff and Tate	Green Creek
Prasiola	3.5	20	0.17	NR	*Howard-Williams and Vincent, 1989*	Canada Stream, Taylor Valley
Binuclearia	2.8	70	0.04	NR	*Howard-Williams and Vincent, 1989*	Canada Stream, Taylor Valley
Gloeocapsa	0.4	40	0.01	NR	This study	Onyx River

P_{max} in each case is the maximum gross rate of photosynthesis, normalized to unit area. R is the dark respiration rate, E_k the saturation onset irradiance parameter, and α the slope of the P-E curve at low irradiance. NR indicates not reported.

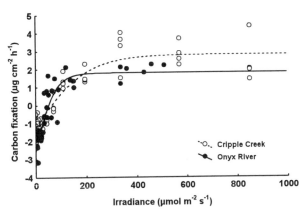

Fig. 2. The relationship between irradiance and photosynthesis for Oscillatoriacean mats from Cripple Creek and the Onyx River. The curves fitted use the *Jassby-Platt* [1976] hyperbolic tangent function.

values remained at 0.01 µg cm^{-2} (Figure 5). We noted the occurrence of accumulations of small fragments of mat on many hemispheres at sites three and four in summer which appeared to have been washed off of rocks higher up and to have attached to the artificial substrates. We also noted that mat fragments were common in early meltwater flows. Mats tended to flake while desiccated over winter, and early flows washed these flakes off of their substrata. In late season flow, such flake transport was less obvious. This mechanism of bulk colonization, in addition to settlement and growth of individual cells and trichomes, may explain the unusual distribution of chlorophyll-*a* concentration at the second sampling. Species composition on the artificial substrates was, like that of natural surfaces, dominated by oscillatoriacean trichomes.

DISCUSSION

There have now been several examinations of the relationship between photosynthesis and irradiance in stream algae from the McMurdo Dry Valley region (Table 1), as well as other parts of Antarctica. These data show a remarkable convergence of values for both

TABLE 2. Rates of Light Saturated Photosynthesis (P_{max}) and Respiration (R) normalized to area, chlorophyll-*a* (Chl*a*), and ash free dry mass (AFDM) for the Onyx River and Cripple Creek.

	Onyx River	Cripple Creek
P_{max} (μg C cm^{-2} h^{-1})	4.28	4.04
P_{max} (μg C μg^{-1} Chl*a* h^{-1})	0.33	0.19
P_{max} (μg C mg^{-1} AFDM h^{-1})	0.41	0.92
R (μg C cm^{-2} h^{-1})	2.51	1.22
R (μg C μg^{-1} Chl*a* h^{-1})	0.19	0.06
R (μg C mg^{-1} AFDM h^{-1})	0.24	0.28

P_{max} (normalized to unit area) and, to a lesser extent, E_k. Values of E_k are consistently low relative to incident radiation, and suggest that light saturation of photosynthesis is attained at irradiances less than incident for most of the growing season. Indeed, the irradiance data for the Onyx River presented in Figure 1a showed that the saturating irradiances of 75–100 μmol photons m^{-2} s^{-1} recorded in benthic communities from this system was exceeded for all but a few hours each day.

These mats comprise compressed photic zones, with a high biomass of chlorophyll-*a* compressed into a thin film [*Hawes,* 1993], and in which light declines rapidly with depth [*Vincent et al.,* 1994]. Mats may consequently be strongly self-shading and individual cells will be adapted to very different irradiance regimes than those incident to the surface, and they may be light limited or even photoinhibited much more frequently. Our data refer only to community rates of activity. As such, calculations of rate of photosynthesis per unit chlorophyll-*a* are misleading, since no cells may in fact show the rate calculated.

The light saturated rates of gross photosynthesis per unit area observed in this study spanned a range of 0.4 to 4.3 μg C cm^{-2} h^{-1}. These values tended to be lower than those for broadly comparable, temperate locations [e.g., data in *Boston and Hill,* 1991; *Guasch and Sabater,* 1995]. This may have reflected the low temperature, but there is an enormous range of community types and productivities within stream ecosystems, even within single rocks, and cross-site comparisons are of questionable value unless planned with this in mind. However general features of photosynthesis and respiration in the dry valley streams could be usefully

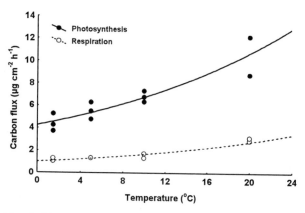

Fig. 3. The effect of temperature on light saturated photosynthesis and respiration in material collected from the Cripple Creek site.

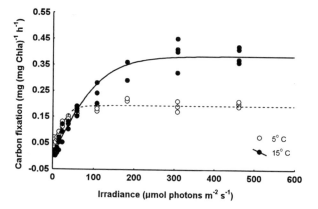

Fig. 4. Photosynthesis-irradiance curves for *Gloeocapsa* from the Onyx River at 5°C and 15°C. Curves are fitted using the *Jassby-Platt* [1976] hyperbolic tangent function.

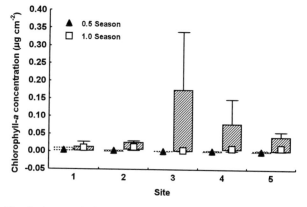

Fig. 5. Accumulation of chlorophyll-*a* on artificial substrata deployed at five sites in the Onyx River. Substrates were deployed in early January 1995, and sampled in September 1995 and again in early January 1996. The samples thus represent 0.5 and 1 seasons accumulation respectively. Points indicate median values, boxes 75 percentile, and whiskers the range of values at each site.

compared with those of temperate locations. For example, the low value of E_k relative to incident radiation is not a property specific to Antarctic mats, but appears to be generally found in thick periphyton films. There have been a number of studies from warm temperate streams that have demonstrated this and shown how this feature can develop as a result of increased self shading as mat thickness and chlorophyll-*a* concentration increases [*Hill and Boston*, 1991; *Guasch and Sabater*, 1995].

Net photosynthesis in our data was much lower than gross photosynthesis, at times approaching zero even at saturating light. This has been reported previously from Antarctic streams [*Vincent and Howard-Williams*, 1986a] and unpublished data from Green Creek in the Taylor Valley also shows this (J. Duff and C. Tate, personal communication). Zero net production implies that, except at specific times of high stress or structural failure, most organic material generated within the mat matrix was respired there.

The possibility for mature, high biomass communities to tend towards zero net production is also seen in temperate streams. However this is often not able to occur in temperate systems even under stable flow regimes because of disturbance by grazing, which tends to maintain low biomass [*Steinman*, 1992; *Hill et al.* 1995]. As stressed by previous authors [summarized in *Vincent et al.* 1993], the ability of dry valley stream communities to develop to high biomass, and approach zero net production, despite low temperatures, reflects the extremely low disturbance, i.e., absence of grazers, in many habitats, thus permitting a balance between microbial heterotrophs and autotrophs to develop. In this context, it is not surprising that in the two mat communities examined in this paper, respiration normalized to AFDM was much more similar than when normalized to area or chlorophyll-*a*, while photosynthesis was similar on an areal basis, but not on an AFDM basis. A similar constancy of P_{max} per unit area (3.0–3.6 µg C cm^{-2} h^{-1}), but not per unit AFDM, was found in two maritime Antarctic streams with much bigger AFDM differences [*Hawes*, 1993].

The low irradiance at which photosynthesis saturated suggests that the performance at light saturation, rather than efficiency at low irradiance will have been most important in terms of determining stream production. This contrasts strikingly with benthic mats from the lakes of the McMurdo Dry Valleys which experience ambient irradiance below saturation for all or most of each year [*Hawes and Schwarz*, in press]. Since both the saturated rate of gross photosynthesis and respira-

tion rate were temperature dependent, temperature rather than irradiance may have been the dominant variable in determining stream production. Previous observations that Q_{10} values for Antarctic stream communities are close to 2 at ambient temperature [*Vincent and Howard-Williams*, 1989] were confirmed by data in this paper. This, and the common observation that experimentally derived temperature optima are well above ambient [*Seaburg et al.*, 1981; *Castenholz and Schneider*, 1993], has been interpreted as evidence of lack of specific low temperature adaptation of these communities. Temperature was clearly an important factor in controlling rate of production in these systems.

We investigated the likely importance of temperature-dependant changes in P_{max} and R on photosynthesis in situ for the Onyx River Oscillatoreacean mat by constructing a simple model. This model calculated photosynthesis at 15 min intervals using the hyperbolic tangent function. The physiological parameters P_{max} and R were obtained from Table 1, and corrected for temperature using the Q_{10} relationships derived from Figure 3. We assumed that α was temperature independent and used a constant value of 0.06. Temperature and irradiance data used to drive the model were those shown in Figure 1a. To determine how this rate of net production should have translated to biomass, we calculated the resulting increase in AFDM assuming a carbon content of 40% AFDM. AFDM at the start of the simulation was set to 10 mg cm^{-2}, the measured value. The model predictions of net photosynthesis and biomass accumulation are shown in Figure 6a. The prediction was for a smooth, temperature related rise and fall in rate of photosynthesis over each day, with a minimum around midnight, when irradiance briefly dropped below saturation. Biomass was predicted to accumulate almost threefold over the month of January.

There have been few studies of cyanobacterial mats in Antarctic streams where biomass as AFDM has been measured repeatedly over the growing season. *Hawes* [1993] showed that, in maritime streams with perennial cyanobacterial mats, there was little increase in AFDM over a six week period, though chlorophyll-*a* increased by 50%. *Howard-Williams et al.* [1986] showed increases in chlorophyll-*a* of 50–100% between overwintering and mid-season biomass in a range of dry valley streams. However the model prediction of a threefold biomass increase is not supported by the limited data available nor by our own casual observations.

As discussed above, respiration appears to have

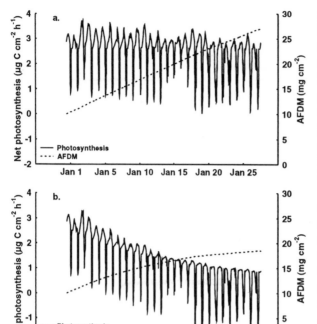

Fig. 6. Output from a model predicting areal net photo-synthesis and ash-free dry mass (AFDM) in an Oscillatoriacean mat from incident irradiance and water temperature. The model uses relationships between light, temperature, photosynthesis, and respiration established in Figures 2 and 3 and raw data from Figure 1. Simulation "a" assumes a constant rate of photosynthesis and respiration per unit area, while simulation "b" assumes that respiration rate is linearly related to AFDM.

been more closely related to AFDM than area, while photosynthesis tended to show the opposite pattern. When we reran the model letting R increase linearly as AFDM increased, what appeared to be a more realistic simulation resulted (Figure 6b). This predicted a smaller change in AFDM, which tended towards a maximum value of 20 mg cm^{-2}, and a gradual decline in net photosynthesis as respiration increased. This simulation predicted respiration rate to have increased over the course of a season's growth, until it balanced photosynthesis. To date we have no data to test these predictions, and the colonization data obtained using the artificial substrates is inappropriate, since it covers only the colonization phase. The importance of measuring photosynthesis and respiration on material at a range of biomass both as AFDM and chlorophyll-a, as well as seasonally, are necessary for better understanding stream production processes.

These models did not allow for the many other factors that might affect production, such as changes in biomass, changes in water level or nutrient concentrations. Neither did they allow for export of dissolved organic material from the mat communities, which is known to occur [*Moorhead et al.,* this volume], particularly under freeze-thaw stress [*Howard-Williams and Hawes,* in press]. However they illustrate the prevailing importance of temperature in hour-to-hour determination of net photosynthesis, and the potential for net photosynthesis to decline towards zero as mat biomass increases.

As similarly low values of saturating irradiance and Q_{10} relationships for photosynthesis were seen in encrusting communities as in mats, it is likely that similar temperature-dominated constraints to produc-tion will occur. Production in all stream communities therefore appears to be temperature limited. The only communities where light may play a more significant role may be the "sub-lithic," shade adapted *Prasiola calophylla* [*Vincent and Howard-Williams,* 1986a]. Nutrient limitation appears to play little role in determining growth rates [*Howard-Williams and Vincent,* 1989]. The nutrient contents of active trichomes extracted from Onyx River mats yielded C:N:P ratios of 103:7.9:1 (I. Hawes, unpublished data) which closely matched the "Redfield ratio," indicating an adequate, balanced supply of nutrients relative to growth requirements [*Tett et al.* 1985].

Accumulation of biomass on newly exposed substrata will depend on primary colonization, fol-lowed by a period of exponential growth until self-shading effects begins to limit growth rate. Distribution of chlorophyll-a on artificial substrates during the early stages of this process might be expected to follow a log-normal distribution, since values are constrained by zero. In the first sampling this was the case, but the second sampling contained a number of high outliers. This is consistent with our observations of flaking of mats during winter freeze-drying, followed by washing downstream during the first flows. Early flows of water from the Onyx River into Lake Vanda contain visibly high concentra-tions of particulate mat material. This may be an important mechanism of colonization in these streams. However it is clear that several years of biomass accumulation and growth would be required to reach that seen in mature communities.

Accumulation of microbial material to the high biomass observed in many McMurdo Dry Valley streams requires an effective overwintering strategy. While observations of rapid recovery of epilithon, following wetting after winter, have been made [*Vincent and Howard-Williams*, 1986a, b], the mechanisms of overwinter survival are poorly understood. We have observed that immediately following wetting in spring microbial communities in the Onyx River lose a significant amount of organic material and that freeze-thaw to $-8\,^{\circ}C$ produces a similar response. Despite this loss of organics, the mats have remained viable (I. Hawes and C. Howard-Williams unpublished data). *Davey* [1989] showed that *Phormidium* and *Prasiola* communities growing on damp Antarctic soil survived repeated excursions to $-15\,^{\circ}C$. He also demonstrated that survival was enhanced in *Prasiola* by high ambient light, and in *Phormidium* by freezing in water rather than in air. Both of these advantages are likely to be available to stream communities, since they will freeze well before winter darkness. *Roser et al.* [1992] found no accumulation of cryoprotectant compounds in cyanobacterial mats, and further research is clearly required to determine the overwintering physiology of these organisms.

Production in streams of the McMurdo Dry Valleys is dominated by microbial communities, particularly the mat forming Oscillatoriacean cyanobacteria. In the absence of disturbance, these communities can develop over many years to high biomass. Areal rates of gross photosynthesis are similar across a range of streams, but as mats accumulate organic material, respiration rate appears to increase until the mats may become balanced with respect to gas exchange. Export of organic material from these mats occurs in the particulate form when they undergo catastrophic detachment, and in dissolved form with leaching of organics, particularly following stress. However the close balance found in some communities between gross photosynthesis and dark respiration suggests that internal cycling processes may be important, though this has yet to be demonstrated. Irradiance over the growing season is normally in excess of the low saturation intensities for photosynthesis. Both photosynthesis and respiration are therefore largely controlled by temperature. The key to the dominance of these streams is the ability to overwinter. While cyanobacteria are evidently capable of overwintering, the adaptations and mechanisms by which they do this have yet to be adequately described.

Acknowledgements. This work was funded by the New Zealand Foundation for Research, Science and Technology under contract CO1506 and CO1614. We thank Anne-Maree Schwarz and Rob Smith for help with field work, and Daryl Moorhead, John Priscu and two anonymous reviewers for valuable comments on the manuscript.

REFERENCES

Boston, H. L., W. R. Hill, Photosynthesis-light relations of stream periphyton communities, *Limnol. Oceanogr. 36*, 644-656, 1991.

Broady, P. A., Taxonomy and ecology of algae in a freshwater stream in Taylor Valley, Victoria Land, Antarctica, *Archiv für Hydrobiol. Suppl. 63*, 331-339, 1982.

Broady, P. A., The distribution of *Prasiola callophylla* (Carmich.) Menegh. (Chlorophyta) in Antarctic freshwater and terrestrial habitats, *Antarctic Sci. 1*, 215-224, 1989.

Castenholz, R. W. and A. J. Schneider, Cyanobacterial dominance at high and low temperatures: Optimal conditions or precarious existence? *Trends in Microbial. Ecol.* ,19-24, 1993.

Chinn, T. J., Physical hydrology of the dry valley lakes, In *Physical and Biochemical Processes in Antarctic Lakes, Antarctic Research Series*, Vol 59, edited by W.J. Green and E.I. Friedmann, American Geophysical Union, Washington D.C. 1-51, 1993.

Conovitz, P.A., D. M. McKnight, L. M. McDonald, A. Fountain, and H. R. House, Hydrologic processes influencing streamflow variation in Fryxell basin, Antarctica, this volume.

Davey, M. C., The effects of freezing and desiccation on photosynthesis and survival of terrestrial antarctic algae and cyanobacteria, *Polar Biol. 10*, 9-36, 1989.

Demmig-Adams, B., W. W. Adams, F-C. Czygan, U. Schreiber, and O. L. Lange, Differences in the capacity for radiationless energy dissipation in the photochemical apparatus of green and blue-green algal lichens associated with differences in carotenoid composition, *Planta 180*, 582-589, 1990.

Garcia-Pichel, F., and R.W. Castenholz, Characterisation and biological implications of scytonemin, a cyanobacterial sheath pigment, *J. Phycol. 27*, 395-409, 1991.

Garcia-Pichel F., N. D. Sherry, R. W. Castenholz, Evidence for an ultraviolet sunscreen role of the extracellular pigment scytonemin in the terrestrial cyanobacterium *Chloroeopsis* sp., *Photochem. Photobiol. 56*,17-23, 1992.

Guasch, H. and S. Sabater, Seasonal variation in photosynthesis-irradiance responses by biofilms in Mediterranean streams. *J. Phycol. 31*, 727-735, 1995.

Hawes, I., Filamentous green algae in freshwater streams on Signy Island, Antarctica, *Hydrobiologia 172*, 1-18, 1989.

Hawes, I. Photosynthesis in thick cyanobacterial films: a comparison of annual and perennial antarctic mat communities, *Hydrobiologia 252*, 203-209, 1993.

Hawes, I. and P. Brazier, Freshwater stream ecosystems of James Ross Island, Antarctica. *Antarctic Sci., 3*, 265-271, 1991.

Hawes, I. and A.-M. Schwarz, Epiphytes from a deep-water characean meadow in an oligotrophic New Zealand Lake: species composition, biomass and photosynthesis, *Freshwater Biol., 36*, 101-117. 1996

Hawes, I and A-M. J. Schwarz, Photosynthesis in benthic mats from Lake Hoare, Antarctica, *Ant. J. U.S.* , In Press.

Hill, W. R., and H. L. Boston, Community development alters photosynthesis-irradiance relations in stream periphyton, *Limnol. Oceanog. 37*, 1375-1389, 1991.

Hill, W. R., M. G. Ryan, and E. M. Schilling, Light limitation in a stream ecosystem: Responses by primary producers and consumers, *Ecology 76*, 1297-1309, 1995.

Howard-Williams, C., I. Hawes, Sources and sinks of nutrients in a polar desert stream, the Onyx River,

Antactica, In *Ecosystem processes in Antarctic ice-free landscapes,* edited by W. B. Lyons, C. Howard-Williams and I. Hawes, Balkema, Rotterdam, In Press.

Howard-Williams, C., and W. F. Vincent, Antarctic stream ecosystems: Physiological ecology of a blue-green algal epilithon, *Freshwater Biol. 16*, 219-233, 1986.

Howard-Williams, C., and W. F. Vincent, Microbial communities in southern Victoria Land streams. I. Photosynthesis, *Hydrobiologia 172*, 27-38, 1989.

Howard-Williams, C., C. L. Vincent, P. A. Broady, and W. F. Vincent, Antarctic stream ecosystems: variability in environmental properties and algal community structure, *Internat. Rev. ges. Hydrobiol., 7,* 511-544, 1986.

Howard-Williams, C., J. C. Priscu, W. F. Vincent, Nitrogen dynamics in two antarctic streams, *Hydrobiologia, 172,* 51-61. 1989a.

Howard-Williams, C., R. Pridmore, M. T. Downes, and W. F. Vincent, Microbial biomass, photosynthesis and chlorophyll *a* related pigments in the ponds of the McMurdo Ice Shelf, Antarctica, *Antarctic Sci., 2,* 125-131, 1989b.

Jassby, A. D. and T. J. Platt, Mathematical formulation of the relationship between photosynthesis and light for phytoplankton, *Limnol. Oceanogr., 21,* 540-547, 1976.

Marker, A. F. H., E. A. Nusch, H. Rai, B. Reimann, The measurement of photosynthetic pigments in freshwaters and standardisation of methods: conclusions and recommendations, *Ergebnisse Limnol., 14,* 91-106, 1980.

McKnight, D. M., A. Alger, C. Tate, G. Shupe, S. Spaulding, Longitudinal patterns in Algal abundance and species distribution in meltwawter streams in Taylor Valley, southern Victoria Land, Antarctica, this volume.

Moorhead, D., D. M. McKnight, C. Tate, Modelling nitrogen transformations in Antarctic Streams, this volume.

Niyogi, D. K., C. M. Tate, D. M. McKnight, J. H. Duff, A. S. Alger, Species composition and primary production of algal communities in Dry Valley streams in Antarctica: examination of the functional role of biodiversity, In *Ecosystem processes in Antarctic ice-free landscapes,* edited by W. B. Lyons, C. Howard-Williams and I. Hawes, Balkema, Rotterdam, In Press.

Roser, D. J., D. R. Melick, H. U. Ling, and R. D. Seppelt, Polyol and sugar content of terrestrial plants from continental Antarctica, *Antarctic Sci., 4,* 413-420, 1992.

Seaburg, K. G., B. C. Parker, R. A. Wharton, and G. M. Simmons, Temperature-growth responses of algal isolates from antarctic oases, *J. Phycol., 17,* 353-360, 1981.

Steinman, A. D., Does an increase in irradiance influence periphyton in a heavily-grazed woodland stream? *Oecologia, 91,* 163-170, 1992.

Tett, P., S. I. Heaney, M. R. Droop, The Redfield ratio and phytoplankton growth rate, *J Mar. Biol. Ass. U.K., 65,* 487-504, 1995.

Vincent, W. F. and C. Howard-Williams, Antarctic stream ecosystems: the physiological ecology of a blue-green algal epilithon. *Freshwater Biol., 16,* 219-233, 1986a.

Vincent, W. F. and C. Howard-Williams, Microbial ecology of Antarctic streams. Proceedings of the IV International Society of Microbiol Ecology, Lubljiana p201-206, 1986b.

Vincent, W. F. and C. Howard-Williams, Microbial communities in southern Victoria Land streams II. The effect of low temperature, *Hydrobiologia, 172,* 39-49, 1989.

Vincent, W. F., C. Howard-Williams, and P. A. Broady, Microbial communities and processes in antarctic flowing waters, *In* Antarctic Microbiology, Wiley-Liss Inc., 543-569, 1993.

Vincent, W. F., R. W. Castenholz, M. T. Downes, and C. Howard-Williams, Antarctic cyanobacteria: light, nutrients and photosynthesis in the microbial mat environment, *J. Phycol., 29,* 745-755. 1994.

Wilson, A. T., Geochemical problems of the Antarctic dry areas, *Nature (London) 280,* 205-208, 1979.

I. Hawes, National Institute for Water and Atmospheric Research, Kyle Street, P.O. Box 8602, Christchurch, New Zealand.

C. Howard-Williams, National Institute for Water and Atmospheric Research, Kyle Street, P.O. Box 8602, Christchurch, New Zealand.

(Received September 12, 1996;
accepted February 25, 1997)

MODELING NITROGEN TRANSFORMATIONS IN DRY VALLEY STREAMS, ANTARCTICA

Daryl L. Moorhead

Department of Biological Sciences, Texas Tech University, Lubbock, Texas

Diane M. McKnight

Civil, Environmental, and Architectural Engineering Department, INSTAAR, Boulder, Colorado

Cathy M. Tate

United States Geological Survey, Water Resources Division, Denver, Colorado

Concentrations of ammonium, nitrate, and urea decline along a glacial meltwater stream in Taylor Valley, southern Victoria Land, Antarctica. These reductions accompany increasing concentrations of particulate and dissolved organic nitrogen (other than urea), suggesting that benthic microbial mats present in these systems may be responsible for transforming dissolved inorganic nitrogen into dissolved and particulate organic compounds. A mathematical model of primary production of microbial mats was used to estimate nitrogen transformation, assuming that nitrogen uptake balanced carbon fixation. Export of organic nitrogen was set equal to inorganic uptake driven by net primary production, based on the assumption of steady-state biomass for mat communities. Model results were comparable to observations although transformation rates generally were lower than observed. The model was sensitive to water retention time in the stream, illustrating the critical importance of accurate assessments of stream geometry and hydrology. Application of this model to three other streams feeding Lake Fryxell (Taylor Valley) suggest that dry valley streams have a large potential to transform mineral nitrogen into organic forms.

INTRODUCTION

The McMurdo Dry Valleys are among the most climatically extreme environments of the world. In Taylor Valley, southern Victoria Land, Antarctica, the mean annual temperature is about –20°C, [*Clow et al.*, 1988] and total annual precipitation is about 10 cm, all of which is received as snow in winter [*Keys*, 1980]. Low humidity and dry föhn winds descending from the polar plateau further enhance the overall aridity of the dry valley system. Nevertheless a number of freshwater streams exist in Taylor Valley, fed by glacial melt.

Streams in the McMurdo Dry Valleys lack most of the flora and fauna of their more temperate counterparts. However stream mats often have standing stocks of biomass exceeding 40 mg ash-free dry mass cm^{-2} [*Alger et al.*, 1996; *McKnight et al.*, this volume], despite the dominance of cyanobacteria (e.g., *Phormidium* and *Nostoc*) [*Vincent*, 1988; *Howard-Williams and Vincent*, 1989; *Alger et al.*, 1996]. Previous studies revealed substantial rates of primary production for mat communities in Canada Stream, a representative meltwater stream in Taylor Valley [*Vincent and Howard-Williams*, 1986; *Howard-Williams and Vincent*, 1989]. Concentrations of nitrate and urea in stream water were reported to decline significantly along the length of Canada Stream, concurrent with significant increases in the concentrations of other forms of organic nitrogen [*Howard-Williams et al.*, 1989]. However few studies have examined the dynamics of carbon and nitrogen in Antarctic streams, and we are aware of no previous attempt to link carbon and nitrogen flows in a functional context for these streams.

BACKGROUND

Taylor Valley is the primary location of the USA-sponsored, McMurdo Dry Valley Long Term Ecological Research (LTER) project (77°00'S, 162°52'E). The valley is approximately 33 km long by 12 km wide and contains three major lakes (Bonney, Hoare, and Fryxell) fed by 15 glaciers (see accompanying CDROM). The streams in Taylor Valley originate primarily from glacial melt, with virtually no inputs of water from the adjacent land surface [*Conovitz et al.,* this volume]. Thus streams are usually short (< 5 km) and flow directly from glacial sources into the lakes.

Microbial mats in dry valley streams are composed primarily of filamentous cyanobacteria and exhibit low rates of photosynthesis [< 1 mg C m^{-2} h^{-1}; *Hawes and Brazier*, 1991; *Davey*, 1993; *Hawes*, 1993] despite high light intensities and adequate nutrients [*Howard-Williams and Vincent*, 1989]. The general lack of herbivores permits accumulations of biomass that may exceed 100 mg ash-free dry mass cm^{-2} [*Alger et al.*, 1996], and *Moorhead et al.* [in press] estimated net annual primary production equivalent to 16–35% of standing stocks (based on 8–28 mg C cm^{-2} standing stock). Thus rates of nutrient uptake may be substantial if they are proportional to net primary production. In fact *Howard-Williams et al.* [1989] found nitrogen uptake rates to be comparable to those of more temperate streams.

A mathematical model of primary production recently was developed to examine patterns of annual biomass accumulation of microbial mats in dry valley lakes and streams [*Moorhead et al.*, in press]. In the present study we used this model to link nitrogen dynamics to primary production for benthic mat communities in Canada Stream, which flows into Lake Fryxell, Taylor Valley, Antarctica. The resulting model also was used to estimate potential uptake of inorganic and urea forms of nitrogen and production of other forms of dissolved and particulate organic nitrogen in streams of the Fryxell basin.

MODELING APPROACH

In this study we simulated the reduction in concentrations of inorganic and urea nitrogen in water of dry valley streams as a process that was directly proportional to the increase in concentrations of other forms of organic nitrogen. We assumed that this process was accomplished by

the biological activities of microbial mat communities found in these streams, occurred as water flowed over benthic communities, and at a rate that was proportional to net primary production. Although many biotic and abiotic processes are responsible for nitrogen dynamics within streams, we considered only the potential for mat communities to fix dissolved inorganic and urea nitrogen into other organic forms. The single exception to this limitation was that atmospheric nitrogen fixation by *Nostoc* mats was estimated for Canada Stream (see below). We assumed that mat biomass was constant (steady state), so that uptake of inorganic and urea nitrogen was equalled by production of other forms of dissolved and particulate organic nitrogen. This progressive substitution of forms of nitrogen was estimated to occur within a parcel of water as it flowed from source to mouth of Canada Stream, Green Creek, Delta Stream, and Von Guerard Stream in Taylor Valley, Antarctica. For convenience, we refer to this process simply as "nitrogen transformation" throughout this manuscript. The availability of detailed, published information allowed a more thorough examination of nitrogen transformation in Canada Stream than in the other streams [*Howard-Williams et al.*, 1989]. Potential transformation of nitrogen was estimated in other streams based on mat characteristics and simulated net primary production.

Net Primary Production

Net primary production of microbial mats was calculated as the difference between photosynthesis and respiration:

$$NPP = P - R \qquad (1)$$

where *NPP* is net primary production, *P* is photosynthesis, and *R* is respiration. Photosynthesis was estimated as a rectangular hyperbolic (Michaelis-Menten) function of light intensity:

$$P = (P_{max} \cdot I) / (\beta + I) \qquad (2)$$

where P_{max} is the maximum photosynthetic rate, β is the half-saturation coefficient, and *I* is light intensity. Data reported by *Howard-Williams and Vincent* [1989] were used to estimate parameter values for the major mat types in Canada Stream, one dominated by *Nostoc* and the other dominated by *Phormidium*. Respiration rates for mat communities were expressed as a function of standing biomass:

$$R = B \cdot \gamma \qquad (3)$$

TABLE 1. Values of Parameters Used in Stream Model

Parameter	Units	Value	Equation
P_{max}	$\mu gC\ cm^{-2}\ h^{-1}$ (*Nostoc*)	3.50	2
	$\mu gC\ cm^{-2}\ h^{-1}$ (*Phormidium*)	2.25	2
ß	$\mu mol\ m^{-2}\ s^{-1}$ (*Nostoc*)	110.0	2
	$\mu mol\ m^{-2}\ s^{-1}$ (*Phormidium*)	80.0	2
γ	$\mu gC\ cm^{-2}\ h^{-1}$ (*Nostoc*)	0.019	3
	$\mu gC\ cm^{-2}\ h^{-1}$ (*Phormidium*)	0.053	3
A	$\mu mol\ m^{-2}\ s^{-1}$	710.0	4
Ø	d	10.5	4
a_0	$\mu mol\ m^{-2}\ s^{-1}$	264.9	5
a_1	$\mu mol\ m^{-2}\ s^{-1}\ d^{-1}$	2.1978	5
a_2	$\mu mol\ m^{-2}\ s^{-1}\ d^{-2}$	2.7437×10^{-1}	5
a_3	$\mu mol\ m^{-2}\ s^{-1}\ d^{-3}$	1.6106×10^{-3}	5
a_4	$\mu mol\ m^{-2}\ s^{-1}\ d^{-4}$	2.3059×10^{-6}	5

where B is biomass and γ is a respiratory coefficient. *Vincent and Howard-Williams* [1986] report respiratory coefficients for mats in Canada Stream. Parameter values are listed in Table 1.

Sunlight Intensity

Hourly sunlight intensities were calculated to drive simulations [*Moorhead et al., in press*], based on meteorological data obtained from the McMurdo LTER project. Hourly sunlight intensity was estimated as a function of time:

$$I = A \cdot cos(2\pi \cdot (H + \text{Ø}) / 24) + S \qquad (4)$$

where H is the hour of the day, Ø is phase shift, A is amplitude, and S is a seasonal effect. The effect of season also was calculated as a function of time:

$$S = a_0 + a_1 \cdot D + a_2 \cdot D^2 + a_3 \cdot D^3 + a_4 \cdot D^4 \qquad (5)$$

where D is the day of year (1-365) and values of a were best-fit estimates (Table 1) based on measurements of incident photosynthetically active radiation (PAR) at Lake Hoare between November 1993 to November 1994.

Stream Geometry and Flow Characteristics

We assumed that the transverse cross-section of Canada Stream was triangular, with a base equal to 3 m, the average width of the wet zone described by *Howard-Williams et al.* [1989]. Maximum depth was set equal to 0.134 m (Figure 1), which provided a cross-sectional area ($0.20\ m^2$) equal to estimates based on depth measurements reported by *Alger et al.* [1996]. *Alger et al.* [1996]

reported stream discharge of ca. $0.054\ m^3\ s^{-1}$, similar to the average value of daily discharge reported by *Howard-Williams et al.* [0.06 $m^3\ s^{-1}$; 1989] during their study. Dividing discharge by cross-sectional area of the stream provided an estimated current velocity of about 0.3 m s^{-1} (1080 m h^{-1}). Thus it takes about 1.9 h for water to travel 2000 m from Canada Glacier to Lake Fryxell.

Our model is structured to follow a volume of water as it moves along the length of Canada Stream in 1 m increments. The volume of water in a 1 m segment of stream is approximately $0.20\ m^3$, given the transverse geometry in Figure 1. The amount of time required for this parcel of water to travel one meter is given by the inverse of the average current velocity (ca. 3.3 s). For the purposes of this study we assumed that stream geometry was constant. Differences in discharge were assumed to affect only current velocity (Table 2).

Alger et al. [1996] provided similar information for Green Creek, Delta Stream, and Von Guerard Stream (Table 3). For simulations, widths of these streams were set equal to 3 m, with depths calculated to match measured stream cross-sectional areas. Current velocities were estimated according to discharge (described above).

Microbial Mat Characteristics

Benthic microbial mats in streams of Taylor Valley are dominated by two groups of algae: *Nostoc* and *Phormidium*. Estimates of cover and standing stock used in simulations were taken from published values. For Canada Stream *Howard-Williams et al.* [1989] reported that *Nostoc* mats covered about 10% of the streambed area while *Phormidium* mats covered 32%. *Vincent and Howard-Williams* [1986] reported standing stocks of 28.1 and 7.9 mg C cm^{-2} for these *Nostoc* and *Phormidium* mats, respectively. For other Taylor Valley streams *Alger et al.* [1996] describe the cover of mat types at particular points [*McKnight et al., this volume*]. Table 4 presents an extrapolation of cover information for the lengths of Green Creek, Delta Stream, and Von Guerard Stream. Biomass carbon varied considerably between and within streams, so an average of values reported by *Alger et al.* [1996] was used in simulations, i.e., 36.7 and 46.2 mg ash-free dry weight cm^{-2} for *Nostoc* and *Phormidium* mats, respectively. We assumed that carbon represented 45% of the mass.

For simulations the total surface area of each mat type in each stream was calculated as the product of (1) total microbial mat cover in the stream, (2) fraction of cover for mat type, (3) stream length, and (4) stream width. Stream length and width were used in calculating total mat area (Figure 1). The standing stock of mat carbon in each stream was estimated as the product of the total areal cover and

TABLE 2. Discharge and Current Velocity for Simulations of
of Canada Stream Nitrogen Transformation

Discharge ($m^3 s^{-1}$)	Current Velocity ($m s^{-1}$)	Comments
0.010	0.049	Minimum [*Howard-Williams et al.*, 1989]
0.017	0.084	Average[*], 1993-5 [McMurdo LTER program]
0.060	0.296	Average [*Howard-Williams et al.*, 1989]
0.083	0.410	Three-day Average [*Alger et al.*, 1996]
0.140	0.691	Maximum [*Howard-Williams et al.*, 1989]

[*]average of values > 0

TABLE 3. Depth and discharge values for simulations of other Taylor Valley streams

Stream	Depth (m)	Discharge ($m^3 s^{-1}$)	Comments
Green	0.0413	0.007	Minimum Record [*Alger et al.*, 1996]
		0.014	Average[a], 1993-5 [McMurdo LTER program]
		0.038	Maximum Record [*Alger et al.*, 1996]
Delta	0.1281	0.011	Average[a], 1993-5 [McMurdo LTER program]
		0.038	Three-day Average [*Alger et al.*, 1996]
		0.046	Maximum Record [*Alger et al.*, 1996]
Von Guerard	0.0242	0.011	Three-day Average [*Alger et al.*, 1996]
		0.012	Average[a], 1993-5 [McMurdo LTER program]
		0.036	Maximum Record [*Alger et al.*, 1996]

[a]average of values > 0

TABLE 4. Biomass and Cover of Microbial Mats in Taylor Valley Streams

Stream	Length (m)	Cover (%)	Fraction of Cover:	
			Phormidium	*Nostoc*
Green	1200	100	0.8	0.2
Delta	5004	100	0.0	1.0
	6196	100	0.8	0.2
Von Guerard	1083	21	0.5	0.5
	1754	100	0.8	0.2
	877	50	0.8	0.2
	1186	0	0.8	0.2

mass per unit area of each mat type. We assumed that mat distributions were uniform within described sections of streams.

Nitrogen Transformation

Uptake of inorganic and urea nitrogen was assumed to balance net carbon fixed via primary production, to maintain constant N:C ratios (g:g) of mat types reported by *Howard-Williams et al.* [1989]:

$$U = r \cdot P \qquad (6)$$

where U is nitrogen uptake and r is the N:C ratio of *Phormidium* ($r = 0.051$) or *Nostoc* ($r = 0.064$) mats, respectively. Standing stocks of microbial mats were con-

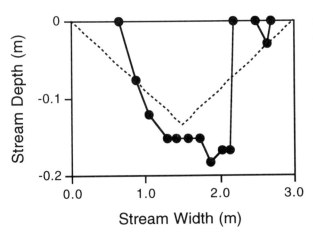

Fig. 1. Width and depth characteristics of Canada Stream. Dotted line: model approximations; solid line: observations from *Alger et al.* [1996].

sidered to remain constant (steady state assumption). In Canada Stream atmospheric nitrogen fixation was estimated to occur at a rate of 45.6 ng N mg⁻¹ mat N h⁻¹ for *Nostoc* mats, with fixed nitrogen added to the stream water [*Howard-Williams et al.*, 1989]. Losses and gains of nitrogen were assumed to occur within each parcel of water occupying a 1 m segment of stream as that parcel traveled the length of the stream. For simulations uptake rates were assumed to be the same for all forms of inorganic nitrogen and urea, proportional to concentrations in stream water. Similarly no differentiation was made among organic forms of nitrogen produced by stream mats. Concentrations of inorganic and urea nitrogen reported by *Howard-Williams et al.* [1989] were used in simulations of Canada Stream. For other Taylor Valley streams, no detailed measurements of various nitrogen compounds were available. Therefore simulations for other streams estimated only the potential for benthic mats to transform inorganic nitrogen and urea into other forms of organic nitrogen.

Simulations

Nitrogen transformations were calculated at various discharge for Canada Stream, Green Creek, Delta Stream, and Von Guerard Stream (Tables 2 and 3). Each simulation was started at the solar equivalent of 12:00 PM on January 1 with the temporal duration of the simulation period defined by the time required for a parcel of water to travel from the source to the mouth of each stream. This time period was determined by dividing stream length by the velocity of the water. Net primary production of microbial mats for each meter of stream length was estimated

for the time period over which a parcel of water traveled a distance of 1 m (Equations 1-3), given the sunlight intensity during the period (Equations 4-5). As previously discussed, nitrogen was transformed from inorganic and urea forms to other organic compounds in proportion to net primary production by assuming a constant N:C ratio of the microbial mat (Equation 6). This calculation was repeated for each 1 m segment of the entire length of each stream and assumes no resistence to uptake or release of nitrogen from mat communities.

RESULTS

Nitrogen transformation in Canada Stream was inversely proportional to discharge (Table 5). In part this was because changes in discharge were assumed to affect only the velocity of moving water, not the geometry of the stream. This is illustrated in Figure 2 where the total change in concentration of inorganic and urea nitrogen, over the entire length of Canada Stream, was plotted against the inverse of discharge (a measure of the amount of time that a parcel of water remains in contact with a particular segment of streambed). The same relationship also existed for the other streams (Table 5). The reason for this pattern was simple; nitrogen transformation was driven by net primary production, which is a rate process. There was less time for nitrogen interaction between a parcel of water and the benthic mats in a segment of stream when current velocity was high. Because it took less time for water to travel the length of a stream at high velocity, increasing the discharge reduced nitrogen transformation. In reality changes in stream geometry probably would accompany changes in discharge. For example increasing depth and width would tend to ameliorate the effects that increasing discharge would have on current velocity. However neither the cover nor biomass of microbial mats could change rapidly enough to track short-term fluctuations in stream flow (except for reductions due to sedimentation or scouring). Thus nitrogen transformation likely would decrease with increasing discharge although the decline may not be as rapid as simulated.

In Canada Stream, the observed decline of inorganic and urea nitrogen was greater than the increase in other forms of dissolved organic nitrogen (DON) [*Howard-Williams et al.*, 1989]; export of particulate organic nitrogen (PON) was about 11% of DON export. However the sum of DON and PON export still would account for only 67% of the total loss of inorganic and urea nitrogen. This suggests that losses of mineral and urea nitrogen from stream water exceeded gains of organic nitrogen, particularly in the upper 1000 m of stream (Figure 3). Model results con-

TABLE 5.   Simulated of Inorganic and Urea Nitrogen to Other Forms of Organic Nitrogen

Stream	Length (m)	Discharge (m³ s⁻¹)	N Transformation (mg N m⁻³)	N Transformation (μg N m⁻¹)
Canada	2000	0.010	130.59	65.3
		0.017	126.91	63.5
		0.060	123.42	61.7
		0.083	91.78	45.9
		0.140	63.33	31.7
Green	1200	0.007	113.16	94.3
		0.014	57.47	47.9
		0.038	21.15	17.6
Delta	11200	0.011	1007.39	89.9
		0.038	283.48	25.3
		0.046	241.53	21.6
Von Guerard	4900	0.011	154.27	31.5
		0.012	141.60	28.9
		0.036	47.11	9.6

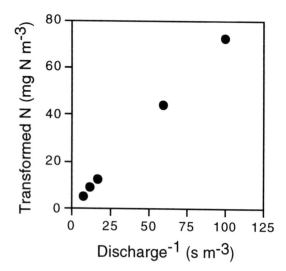

Fig. 2. Estimated quantities of nitrogen transformed from inorganic and urea nitrogen into other organic forms plotted against the reciprocal of discharge.

sistently underestimated the decline in inorganic nitrogen and urea (Figure 3a), although simulations and observations were similar at the lowest discharge (63 and 29 mg N m⁻³, respectively). In contrast the observed increase in organic nitrogen concentrations (54 mg N m⁻³) fell within values estimated by the two lowest discharge scenarios used in the model (44-73 mg N m⁻³).

Howard-Williams et al. [1989] found that concentrations of DON + PON actually declined by about 7 mg N m⁻³ in the first 1000 m of Canada Stream concurrent with a decline of about 74 mg N m⁻³ in mineral and urea nitrogen content. However DON + PON concentrations increased by 61 mg N m⁻³ in the lower half of the stream, almost twice the reduction in mineral and urea nitrogen concentration in this reach (34 mg N m⁻³). Nitrogen tranformation characteristics of Canada Stream varied with distance from Canada Glacier; the first 1000 m represented a nitrogen sink while the last 1000 m served as a nitrogen source (Figure 3a). In contrast our model held stream and mat characteristics constant over the entire length of the

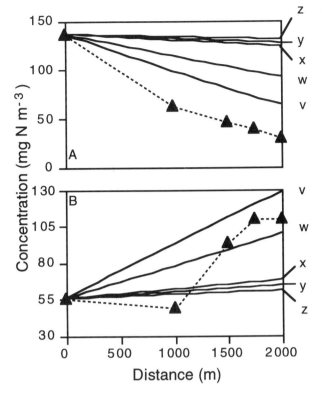

Fig. 3. Changes in concentrations of inorganic and organic nitrogen along the length of Canada Stream. Solid lines: model output; dotted lines: field observations. A : decline in inorganic + urea nitrogen concentration; B : increase in other forms of organic nitrogen. Simulated discharge is v = 0.010, w = 0.017, x = 0.060, y = 0.083, and z = 0.140 $m^3 s^{-1}$.

stream so simulated reductions in mineral and urea concentrations exactly matched increases in concentrations of organic nitrogen at all points (Figure 3). Differences between observations and model results suggest the likely importance of variations in mat distributions and streambed geometry along the length of Canada Stream. It also is likely that exchange of water between stream channel and hyporheic zones could alter concentrations of DIN and DON [Conovitz et al., this volume].

Fewer data were available to evaluate model output for Green Creek, Delta Stream, and Von Guerard Stream. In these simulations we simply estimated the total uptake potential of mineral and urea nitrogen along the length of each stream, given characteristics of discharge, mat distributions, and streambed geometry. Model results (Table 5) suggest that potential uptake was highest for Delta Stream (the longest stream examined) and lowest in Green Creek (the shortest stream). However Alger et al. [1996] reported similar nitrate concentrations (11–13 mg N m^{-3}) for all three streams at points near their outlets in Lake Fryxell, which may represent concentrations too low for significant removal via immobilization. Measures of mineral nitrogen in glacial meltwater feeding these streams, or concentrations of DON or PON leaving these streams, would help to elucidate nitrogen transformations. Such data are not available, but simulations suggest that dry valley streams have a substantial capacity to utilize mineral nitrogen.

DISCUSSION

Nutrient dynamics in flowing water has at least one feature lacking in other ecosystems: flowing water is a moving medium in which nutrients are physically transported in particulate and soluble forms while biogeochemical processes of nutrient cycling occur along the flow path [Newbold et al., 1982; 1983; Maltchik et al., 1996]. This adds a linear dimension to the nutrient cycle, defining an idealized spiral [Webster and Patten, 1979]. Given the nitrogen characteristics of Canada Stream used in the current study [Howard-Williams et al., 1989], biological processes should account for some of the changes in concentrations of mineral and organic forms of nitrogen along the stream. However dry valley streams have some unusual features that may affect nutrient spiralling.

The bulk of research on nutrient spiralling has focused on permanent streams in temperate regions where invertebrate grazers have substantial impacts on periphyton communities and nutrient dynamics [Newbold et al., 1982; 1983; Mulholland et al, 1991; 1994]. However cyanobacterial mats seldom exist in the presence of significant grazing [Stal, 1995] and dry valley streams appear to be devoid of grazers. Newbold et al. [1983] and Mulholland et al. [1985] also noted a dominant role of coarse particulate organic matter in nutrient cycling within a woodland stream in Tennessee, USA (Walker Branch). This material was of detrital origin but dry valley streams receive no allochthonous inputs of organic matter [McKnight et al., 1993]. Thus some of the controls on nutrient cycling in streams of other regions are not present in the dry valleys.

Other factors affecting nutrient spiralling include abiotic retention in stream sediments [Meyer, 1979] and water velocity [Bencala, 1983]. Our simulations indicated an inverse relationship between discharge and likely loss of mineral and urea nitrogen from water in Canada Stream (Figure 3a). This is consistent with an inverse relationship between discharge and inorganic nitrogen concentrations in Canada Stream observed by Howard-Williams et al. [1989] and observed in another Antarctic stream by

Hawes and Brazier [1991]. *Bencala* [1983] noted a similar pattern for a mountain stream in Colorado, USA, where more solute (strontium) was lost from water moving at low velocity along a stream reach than at high velocity. Whether solutes are lost from water due to chemical reactions with streambed sediments or via uptake by biota, greater opportunity for such losses exist in a given stream reach at lower current velocities [*Bencala,* 1983].

Meyer [1979] noted that the phosphorus dynamics of a headwater stream of the Hubbard Brook watershed (New Hampshire, USA) was dominated by interactions with sediments. The role of abiotic processes in the nitrogen dynamics of Canada Stream is uncertain, but studies reveal the existence of substantial hyporheic zones underlying dry valley streams [*Conovitz et al.,* this volume] in which solutes could interact with streambed materials. Some of the fluctuations in stream solute concentrations in Antarctic streams also can be related to other physical factors. For example, early season glacial melt would be expected to contain high levels of materials deposited on glacial surfaces over winter, additionally concentrated by ablation of the ice surface. Diurnal or short-term reductions in melting would tend to freeze-concentrate solutes in remaining water [*Howard-Williams et al.,* 1989; *Vincent and Howard-Williams,* 1986]. However increases in concentrations of organic forms of nitrogen in Canada Stream, in conjunction with decreases in mineral and urea nitrogen concentrations, suggest that biological mechanisms could account for much of the nitrogen dynamics observed for this stream.

Microorganisms also play substantial roles in nutrient spiralling [*Elwood et al.,* 1981]. For example, *Mulholland et al.* [1991; 1994; 1995] examined interactions between periphyton communities and nutrient cycling in a set of artifical streams in the absence of grazers. Decreases in streamwater nitrogen and phosphorus concentrations along the length of these streams accompanied reductions in chlorophyll-*a* and increases in densities of cyanobacteria. No differences in standing biomass, gross primary productivity, or respiration existed along streams despite declining rates of nitrogen and phosphorus uptake per unit primary productivity at downstream locations [*Mulholland et al.,* 1995]. The most conclusive demonstrations of microbial impacts on nitrogen dynamics in Antarctic streams are provided by the studies of *Howard-Williams et al.* [1986; 1989] and *Davey* [1993]. *Howard-Williams et al.* [1986] found that nitrate concentrations in several dry valley streams were lower at downstream locations than near their sources. In particular, nitrate concentrations in water of the Onyx River fell from 47.4 to 3.1 mg N m^{-3} as it passed through a boulder pavement area supporting a high biomass of microbial mat. This is similar to patterns reported by *Howard-Williams et al.* [1989] in which nitrate concentrations of water declined with distance in streams containing microbial mats. However only *Davey* [1993] has demonstrated that rates of dissolved inorganic nitrogen uptake were consistent with the observed growth rates of mats. His carbon and nitrogen budgets for microbial mats in Antarctic streams demonstrated a functional linkage, corroborating the suggestion made by *Hawes and Brazier* [1991] that nitrogen uptake in streams was driven by photosynthetic activity of microbial mats.

The work by *Davey* [1993] formed the conceptual basis of the present study, but concentrations of mineral and urea nitrogen in Canada Stream declined more rapidly at upstream locations (< 1000 m) than could be explained by simulated or observed production of other organic forms (Figures 3a, b). In contrast production of organic forms of nitrogen at downstream locations (> 1000 m; Figure 3b) exceeded uptake of mineral and urea forms in this reach (Figure 3a). The exchange of solutes between channel and hyporheic volumes of water could alter apparent spatial and temporal relationships between uptake and production of nitrogenous compounds. Losses of nitrogen from channel water could occur at upstream locations due to mixing with hyporheic water. Subsurface movement of hyporheic water also could provide a source of nitrogen to downstream communities not observed by *Howard-Williams et al.* [1989]. Finally the steady state assumption of constant mat biomass used in simulations may not be correct. A net growth in upstream communities and net reduction in downstream communities could account for some of the discrepancies between simulations and observations (Figures 3a, b).

Another assumption included in this modeling study was that nitrogen uptake from stream water was driven by photosynthetic carbon fixation with no provision for cycling of nitrogen within mats. However internal cycling appears to be important in some cases [*Hawes et al.,* 1993; *Mulholland et al.,* 1995] and may be enhanced by structural features of cyanobacterial mats [*Stal,* 1995]. Mats usually have a surface layer of sand, silt, or dense pigmentation [*Doemel and Brock,* 1977; *Vincent et al.,* 1993a; 1993b; *Stal,* 1995], embedded within a consolidated, mucilage matrix. Beneath this layer exists a complex community of filamentous cyanobacteria (e.g., *Oscillitoria*), photosynthetic bacteria (e.g., *Chloroflexus*), chemoautotrophic and heterotrophic bacteria [*Stal,* 1995]. Interstitial concentrations of oxygen, nitrogen, and phosphorus can be much higher within mats than in overlying water, demonstrating resistence to free exchange [*Revsbech and*

Ward, 1984; *Hawes et al.*, 1993; *Vincent et al.*, 1993a; *Stal*, 1995]. Large temporal fluctuations in oxygen concentrations and pH can occur within mats and abrupt changes in chemical characteristics also occur over small spatial distances [*Revsbech and Ward*, 1984; *Vincent et al.*, 1993a; 1993b; *Stal*, 1995]. For these reasons nitrogen supporting primary production of benthic mats in Canada Stream could be supplied partly from internal cycling. However our simulations underestimated losses of mineral nitrogen from stream water (Figure 3a) and including a mechanism for internal cycling would further reduce the requirement for uptake from stream water to meet nitrogen demand.

Further insights to the functioning of cyanobacterial mats in dry valley streams may be gained by comparisons to those of geothermal streams, which have been the focus of detailed studies for several decades [e.g, *Brock and Brock*, 1966; 1967; 1969a; 1969b; *Winterbourne*, 1969; *Castenholz*, 1976; 1984; *Forsyth*, 1977]. Microbial mats in these two ecosystems are similar in that both are dominated by bacteria and cyanobacteria, and have few grazers. Studies of mats in thermal springs of Yellowstone National Park, USA [*Brock*, 1967a; 1967b; *Brock and Brock*, 1969a; 1969b; *Wiegert and Fraleigh*, 1972], led to a modeling synthesis of community succession [*Fraleigh and Wiegert*, 1975]. One of the more interesting results of the study was that rates of gross production of geothermal mats were directly related to day length but relatively independent of light intensity. The same relationship has been observed for mats in Antarctic streams [*Moorhead et al.*, in press] and appears to result from saturation of photosynthesis at very low light intensities. Thus the duration of sunlight has a stronger influence on primary production than the intensity of sunlight above the low saturation threshold that is characteristic of cyanobacterial mats (ca. $\leq 200 \ \mu mol \ m^{-2} \ s^{-1}$).

There may be several reasons for such low rates of gross primary production in cyanobacterial mats, including self-shading or limitations imposed by lack of nutrients. *Fraleigh and Wiegert* [1975] noted that rates of gross primary production were directly related to chlorophyll-*a* content of mats, but they also noted that rates of photosynthesis per unit of chlorophyll-*a* declined as standing stocks of chlorophyll-*a* increased. They concluded that self-shading was an important factor limiting rates of primary production. In contrast to geothermal mats there appears to be little relationship between rates of carbon fixation and chlorophyll content of Antarctic mats [*Vincent and Howard-Williams*, 1989], but light intensities within these mats are reduced by surface debris, pigmentation, and dense accumulations of chlorophyll [*Vincent et al.*, 1993a; 1993b].

The role of nutrient limitation in mat productivity is unclear but *Vincent and Howard-Williams* [1989] found little evidence of nutrient limitations to Antarctic mats. *Mulholland et al.* [1995] reported considerable flexibility in the ability of periphyton communities in temperate streams to utilize external and internal sources of nutrients, and found that internal cycling was more important for communities having higher cyanobacterial densities. *Hawes and Brazier* [1991] noted that biomass of microbial mats in an Antarctic stream actually increased downstream, concurrent with decreasing concentrations of total dissolved nitrogen, and *Hawes et al.* [1993] suggested strong internal cycling of nitrogen in cyanobacterial mats in Antarctic ponds. In hot spring communities *Doemel and Brock* [1977] suggested that in situ biomass turnover and mineralization approximated rates of primary production, suggesting a balance between mineralization and immobilization of nutrients. *Wiegert and Fraleigh* [1972] concluded that CO_2 concentrations were limiting productivity of mats in a geothermal stream, but other elements were not limiting. Finally many species of heterocyst and nonheterocyst cyanobacteria fix nitrogen in microbial mats [*Stal*, 1995] and we included fixation rates reported for *Nostoc* mats [*Howard-Williams et al.*, 1989] in our model of nitrogen dynamics of Canada Stream. However levels of mineral and urea nitrogen in Canada Stream used in the present study suggest that nitrogen is not a limiting nutrient.

In summary the nitrogen dynamics of Antarctic streams vary widely and are influenced by many factors. A recent review reported dissolved inorganic nitrogen concentrations as low as 2 mg N m^{-3} and as high as 520 mg N m^{-3} in Antarctic streams [*Vincent et al.*, 1993c]. Sources of inorganic nitrogen include aerial deposition, soil mineral leaching, erosion of ancient lake sediments, and inputs from penguin rookeries [*Howard-Williams et al.*, 1986; *Vincent et al.*, 1993c]. In the stream, other factors influence concentrations of nitrogen, including freeze-concentration, current velocity, and interactions with sediments or biota. Cyanobacteria dominate many of the microbial communities of Antarctic streams and apparently play significant roles in nitrogen dynamics. Environmental conditions permit substantial accumulation of mat bio-mass which immobilizes nitrogen from passing water [*Davey*, 1993], fixes nitrogen from the atmosphere [*Howard-Williams et al.*, 1989], and cycles nitrogen internally [*Hawes et al.*, 1993]. Insights gained from detailed examinations of the internal structural and functional features of cyanobacterial mats found in geothermal streams suggest the need for expanding investigations of similar Antarctic communities [*Vincent et al.*, 1993a;

1993b]. As research within the McMurdo LTER program continues to elucidate the distribution, composition, biomass, and productivities of mat communities in Taylor Valley streams, in conjunction with monitoring hydrologic discharge and nutrient concentrations in stream outflow, glacial melt and exchange between channel and hyporheic volumes, we will be able to better predict the role of microbial mats in the nitrogen dynamics of Antarctic streams.

Acknowledgments. We wish to thank I. Hawes, C. Howard-Williams, and A.-M. Schwarz for their insights and enthusiastic support. Meterological data were provided by the USA McMurdo Long-Term Ecological Research program. Funding for this work was provided by National Science Foundation Office of Polar Programs research grant OPP-9211773.

REFERENCES

Alger, A. S., D. M. McKnight, S. A. Spaulding, C. M. Tate, G. H. Shupe, K. A. Welch, R. Edwards, E. D. Andrews, and H. R. House, Ecological processes in a cold desert ecosystem: The abundance and species distribution of algal mats in glacial meltwater streams in Taylor Valley, Antarctica. 102 pp., *United States Geological Survey.* Boulder, Colorado, 1996.

Bencala, K. E., Simulation of solute transport in a mountain pool-and-riffle stream with a kinetic mass transfer model for sorption. *Wat. Resour. Res.*, 19, 732–738, 1983.

Brock, T. D., Micro-organisms adapted to high temperatures. *Nature*, 214, 882–885, 1967a.

Brock, T. D., Relationship between standing crop and primary productivity along a hot spring thermal gradient. *Ecology*, 48, 566–571, 1967b.

Brock, T. D., and M. L. Brock, The measurement of chlorophyll, primary productivity, photophos-phorylation, and macromolecules in benthic algal mats. *Limnol. Oceanogr.*, 12, 600–605, 1967.

Brock, T. D., and M. L. Brock, Effect of light intensity on photosynthesis by thermal algae adapted to natural and reduced sunlight. *Limnol. Oceanogr.*, 14, 334–341, 1969a.

Brock, T. D., and M. L. Brock, Recovery of a hot spring community from a catastrophe. *J. Phycol.*, 5, 75–77, 1969b.

Castenholz, R. W., The effect of sulphide on the blue green algae of hot springs. I. New Zealand and Iceland. *J. Phycol.*, 12, 54–68, 1976.

Castenholz, R. W., Composition of hot spring microbial mats, in *Microbial mats: Stromatolites*, edited by Y. Cohen, R. W. Castenholz and H. O. Halvorson, pp. 101–119, Alan R. Liss, Inc., 1984.

Clow, G. D., C. P. McKay, G. M. Simmons, Jr., and R. A. Wharton, Jr., Climatological observations and predicted sublimation rates at Lake Hoare, Antarctica. *J. Clim.*, 1, 715–728, 1988.

Conovitz, P. A., D. M. McKnight, L. M. McDonald, A. Fountain, and H. R. House, Hydrologic processes influencing streamflow variation in Fryxell Basin, Antarctica, this volume.

Davey, M. C., Carbon and nitrogen dynamics in a maritime Antarctic stream. *Freshwat. Biol.*, 30, 319–330, 1993.

Doemel, W. N., and T. D. Brock, Structure, growth, and decomposition of laminated algal-bacterial mats in alkaline hot springs. *Appl. Environ. Microbiol.*, 34, 433–452, 1977.

Elwood, J. W., J. D. Newbold, A. F. Trimble, and R. W. Stark, The limiting role of phosphorus in a woodland stream ecosystem: Effects of P enrichment on leaf decomposition and primary producers. *Ecology*, 62, 146–158, 1981.

Fraleigh, P. C., and R. G. Wiegert, A model explaining successional change in standing crop of thermal blue-green algae. *Ecology*, 56, 656–664, 1975.

Forsyth, D. J., Limnology of Lake Rotokawa and its outlet stream. *NZ J. Mar. Freshwat. Res.,* 11, 525–539, 1977.

Hawes, I., Photosynthesis in thick cyanobacterial films: a comparison of annual and perennial Antarctic mat communities. *Hydrobiol.*, 252, 203–209, 1993.

Hawes, I., and P. Brazier, Freshwater stream ecosystems of James Ross Island, Antarctica. *Ant. Sci.*, 3, 265–271, 1991.

Hawes, I., C. Howard-Williams, and R. D. Pridmore, Environmental control of microbial biomass in the ponds of the McMurdo Ice Shelf, Antarctica. *Arch. Hydrobiol.*, 127, 271–287, 1993.

Howard-Williams, C., J. C. Priscu, and W. F. Vincent, Nitrogen dynamics in two antarctic systems. *Hydrobiol.*, 172, 51–61, 1989.

Howard-Williams, C., C. L. Vincent, P. A. Broady, and W. F. Vincent, Antarctic stream ecosystems: Variability in environmental properties and algal community structure. *Int. Rev. Gesamten. Hydrobiol.*, 71, 511–544, 1986.

Howard-Williams, C., and W. F. Vincent, Microbial communities in southern Victoria Land streams (Antarctica) I. Photosynthesis. *Hydrobiol.*, 172, 27–38, 1989.

Keys, J. R, Air temperature, wind, precipitation, and atmospheric humidity in the McMurdo region. 57 pp., Dept. of Geology Publ. 17 (Antarctic Data Ser. No. 9), Victoria University of Wellington, New Zealand, 1980.

Maltchik, L., S. Mollá, C. Montes, and C. Casado, Measurement of nutrient spiralling during a period of continuous surface flow in a mediterranean temporary stream (Arroyo de La Montesina, Spain). *Hydrobiol.*, 335, 133–139, 1996.

McKnight, D. M., G. R. Aiken, E. D. Andrews, E. C. Bowles, and R. A. Harnish, Dissolved organic material in dry valley lakes: A comparison of Lake Fryxell, Lake Hoare, and Lake Vanda. Am. Geophys. Union, Ant. Res. Series, 59, 119-133, 1993.

McKnight, D. M., A. Alger, C. Tate, G. Shupe, and S. Spalding, Longitudinal patterns in algal abundance and species distribution in meltwater streams in Taylor Valley, southern Victoria Land, Antarctica, this volume.

Meyer, J. L., The role of sediments and bryophytes in phosphorus dynamics in a headwater stream ecosystem. *Limnol. Oceanogr.*, 24, 365–375, 1979.

Moorhead, D. L., W. S. Davis, and R. A. Wharton, Jr., Carbon dynamics of aquatic microbial mats in the antarctic dry valleys: A modelling synthesis. *Proc. of the Polar Desert Confr.* Christchurch, New Zealand. Balkema Publications, in press.

Mulholland, P. J., E. R. Marzolf, S. P. Hendricks, R. V. Wilkerson, and A. K. Baybayan, Longitudinal patterns of nutrient cycling and periphyton characteristics in streams: a test of upstream-downstream linkage. *J. N. Am. Benthol. Soc.*, 14, 357–370, 1995.

Mulholland, P. J., J. D. Newbold, J. W. Elwood, L. A. Ferren, and J. R. Webster, Phosphorus spiralling in a woodland stream: Seasonal variations. *Ecology*, 66, 1012–1023, 1985.

Mulholland, P. J., A. D. Steinman, E. R. Marzolf, D. R. Hart, and D. L. DeAngelis, Effect of periphyton biomass on hydraulic characteristics and nutrient cycling in streams. *Oecologia*, 98, 40–47, 1994.

Muholland, P. J., A. D. Steinman, A. V. Palumbo, and J. W. Elwood, Role of nutrient cycling and herbivory in regulating periphyton communities in laboratory streams. *Ecology*, 72, 966-982, 1991.

Newbold, J. D., J. W. Elwood, R. V. O'Neill, and A. L. Sheldon. Phosphorus dynamics in a woodland stream ecosystem: A study of nutrient spiralling. *Ecology*, 64, 1249–1265, 1983.

Newbold, J. D., R. V. O'Neill, J. W. Elwood, and W. Van Winkle, Nutrient spiralling in streams: Implications for nutrient limitation and invertebrate activity. *Am. Nat.*, 120, 628–652, 1982.

Revsbech, N. P., and D. M. Ward, Microelectrode studies of interstitial water chemistry and photosynthetic activity in a hot spring microbial mat. *Appl. Environ. Microbiol.*, 48, 270–275, 1984.

Stal, L. J., Physiological ecology of cyanobacteria in microbial mats and other communities. Tansley Review No. 84. *New Phytol.*, 131, 1–32, 1995.

Vincent, W. F., *Microbial ecosystems of Antarctica.* 304 pp., Cambridge University Press, New York, 1988.

Vincent, W. F., R. W. Castenholz, M. T. Downes, and C. Howard-Williams, Antarctic cyanobacteria: Light, nutrients, and photosynthesis in the microbial mat environment. *J. Phycol.*, 29, 745–755, 1993a.

Vincent, W. F., M. T. Downes, R. W. Castenholz, and C. Howard-Williams, Community structure and pigment organisation of cyanobacteria-dominated microbial mats in Antarctica. *Eur. J. Phycol.*, 28, 213–221, 1993b.

Vincent, W. F., C. Howard-Williams, and P. A. Broady, Microbial communities and processes in Antarctic flowing waters, in *Antarctic Microbiology*, edited by E. I. Friedmann, pp. 543–569, Wiley-Liss, Inc., 1993c.

Vincent, W. F., and C. Howard-Williams, Antarctic streams ecosystems: Physiological ecology of a blue-green algal epilithon. *Freshwat. Biol.*, 16, 219–233, 1986.

Vincent, W. F., and C. Howard-Williams. Microbial communities in southern Victoria Land streams (Antarctica) II. The effects of low temperature. *Hydrobiol.*, 172, 39–49, 1989.

Webster, J. R, and B. C. Patten, Effects of watershed perturbation on stream potassium and calcium dynamics. *Ecol. Monogr.*, 49, 51–72, 1979.

Winterbourne, M. J., The distribution of algae and insects in hot spring thermal gradients at Waimangu, New Zealand. *NZ J. Mar. Freshwat. Res.*, 3, 459–465, 1969.

Wiegert, R. G., and P. C. Fraleigh, Ecology of Yellowstone thermal effluent systems: Net primary production and species diversity of a successional blue-green algal mat. *Limnol. Oceanogr.*, 17, 215–228, 1972.

Diane M. McKnight, Civil, Environmental, and Architectural Engineering Department, INSTAAR, 1560 30th St., Boulder, Colorado 80309

Daryl L. Moorhead, Department of Biological Sciences, Texas Tech University, Lubbock, Texas 79409-3131

Cathy M. Tate, United States Geological Survey, Water Resources Division, PO Box 25046, Denver, Colorado, 80225-0046

(Received September 13, 1996; accepted April 14, 1997)

PHYSICAL LIMNOLOGY OF THE MCMURDO DRY VALLEYS LAKES

Robert H. Spigel

University of Canterbury, Christchurch, New Zealand

John C. Priscu

Department of Biological Sciences, Montana State University, Bozeman, Montana

We present high-resolution measurements of conductivity and temperature made from January 1990 to December 1993 in the east and west lobes of Lake Bonney and in Lakes Vanda, Hoare, Fryxell, Joyce, and Miers. These measurements were used to calculate profiles of density and stability, and thereby infer mechanisms and strengths of mixing in the water columns of the lakes. Transects along the length of Lake Bonney allowed estimates of horizontal exchanges in and between the two lobes of that lake and help to explain some of the characteristics of single profiles measured in other lakes. Stratification in all the lakes is controlled mainly by concentration of dissolved solids ("salinity"), with temperature exerting such a minor influence as to act virtually as a passive tracer. An exception is in the upper two-thirds of Lake Vanda and at the bottom of Lake Miers, where solar heating in the presence of weak salinity gradients gives rise to thermohaline convection. The distinctive and relatively invariant shapes of the density profiles in the different lakes is due to distinctive distributions of salts in the water columns of these lakes, distributions that can only be explained in terms of geochemical processes acting over time scales much longer than the annual overturning cycle that dominates patterns of stratification and mixing in temperate, freshwater lakes. Temperatures in the McMurdo Dry Valleys lakes, in contrast to salinities, do respond to changes in weather, climate, and water levels on a seasonal and annual basis, although to a much smaller extent than in temperate lakes. Stability reaches extremely high levels in the chemoclines of the two lobes of Lake Bonney, being slightly lower in the bottom waters of Lake Vanda. Stabilities in Lakes Fryxell and Joyce, although still very high in comparison with freshwater lakes, are much lower than in Bonney and Vanda. Maximum stabilities in Lakes Hoare and Miers are similar to those found in the summer thermoclines of freshwater lakes. With the exception of thermohaline convection cells in Lake Vanda and Lake Miers, our measurements do not support the presence of turbulent diffusion in the main bodies of the lakes; however, profiles did document mechanically generated turbulence just below the ice in Lake Miers (probably associated with the meltwater stream through-flow in that lake, the only lake with a stream outlet) and much weaker turbulence in the narrows connecting the two lobes of Lake Bonney (probably associated with the exchange flows between these basins).

INTRODUCTION

The quiet waters of ice-covered lakes provide a refuge for life in the harsh surroundings of the McMurdo Dry Valleys, and thus form an important part of the dry valleys ecosystem. The thermal and salinity characteristics of these lakes are so unusual and varied that they have puzzled and fascinated scientists ever since the first extensive measurements were made in the early 1960s. Our goal in this chapter is to combine what has been learned from previous studies about the physical limnology of these lakes with results from high-resolution conductivity and temperature measurements that we made recently in the east and west lobes of Lake Bonney, and in Lakes Vanda, Fryxell, Hoare, Joyce, and Miers. Figure 1

153

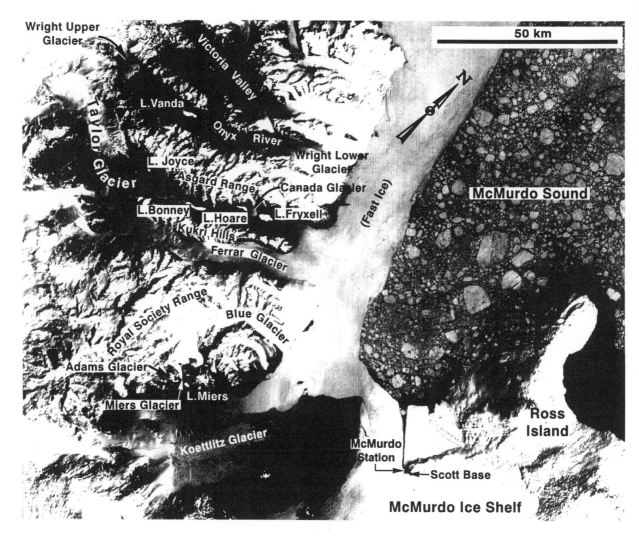

Fig. 1. Annotated satellite photo of the McMurdo Sound region; the dry valleys are in the left half of the picture. The lakes have been highlighted to appear white against the dark background of the valleys. The illustration is based on a Landsat 1 MSS digitally enhanced composite image (NASA image 1174-19433, bands 4, 5, and 7; Jan. 13, 1973; path 56, row 116), supplied by Landcare Research, New Zealand.

shows the locations of the lakes in southern Victoria Land, in the ice-free valleys between the peaks of the Transantarctic Mountain range and the coastline of McMurdo Sound. The recent conductivity and temperature measurements are used to infer patterns of water column stability and, in some cases, mixing and circulation. It has been hypothesized that the lakes of the dry valleys provide a highly stable environmnent for the microorganisms that inhabit them, largely free of turbulent mixing and basin-scale stirring that characterizes ice-free lakes; this is a consideration of major importance for lake biology [*Vincent*, 1981; *Hawes*, 1983]. The main reason for making high-

resolution measurements was to check the stability hypothesis, i.e., to determine whether turbulence does exist in these lakes, and, if so, in what strengths and with what effects on mixing. Many of our measurements were originally made to support an investigation of plankton ecology in Lake Bonney; as a result more examples from Lake Bonney appear in this paper than from the other lakes. But we present data on all major lakes to illustrate the diversity (and common themes) of the physical limnology in these systems. (Optical properties of the lakes are not discussed in detail here; they are treated elsewhere in this volume by *Howard-Williams et al.*)

One common theme is density stratification, present to a greater (extreme, in the pycnocline of the west lobe of Lake Bonney) or lesser (almost absent in Lake Hoare, but still there) degree in all of the lakes, and the importance of stratification in controlling the transport of nutrients from enriched and sometimes anoxic bottom waters to higher elevations, where the nutrients are accessible to photosynthesizing plankton. The insulation provided by 3 to 5 meters of ice covering almost all of the water surface year-round precludes turbulent mixing by wind or by convective overturning due to loss of heat from the water directly to the atmosphere. Heat is lost through the ice from 0°C water at the base of the ice when air temperatures remain less than 0°C for an extended period of time, but this heat loss occurs via molecular conduction through the ice, and the result of the heat loss is not convective overturn of the water but freezing onto the underside of the ice cover. Also, some water is exposed to wind during summer months when a moat melts around the edges of the lakes, but the relative surface area thus exposed to wind, compared to the entire lake surface area, is probably on the order of 1–3% for most of the lakes. The mixing efficiency of wind blowing over water surfaces with severely limited fetches is itself quite small, on the order of 1% [Wu, 1973; Richman and Garrett, 1977], so that the fraction of wind energy that is eventually available to directly mix the stratified water column in the lakes is negligible.

The moats that form in summer around the edges of the lakes do play a role in transferring substances from the land boundaries of the lakes, where they are introduced in surface waters by meltwater runoff from glaciers, to the interiors of the lakes. The moats provide one example of a way that very localized turbulence at the boundary of a lake can generate laminar (non-turbulent) water movements in the main body of that lake. Turbulence at the lake's edge causes localized mixing, which in turn leads to an imbalance in the buoyancy field, creating pressure gradients to drive motions that eventually redistribute mass. The most important effects of turbulence may therefore not be felt where the turbulence is generated. Because the water columns of most of the dry valleys lakes are so stably stratified, motions in the main bodies of the lakes are predominantly horizontal with very limited vertical extent. These motions lead to intrusions, interleaving, and layering so characteristic of density stratified systems, and to an "apparent horizontal diffusion" that is orders of magnitude more efficient than molecular diffusion. The limited vertical extent of these motions

contrasts with observations in some freshwater ice-covered Arctic lakes, where shore-bound density currents, generated by release of heat from sediments, may keep the entire water mass in weak but continuous circulation in winter [Mortimer and Mackereth, 1958; Welch and Bergmann, 1985]. The currents generated in the dry valleys lakes are probably even slower and weaker than such freshwater counterparts, and mixing times have more in common with geological time scales than the seasonal cycle of stratification and overturn that renews the water column annually in most other lakes.

Another example of localized mixing is the generation of weak density currents during moat freezing. This occurs when enough salt is rejected from freezing moat ice to form a saline surface layer of water just under the ice that is denser than the fresher water below it, and therefore sinks. Miller and Aiken [1996] recently suggested that such a process could explain the source of a lens of relatively "young" water at depth in the interior of Lake Fryxell. Miller and Aiken also pointed out that horizontal redistribution of surface water on the scale of the entire lake basin can be caused by freezing and melting processes on the underside of the lake's permanent ice cover. Still another example of localized mixing is the melting of glaciers that are in direct contact with the water column, with ice faces extending over depths of several meters, submerged in relatively "warm" (i.e., temperatures greater than 0°C) lake water, as in Lake Bonney's west lobe, Lake Hoare, and Lake Joyce. The meltwater plumes that develop along such near-vertical ice faces may be laminar or weakly turbulent [Huppert and Josberger, 1980]. Other examples include episodic strong meltwater inflows just below the ice-cover during periods of warm weather; stream throughflow under the ice (this occurs in Lake Miers, the only lake of the seven considered here that has a stream outflow); mixing of water in contact with the rough sides and bottom when lake water is forced, for example, through the narrow, shallow passage that connects the east and west lobes of Lake Bonney, or back and forth over a lake's boundary in response to movement of the ice-cover when it shifts under strong winds within the confines of a moat (Priscu, personal observation).

It would be misleading to give the impression that analogous processes do not occur in "normal," ice-free lakes. Imberger [1985a] and Imberger and Patterson [1990] have pointed out that in most small to medium-sized lakes turbulence below the thermocline occurs only at boundaries and in isolated patches in the interior; most of the flows in the interior are laminar and the diffusion molecular. Advection by buoyancy driven flows, combined with diffusion, together give rise to an apparent vertical eddy diffusion that may be 10 to 100 times molecular. In ice-free lakes, energy to generate localized mixing comes mainly from the wind and plunging river inflows [*Imberger and Patterson*, 1990; *Imberger*, 1979], sources all but completely absent in the ice-covered lakes of the dry valleys. Hence, while apparent vertical mixing rates in the ice-covered lakes may be somewhat higher than molecular, they cannot be 10 or 100 time higher. There is an exception; mixing is probably turbulent inside thermohaline convection cells that are found over much of the water column of Lake Vanda, and to a much more limited extent in Lake Miers. However, even in these lakes, the overall "effective eddy diffusivity" is limited by molecular diffusion in the regions that separate the cells.

In addition to their year-round ice covers, the lakes of the dry valleys feature distinctive salinity gradients that control density stratification; concentrations of total dissolved solids increase with depth in all of the lakes. Maximum salinities range from almost a saturated brine at the bottom of Lake Bonney's east lobe, to within an order of magnitude of values thought to be typical for freshwaters, as in Lake Hoare, to just above the typical freshwater salinity range, as in Lake Miers. (*Hutchinson* [1957, pp. 552–563] gives a range 0.01 to 0.25 grams liter^{-1} total dissolved solids concentration for most freshwater lakes and rivers, with "average" values around 0.1 grams liter^{-1}.) In most of the lakes, temperature plays a secondary role in water column stability, both because the temperature range is so small (except in Vanda), and because the value of thermal compressibility, the rate at which density changes with temperature, is so small near 0°C, and equals zero at 4°C for pure water at atmospheric pressure. In fact temperature is so "passive" in most of the lakes that it can be used as a tracer to identify water movements; this is the reverse of the situation in freshwater lakes, where stratification is dominated by temperature effects, and slight variations in dissolved solids concentrations can sometimes be used to track water motions. The role of temperature is further confused in the dry valleys lakes because of the occurrence of a density maximum in water at a temperature just above the freezing point, depending on salinity and pressure. This effect disappears in salty water; for example, in seawater with salinities greater than around 24, density increases monotonically to the freezing point as temperature drops. Empirical formulae that quantify this behavior are given by *Caldwell* [1978] for the temperature of maximum density, and *Millero and Leung* [1976] for the freezing point of seawater [quoted by *Fofonoff and Millard*, 1983]. The temperature range in several of the dry valleys lakes overlaps the point of maximum density, so that over part of the water column density increases with increasing temperature, while over the remaining part density decreases with increasing temperature. In all of the lakes temperatures increase with depth over at least part of the water column, sometimes to the bottom. But in almost all cases, salinity, not temperature, determines stability.

We have used the term "salinity" in a sense that implies equivalence with total dissolved solids concentration, but have also had occasion to refer to salinity in an oceanographic sense as defined for seawater. As described in more detail in the section on methods, we have found the use of oceanographic salinity to be helpful, partly because of the now widespread use of the Practical Salinity Scale 1978 (PSS78) for converting conductivity, temperature, and pressure measurements to salinity [*Lewis*, 1980] and then to density via the International Equation of State for Seawater [*Millero et al.*, 1980; *Millero and Poisson*, 1981; *Fofonoff and Millard*, 1983]. Although strictly applicable only to standard seawater with salinity less than 42, the equation and its predecessors have been successfully used to assess stability effects in laboratory and field studies of mixing, and can be adapted to include effects of higher concentrations and varying ionic composition [e.g., *Millero et al.*, 1976; *Spigel and Priscu*, 1996]. We will address these issues further in the section on methods, but for now it is sufficient to note that by "salinity" we mean salinity in the oceanographic sense (PSS78); but note that for salinities less than about 50, PSS78 salinity can often be used as a reasonable estimate for total dissolved solids concentration (TDS, grams per liter). When TDS exceeds 50 grams liter^{-1}, we have found that PSS78 predictions for TDS are too low.

STUDY SITES, AND SUMMARY AND RELEVANCE OF EARLIER WORK

Locations of the lakes are shown in Figure 1; Figure 2 presents a comparison of basin outlines (with depth contours where available) drawn to the same horizontal scale. Comparative information on lake morphometry and parameters associated with

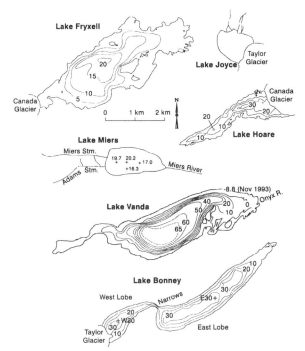

Fig. 2. The individual lakes, with depth (meters) information (where known). Bathymetry of Lake Fryxell is from *Lawrence and Hendy* [1985], modified by Wharton (unpublished data); bathymetry of Lake Hoare is from *Wharton et al.* [1986], modified by Wharton (unpublished data); bathymetry of Lake Bonney is from *Angino et al.* [1964], modified by Wharton (unpublished data) and Spigel and Priscu (unpublished data). Spot depths for Lake Miers are from *Bell* [1967]. Bathymetry for depths in Lake Vanda of zero meters and deeper are from *Chinn and Woods* [1984]; the zero meter contour coincides with the shoreline shown by *Chinn and Woods* [1984] and is assumed to coincide with the average water surface elevation of 84 m asl from 1971 to 1983 [*Chinn*, 1993]. The –8.8 m contour is based on a Nov. 1993 survey of the lake's ice edge by the New Zealand Department of Survey and Land Information (ref. No. 37/201) that recorded an elevation for the ice surface of 93.05 m asl; we have assumed the free water level to be 0.3 m below the ice surface [*Spigel and Priscu*, 1996].

stratification (estimates of maximum and minimum temperatures, maximum salinities, and estimates of light attenuation parameters) are given in Table 1. The work of earlier investigators, cited as the sources for information in Table 1, reveals that Lakes Bonney, Hoare, Joyce, and Fryxell have temperature maxima in the upper third of their water columns; temperatures then decline steadily to minimum values at the bottom. These minima are below 0°C in both lobes of Lake Bonney, but still well above the freezing point for the very saline waters found at the bottom of these lakes. Temperature and salinity profiles in the east lobe of Lake Bonney [*Hoare et al.,* 1964; *Shirtcliffe*, 1964; *Shirtcliffe and Benseman*, 1964], in Lake Fryxell

[*Hoare et al.,* 1965], and in the bottom 15 m of Lake Vanda [*Wilson and Wellman*, 1962; *Wilson*, 1964] are very smooth, suggesting absence of turbulence with transport dominated by molecular processes. Profiles in the upper 55 m of Lake Vanda [*Hoare*, 1966 and 1968] consist of a series of steps, the vertical portions being of widely varying heights and of such complete uniformity as to appear almost artificial, implying a great efficiency of mixing. The vertical portions are separated by strong gradients in heat and salt, gradients that are sharply discontinuous where they join the vertical parts of the profile, but quite smooth between vertical parts, indicating absence of turbulence in the strong-gradient regions. Lakes Joyce, Hoare, and the west lobe of Lake Bonney exhibit yet another distinctive pattern, but one that has only become clear from the higher resolution profiles measured for this study (see Figures 3, 5, 8, and 9). All of these lakes contain submerged faces of glaciers at one end, and, while their salinity profiles are smooth, their temperature profiles exhibit considerable variability (for Lake Bonney's west lobe this is most noticeable in profiles made near the glacier).

The climate of the dry valleys is extremely arid and cold; precipitation averages 10 mm yr^{-1} while potential ablation exceeds 300 mm yr^{-1} [*Chinn*, 1993]; average annual air temperature is around –20°C, and winter temperatures frequently drop below –30°C [*Clow et al.*, 1988]. The scientific excitement created by the discovery of warm saline lakes in such an environment led researchers in the 1960s to seek answers to questions that one might categorize in terms of either hydrology, energy balance or salt balance: (1) What maintains the hydrological balance of these lakes and their equilibrium ice thickness? And why does one also find in the dry valleys, in addition to the lakes containing liquid water, "lakes" frozen (practically) solid year-round? (2) Where does the heat come from to sustain the warm temperatures? And why do the lakes, all of such similar general shape and exposed to roughly the same climatic conditions, have such different profiles of temperature? (3) What is the source of the salts in the various lakes? And why do the concentrations, profiles, and compositions vary so greatly from one lake to another? It is probably safe to say that the answers to these questions are now understood, at least in a general way. Since the questions and answers form a background necessary for understanding the more recent measurements presented later in this chapter, a brief summary of contributions by earlier investigators follows.

TABLE 1. Hydrographic Parameters
(Unless Otherwise Noted, Values For Water Properties Based on Measurements for This Study)

	Bonney east	Bonney west	Hoare	Fryxell	Joyce	Vanda	Miers
Altitude	57 m	57 m	73 m	18 m	325 m	143 m	240 m
Distance to sea	25 km	28 km	15 km	9 km	44 km	47 km	20 km
Latitude	77°43'S	77°43'S	77°38'S	77°37'S	77°43'S	77°32'S	78°06'S
Longitude	162°26'E	162°17'E	162°55'E	163°09'E	161°37'E	161°34'E	163°58'E
Max Depth	37 m	40 m	34 m	20 m	35 m	75 m [16]	21 m
Length	4.8 km	2.6 km	4.2 km	5.8 km	1 km	8 km	1.5 km
Width (max)	0.9 km	0.9 km	1.0 km	2.1 km	1 km	2 km	0.7 km
Surface area	3.32 km^2	0.99 km^2	1.94 km^2 [20]	7.08 km^2 [19]	0.83 km^2	5.2 km^2	1.3 km^2
Ice thickness (m)	3–4.5[1]	2.8–4.5[1]	3.1–5.5[1]	3.3–4.5[1]	3.9–5.6[1]	2.8–4.2[1]	3.4–6[1]
%Transmission through ice	1–5%[2]; 3.3%[7] 1.7–3.3%, 2.73% avg[17]	1–5%[2]; 2.8%[7]	2%[3]; 0.5%[7] 0.5–2.8%, 1.59% avg[17]	1–5%[2]; 3.2%[7] 0.5–3.2%, 1.34% avg[17]	1.2%[7]	5–8%[4,5]; 5.2%[7] 5.2–20%, 13.2% avg[17]	1–2%[6]; 0.3%[7]
Extinction Coefficient (m^{-1})	0.1–0.2[2] 0.19[9] 0.12[12] 0.14[7]	0.1–0.2[2] 0.19[9] 0.16[7] 0.095–0.127[17]	0.2–0.3[8] 0.16[7] 0.17[17]	0.4–0.6[2] 1.0[7] 0.074–0.25[16]	0.17[7]	0.034–.05[4] 0.05[5] 0.045[10] 0.056[7] 0.04–0.055[17]	0.07–0.1[6] 0.19[7] 0.11[17]
Maximum Temperature, °C	7.0[9,11] 7.5[12] 7.9[4] 6.3–6.4	1.35[9] 2.0[11] 3.1–3.2	1.0[7] 1.2	2.2[13] 2.3[14] 3.5	2.5[7] 0.7	25[4,5,10] 21	5.25[6] 5.8[7] 5.5
Depth	14–16 m	9–10 m	4.5–5 m	10.5 m	10 m	bottom	bottom
Minimum Temperature, °C	–2.8[4]–2.6[9] –2.5[11]–2[12] –2.0	–5.35[9] –4.3[11] –5.0	0	0	0	0	0
Depth	bottom	bottom	surface	surface	surface	surface	surface
Max Con. @15°C	14.8 S m^{-1}	12.6 S m^{-1}	0.113 S m^{-1}	0.876 S m^{-1}	0.515 S m^{-1}	10.5 S m^{-1}	0.039 S m^{-1}[6]
Max Sal (PSS78)	150	125	0.70	6.23	3.50	96.1	
Max TDS (g L^{-1})	239	149	0.70 est	6.2 est	3.5 est	123[15]	0.287[6]
Depth	bottom	bottom	bottom	bottom	bottom	bottom	bottom

Sources: [1] *Wharton et al.*, 1993; [2] *Lizotte and Priscu*, 1992; [3] *McKay et al.*, 1994; [4] *Ragotzkie and Likens*, 1964; [5] *Wilson and Wellman*, 1962; [6] *Bell*, 1967; [7] *Parker et al.*, 1982; [8] *Palmisano and Simmons*, 1987; [9] *Hoare et al.*, 1964; [10] *Hoare*, 1966; [11] *Angino et al.*, 1964; [12] *Shirtcliffe and Benseman*, 1964; [13] *Angino et al.*, 1962; [14] *Hoare et al.*, 1965; [15] *Vincent*, 1987; [16] *Hawes et al.*, 1996; [17] *Howard-Williams et al., this volume*; [18] *Vincent, 1981*; [19] *Lawrence and Hendy, 1985*; [20] *Wharton et al., 1986. U.S. Geological Survey*, Topographic Maps, Antarctica. 1:250,000 series, 1970: Ross Island; Taylor Glacier; Mount Discovery. 1:50,000 series, 1977: Lake Vanda; Lake Bonney; Lake Fryxell; Joyce Glacier (provisional).

In his 1981 review, *Wilson* pointed out that the existence of perennially ice-covered lakes, unknown outside the arid areas of Antarctica, depends on the coincidence of a number of rather special hydrologic and climatic conditions. Conditions must be:

(1) sufficiently warm during the heights of summer for some liquid water to flow into the lakes, to

6 CTD casts Nov 24, 90 - Dec 9, 93, W20 Lake Bonney

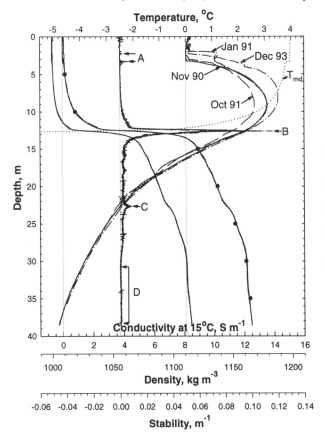

Fig. 3. Six CTD casts (fine-structure) measured at the central site (W20) in the west lobe of Lake Bonney on Nov. 24 and 29, 1990, Jan. 11 and 12, 1991, Oct. 21, 1991, and Dec. 9, 1993. Temperature profiles are labeled with dates; conductivity profiles are overplotted with solid circles that represent data from samples collected on Jan. 14, 1990; density profiles are solid curves with no labeling that have the same shape as the conductivity profiles; stability profiles are labeled with letters A to D. T_{md} denotes temperature of maximum density, plotted as a dotted line, calculated as a function of salinity and pressure from the observed temperature and conductivity profiles using equations from *Caldwell* [1978].

replace evaporation, (2) insufficiently warm during the summer for the ice to melt out completely, and (3) sufficiently arid for significant water to ablate off the surface ice cover during the year so that when it is replaced by new ice being frozen on the bottom of the ice during winter the latent heat of freezing produced will match that conducted through the ice cover during winter. . . Essentially such lakes use the latent heat of fusion of water to make up for the heat that escapes through their surfaces in winter. . .Clearly the resulting

equilibrium [ice] thickness must not exceed the depth of the lake.

Thus it can be seen why perennially ice-covered lakes are common in the Taylor Valley at low altitude. . .whereas the lakes in the Victoria Valley, which is at a much greater altitude and consequently closer to the snow line, are all frozen to their base. . .

The hydrologic balances implied in *Wilson's* summary have been quantified for several lakes by *Chinn* [1993], while *Conovitz et al.* [this volume] provide details of ongoing work at Lake Fryxell. *McKay et al.* [1985] succinctly quantified the energy balance for the dry valleys and explained the observed range in equilibrium ice thickness for the lakes containing liquid water. Detailed meteorological measurements made at Lake Hoare were used by *Clow et al.* [1988] to calculate a sublimation rate for 1986 of 350 ± 63 mm y[-1].

Wilson [1981] further pointed out that while the strong sunlight of the austral summer can heat glacial meltwater streams to several degrees above freezing (*Bell* [1967] observed temperatures as high as 8°C in the Adams Stream flowing to Lake Miers, although temperatures of 3°C were more typical), the latent heat plus the sensible heat (above 0°C) carried by these streams could not explain the elevated temperatures found in Lake Bonney and, to an astonishing degree, in Lake Vanda, and alternative energy sources would have to be found. The search proved interesting.

After some initial conjecture that most or all of the energy to heat Lakes Vanda and Bonney came from geothermal sources [*Armitage and House*, 1962; *Angino and Armitage*, 1963; *Ragotzkie and Likens*, 1964; *Angino et al.*, 1964], several studies were published using energy budget calculations showing that the warm temperatures could be explained completely in terms of solar energy that penetrated the ice and was stored at depth within a water column stabilized by salt gradients [*Wilson and Wellman*, 1962, and *Hoare*, 1966, for Lake Vanda; *Shirtcliffe and Benseman*, 1964, and *Hoare et al.*, 1964, for Lake Bonney; *Hoare et al.*, 1965, for Lake Fryxell; and *Bell*, 1967, for Lake Miers]. *Wilson and Wellman* [1962] and *Shirtcliffe and Benseman* [1964] noted that this is the mechanism at work in solar ponds, one that allows the extremely saline water in the bottom of these ponds to reach temperatures well in excess of 60°C [*Tabor*, 1981]. *Wilson* [1981] identified the higher transparency of the ice cover of Lake Vanda (see Table 1) as being responsible for the warmer temperatures found there compared with the other lakes, but to this must be added the remarkable transparency of Vanda's water, which approaches that for the purest natural waters (compare extinction coefficients for the different lakes in Table 1; also *see Howard-Williams et al.* [this volume] and *Lizotte and Priscu*'s [1992] Figure 6) and allows radiant energy to penetrate to greater depths, where it can be captured by the salty bottom waters. Deep chlorophyll maxima, accompanied by dramatic increases in light attenuation, are found in many of the dry valleys lakes [*Howard-Williams et al.*, this volume] and provide for further capture of solar energy at depth.

Further rejection of the geothermal hypothesis occurred when, as part of the dry valleys Drilling Project in 1973-1974, a hole was drilled in the deepest part of Lake Vanda, through 12 m of sediments and 5 m into the underlying rock. Measurements made in this hole showed temperature decreasing with depth, implying that heat flux in the sediments was downward from the lake to the sediments, rather than the other way around [*Wilson et al.*, 1974b; *Bydder and Holdsworth*, 1977].

The unusual shapes of the temperature and salinity profiles in the lakes also attracted considerable attention. *Wilson and Wellman* [1962] concluded that the regions of uniform temperature and salinity in Lake Vanda were undergoing strong convection. *Hoare* [1968] recognized the vertical steps as a series of thermohaline convection cells, and a similar process was identified by *Bell* [1967] near the bottom of Lake Miers. It remained to *Yusa* [1975, 1977] to apply existing theoretical linear stability criteria that salinity and temperature profiles must satisfy in order for such cells to develop in the presence of small disturbances. The phenomenon is now known to be fairly widespread in the ocean [*Gregg*, 1973; *Turner*, 1973] and requires that there be two components present that independently affect density, in this case heat and salt. One component must have gradients that tend to destabilize the water column (in both Vanda and Miers, the cells occur where temperatures, above the temperature of maximum density, increase with depth) while the other component stabilizes the water column (salinity increasing with depth), so that overall the density gradient is statically stable. The motions can develop in spite of this overall stability, providing the salt gradient is not too strong (or the temperature gradient not too weak), *and* the molecular diffusion coefficients for the two components differ markedly (in the case of heat and salt, heat diffuses approximately 100 times

faster than salt). Synonymous terms for the phenomenon are double-diffusive convection or thermosolutal convection, and the type observed in Vanda and Miers is one in which the component with the smaller diffusion coefficient is the stabilizing component. The reverse situation (e.g., cold, fresh water underlying warmer, salty water, with heat being the stabilizing component) leads to a somewhat different result ("salt fingering"), although homogeneous layers can also be produced. The situation in Vanda was also the subject of an analysis by *Huppert and Turner* [1972], who were interested to see if results from laboratory-scale experiments could be applied to natural systems on a much larger scale (their answer was "yes"). *Yusa* [1977] showed that salt gradients in Bonney and Fryxell were too strong to permit convection, thereby explaining the absence of such cells in those lakes (he did not consider Lake Joyce or Lake Hoare, but we will do so later).

Mathematical models to explain the shapes of temperature and salinity profiles in Lake Bonney (east lobe), Lake Fryxell, and the bottom of Lake Vanda all relied on the one-dimensional diffusion equation and assumed that vertical heat and mass transfer in the interiors of these lakes is controlled by molecular diffusion and (for heat) volumetric heating by solar radiation at depth. The one-dimensional approach was well justified by data showing extremely flat isotherms extending practically over the entire basin in all three lakes. Unfortunately insufficient data were available to specify all the necessary physical parameters (solar flux penetrating the ice, extinction coefficients throughout the water column, conductive heat loss or gain at the boundaries) independently of the calculations, so some fitting was involved, usually accompanied by discussion to show that calibration parameter values were physically reasonable. A steady state solution for Fryxell [*Hoare et al.*, 1965] gave good results, but a similar approach for the west lobe of Lake Bonney [*Hoare et al.*, 1964] yielded poor results over much of the water column. Solutions for the east lobe of Lake Bonney gave results that did not match observed temperatures in the lower part of the water column [*Hoare et al.*, 1964; *Shirtcliffe and Benseman*, 1964], leading *Shirtcliffe* [1964] to undertake more elaborate calculations. It is of interest to note that measurements of *Hoare et al.* [1964; measurements made in December 1963] and *Shirtcliffe and Benseman* [1964; measurements made in January 1963] both show a decrease in temperature gradient in the east lobe as the bottom is approached, with an inflection point in the profile at

10–15 m above the bottom. Our profiles do not show this behavior. *Shirtcliffe* [1964] pointed out that the decrease in gradient implied a decrease in downward heat flux, hence a convergence of thermal energy, and concluded that under these circumstances (accumulation of heat) a steady state was not possible. His unsteady solution for the east lobe gave an excellent fit to the measured temperature profile, but involved still more parameters that required calibration. His optimum parameter set corresponded to (among other things) a geothermal heat flux of 0.167 ± 0.084 W m^{-2}, roughly 2.7 times the global average of 0.062 W m^{-2} [*Williams and von Herzen*, 1974], but overlapping (and hence not significantly higher than) the range of the geothermal heat flux measurements that have since been made in the dry valleys of 0.059 to 0.142 W m^{-2} [*Decker and Bucher*, 1977].

Of perhaps more significance was the time scale for diffusion that *Shirtcliffe* [1964] derived from his calculations on profiles of both heat and salt, as 60 ± 20 years. In 1964, as now, the bottom waters of the east lobe consisted of a highly concentrated salt solution (dominated by sodium chloride) that has an almost uniform concentration for the first 8 m above the bottom; above 8 m salinity drops steadily, slowly at first but then more steeply, the gradient reaching a maximum at about 18 m (*Shirtcliffe* gives 19 m) above the bed. *Shirtcliffe* [1964] "inferred that at one time the lake was somewhat shallower than now and of uniformly high salinity, and that freshwater ran in over the salt solution to bring the lake up to its present level"; he felt that the density gradient that formed at the interface between fresh and salt water would have had sufficient stability to suppress any subsequent turbulent mixing. *Shirtcliffe* actually said very little about climate or ice cover, but if at some time in the past, water levels were much lower in the east lobe, perhaps as a result of an extended period of cold, dry weather, and lake salinities became so high as a result of concentration by evaporation that the ice cover disappeared (the freezing point having been sufficiently depressed by the high salinities), the lake water would then have been exposed to wind, as Don Juan Pond (near Lake Vanda) is today. Strong Antarctic winds would have kept the lake well mixed, the depth of this mixed layer probably corresponding to the 18 m elevation of the present peak salinity gradient (this *Shirtcliffe* also assumed). Climate warming would eventually have led to increased meltwater runoff from glaciers and a refilling of the east lobe. Fresh-water inflow from the glaciers would have overflowed the

denser saline waters already occupying the basin, and, as lake levels rose, an ice cover would have been able to form. Once the ice cover formed, diffusion would control any further evolution of the salinity gradient. Regardless of what allowed diffusion to take over (density gradient between salt and fresh-water, or ice cover) *Shirtcliffe* made it clear that he associated the 60 ± 20 year diffusion time scale with the onset of a freshwater overflow event that signaled the beginning of a rise in the lake to present levels. His figure of 60 years coincided almost exactly with the earliest measurement from which lake elevation can be inferred, that by the western party of Scott's *Discovery* expedition in 1903 of the contraction separating Lake Bonney's east and west lobes as being "only seventeen feet (5.2 m) wide" [*Scott*, 1905; quoted by *Chinn*, 1993]. Based on soundings he made in the narrows in February 1964, *Shirtcliffe* [1964] estimated a rise in water level between 1903 and 1964 of 9.2 m, assuming that the thickness of the ice cover did not change appreciably over that time. *Chinn* [1993] extrapolated this estimate further, based on measurements of the side-slopes of the channel cross-section and water level measurements begun in 1971-1972, to arrive at an estimated water level rise of 13.5 m between 1903 and January of 1990. Our measurements in the narrows (soundings made along a transect of holes drilled through the ice across the width in November 1990) together with temperature and salinity transects made through the narrows in November 1990, indicated a maximum water depth over the sill in the narrows of approximately 12.5 m. If this latter estimate is accurate, it implies that the estimate of 13.5 m rise since 1903 is too high by about 1 m, assuming that water level was just at the height of the sill in the narrows in 1903. In any event, it appears that around 1900 water levels in both lobes must have been very close to the top of the sill that separates the two basins, and that by then an ice cover had been established in the east lobe. Since most of the meltwater flow to the east lobe apparently comes over this sill, from the Taylor Glacier (Priscu unpublished current meter data), this lends support to *Shirtcliffe's* hypothesis that the evolution of the chemocline in the east lobe did not begin until around 1900.

Calculations for Lake Vanda led along a different path. Use of the unsteady diffusion equation (with a molecular diffusion coefficient) to model the salt gradients in the bottom waters of Lake Vanda led *Wilson* [1964] to "speculate on the origin of the salt concentration gradient in the lake:"

The only reasonable explanation seems to be that at some period in the past the climate was such that the Onyx River did not supply appreciable water to Lake Vanda. Under these conditions the lake level would have dropped until only a few feet of concentrated calcium chloride remained. When the climate changed, the Onyx would flow during the summer and fresh water would have flowed on top of this strong salt solution. Since that time the calcium chloride has been diffusing upwards. If such a model is assumed, it is possible to calculate the time in the past when this climatic change occurred.

The number that Wilson arrived at was 1,200 years.

Since 1964, a number of papers have been published utilizing the diffusion equation, supplemented by other geochemical information, in an effort to determine whether, and when, a major drying event could have occurred that caused one or more of the lakes to contract into saturated brine ponds, or possibly dry up altogether [Hendy et al., 1977; Matsubaya, 1979; Lawrence and Hendy, 1985]. Hendy et al. [1977] suggested adjustments to some of the parameters used by Shirtcliffe [1964] that led to an upward revision of the 60 years figure for the time origin of diffusion to 400–750 years. Lyons et al. [in press] reviewed all this work, as well as more recent geological and geochemical findings, and concluded that there is a reasonable amount of evidence to support a scenario in which a relatively warm, wet period allowed the dry valleys lakes to reach high levels 3000–2000 years ago, followed by a cooling, drying trend lasting 500–1000 years, that caused the east lobe of Lake Bonney, Lake Vanda, and Lake Fryxell to lose their ice covers and contract to hypersaline ponds. Lake Hoare may have evaporated completely. The same fate did not befall the west lobe of Lake Bonney, presumably because of its direct proximity to the Taylor Glacier, which, unlike the much smaller mountain and cirque glaciers that supply meltwater to Lakes Hoare, Fryxell, Vanda, and Bonney east lobe, is a terminal glacier of the polar ice cap. The west lobe must have fallen below the level of the sill separating the two basins, however, and once this happened, the east lobe would have been deprived of inflow from the Taylor Glacier. When warmer, wetter conditions returned (from 1000-1500 years ago, in agreement with Wilson's estimate), the lakes began to refill with less saline, relatively fresh meltwater. It seems likely, in fact, that the warming trend has persisted until now.

We will return later to the apparent contradiction between Shirtcliffe's and Wilson's time scales for diffusion; for now we note that patterns of stratification observed in all of these lakes are intimately connected with the geology and geochemistry of their terrestrial surroundings in a markedly different way from most lakes. Furthermore, time scales for mixing in these lakes (other than within thermohaline convection cells) are very long, with substantial modifications in density structure only occurring as a result of climate change. This much follows from the diffusion calculations and related events described above. Other examples appear later in this chapter.

Questions of where the salts in the lakes originated, and why the lakes have such different compositions and concentrations, are also intimately connected with the geology and geochemistry of the region. Proximity to the sea, extreme aridity, fractionation of solutions accompanying evaporation and freezing, and soil- or rock-water interactions feature in most explanations [Burton, 1981]. Matsumoto [1993] concisely sum-marizes the general situation: "These salts can be explained by the mixing of a common source of atmospheric fallout with various local sources, such as rock weathering, groundwater (including hydrothermal activity), and trapped seawater. Also, the fractional crystallization of certain Na, K, Ca, and SO_4 salts during the concentration of lake and pond waters plays an important role in the distribution of dissolved salts." Details, which are outside the scope of this chapter, vary from lake to lake, and continue to be the focus of active research [e.g., Carlson et al., 1990; Matsumoto, 1993; Welch and Lyons, in press].

The work of other investigators summarized in this section has relied on measurements and samples taken mostly at meter-scale intervals in the water column. In the remainder of the chapter, we will see what light measurements acquired with an accurate high-resolution, continuous-profiling conductivity-temperature-depth probe(CTD) can shed on questions related to mixing and transport.

METHODS

Instrumentation

An SBE 25 Sealogger CTD (s/n 252247-6), manufactured by Sea-Bird Electronics according to our specifications, was deployed through 26 cm diameter holes in the ice at drop speeds between 0.2 and 0.5

m s $^{-1}$ (the drop speed for any one profile was held as constant as possible) using a hand-operated, adjustable-ratio winch that could be set up directly over an ice-hole and was fitted with a TSK metering block. The winch was designed and fabricated with special attention to ease and smoothness of operation and low weight so it could be carried in a backpack. The CTD carried two sets of conductivity and temperature probes, mounted outside the main instrument casing. We will denote one temperature-conductivity pair as fine-structure sensors, and the second pair as microstructure sensors. The fine-structure temperature sensor is an SBE 3-01/F Oceanographic Thermometer with a response time of 0.084 s at a drop speed of 0.5 m s $^{-1}$. Accuracy and stability are ±0.01°C per 6 months (guaranteed), ±0.004°C per year (typical); resolution is better than 0.001°C [Sea-Bird Electronics, 1989a]. The fine-structure conductivity cell is an SBE 4-01/0 Conductivity Sensor with a response time of 0.085 s at a drop speed of 0.5 m s $^{-1}$, with a submersible, constant-speed pump (SBE-5LD, operating at 1800 rpm) to draw water through the conductivity cell. Pump connections are designed to eliminate the effects of any probe accelerations or decelerations that can affect pump performance and introduce noise into the temperature and conductivity measurements [Sea-Bird Electronics, 1989b]. Water is drawn first through the duct that houses the temperature sensor and then through the conductivity cell, so that temperature and conductivity measurements are made on the same water. The SBE 3 and SBE 4 act as variable resistances in Wien oscillating bridge circuits, their outputs being frequencies rather than voltages. Frequency for each sensor is measured by two 12-bit counters, one for whole-period counts and the second for fractional count. Fine-structure sensors sample at a rate of 8 Hz, giving resolution of 6.2 cm at a drop speed of 0.5 m s $^{-1}$, to 2.5 cm at a drop speed of 0.2 m s $^{-1}$. Both SBE 3 and SBE 4 sensors are well-proven and widely used [Pederson, 1973; Pederson and Gregg, 1979; Gregg and Hess, 1985], and provided both accuracy and stability to which the faster microstructure sensors' output could be anchored.

The microstructure temperature sensor (SBE 8-02) contains a high-speed Thermometrics (type FP07) thermistor, with a response time of 0.007 s. The microstructure conductivity sensor (SBE 7-02) contains a two-terminal, dual-needle platinized-electrode cell, with response time limited only by flushing. The design and operating characteristics of the microstructure sensors, which were originally developed in the Applied Physics Laboratory and School of Oceanography, University of Washington, in response to a need by the U.S. Naval Research Laboratory for an open-ocean microstructure measurement system, have been documented by Gregg et al. [1978] and Meagher et al. [1982]; the low-noise dual-needle electronics of the microconductivity sensor have been used to drive several configurations of Neil Brown conductivity cells. Outputs from both microstructure sensors are voltages that are appropriately filtered prior to digitizing. Signal processing for both sensors includes circuitry to "pre-emphasize" the output signals in proportion to frequency. The circuit contains a low-noise amplifier and associated resistance and capacitance components that give a voltage gain proportional to variations in the input voltage, thereby increasing the sensitivity of the instrument to signals that are changing rapidly. Pre-emphasis is designed to overcome the digital threshold imposed by the 12-bit analog-to-digital conversion, which would otherwise prove a serious limitation to the sensors' ability to resolve microstructure gradients. The SBE 7's pre-emphasis response magnifies a 100 Hz signal by a factor of 1000, and the SBE 8's pre-emphasis response magnifies a 20 Hz signal by a factor of 200. The microstructure sensors sample at a rate of 72 Hz, giving a resolution of 6.9 mm at a drop speed of 0.5 m s $^{-1}$, to 2.8 mm at a drop speed of 0.2 m s $^{-1}$. Preliminary calculations indicated that this resolution would be adequate for the low levels of energy dissipation we expected to encounter in Lake Bonney; we will return to this consideration later.

Pressure is measured with a Paine absolute pressure transducer, housed inside the main body of the CTD and ported to the outside with an oil-filled capillary tube that protrudes slightly through the bottom of the CTD end-cap. The sensor has a range of 0 to 1.168×10^6 pascals absolute (300 psia), corresponding to a maximum gauge pressure of 1.067×10^6 pascals at standard atmospheric sea level pressure (0.1013×10^6 pascals); the maximum gauge pressure is equivalent to 106.7 decibars, corresponding to a maximum depth range in seawater of a little less than 106.7 m. Total error band is specified as ±0.5% of full range, or ±5840 pascals (±0.584 decibars, i.e., approximately ±0.6 m in terms of water depth). This may seem to impose a limitation on the depth resolution that can be achieved in profiling, but the "error" refers to accuracy, not precision, and in practice satisfactory resolution is achieved by maintaining a steady drop speed and then smoothing the pressure signal with a Gaussian filter.

For every cast, the CTD was switched on manually and then lowered to a preset level in a hole, with the cage's top ring just below the water surface; the CTD was held stationary for 45–60 s before lowering. This allowed ample time for the pumping system to prime and purge all air (this could be confirmed by inspection) and for the probe assembly of the CTD (which had to be maintained as "warm" as possible between casts to prevent freezing of water inside the fine-structure temperature-conductivity duct or conductivity cell) to come to equilibrium with conditions in the water. It also allowed subsequent correction for any pressure offset, as the pressure sensor was at a known depth (0.965 m) below the water surface during this initialization stage.

The CTD is surrounded by a cage that offers protection from the ice when the instrument is being lowered or raised in a hole. The bottom of the cage is open and flush with the bottom of the instrument (the level at which the microstructure sensors are exposed) and therefore does not interfere with sampling during a drop. Only measurements recorded as the CTD is dropping are subsequently used.

Instrument calibration was carried out before each field season by SeaBird Electronics, Inc. Calibration coefficients were then incorporated in the software supplied by SeaBird Electronics; the software was used to download cast data from the CTD and transform the measured voltages and frequencies to conductivities, temperatures and depths.

Data Reduction

Conversion of pressures, temperatures and conductivities recorded during CTD casts to depths, densities, and measures of stability rests on the analyses of Lake Bonney water samples presented by *Spigel and Priscu* [1996]. They concluded that density could be computed by first calculating PSS78 salinity (S) from in situ conductivity, temperature and pressure [*Lewis*, 1980; *Fofonoff and Millard*, 1983], and then calculating density from salinity, temperature and pressure using a modified form of the UNESCO International Equation of State for Seawater. Denoting the unmodified density calculated from the UNESCO Equation of State [*Millero et al.*, 1980; *Millero and Poisson*, 1981; *Fofonoff and Millard*, 1983] as ρ_{UN}, the in situ density is given by ρ_{UN} for $S \leq 42$ and by:

$$\rho = \rho_{UN} + a(S-42)/[b - 1(S-42)], \text{ east lobe} \quad (1)$$

for $S \geq 42$ where a = 15.299 and b = 133.6 (note that there is an error in *Spigel and Priscu* [1996], where values for a and b have been interchanged), and:

$$\rho = \rho_{UN} + a(S-42), \text{ west lobe} \quad (2)$$

for $S \geq 42$ where a = 0.13696. The correction for salinity is nonlinear in the east lobe because of the much higher salinities found there in comparison to the west lobe, where a linear correction was found to be adequate. The unmodified UNESCO equations were used for all the other lakes, although this may introduce some error for data from Lake Vanda. Lake Vanda is the only lake other than Lake Bonney in which salinities exceed those of seawater, but we do not have information on density-conductivity-temperature relations for its waters.

It was found that conductivities (and hence salinities and densities) in the very saline waters of Lake Bonney were overestimated by the CTD, based on results found by *Spigel and Priscu* [1996]; conductivity calibration in their experiments covered the entire range of conductivities found in Lake Bonney, while calibration of the CTD covers only the range normally found in seawater. Hence, for salinities over 42 calculated from CTD measurements, a correction was made that incorporated the laboratory calibration for high salinities, as:

$$S = 1.0609S_o - 8.9918 \times 10^{-4} S_o^2$$
$$- 2.8325 \times 10^{-7}S_o^3, \quad S_o > 42 \quad (3)$$

where S is the corrected salinity and S_o is the salinity predicted from CTD measurements. Equation 3 is based on east lobe data, but also provides good results when used with west lobe measurements. If the correction is applied instead to conductivity adjusted to the relatively warm temperature of 15°C, however, separate equations for east and west lobe are necessary, as:

$$C = 1.0433C_o - 4.2288 \times 10^{-3}C_o^2$$
$$-2.2600 \times 10^{-4}C_o^3, \text{ east lobe}, C_o > 5 \text{ S m}^{-1}$$
$$C = 1.0256C_o - 0.010184C_o^2$$
$$+ 2.1141 \times 10^{-4}C_o^3, \text{ west lobe}, C_o > 5 \text{ S m}^{-1} \quad (4)$$

where C, C_o are the corrected and uncorrected

conductivities at a reference temperature of 15°C, and T is temperature; the formulas of *Hewitt* [1960] were used to adjust conductivity for temperature.

Equations (1) and (2) give density as functions of salinity and (through the unadjusted equations for ρ_{UN}) of temperature and pressure; they can be differentiated to give expressions for thermal compressibility, $\alpha = (1/\rho)(\partial\rho/\partial T)$, and compressibility due to salt, $\beta = (1/\rho)(\partial\rho/\partial S)$. Values of α and β were calculated for each data point. Adiabatic lapse rate,

$$\partial T/\partial p \mid_{adiabatic} = (T_{abs}/c_p)(\partial v/\partial T) = -\alpha\, T_{abs}/(\rho c_p) \quad (5)$$

[*Gill*, 1982, p. 50, although note the different sign convention for α], where T_{abs} is absolute temperature, $v = 1/\rho$ is specific volume and c_p is specific heat capacity, was also calculated for each data point, with c_p calculated as a function of S, T, p using the empirical formula of *Millero et al.* [1973; quoted in *Fofonoff and Millard*, 1983]. Potential temperature, θ, could then be calculated at any depth by integrating (numerically) Equation 5 over pressure from the depth in question to the free surface. A single-step, fourth-order Runge-Kutta algorithm was used; the results were very little different from those produced by an empirical formula for potential temperature as a function of S, T, and p presented by *Fofonoff and Millard* [1983], based on their results for similar integrations for standard seawater. Potential density was calculated from the appropriate density expression (Equation 1 or 2, or as ρ_{UN}) using the potential temperature, θ, in place of the in situ temperature, T, and with p = 0. Potential temperature is the temperature of a fluid particle if reduced adiabatically and with constant salinity from its in situ temperature and pressure to a reference pressure, taken here as atmospheric pressure. Because water is slightly compressible, a sample brought from depth to the surface will expand and therefore tend to cool if it is above the temperature of maximum density; if it is below the temperature of maximum density it will warm. Potential temperatures below the water surface are therefore slightly cooler than in situ temperatures for temperatures above the temperature of maximum density, and slightly warmer for temperatures below the temperature of maximum density. Use of potential density is one way of accounting for the effects of pressure on static stability. In limnological work, except in very deep lakes, it is customary to neglect the effects of pressure on density for purposes of assessing stability, density being treated as a function of S,T

only, not S, T, p. However we included pressure effects in all our calculations from the outset to rule out any errors that their neglect might introduce, particularly in regions where temperature and salinity gradients are small and stability approaches neutral conditions.

Depths were calculated by numerically integrating the hydrostatic pressure equation, $\partial p/\partial z = \rho g$, where p is pressure, z is depth below the free water surface in the ice hole (z is positive downwards), and g = 9.81 m s^{-2} is the acceleration of gravity. Stability was assessed by calculating the oceanographic stability, E, [*Pond and Pickard*, 1983, p. 27]

$$E = \beta\,\partial S/\partial z + \alpha(\partial T/\partial z - \Gamma) \quad (6)$$

where $\Gamma = \rho\, g(\partial T/\partial p)\mid_{adiabatic}$; positive values of E correspond to statically stable stratification, negative values to unstable stratification, and a value of zero to neutral stability. A check was provided by also calculating the square of the buoyancy frequency based on potential density, $N_\theta^2 = (g/\rho_\theta)(\partial\rho_\theta/\partial z)$. Differences between values of N_θ^2/g and those of E were negligible.

Other calculations for each data point included viscosity as a function of S,T,p using equations and data presented in *Riley and Skirrow* [1975, pp. 576–577]; diffusivity of salt in water, using data for NaCl solutions in *Robinson and Stokes* [1955, pp.494–495] and for seawater by *Montgomery* [1957]; and thermal conductivity, using data for pure water from *Kestin and Wakeham* [1988, p.215] with an adjustment for salinity proposed by *Reid et al.* [1977, pp. 516, 536]. These transport properties were used to evaluate parameters that determine susceptibility of the water column to double diffusion.

RESULTS AND DISCUSSION

Lake Bonney

The extensive data from Lake Bonney provide good examples of the processes discussed earlier, and serve as a guide to interpreting the less comprehensive data from other lakes.

Vertical structure of conductivity, temperature and stability. Profiles of conductivity (adjusted to 15°C), temperature, density and stability (E in Equation 6), measured at the central sampling sites in the west and east lobes of Lake Bonney over a three-year period, are shown in Figures 3 and 4. Circles on the conductivity profiles are from samples collected with a Niskin bottle on November 28, 1989 in the east

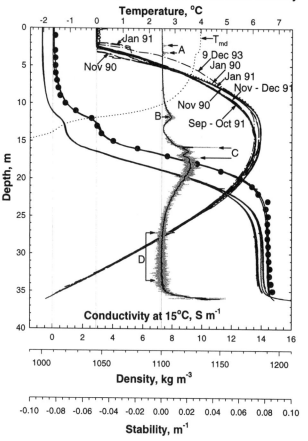

20 CTD casts Jan 9, 90 - Dec 9, 93, E30 Lake Bonney

Fig. 4. Twenty CTD casts (fine-structure) measured at the cen-tral site (E30) in the east lobe of Lake Bonney on Jan. 9 and 10, 1990; Nov. 17, 23, and 27, 1990; Jan. 10 and 13, 1991; Sept. 20 and 27, 1991; Oct. 5, 11, 18, and 25, 1991; Nov. 1, 7, 14, 20, and 26, 1991; Dec. 2, 1991; and Dec. 9, 1993. See the caption to Figure 3 for further explanation. Solid circles represent data from samples collected on Nov. 29, 1989.

lobe and January 14, 1990 in the west lobe, as described by *Spigel and Priscu* [1996]. Zero depths correspond to water surface elevations on those dates. Lake levels rose by more than a meter over the period of measurement covered by Figures 3 and 4, so water surface elevations for CTD casts made after January 1990 correspond to negative values of depth in terms of the depth scale shown in the graphs; these adjusted negative depths are inside the ice holes and the corresponding sections of the profiles are not shown in Figures 3 and 4. Reference elevations for profiles were fixed relative to the chemoclines in both figures (the region of steepest gradient in the conductivity profiles), by translating the depth scales for each conductivity profile upward by trial and error until chemocline

position matched the reference chemocline to within ±0.05 m. The reference chemocline in the east lobe was that of the November 1989 samples; in the west lobe, no samples were available within the chemocline because of the large sampling interval, so the chemocline of the first CTD cast (November 24, 1990) served as the reference.

The extraordinary consistency of the conductivity profiles from year to year in both lobes is proof of the stability of these systems and of the absence of turbulence in the interiors of the two basins. Density profiles in both figures appear to the left of the conductivity profiles, their shapes being similar to that of the conductivity profiles because of the control that dissolved solids concentrations exert on density variations. In the east lobe below the chemocline it is possible to distinguish a very slight flattening of curvature in the conductivity profiles over time, resulting in an apparent shift to the left of the profiles where profile curvature is strongest. Profiles converge again as the bottom is approached. The small differences in conductivity are magnified considerably in the density profiles, because of the strongly nonlinear, increasing effect of high salinities on density in the east lobe (the denominator in Equation 1). Although the change in curvature in conductivity profiles is consistent with the effects of diffusion, it would be best to obtain further information from subsequent years, paying particular attention to higher salinities in CTD calibration, before speculating further on the east lobe conductivity changes. Similar changes are not seen in the west lobe profiles; maximum west lobe salinities are lower than in the east lobe, and it is likely that saline meltwater intrusions from the submerged face of the Taylor Glacier mask the effects of diffusion in the bottom waters of the west lobe.

Temperature profiles, although retaining their general shape from year to year, show much greater variability than do conductivity profiles. This is to be expected, as temperatures respond to variations in heat transfer between the lake and the atmosphere. In both basins, changes in temperature occur mainly in the upper 25 m of the water column. At this depth, only about 0.11% of solar radiation incident on the ice-cover remains (based on typical values in Table 1 for an extinction coefficient of 0.15 m^{-1}, 3% transmission through the ice and 3 m thick ice cover). Response to solar radiation is evident in the east lobe record, where more extensive measurements are available and where meltwater inflow does not occur at depth. For seasons in which profiles are available over a period of several months, small but definite warming trends can be seen. From mid-November 1990 to mid-January 1991, for example, maximum east lobe temperatures increased by 0.085°C at the main sampling site; similar increases were observed at all east lobe sites. November–December 1990 included some extended periods of warm air temperatures that led to large meltwater inflows. Although we do not have flow records for Lake Bonney, this observation is supported by streamflow measurements made at Lake Fryxell, in which total meltwater volumes were largest for the 1990-1991 season [see *Conovitz et al.*, this volume, and the accompanying CDROM]. The effects of relatively warm, fresh inflows immediately under the ice are clearly visible as a "step" at the top of the January 1991 temperature profiles for both the east and west lobes (see also the temperature transect in Figure 5, where the step can be clearly seen in all temperature profiles). The greatest total heat content (proportional to the area under the temperature profile) in both lobes was observed in December 1993 (Figures 3 and 4). We have no profiles to fill the gap between October 1991 and December 1993, so we cannot follow the development of the temperature profile over this period. Neither do we have meteorological data spanning this period, although the change in temperatures must be a response to changes in meteorological conditions.

As mentioned earlier, the solar heating and molecular diffusion model of *Shirtcliffe* [1964] was successful in explaining the form of the east lobe temperature and conductivity profiles, while a similar (somewhat less elaborate) model presented by *Hoare et al.* [1964] gave unsatisfactory results in the west lobe. Our explanation for this is that meltwater influences a much greater part of the water column in the west lobe than in the east lobe, because of the direct contact maintained between the water column and the face of the Taylor Glacier, providing a source for heat (or cold), water and salt in the west lobe that cannot be accounted for in a model that balances absorption of solar energy only with vertical molecular diffusion. A complete balance must include lateral transport by horizontal advection. The depth to which the glacier extends is uncertain, but variability in the temperature and stability profiles (Figure 3) indicate that the glacier exerts an influence over a depth of nearly 25 m. The local maximum in stability between 20 and 25 m (see the cusp-like feature labeled C in Figure 3) is probably associated with a saline intrusion of meltwater originating from the glacier. Surface discharges of salty water from the terminus of the Taylor Glacier, rich

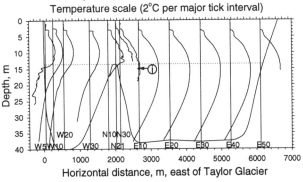

Lake Bonney temperature transect Jan 10-11, 1991
Temperature scale (2°C per major tick interval)

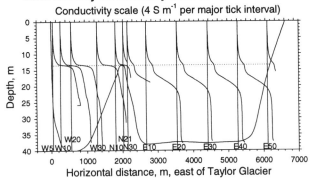

Lake Bonney conductivity transect Jan 10-11 1991
Conductivity scale (4 S m^{-1} per major tick interval)

Fig. 5. Fine-structure temperature and conductivity (adjusted to 15EC) profiles measured along a west-to-east transect through Lake Bonney, Jan. 10 and 11, 1991. The profile of the lake bed is shown, with vertical lines marking the positions of the sampling sites (W5-W30 in the west lobe, N10-N30 in the narrows, and E10-E50 in the east lobe). The vertical lines also serve as references for temperature T = 0 EC in the temperature transect and conductivity C_{15} = 0 S m^{-1} in the conductivity transect.

enough in iron oxides to have stained a section of the glacier face red, have been described in detail by *Black* [1969], *Black et al.* [1965], *Black and Bowser* [1968], and *Keys* [1979]. These discharges were so large during the summer of 1990-1991 that by January 1991 reddish-orange slush covered extensive areas of shoreline next to the glacier and was conspicuous from a distance of several kilometers. Water samples collected in the west lobe from depths between 20 and 25 m when sampling resumed in October 1991 contained turbid, reddish iron-oxide-rich water that had not been previously observed there and that created difficulties in routine chemical and biological assays (Priscu unpublished data). This must have been water from an intrusion layer that originated at the face of the Taylor Glacier, its source being either an underflow from the well-documented red, saline surface discharge, or possibly a submerged outlet in the glacier face. Regardless of the exact location of the intrusion's source, it was a striking manifestation of a submerged saline inflow that may not be energetic enough every year to transport suspended sediments into the interior of the lake, but that nevertheless probably supplies some saline water at depth to the west lobe in most years.

The noisy appearance of the twenty superposed east lobe stability profiles, shown in grey in Figure 4, is partly a consequence of nonlinear amplification of conductivity differences, and of the effect that this amplification has in the calculation of density differences at high salinities (as discussed above in connection with shifts in conductivity and density profiles in the bottom of the east lobe). The magnification applies not only to real differences in conductivity, but also to any noise in the conductivity signal; noise from the temperature signal is amplified as well because of the strong dependence of conductivity on temperature. The problem is compounded because stability is the derivative of the density profile, involving calculation of small differences between large numbers, an operation that is inherently numerically unstable. Rather than smooth any of the original signals, we have presented the average of the twenty profiles (black curve) as the least disruptive way of filtering the noise. This was not necessary in the west lobe, where the magnification effect is much less pronounced because salinities are lower and the correction to the salinity-density relation is linear (compare Equations 1 and 2).

All profiles for both lobes show a step increase in density immediately below the ice, as indicated by the spikes in stability labeled A in Figures 3 and 4 (recall that the free water surface rose by more than 1 m from 1990 to 1993; this is reflected in upward movement of the ice-water interface over time and hence of the positions of the spikes in stability). Above these spikes, the data are from water inside the confines of the 26 cm diameter ice hole, water that is generally well mixed and contains turbulence caused by disturbances accompanying the clearing of the ice hole and lowering of the CTD. These disturbances do not propagate below the ice because of the stability of the water column. A spike in stability occurs as the CTD begins to sample water immediately below the ice, and is caused by a jump in salinity from the fresher water in the ice hole to the slightly saltier lake water just underneath the ice. (The step-increase is small relative to the very high salinities at depth and is not clearly visible in Figures 3 and 4 because of the overlapping of so many curves; it can be clearly seen in individual profiles, however.) This pattern of stability just below the ice contrasts with ice-covered freshwater lakes, where solar heating of water in contact with the ice cover, and thus below the temperature of maximum density, can cause a convecting mixed layer to form immediately beneath the ice, extending to depths of several meters in some cases [*Matthews and Heaney*, 1987; *Patterson and Hamblin*, 1988]. No such surface mixed layer was found in any of our profiles.

The exceptionally high peak in stability in the west lobe between depths of 12 and 13 m (B, Figure 3) coincides with the top of the sill in the narrows (see Figure 5) that blocks almost all flow of salty west lobe water to the east lobe. Streamflow gauging in the Lake Bonney catchment by the U.S. Geological Survey since 1992 indicates that meltwater inflow to Lake Bonney is dominated by the Taylor (and neighboring) Glaciers, entering at the western end of the west lobe (H. House, personal communication; see also accompanying CDROM). Current meters moored in the narrows directly over the sill have shown a net flow of relatively fresh meltwater from west to east over the sill (Priscu, unpublished data). Recall that it was this flow that *Shirtcliffe* [1964] invoked in his explanation of the evolution of the east lobe salinity profile. This flow must be partly responsible for maintaining the sharpness of the salinity gradient between 12 and 13 m in the west lobe by sweeping away salt that diffuses upward from below this level. Sharpening of the gradient would be reinforced by any inflow of saline meltwater to the west lobe below the chemocline. Because the west lobe water column is density stratified over its entire depth, any such inflows will form horizontal intrusions of limited vertical extent at their level of neutral density (the iron-oxide-rich layer described above being one such example), lifting all of the water above this depth to higher levels. The net west-to-east flow through the narrows must dilute and skim off salty west lobe bottom water as it rises above the sill. Hence the gradient is intermittently resharpened in the west lobe, whereas no such mechanism exists in the east lobe. The main east lobe gradient is controlled by vertical molecular diffusion, while that of the west lobe depends on both advection (the horizontal west-to-east flow and the vertical lifting of water above saline intrusions) and vertical diffusion. We believe that the distinctive "shoulder" in the east lobe conductivity and density profiles just below sill level, corresponding to the minimum in stability below B in

Figure 4, is the signature of salty west lobe water that is washed over the sill and sinks until it reaches its level of neutral density on top of the east lobe chemocline. *Yusa* [1977] concluded that the shoulder marked a thermohaline convection cell, but gradients are too smooth, and even the stability minimum is far too strong, for this to be the case. Support for our hypothesis will be presented in connection with discussion of Figure 5 and horizontal variability.

Before doing so, it is helpful to put the Lake Bonney profiles of conductivity and stability in perspective by comparing their magnitudes with values typical of more familiar aquatic systems. For the range in salinities of 0.01 to 0.25 (10 to 250 mg liter $^{-1}$) for "normal fresh waters" quoted earlier, corresponding conductivities at 15°C are 0.0018 to 0.036 S m^{-1}. Conductivity of seawater at 15°C and salinity 35 is 4.914 S m^{-1}. Peak conductivity in the east lobe is nearly three times the value for seawater. Stability defined in Equation 6 is essentially the gradient of potential density divided by potential density, $(1/\rho_\theta)(\partial\rho_\theta/\partial z)$, where ρ_θ is potential density, and can be approximated for a freshwater thermocline of thickness z as $(1/\rho)(\Delta\rho/\Delta z)$, where ρ is the average density in the thermocline and $\Delta\rho$ is the difference in density across the thermocline. For a temperature change of 20°C to 4°C the relative density difference $\Delta\rho/\rho = 1.8 \times 10^{-3}$, giving stability for a 15 m thick thermocline of E = 1.2×10^{-4} m^{-1}. Peak stability of the west lobe chemocline is roughly 1000 times this size, while that in the east lobe is approximately 200 times larger.

In both lobes water in the first 5–10 m above the bottom appears to be of almost uniform density, with negligible stability (the segments labeled D in Figures 3 and 4). In fact these regions are quite stable, although they do not appear that way on the scale necessary to show the much larger peak values. Average stability below 30 m in the west lobe is 3.7×10^{-4} m^{-1}, while that between 28 and 34 m in the east lobe is 3.3×10^{-4} m^{-1}; in both cases about 90% of the stability is contributed by the salt gradient, and 10% by the temperature gradient. The region of nearly uniform conductivity continues to the bottom in the west lobe, while in the east lobe there is a boundary layer about 1 m thick in which conductivities increase sharply to larger values near the lake bottom. This is consistent with what is known of the lake beds in the two lobes. The bottom in the east lobe contains large crystals of halite (NaCl) overlying beds of dihydrohalite, halite, aragonite and gypsum to a depth of at least 1.6 m [*Wilson et al.*,

1974a, *Craig et al.*, 1974], possibly the remains of the hypothetical drying event that reduced the east lobe to a hypersaline pond more than 1000 years BP. One would expect there to be a diffusive sublayer above the salt crystals in which concentrations approached saturation. In the west lobe, however the lake bed is more conventional, consisting of a thin layer of gypsumniferous silts overlying subaerially weathered tills [*Gumbley*, 1975, cited in *Hendy et al.* 1977]. There is no salt-saturated sublayer, hence the observed zero slope of the salt gradient at the bed, the condition necessary for there to be no flux of salt from the lake bed to the water column.

The sharp local maxima in stability in the east lobe, labeled B and C in Figure 4, are associated with transport from the west lobe, and are best discussed in connection with horizontal transport processes.

Horizontal variability. Temperature and conductivity profiles shown in Figure 5 are representative of profiles that have been measured on several longitudinal transects through the lake. Using temperature as a tracer to track water movements, it is possible to identify horizontally distinct subregions along the transect: water immediately bordering the Taylor Glacier at the western end of the west lobe (site W5, Figure 5), the main body of the west lobe (sites W10–W30), the narrows (N10–N30), and the main body of the east lobe (E10–E50). The above classification is based on fine-scale variability (especially temperature inversions) that can be observed in the temperature profiles even at the scale of Figure 5. We interpret this variability as the result of interleaving of water layers that have sources of different temperature. (Micro-scale variability, which we interpret as turbulence, cannot be seen at the scale of Figure 5 and will be discussed later). For such interleaving there can be no corresponding variability or inversions in conductivity, because salinity controls density, and buoyancy forces dictate that interleaving will occur where densities, and hence salinities and conductivities, match; inversions in conductivity would signal instability. Depending on the age and vertical extent of an interleaving layer, and whether there is sufficient circulation or turbulence within the layer to keep it mixed, one might observe uniform conductivity through the layer. However such uniform steps do not appear in the fine-scale conductivity profiles in Lake Bonney. In Lake Bonney the causes of the interleaving are intrusions of meltwater from the Taylor Glacier at the west end of the west lobe and exchange flow between basins through the narrows. Meltwater intrusions from the Taylor Glacier have already been discussed.

Conductivity profiles in the two lobes are virtually identical above the level of the narrows' sill (see dotted horizontal line in Figure 5), as a result of the net west-to-east flow of meltwater. However temperature differences between lobes above sill level, and any small salinity differences that do exist, create horizontal density gradients across the sill, resulting in a complex exchange flow between basins through the narrows that is superimposed on the net west-to-east flow. The net west-to-east flow is driven by barotropic (i.e., external) pressure gradients caused by a higher piezometric surface in the west lobe when inflows to the west lobe are large enough to raise the (imaginary) free water surface level there above that of the east lobe. (The "imaginary" level is the elevation to which the water rises in a hole drilled through the ice, or the elevation of the water surface in a moat, when a moat exists; the terms "free surface" and "piezometric surface" are synonymous when pressures are hydrostatic, an assumption that can be safely made here.) Note that it is the elevation of the free surface that matters in determining flow direction, not the actual volume of water. Because the west lobe is smaller than the east lobe, inflows into the west lobe do not necessarily need to be greater in volume than those to the east lobe, although this does seem to be the case, in order to create a west-to-east barotropic pressure gradient. The barotropic component of pressure gradient causes a one-way flow, down the gradient. Bi-directional or true exchange flow, on the other hand, is driven by baroclinic (internal) pressure gradients that arise from horizontal variations in density. While steady two-layer exchange flows between basins of uniform densities have been analyzed successfully (e.g., *Wood* [1970], *Farmer and Armi* [1986]), the flow between the two lobes of Lake Bonney is unsteady, multi-layered, and very complex, and no attempt will be made here to analyze it. (Analogous, but more energetic, flows have been observed recently at depths greater than 4000 m in the Romanche Fracture Zone between the Brazil and Sierra Leone Basins of the Atlantic Ocean by *Polzin et al.* [1996], as dense Antarctic Bottom Water makes its way north.) An aspect of the Lake Bonney exchange that is of particular interest is the intermittent overflow of salty water from west to east hypothesized to be the source of the "shoulder" in all the east lobe conductivity profiles. That west lobe water does reach this depth can be seen from temperature inversions below sill level in the profile for N30, Figure 5; in particular the notch noted

with an "i" marks an intrusion of colder water that can only have come from the west lobe and is found in all N30 profiles at an elevation that coincides with the middle of the shoulder in the east lobe conductivity profiles. At the bottom of the shoulder there are sharp, local stability maxima that appear in all east lobe profiles (C in Figure 4) and that we interpret as marking the lower depth limit for the influence of west lobe water in the east lobe. The local stability maximum at B, Figure 4, lies just above the top of the sill and is a greatly truncated (by the blocking effect of the sill on flow from west to east) version of the top portion of the west lobe stability profile.

Thermohaline convection (double-diffusive processes) and meltwater intrusions. There are no basin-scale thermohaline convection cells in Lake Bonney like those found in Lake Vanda. Thermohaline convection, if in fact any occurs at all in Lake Bonney, must be limited to the vicinity of the submerged Taylor Glacier face in the west lobe. Recall that in order for "diffusive type" convecting cells to form, the water column must be (1) stabilized by a salt gradient (but not too stable), and (2) destabilized by a heat flux from below or, equivalently, a temperature gradient with temperature increasing with depth. (If temperatures are below the temperature of maximum density, then the directions corresponding to a destablizing heat flux or temperature gradient are reversed.) There is a rather narrow range of conditions over which the cells will form; if the salt gradient is too strong, then convection cannot occur; if the temperature gradient is too strong it will make the water column statically unstable overall, leading to overturn. Mathematically, the range of conditions that will allow cells to form is usually expressed as [*Turner*, 1973, pp.255-256]:

$$Rs > Ra > \frac{Pr+\tau}{Pr+1} Rs + \left(1+\frac{\tau}{Pr}\right)\frac{27\pi^4}{4} \qquad (7)$$

where $Rs = g\beta\Delta Sd^3/(\nu\kappa)$ is the salt Rayleigh number, g is acceleration of gravity, β is the compressibility for dissolved salt (defined earlier), ΔS is the change in salinity that occurs over depth d, d being the extent of the water column being investigated for conditions of instability, ν is the (molecular) kinematic viscosity of the salt solution, and κ is the (molecular) diffusivity for heat of the salt solution; $Ra = g\alpha\Delta Td^3/(\nu\kappa)$ is the thermal Rayleigh number, α is thermal compressibility (defined earlier), and ΔT is the change in temperature that occurs over depth d; $Pr = \nu/\kappa$ is the Prandtl number; and $\tau = D/\kappa$ is the ratio of diffusivity for salt to that for heat. The left-hand inequality assures overall static stability ($\beta\Delta S > \alpha\Delta T$, i.e., that the stabilizing density difference due to salt is greater than the destabilizing density difference due to heat), while the right hand inequality assures that the destabilizing temperature gradient is strong enough to generate some convective motion. The value of $27\pi^4/4 \approx 657$ is the critical Rayleigh number for free convection to occur in a single component (i.e., heat only, no salt) fluid; the other terms on the right-hand side of the inequality are associated with the effects of a salt gradient. The right-hand inequality in Equation 7 is based on assumptions of linear heat and salt gradients and idealized (stress-free, constant temperature, constant salinity) upper and lower boundaries, conditions that are not met when applying the criterion to the dry valleys lakes. For our purposes these restrictions are minor considerations and Equation 7 serves as an adequate guide to whether or not thermohaline convection is likely to occur. Although diffusivities for momentum, heat and salt (ν, κ, and D) vary with salinity, temperature, and pressure (values are calculated for every data point as part of the CTD data processing), the variation is not great and it is possible to simplify Equation 7 by using typical values for Lake Bonney. Moreover inspection of Figures 3 and 4 (which show the variation of temperature of maximum density with depth) indicates that in the west lobe the temperature profile is stable over almost the entire water column, ruling out the possibility of thermohaline convection anywhere in the interior of the basin, while in the east lobe an unstable temperature gradient exists only in a range of depths from 6.2 m to 12 m, just above the stability maximum at B (Figure 4). Typical values for diffusivities in this depth segment are $\nu = 1.5\times10^{-6}$ m^2 s^{-1}, $\kappa = 1.4\times10^{-7}$ m^2 s^{-1}, and D = 8.8×10^{-10} m^2 s^{-1}. Substituting these values into Equation 7 gives:

$$Rs > Ra > 0.92Rs + 660 \qquad (8)$$

and since both Ra and Rs are very large (of order 10^9 to 10^{11}) compared with 660, this can be simplified still further to:

$$1 > \frac{\alpha\Delta T}{\beta\Delta S} > 0.92 \qquad (9)$$

Considering 1 m intervals over the depths 6–12 m in the east lobe, typical values for the relative density differences are $\alpha\Delta T \approx 1.8\times10^{-5}$ and $\beta\Delta S \approx 1.8\times10^{-3}$,

giving the ratio $\alpha\Delta T/(\beta\Delta S) \approx 0.010$, only about 1% of the value needed to allow cells to form. The maximum value that the ratio reaches over the depths of interest is only 0.015. Hence density stratification in the interior of Lake Bonney is far too stable to allow thermohaline convection to develop.

Thermohaline convection can also be generated at boundaries under some circumstances. The effects of heating the side walls of solar ponds or laboratory tanks containing salt-stratified water have been extensively investigated [e.g., *Schladow and Imberger*, 1987; *Jeevaraj and Imberger*, 1991; *Schladow et al.*, 1992]. Much less attention has been devoted to the effects of vertical ice walls melting in salt water [e.g., *Huppert and Turner*, 1978, 1980; *Huppert and Josberger*, 1980; *Josberger and Martin*, 1981], and of these studies none has considered the exact situation relevant to the McMurdo Dry Valleys lakes–the ice wall bounding a mass of ice on one side that can act as a sink for heat at temperatures well below 0°C, and on the other side water containing gradients of both heat and salt, with water temperatures extending several degrees below 0°C. But the sidewall heating and ice-melting studies do have the following phenomenon in common with the dry valleys lakes: introducing a source of buoyancy, such as heat or fresh water, at the lateral boundary of a salt-stratified solution will almost always give rise to one or more horizontal (or near-horizontal) intrusions from the sidewall into the fluid. Double-diffusive effects may or may not enter into the process; this depends on the details of the boundary conditions, the existing gradients in the main body of the fluid, and the thermodynamics of melting and subsequent dilution of meltwater with ambient lakewater. In the west lobe of Lake Bonney, temperatures measured below the chemocline at site W5, approximately 50 m from the shoreline face of the Taylor Glacier, are 1°C to 2°C cooler than in the interior of the basin (Figure 5), indicating considerable heat loss from the water at temperatures between 0°C and –4°C to the ice. Some of this heat must be taken up as latent heat for melting, and the highly irregular shape of the temperature profile provides evidence of several intrusions propagating from the boundary. Double diffusive effects may be of minor importance in the formation and propagation of these intrusions.

Turbulence. We interpreted the presence of any extensive microstructure activity, at scales smaller than that of the resolution of the fine-structure probes, as evidence of turbulence. We checked for this by stretching the fine-structure profiles with spline interpolation to give them the same number of data points as the microstructure profiles, correcting the microstructure for any offsets from the fine-structure at the 8-Hz sampling frequency of the fine-structure, and then computing differences T′ between the adjusted microstructure and the smoothly interpolated fine-structure. An example of the results is shown in the left-hand graph of Figure 6, a profile of the temperature differences T′ between fine- and microstructure scales at site N21 in the narrows. (Note that temperature gradient could equally well have been plotted; the general appearance of the graph would not change. We felt that values of temperature differences, rather than gradients, were easier to interpret. Temperature gradients were used in calculation of the sample spectra shown in Figure 6, as discussed below.) Turbulence is present at the top and bottom of the profile (segments A, B, and D in Figure 6), while the middle of the water column (segment C) is essentially devoid of turbulent activity.

Except in the narrows, all our profiles for T′ appeared much the same as segment C in Figure 6. This indicates lack of turbulence throughout the main bodies of both basins, with turbulence occurring only where inter-basin flows must reach their maximum velocities and closest proximity to solid boundaries. Undoubtedly turbulence occurs near shore and under the ice during energetic episodes of meltwater runoff, but we never had the opportunity to capture one of these episodes with a CTD cast. We did not carry out any casts in moats.

Turbulence in segment A, Figure 6, is inside the ice hole, and was probably generated when clearing the hole of any surface ice and then lowering the CTD into the hole for initialization. Similar disturbances are present in the ice hole segments of all casts, but the disturbances do not propagate below the bottom surface of the ice. We interpret segment B as turbulence produced by flow in a boundary layer next to the ice (the bottom of the ice is far from flat and smooth), while turbulence in segment D is produced in a boundary layer next to the rough bed. The channel cross-section is quite narrow over the depths spanned by D, making it likely that turbulence found within this segment is generated at both the bottom and sides of the cross-section.

If the variance of T′ is essentially constant within a segment (and this is the case for segments B and D), then scales of motion and levels of energy dissipation can be quantified by fitting a theoretical model, known as the Batchelor spectrum, to the power spectra calculated for microstructure temperature gradients in

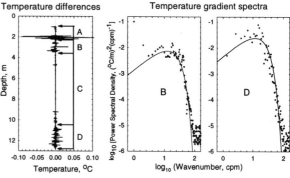

Temperature microstructure, Lake Bonney, N21, Jan 12, 1991

Fig. 6. Temperature microstructure from a CTD cast on the sill in the narrowest section of the connecting channel between east and west lobes, Lake Bonney, Jan. 12, 1991. The first graph is a profile of differences between adjusted microstructure and fine-structure profiles; the regions of the profiles labeled with letters A to D are described in the text. The second two graphs show power spectra of microstructure temperature gradients for profile segments B and D; the crosses are the sample spectra, and the solid curves show the fitted Batchelor spectra. Dissipation values associated with the fitted spectra are 3.8×10^{-10} m^2 s^{-3} for B and 3.0×10^{-10} m^2 s^{-3} for D; corresponding Batchelor lengths are 1.9 cm and 2.0 cm, respectively.

each segment [*Caldwell et al.*, 1980; *Imberger*, 1985b; *Imberger and Ivey*, 1991]. The Batchelor spectrum for vertical gradients of a passive, scalar contaminant in homogeneous, isotropic turbulence, is given as [*Gibson and Schwarz*, 1963; *Dillon and Caldwell*, 1980]:

$$S(k) = \sqrt{q/2} \; 2\pi \, Vf(a)/(k_B \kappa)$$

$$V = 12\pi \; \kappa \int_0^\infty S(k)dk = 6\kappa \operatorname{var}(\partial T'/\partial z),$$

$$f(a) = a \left\{ e^{-a^2/2} - a \int_0^\infty e^{-x^2/2} dx \right\},$$

$$a = \sqrt{2q} \; k/k_B \qquad (10)$$

S (k) is the power spectral density (in units of $(^\circ C\, m^{-1})^2$ $(cpm)^{-1}$, where cpm denotes cycles per meter) of the microstructure temperature fluctuation gradients, $\partial T'/\partial z$; k is wavenumber (cpm); q is a universal constant, taken as $2\sqrt{3} \approx 3.464$; V is 6κ times the variance of the microstructure gradients (equal to $12\pi\kappa$ times the area under the spectrum); κ is molecular thermal diffusivity; and a is wavenumber scaled by $\sqrt{2q}$ and nondimensionalized by k_B, where k_B is the Batchelor wavenumber (cpm). The Batchelor wavenumber (or more precisely, the Batchelor length, which is 2π over the Batchelor wavenumber) is the length scale for the smallest temperature fluctuations that one would expect to find in a fluid, given the intensity of turbulence that exists in the fluid. Smaller-scale fluctuations would be smoothed by molecular diffusion; the more energetic the turbulence, the smaller the size of the smallest temperature fluctuation that can be maintained against the effects of diffusion. The Batchelor wavenumber is related to the rate at which turbulent kinetic energy is being dissipated in the fluid by [Dillon and Caldwell, 1980; Caldwell et al., 1981]:

$$2\pi k_B = \left[\varepsilon / (\nu \kappa^2) \right]^{1/4} \qquad (11)$$

where ε is the rate of turbulent kinetic energy dissipation per unit mass ($m^2 s^{-3}$) and ν is molecular kinematic viscosity. In practice, one uses Equations 10 and 11 with k_B (or ε) and V as parameters to fit the Batchelor spectrum to a measured spectrum, having first determined that the corresponding segment of the temperature difference profile is stationary and that the sample spectrum resembles the Batchelor spectrum. The best fit gives estimates of k_B and ε for that segment of the profile. Automated segmentation algorithms and nonlinear fitting procedures to obtain the best fit have been described by Imberger and Ivey [1991], but we have carried out the process for segments B and D in Figure 6 in a series of separate stages, "by hand." The sample spectrum of segment D is based on 683 points and was calculated by averaging results from fast Fourier transforms of nonoverlapping 256-point sections using a Hanning window. For segment B only 256 points were available and the sample spectrum is based on a single Fourier transform. In both cases the fit to the shape of the Batchelor spectrum (the solid curve) is reasonable. Some discrepancy at either end of the spectrum is to be expected, by "contamination" of the sample spectrum at lower wavenumbers by nonturbulent (e.g., internal wave) motions, and at the high wavenumber end by noise [Caldwell et al., 1980]. Fitting was done by trial and error, the "best" fit having been judged by inspection. This was adequate for estimating dissipation to within two significant figures. Estimates for dissipation and Batchelor wavenumber are 3.8×10^{-10} $m^2 s^{-3}$ and 53 cpm for segment B, and 3.0×10^{-10} $m^2 s^{-3}$ and 50 cpm for segment D. These energy levels are very low, at least an order of magnitude smaller than the lowest values that have been reported from active turbulent patches in and below the thermocline in lakes without ice cover by Imberger [1985b] and Imberger and Ivey [1991], and of the same order (or less) than the detection limits for dissipation given by those authors. We believe the results shown in Figure 6 to be real, however, and not due to instrument noise. Turbulence was rarely detected in our CTD casts, but when a patch was detected, the signal was very clear. The estimated Batchelor lengths of 1.9 cm to 2.0 cm are six to seven times larger than the sampling resolution of the microstructure temperature sensors at the drop speed of approximately 0.24 m s^{-1} used for the cast. The length of the data segments are sufficient to estimate a spectrum, and although energy levels are low, it should be noted that density stratification was also very weak in regions where turbulence was detected, making very weak turbulence possible. Even if the total energy in the spectrum is not accurately resolved (i.e., the variance V in Equation 10), the fitting method is not sensitive to V, but depends mainly on k_B, the Batchelor wavenumber, which acts as a location parameter in the fitting process, and from which the value of dissipation is estimated (Equation 11). Further smoothing of the sample spectra would improve their resemblance to the theoretical spectrum, but there seems little point in doing so. Finally, the dissipation estimates are consistent with a simple integral scale estimate based on current meter records obtained in the narrows with S4 current meters (Priscu, unpublished data) that show velocities on the on the order of 1 to 2 cm s^{-1}. Assuming a friction factor of f = 0.02 (this is based on the cross-sectional shape in the narrows, with a roughness size of about 10 cm; but the results are not sensitive to reasonable variations in f), a shear velocity may be estimated from [e.g., White, 1979] $u_* = u \sqrt{(f/8)} \approx (0.015$ m s$^{-1})(0.05) = 7.5 \times 10^{-4}$ m s^{-1}, and an estimate for dissipation in a closed conduit with hydraulic radius R $\approx$ 2.4 m (cross-sectional flow area 235 m^2 and wetted perimeter 100 m) is $\varepsilon \sim u_*^3 / R = 1.8 \times 10^{-10}$ m^{2}s^{-3}.

Other Lakes

Lake Fryxell. Lake Fryxell (Figure 7) is the shallowest of the seven basins included in this study; recently it has been the center of intensive hydrological and chemical studies, some of the results of which are presented elsewhere in this volume [e.g., Conovitz et al.; Lyons et al.; Lizotte and Priscu]. Its ice cover is thicker and rougher than that of Lake Bonney, and its water less transparent because of the relatively high

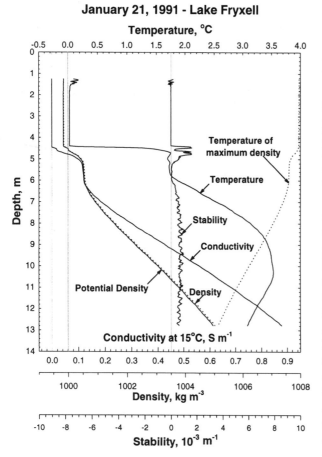

January 21, 1991 - Lake Fryxell

Fig. 7. Results (fine-structure) from a CTD cast in Lake Fryxell at a mid-lake site, Jan. 21, 1991. Zero depth corresponds to the free water surface.

amounts of phytoplankton and dissolved organic carbon it contains [*Lizotte and Priscu*, 1992; *Lizotte and Priscu*, this volume; Vincent, personal communication]. The temperature maximum of 3.53°C is lower than that of the east lobe of Lake Bonney (6.3°C), and occurs at a depth of 10.3 m for the cast shown in Figure 7, almost 5 m less than the 15 m depth of the temperature maximum in the east lobe of Lake Bonney. The cooler maximum temperature and the shallower depth at which it occurs is likely a consequence of Lake Fryxell's thicker ice cover, less transparent water, and greater snow cover resulting from its coastal location. The ice cover of the lake just makes contact with the Canada Glacier in a shallow embayment at the western end of the lake, but there is little direct contact between the Canada Glacier and the water column of the lake. Profiles of temperature and conductivity show no evidence of

turbulence or thermohaline convection, and visual evidence of horizontal intrusions is limited to an approximately 1 m thick layer just below the ice, demarcated by steps in the conductivity and temperature profiles of Figure 7 between 5 m and 6 m, similar to those caused by the high 1990-1991 summer meltwater inflows in the January 1991 Lake Bonney profiles. Below this step, conductivity increases almost linearly to the bottom, although salinities are much lower than those found in Lake Bonney. The water column is stable over its entire depth below the ice, with stabilities ranging between 0.5×10^{-3} m^{-1} and 1.0×10^{-3} m^{-1} below 7.5 m, four to eight times higher than the figure of 1.2×10^{-4} m^{-1} quoted earlier for a typical freshwater thermocline. The temperature gradient is unstable over a small range in depth between 8.8 m (where the profile intersects the temperature of maximum density, Figure 7) and 10.3 m, where temperature reaches its maximum value. The stability ratio $\alpha\Delta T/(\beta\Delta S)$ over this range is of the order 5×10^{-4}, far too small to satisfy the criterion of Equation 9 for the onset of thermohaline convection. In summary, there is nothing in the temperature or conductivity profiles, either at fine-scale or micro-scale, to contradict the assumptions of earlier workers [*Hoare et al.*, 1965; *Lawrence and Hendy*, 1985] that vertical transport in Lake Fryxell is dominated by molecular diffusion. *Miller and Aiken* [1996], however, provide evidence from tritium measurements that bottom water between 14 m and 18 m was "recently at the surface," and they hypothesize that it found its way to depth as a gravity current down the slope of the lake bed after being enriched in solutes either by salt exclusion from moat ice when freezeover occurred, or "by dissolution of soluble salts during subsurface transport" (i.e., as a shallow groundwater input).

Lake Hoare. Very little seems to be known about the physical limnology of Lake Hoare. The few references that we are aware of are from studies in which temperature and salinity of the water column were only of peripheral interest [*Wharton et al.*, 1986, 1987; *Palmisano and Simmons*, 1987; *Priscu*, 1995; *Priscu et al.*, 1996; *Priscu*, 1997], and these generally refer to the water column as being fresh (or nearly fresh) and isothermal. Our profiles (an example is shown in Figure 8) indicate that although salinities are low, they do reach levels nearly three times greater than the upper limit quoted earlier for "normal" freshwaters of 250 mg liter^{-1} (the corresponding conductivity at 15°C is approximately 0.036 S m^{-1}). The salinity structure is similar to that of the other lakes in some respects (there

Nov 23, 1995 - Jan 22, 1996, Lake Hoare

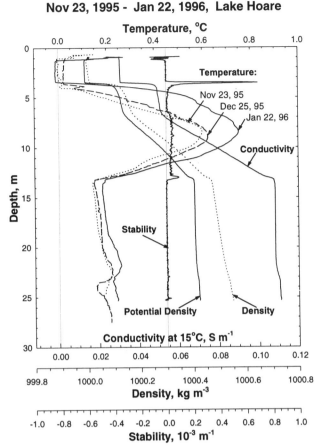

Fig. 8. Results (fine-structure) from a CTD cast in Lake Hoare at a mid-lake site, Jan. 22, 1996; also shown are temperature profiles from casts on Nov. 23, 1995, and Dec. 25, 1995. Zero depth corresponds to the free water surface.

is an overall increase in salinity from top to bottom, including a region with a relatively strong gradient), but there are also some distinctive features. There is a step-increase in salinity immediately below the ice (at 3.5 m below the free water surface in Figure 8); below this step is a stable gradient extending to just below 13 m. From 13.5 m to 23 m there is a layer of nearly uniform salinity, below which salinity again increases slightly. The cast shown in Figure 8 stops well above the bottom, which is at approximately 30 m; but casts to deeper levels show that the salinity increases more rapidly as the bottom is approached.

Temperature varies over a range of less than 1°C, maximum temperatures occurring just below 8 m. The relatively low temperatures of Lake Hoare are probably due to a combination of circumstances; the rough surface and relatively low transparency of the ice [*McKay et al.*, 1994], the direct contact between the

Canada Glacier and the water column over a significant fraction of total lake depth, snow accumulation, and the shading effects of surrounding mountains [*Dana et al.*, this volume]. The three temperature profiles included in Figure 8 provide a good illustration of warming over the summer by solar heating.

Water temperatures in Lake Hoare are all below the temperature of maximum density; hence portions of the temperature profile in which temperature decreases with depth contribute to instability of the density profile, while those in which temperature increases with depth contribute to stability. A destabilizing temperature gradient occurs from 8.5 m to 13 m, but is not strong enough to overcome the stabilizing effects of the salinity gradient. Stability between 7 m and 13 m is in the range 2×10^{-5} m^{-1} to 9×10^{-5} m^{-1}. Although low (the average value of 4×10^{-5} m^{-1} is less than half that quoted earlier for a freshwater thermocline with a temperature drop from 20°C to 4°C over 15 m), these stabilities are more than adequate to weaken any turbulence that might be generated, and in fact no evidence of turbulence was found in microstructure measurements over the entire water column below the ice cover. Thermohaline convection would be possible only over those depths in which there is a destabilizing temperature gradient, but values of the density-difference ratio, $\alpha \Delta T/(\beta \Delta S)$, are in the range 0.1 to 0.3 between 8.5 m and 13 m, not large enough to satisfy the condition of Equation 9 for the onset of instability, although much closer than in either Lake Bonney or Lake Fryxell. There was no evidence of thermohaline convection in either fine-structure or microstructure measurements.

One of the most striking aspects of Figure 8 is the fairly abrupt change in the character of the profiles that occurs just below 13 m. Below 13 m stability is very weak, averaging only 1.6×10^{-6} m^{-1} between 14 m and 21 m. Conductivity between these depths is constant (conductivity at 15°C = 0.1083 S m^{-1}), what little stability there is being provided by a very small increase in temperature with depth. The uniformity of the conductivity profile, coupled with the absence of a smooth, diffusive transition to the stronger gradients above 13 m, leads one to speculate that the water column between 14 m and 21 m has been recently mixed, despite the arguments above that rule out either thermohaline convection or turbulent diffusion at the time the profiles were measured. Profiles of dissolved gases made in this lake [*Priscu*, 1997; Priscu, unpublished data] exhibited a great deal of vertical variability. McKay (personal communication) has interpreted this variability as being caused by water that had initially been in contact with either lake ice or glacier ice, having had its gas content altered by freeze-thaw processes and in some cases biological activity, then moving through the lake with little mixing because of the absence of turbulence. This is consistent with our failure to detect turbulence, and implies that the stirring alluded to above may be the result of convective circulation associated with the submerged vertical ice face of the Canada glacier. Another possibility is that thermohaline circulation occurs below 14 m if there are periods when the temperature gradient below 14 m reverses and becomes unstable due to loss of heat at depth during winter. These questions might be resolved by measuring temperature and conductivity profiles along a transect starting at the Canada Glacier and running the length of the lake. This would help clarify the role that the glacier plays in generating horizontal intrusions and any convective circulations. Additionally collecting profiles at a central site over time would give information on how the vertical distribution of heat changes during the year.

Lake Joyce. Lake Joyce, at an altitude of 325 m in the upper Taylor Valley and dammed along one side by the Taylor Glacier, has the lowest maximum water temperature of any of the lakes included in this survey; its maximum measured ice thickness of 5.6 m is a close second to that of 6 m measured for Lake Miers (Table 1). The large amount of fine-scale variation evident in Lake Joyce's temperature profile (Figure 9) contrasts markedly with the relative smoothness of its conductivity profile, and provides further evidence of horizontal intrusive flows to be expected in lakes that have direct contact with glaciers. Density is controlled by salinity, the density profile being virtually a copy of the conductivity profile. The conductivity profile bears some resemblance to that of the east lobe of Lake Bonney, giving one reason to believe that an increased input of fresher meltwater overflowed higher salinity water at some time in the past. Salinities are not nearly as high as those of Lake Bonney, maximum values being little more than one-tenth those of seawater. The resemblance to the east lobe of Lake Bonney extends to the occurrence of a "shoulder" in conductivity (between 10 m and 14 m in Lake Joyce). We explained the shoulder in the Lake Bonney profile as the result of a colder, saline intrusion from the west lobe. In Lake Joyce the source is also probably an intrusion, this time from the Taylor Glacier, as evidenced by the colder

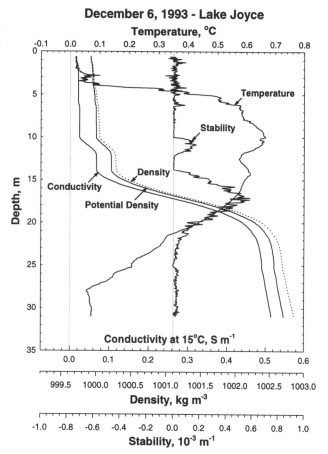

Fig. 9. Results (fine-structure) from a CTD cast in Lake Joyce at a mid-lake site, Dec. 6, 1993. Zero depth corresponds to the free water surface.

temperatures between 10 m and 14 m that appear as a sideways U-shaped groove in the temperature profile. The density profile of Lake Joyce is statically stable, although there are two segments of nearly uniform density (one from 5.5 m to 10 m, the next from 12 m to 13.8 m) in which stability approaches zero (average stability 6.3×10^{-6} m^{-1} in the first segment, 3.2×10^{-6} m^{-1} in the second). There do not appear to be any thermohaline convection cells. The two near-uniform density segments of the profile coincide with stabilizing temperature gradients (temperature increasing with depth below the temperature of maximum density). In regions where the temperature gradient is unstable, the density-difference ratio, $\alpha\Delta T/(\beta\Delta S)$, is too low to satisfy the criterion for onset of double diffusive instability (Equation 9), being of the order of 0.05; the ratio reaches its highest values (approximately 0.2) around 25 m. Microstructure profiles indicate that meltwater intrusions immediately beneath the ice cover

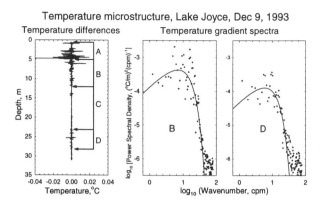

Temperature microstructure, Lake Joyce, Dec 9, 1993

Fig. 10. Temperature microstructure from the CTD cast in Lake Joyce shown in Figure 9. The format of the graphs is the same as that described in the caption for Figure 6. Dissipation values associated with the fitted spectra are 3.7×10^{-11} m^2 s^{-3} for B and 1.8×10^{-11} m^2 s^{-3} for D; corresponding Batchelor lengths are 3 cm and 4 cm, respectively.

and near the bottom may be energetic enough to generate turbulence, although the turbulence is very weak by the time it reaches the sampling station in the center of the lake. The microstructure activity is found in regions of low stability (segments B and D in Figure 10); as with profiles measured in the narrows of Lake Bonney (Figure 6), turbulence also occurs inside the ice hole (segment A, Figure 10), while the middle of the water column is quiet (segment C, Figure 10). Dissipation rates, 3.7×10^{-11} m^2 s^{-3} and 1.8×10^{-11} m^2 s^{-3} for segments B and D, respectively, Figure 10, estimated by the spectrum-fitting method described earlier, are an order of magnitude smaller than those measured in Lake Bonney, and the Batchelor length scales correspondingly larger (3 cm to 4 cm). There appears to be more scatter in the spectra shown in Figure 10, especially at lower wavenumbers, than in those presented for Lake Bonney (Figure 6) and Lake Miers (Figure 13). This is probably due to the very low energy levels of the turbulence by the time it reaches mid-lake, and the turbulence may no longer be active.

Lake Vanda. *Vincent* [1987] has called Lake Vanda in the Wright Valley "probably the best known of all Antarctic lakes." In terms of physical limnology, Lake Vanda is perhaps the most unusual lake among a group of unusual and unique lakes. Of the lakes considered here, Lake Vanda has the greatest depth, the most transparent ice with the smoothest surface, the warmest water temperatures, the largest inflowing meltwater stream (the Onyx River; see *Chinn,* [1981]), water with transparency comparable to that of the

clearest of natural waters, and temperature and conductivity profiles that provide one of the most striking natural examples of thermohaline convection cells found in any lake in the world (Figure 11). While thermohaline convection has been documented in other lakes, for example by *Osborn* [1973] in Powell Lake (a fiord lake in British Columbia, Canada) and *Newman* [1976] in Lake Kivu (an East African rift lake), nowhere (except in Vanda) are convection cells to be found with vertical spans of up to 20 m and temperature differentials of nearly 1°C. Only in 500 m deep Lake Kivu, where heat and salt are supplied at depth by submerged geothermal springs, do convection cells occur over a greater range of depths, though the individual cells are much smaller than in Lake Vanda. As discussed earlier, the much higher temperatures in Lake Vanda compared with the other dry valleys lakes

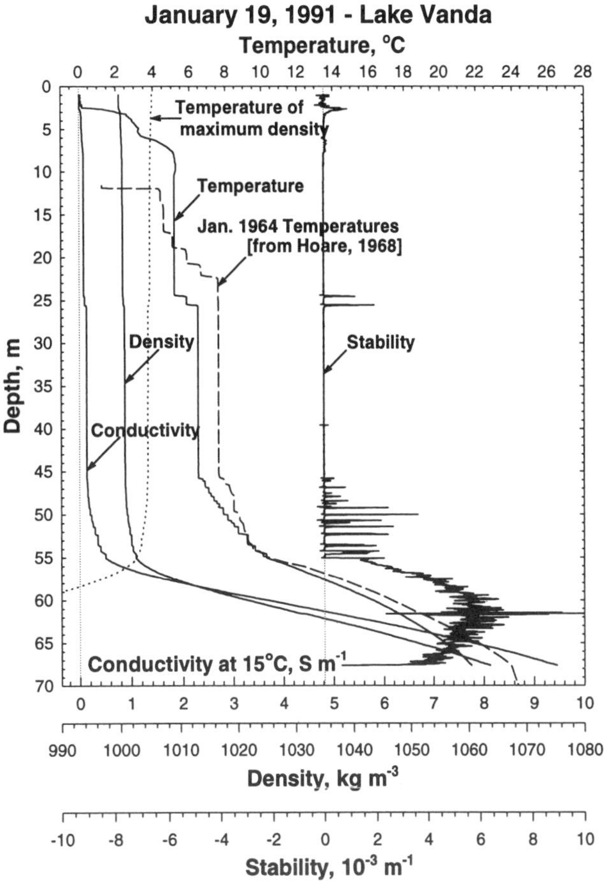

January 19, 1991 - Lake Vanda

Fig. 11. Results (fine-structure) from a CTD cast in Lake Vanda at a mid-lake site, Jan. 19, 1991. Jan. 1964 temperature data from *Hoare et al.* [1968] are shown for comparison, shifted down by 8 m to account for rise in water level since 1964. Zero depth corresponds to the free water surface on Jan. 19, 1991.

must be largely due to the greater transparency of Vanda's water and ice cover, although differences in cloudiness and shading by surrounding hills may also play a role [see *Dana et al.*, this volume]. In Lake Bonney, we found that the influence of solar radiation on water temperatures appears not to extend much below 25 m, a depth at which only 0.11% of incident solar radiation remains. Similar considerations for Lake Vanda (assuming values from *Howard-Williams et al.* [this volume] of 13% transmission through the ice, an extinction coefficient of 0.045 m^{-1} and a 3 m thick ice cover) show that at 25 m depth, 4.2% of the incident light remains, and that at this rate the 0.11% light level would be deeper than 100 m, well beyond the maximum depth (75 m) of the lake. This calculation is unrealistic for another reason: it does not account for the nearly four-fold increase in extinction coefficient associated with the deep chlorophyll maximum below 55 m, as documented by *Howard-Williams et al.* [this volume], or the even larger increase in extinction coefficient in the 6 m above the bottom reported by *Hoare* [1968] to a value of 1.9 m^{-1}. Such increases in the rate at which solar energy is absorbed in the water column contribute to the rapid temperature rise below 55 m in Lake Vanda (see *Lewis et al.'s* [1983] discussion of the effect of chlorophyll maxima on local heating rates in the upper ocean).

In spite of the scientific interest aroused by Lake Vanda's temperature and salinity profiles, no one has yet been able to provide a complete explanation of how they developed. In part this is due to lack of long term data on river inflows, lake levels, and solar radiation, and in part to mathematical difficulties inherent in modeling thermohaline convection. The linear stability criteria given in Equations 7, 8, and 9 only tell whether or not convection is likely to occur, but give no information about the details of the motion that ensues. The onset of instability observed in experiments has been generally found to agree well with the predictions of linear theory [e.g., *Shirtcliffe*, 1967], but detailed calculations to describe the subsequent evolution of initially linear stratification into the staircase structure that is observed, containing not one but a series of convecting layers, has so far eluded applied mathematicians. Scaling arguments and experimental observations suggest that the vertical extent of the layers that form depends directly on the three-quarters power of the destabilizing heat flux and inversely on the strength of the initial stabilizing salt gradient [*Turner*, 1973, p. 265], so that the stronger the salt gradient relative to the heat flux, the smaller the layers

will be. This is in general agreement with the profiles for Lake Vanda (Figure 11), where the layers become smaller as the salinity gradient first begins to strengthen, and then disappear altogether as stability increases still further. The largest layer depths coincide with regions of low salinity gradient and higher heat flux (note that upward heat flux in the lake must decrease with depth if it is to balance, on average, the input of heat from the sun). Most of the variability in the stability profile below 55 m in Figure 11 is noise, for the same reasons discussed earlier in connection with stability calculations at high salinities in the east lobe of Lake Bonney. However the spike in stability just above 62 m is real (it appears in all casts); it is associated with a local steepening in conductivity gradient at that depth, the cause of which is unknown.

Calculations of the density-difference ratio, $\alpha \Delta T / (\beta \Delta S)$, based on overall temperature and salinity differences in the convecting regions, give values in the range $\alpha \Delta T / (\beta \Delta S) \approx 0.12$ to 0.17, too low to satisfy the criterion for the onset of instability given by Equations 7 to 9. This does not imply that convection is not occurring, a contradiction of what is plainly observed. It does imply, however, that if there were no convection (if we were to draw smooth curves through all the profiles that eliminated all the steps), that convection would not start up again, given the new values for temperature and salinity gradients. Similar results for $\alpha \Delta T / (\beta \Delta S)$ were presented by *Yusa* [1977], who was sufficiently disturbed by what he interpreted as a contradiction between theory and observations that he rederived the stability condition (Equation 7) in terms of heat flux instead of temperature difference. Assuming steady state conditions, he equated upward heat flux with average annual downward solar flux (he neglected heat conduction between the lake and the sediments), and was able to demonstrate that his new condition was satisfied over at least part of the convecting region. We reiterate, however, that the linear stability criterion says nothing whatsoever about the subsequent evolution of motion once convection has been initiated. Calculations based on nonlinear theory show that once a cell has been established, the ensuing "strong finite amplitude motions which exist at sufficiently large [thermal] Rayleigh numbers tend to mix the solute and distribute it so that the interior layers of the fluid are more nearly neutrally stratified. When this happens, the inhibiting effect of the solute gradient is greatly reduced and the fluid can convect nearly as much heat as it does in the absence of the solute" [*Veronis*, 1968, p. 327]. Published experimental results for heat fluxes in thermohaline convection cells extend to values of $\alpha\Delta T/(\beta\Delta S)$ as low as 0.14 [*Turner*, 1965], and *Osborn* [1973], who observed thermohaline cells in Powell Lake at $\alpha\Delta T/(\beta\Delta S) = 0.12$, suggests a lower limit for maintenance of convection as $\alpha\Delta T/(\beta\Delta S) = 0.07$.

There is no evidence of any turbulence in the strong gradients that separate the convection cells, or in the stable bottom waters below 55 m, indicating that diffusion is molecular between the cells and below 55 m. Our CTD cannot be used to detect turbulence within the cells because of the complete uniformity of temperature and salinity in the cells. Detection of turbulence with a CTD relies on variations in temperature and conductivity microstructure, created by small-scale, overturning movement of fluid containing gradients of temperature and conductivity, to indicate turbulence. A velocity-shear microstructure probe, which does not rely on temperature or conductivity differences, could be used to quantify turbulence within the cells. There is evidence that the circulation inside the convection cells is turbulent. Values of the thermal Rayleigh number ($Ra = g\alpha\Delta T d^3/(\nu\kappa)$, as defined earlier) calculated for the cells in Lake Vanda average 2.0×10^8 for the smaller cells found between depths of 45 m to 55 m, to 1.8×10^{12} for the larger cells between 10 m and 45 m. These values are much larger than the value of 3.7×10^4 at which turbulence is observed to occur in free convection between horizontal boundaries [*Turner*, 1973, p. 220], and indicates that the flow in Lake Vanda's cells is almost certainly turbulent. That turbulence is present is also consistent with *Ragotzkie and Likens'* [1964] description of the two releases of radioactive iodine-131 they made in the middle of the large, upper convecting layer in Lake Vanda. (Their releases were made at depths of 23 m and 20 m below the ice surface, as it was then. Referring to the profile taken from *Hoare's* [1968] measurements in Figure 11, these correspond to depths of approximately 31 m and 28 m on the scale of that figure.) The first release dispersed so rapidly that no trace of it could be detected within their observation grid covering a 20 m diameter circle when measurements were made three hours after the release. Within two minutes after the second release, made through a rubber tube of unspecified diameter, the tracer had spread through a vertical distance of 6 m, and within 5 minutes had disappeared completely from the release point. After 17 minutes, tracer was detected at the edge of their grid, 10 m from the release point in a 7 m thick column of water. Within 20 minutes the tracer disappeared completely from the grid, and no trace was observed when the grid

was resampled four and one-half hours later. *Ragotzkie and Likens* [1964] point out that "horizontal transport of 10 m in 17 min. indicates a current of 1 cm sec^{-1}. This is a high velocity for any ice-covered lake . . ." Further evidence of activity within the cells, and of their responsiveness to seasonal changes in heat exchange between the lake and the atmosphere, was provided by *Hoare* [1968], who published results from a series of profiles made during January and November 1964. One set of profiles, measured between November 23 and 26 in the upper 10 m of the water column at a central location in the lake, documents the disappearance of a small, intermediate cell between two larger cells over the four day period. A longitudinal transect revealed horizontal variability in the upper 15 m of the water column in terms of cell numbers and cell sizes, but no detectable variation below 15 m. Finally, a comparison of profiles measured at a central site before and after the winter of 1964 showed a cooling trend over the entire depth, temperatures dropping by up to 0.69°C near the surface, to 0.21°C at 65 m. More recently, Hawes (personal communication) also observed amalgamation of cells in the upper water column accompanying summer heating, and cooling of the entire water column during winter.

A longer term cooling trend can be seen in Figure 11, where *Hoare's* [1968] January 1964 profile is compared with our January 1991 profile; *Hoare's* profile has been shifted downward by 8 m, a result of the rising lake levels discussed earlier. This cooling trend is not new and was noted by *Yoshida* [1975], who cited as possible causes "(i) change of lake level, (ii) tendency towards a decrease in insulation, (iii) the generation of convection in the past due to a steep temperature gradient." *Hawes et al.* [1996] give change in lake level as the main reason for cooling, but arguments can be made to support all three causes. Changes in lake level reduce the amount of solar radiation reaching deeper levels. Figure 11 shows that heat stored in the 1991 profile above 19 m (the depth at which 1991 and 1964 profiles cross) offsets the deficit in heat below 19 m (associated with the area between the two profiles); but temperature differences need to be weighted by lake volumes in order to quantify the balance (or imbalance). We assume that "decrease in insulation" refers to a decrease in ice-cover thickness, which would increase heat loss during winter. *Wharton et al.* [1993] give ice thickness measurements for Lake Vanda since December 1960. Although there is considerable variability, there appears to be a thinning of Vanda's ice cover from approximately 4 m to 3 m between 1961 and 1982; after 1982 there is no clear trend. Other things being equal (i.e., thermal conductivity of the ice, winter air temperatures, wind speeds, etc.), the steady-state conduction through a 3 m thick ice sheet would be 4/3 times that through a 4 m thick sheet. The problem is of course not that simple, and extra heat loss during winter might be offset by the extra heat gain in summer from an increase in solar radiation penetrating a thinner ice cover. Finally, we interpret "generation of convection in the past due to a steep temperature gradient" to mean that the heat balance of the lake has not yet reached equilibrium following the onset of thermohaline instability and the creation of convection cells. As we have noted, present temperature gradients are not strong enough to generate the instability necessary to trigger the formation of cells, which implies that gradients were stronger at some time in the past. Following the onset of convection, heat transfer through the water column would have increased dramatically. Of course the system as a whole would tend toward equilibrium by adjusting cell thicknesses, inter-cell diffusion gradients, and ice thickness to match the external forcing conditions of climate, inflow, and water levels. Changes in heat storage (the decrease in lake temperatures) are the transient response, or residual term in the heat balance. We do not know how fast the time scale for the transient response is. Hence we do not know to what extent the cooling trend represents a remnant of the unsteady response to changes in heat exchange that accompanied the onset of thermohaline convection. *Yoshida's* comment that "quantitative treatment is one of the subjects for further discussion" seems as valid today as in 1975.

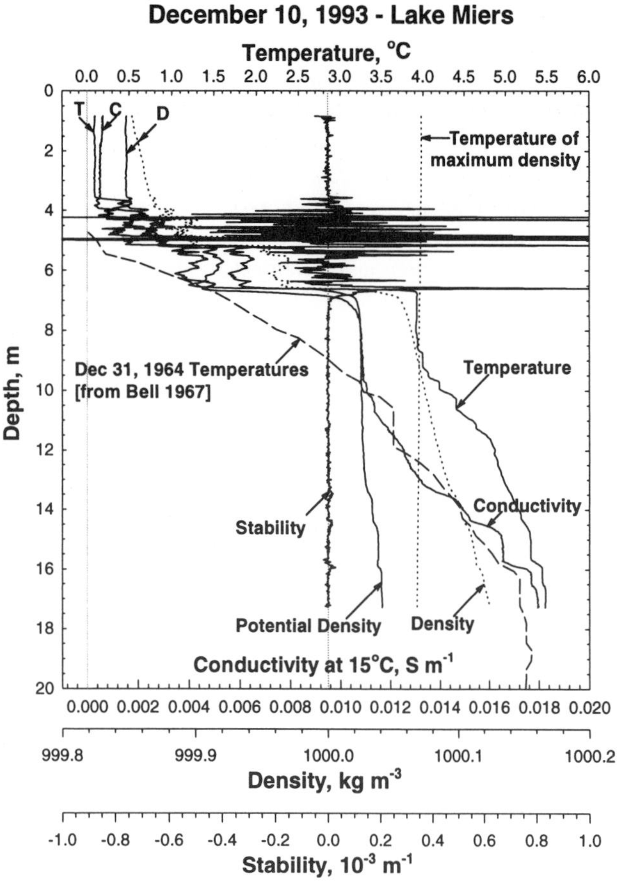

Lake Miers. Lake Miers, in the Miers Valley, receives meltwater at its western end from Adams Stream draining Adams Glacier and Miers Stream draining Miers Glacier. Lake Miers in turn empties into the Miers River at the eastern end of the lake; it is the only lake included in this study that has a stream outlet. The outlet controls lake level, so that, unlike the other closed-basin dry valleys lakes, Lake Miers' depth cannot increase by more than a small amount once water level rises above the outlet. Meltwater thus flows through the lake from west to east, and although the throughflow is confined to the upper half of the water column, it provides a mechanism, not present in the other lakes, for flushing salts from Lake Miers. Perhaps for this reason the waters of Lake Miers are the least saline of all the lakes, reaching a maximum TDS concentration at the bottom (Table 1) that is only

Fig. 12. Results (fine-structure) from a CTD cast in Lake Miers at a mid-lake site, Dec. 10, 1993. Dec. 31, 1964 temperature data from *Bell* [1967] are shown for comparison; no shift of depth axis was required (Lake Miers has a stream outlet and there is no significant difference in water levels). Zero depth corresponds to the free water surface.

slightly greater than the upper limit quoted earlier for "normal" freshwaters. In spite of its relative freshness, however, Lake Miers could hardly be said to resemble a typical freshwater lake. What little salinity gradient it has is strong enough to support the kind of thermal and density stratification characteristic of other dry valleys lakes, although both maximum and average values of stability are much lower than in any of the other lakes (Figure 12); average stability below 7 m is 2.0×10^{-6} m^{-1}, and the maximum stability below 7 m, 2.9×10^{-5} m^{-1}, is almost an order of magnitude less than values typical of freshwater thermoclines.

Comparison with one of the temperature profiles measured by *Bell* [1967] shows that considerable warming has taken place since 1964. The ice cover has also thinned, from a thickness of 6 m reported by *Bell* [1967], to less than 4 m in December 1993. Unlike

Lake Vanda, there has been little increase in water level, so the increase in water temperatures at depths originally surveyed in 1964 is unequivocally associated with an increase in heat content of the lake. This can only have come about as a result of a net increase in fluxes of heat supplied to the lake by the atmosphere and inflowing streams since 1964.

Lake Miers has no direct contact with a glacier, and its maximum water temperature is greater than those of lakes that are in contact with submerged glacier faces (west lobe of Lake Bonney, Lake Hoare and Lake Joyce). Transparency of the ice cover and water column appear to be similar to those of Lake Bonney (Table 1). The temperature profile below the bottom of the ice cover (about 3.5 m in Figure 12) can be divided into three distinct regions, the first two having temperatures colder than the temperature of maximum density, and the third with temperatures increasing to the bottom and warmer than the temperature of maximum density. We associate the first two regions, down to a depth of about 8.5 m, with direct influence from meltwater inflows. The first is a turbulent region that extends from 3.5 m to 6.5 m, and probably corresponds to stream throughflow that was occurring under the ice when the profile was measured. Microstructure activity is evident in this region, and temperature gradients appear to be well described by a Batchelor spectrum (Figure 13) with energy dissipation per unit mass of 3.7×10^{-9} m^2 s^{-3}, about ten times higher than that measured in the narrows of Lake Bonney; the Batchelor length scale is 1.1 cm. Both meltwater streams were flowing on the day we made our CTD cast; judging by the fractions of the channel cross-sections that were occupied by the flows, discharges were not above average. *Bell* [1967] observed streamflows over a wide range of conditions. He describes two days of "exceptional warmth" when stream discharges were high and stream temperatures "appreciably" greater than 4°C. "On reaching the lake, such water would enter immediately beneath the ice cover, melting part of this and cooling. It would then sink into the lake until it reached equilibrium with the surrounding water. This would mean descending almost to the mean 4°C level at 12.4 m, and considerable mixing would be likely in the neighborhood of the inflow water." *Bell* noted that inflows that took place on these two occasions temporarily transformed the normally linear temperature gradient between 10.5 m and 12.2 m into one of uniform temperature at 3.9°C. The uniform layer at about 3.9°C in our profile between 6.7 m and 8.5 m (the second of the three regions

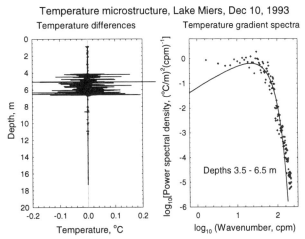

Fig. 13. Temperature microstructure from the CTD cast in Lake Miers shown in Figure 10. The format of the graphs is the same as that described in the caption for Figure 6, except that there is only one segment of the profile that exhibits active microstructure. The dissipation value associated with the fitted spectrum is 3.7×10^{-9} m^2 s^{-3} and the corresponding Batchelor length is 1.1 cm.

mentioned above) is probably associated with similar inflow processes.

Below 8.5 m the increase in temperature with depth acts to destabilize the water column, overall stability being maintained by a weak salinity gradient. *Bell* [1967] found that concentrations of some dissolved salts approached saturation in a thin layer immediately next to the lake bed (our profiles stopped well short of the bottom), and hypothesized that salinity in the lake is maintained by diffusion of salt from the lake bed into the overlying meltwater. Below 8.5 m our temperature and conductivity profiles exhibit step-like structures (layers in which temperature and conductivity are uniform, separated by layers in which gradients are relatively strong). We believe that the regions of uniform temperature and conductivity represent thermohaline convection cells. They are not likely to be associated with intrusions, because they occur below the level of influence of the fresher meltwater inflows, and there is no evidence of submerged saline inflows. *Bell's* [1967] profiles contained similar features that he recognized as thermohaline convection cells, but they were confined to depths below 15 m. The density difference ratio $\alpha\Delta T/(\beta\Delta S)$, averaged over the cells, varies between 0.53 and 0.64, four to five times greater than in Lake Vanda, reinforcing our hypothesis. Values of the thermal Rayleigh number (Ra = g $\alpha\Delta T d^3/(\nu\kappa)$, as defined earlier) calculated for the cells in Lake Miers average 1.8×10^6 for the cells between 9.2 m to 10.5 m, to 1.2×10^8 for the cells between 14.5 m and 17 m. These values are smaller than those for Lake Vanda, but still much larger than the value of 3.7×10^4 at which turbulence occurs in free convection between horizontal boundaries [*Turner*, 1973, p. 220], indicating that flow in Lake Miers' cells is turbulent. No turbulence was detected in the gradient regions between cells.

CONCLUSION

In many ways the physical limnology has changed little in the lakes of the McMurdo Dry Valleys over the 30 years or so since the first measurements were made. A scientist familiar with temperature and conductivity profiles measured in 1964, if presented with a profile measured in 1994 and asked to identify the lake that it came from, would have little trouble in doing so. We have seen how the unique thermal and salinity structure of each lake can only be explained by combining information about the evolution of the lake and its climate over at least the past 1000 years with knowledge of physical processes occurring now.

In the closed basin lakes of Fryxell and Bonney's east lobe, below the depth of influence of recent meltwater inflow, the governing processes are those of molecular diffusion and absorption of solar radiation. The same is true in Lake Vanda below the influence of thermohaline convection. In Lake Fryxell, where bottom salinities are less than 20% that of standard seawater, a mechanism has recently been proposed that short-circuits vertical diffusion by concentrating the salinity of some inflows sufficiently to allow them to sink from the surface layers to the near bottom [*Miller and Aiken*, 1996]. In Lake Bonney's east lobe this would not be possible because of the very high salinities that exist in that basin at depth. We have interpreted fine-structure in Lake Bonney as indicating that cool, saline flows from the west lobe do sink below the level of the sill that separates the two basins, but only to a depth of 15 m.

In lakes where the submerged face of a glacier maintains direct contact with the water column, horizontal and vertical laminar advection must be added to the radiation-diffusion balance, the glaciers providing a sink for heat and sources for buoyancy and meltwater (possibly saline in some cases). Temperatures in these lakes (Bonney's west lobe, Hoare, and Joyce) are cooler than they would otherwise be if the glaciers were not present. Temperature profiles in these lakes show unmistakable evidence of intrusive,

interleaving flows of large horizontal and small vertical extent. Salinity controls stability at almost all depths in the dry valleys lakes, and most of the fine-scale variability in temperature observed in Hoare, Joyce, and Bonney's west lobe near the Taylor glacier has little influence on density, but indicates horizontal movement of water originating from sources at slightly different temperatures. Similarly fine-scale temperature structure in the narrows separating the east and west lobes of Lake Bonney arises from exchange between water at different temperatures in the two basins.

Thermohaline convection dominates most of the upper 55 m of Lake Vanda. At some time in the past, salinity gradients above the salt-concentrated bottom waters became too weak to prevent the onset of convection, convection driven by heat from solar energy that had been stored in the dense bottom brines of the lake. This leads one to speculate that meltwater inflows occurred relatively suddenly in Lake Vanda after the cool, dry climate changed to one of warming about 1200 to 1000 years ago, covering the remaining brines quickly enough and to a great enough depth to insulate the salt-freshwater interface from the effects of wind mixing and allow an ice cover to form. In accordance with this reasoning, inflows occurred more gradually in the east lobe of Lake Bonney, probably because of storage effects of the Taylor Glacier for heat and the west lobe for water, thus delaying the establishment of ice cover in the east lobe and precluding conditions necessary for the onset of thermohaline convection. The apparent contradiction between *Shirtcliffe's* (1964) diffusion time-scale for Lake Bonney of 60 years and *Wilson's* (1964) diffusion time-scale of 1200 years may therefore be resolved if one allows for a long delay between the times that ice covers became established in Lake Vanda and the east lobe of Lake Bonney. This is reasonable considering the much greater responsiveness of the Lower Wright Glacier to warming than the more massive Taylor Glacier, and the lack of correlation that has been observed between meltwater flows from these two systems [*Chinn*, 1993]. In Lake Miers, salinity gradients are sufficiently weak, and heat capture at depth sufficiently great, that thermohaline convection also became established there, as discovered by *Bell* [1967] in 1964. Further warming has resulted in expansion of the region susceptible to convection, and our measurements indicate that new cells have developed in Lake Miers since 1964.

With the exception of flow inside thermohaline convection cells, turbulence is a rare occurrence in the lakes of the dry valleys. We have used microstructure measurements to detect turbulence in the exchange flow in the narrows separating east and west lobes in Lake Bonney, in the stream throughflow in lake Miers, and possibly in intrusions in Lake Joyce. In all cases the turbulence occurred in boundary layers next to the bottom of the ice cover; in the narrows of Lake Bonney, it also occurred in a boundary layer next to the channel bed. Other locations where turbulence would almost certainly be detected include wind-stirred surface water in moats, inflows during an episode of prolonged or intense melting, and buoyant meltwater rising along the face of a submerged glacier. These are localized and intermittent sources. The interiors of the lakes are quiet (except for the thermohaline cells of Vanda and Miers). These findings are consistent with measurements of chlorophyll-*a* fluorescence made during the summer of 1996-1997, using an in situ fluorometer in place of the microstructure probes on the SBE-25 CTD (Priscu, unpublished data). Results from these casts in Lakes Hoare and Fryxell, for example, show considerable detail and structure in fluorescence where stability is greatest (i.e., in regions of strongest conductivity gradient). Such detail could not be preserved if turbulence or vertical circulation were present. In Lakes Miers and Joyce, steps in conductivity and temperature correspond to steps in fluorescence, implying the presence of sufficient motion with the steps to sustain vertical mixing at a rate that must be faster than the production rate of the plankton that are present.

The most readily apparent differences in profiles since the 1960s stem from the warming trend that has caused changes in water temperatures and, in the closed-basin lakes, rises in water levels. The most extreme example is that of Lake Vanda, where levels have risen approximately 10 m from 1960 to 1996 [*Hawes et al.*, 1996]. In Lake Miers, where lake level is controlled by a stream outlet, water levels have not risen significantly and the warming trend has had the straightforward effect of increasing water temperatures over most of the water column. In the closed-basin lakes the results have not been as straightforward because of the effect that lake depth has on the distribution of water temperatures. In Lake Vanda and in the east lobe of Lake Bonney, maximum water temperatures have fallen, while in the west lobe of Lake Bonney and in Lake Fryxell maximum water temperatures have risen. We do not have water level records extending to the 1960s for these lakes, but *Chinn* [1993] presents comparisons for all lakes that show increases in water levels between 1974 and 1990 of approximately 5 m in Lake Vanda, 4 m in Lake

Bonney, and 1.9 m in Lake Fryxell. Some insight into the effect of a change in lake level on water temperature distribution can be gained by considering the simplified case in which there is a steady state balance between vertical diffusion and absorption of solar radiation, the model proposed by *Hoare et al.* [1964, 1965] that successfully explains the shapes of the temperature profiles in Lake Bonney's east lobe and Lake Fryxell, expressed as:

$$\rho c_p \kappa dT/dz = Q_o e^{-\eta z} - G \qquad (12)$$

where Q_o is the solar radiation transmitted through the ice cover, G is the heat flux from the lake to the sediments, and the other symbols are as defined earlier. Equation 12 shows that if there is a flux of heat from the lake to the sediments (G positive) then there will be a temperature maximum in the water column ($dT/dz = 0$) where $G = Q_o e^{-\eta z}$, i.e., where the downward flux of solar radiation is equal to the flux of heat lost to the sediments. If such a balance is disturbed by a sudden increase in water level Δh, and a new level of solar flux, then the system will tend toward a new steady state with the maximum temperature at an elevation above its original level by an amount (assuming no change in extinction coefficient):

$$z_1 - z_2 = \Delta h - \eta^{-1} \ln[(Q_{o2}/G_2)/(Q_{o1}/G_1)] \qquad (13)$$

where z_1 and z_2 are the depths at which the maximum temperature is found in the old and new steady states, and Q_{o1}, Q_{o2} and are the corresponding solar and bottom heat fluxes. Hence the old maximum temperature will be left behind and temperature at the old level (z_1) will drop, while temperature at the new level (z_2) will increase. How much it will increase, whether the new maximum temperature will be greater or less than the old one, and how long it takes to reach equilibrium depends on Δh and the shape of the original temperature profile. Although the above analysis is oversimplified, we note that a small decrease in the depth of maximum temperature has been observed even over the short period 1990-1993 (Figure 4) in Lake Bonney. Furthermore, the above reasoning indicates that the larger the rise in lake level and the deeper the lake, the more likely it is that the new maximum temperature will be lower than the old one. This too is consistent with what is observed.

Although many of the conclusions reached in this chapter, like the argument just presented, have of necessity been based on limited data, we believe that the capability of making high-resolution temperature and conductivity measurements has allowed us to make progress in our understanding of the physical limnology of the lakes in the McMurdo Dry Valleys. We feel that the insights gained from the more extensive data set collected in Lake Bonney have justified our efforts and that similar advances may be possible by applying these techniques in the other lakes. It is also encouraging that there is now an emphasis on more comprehensive coverage of hydrologic and climatic factors throughout the dry valleys, exemplified by other contributions in this volume. These data sets will eventually allow us to take a more holistic view of these globally unique lakes.

Acknowledgments. We thank Ian Sheppard, Ian Forne, Tom Sharp, Rob Edwards, Richard Bartlett, Christopher Woolston, and Vann Kalbach for assistance in the field; Alan Poynter for his design and construction of the winch; and Neil Sutherland and David MacPherson for help with laboratory analyses. We are grateful to Greg Ivey and Warwick Vincent for valuable criticism during the review process. Antarctic Support Associates and the U.S. Navy furnished logistical support. This work was supported by the U.S. National Science Foundation, Office of Polar Programs, under grants DPP-88-20591, OPP 91-17907, OPP 92-11773, and OPP94-19423 to JCP.

REFERENCES

Angino, E. E., K. B. Armitage, and J. C. Tash, Chemical stratification in Lake Fryxell, Victoria Land, Antarctica, *Science, 138*, 34–36, 1962.

Angino, E. E and K. B. Armitage, A geochemical study of Lakes Bonney and Vanda, *J. Geol.,71,* 89–95, 1963.

Angino, E. E., K. B. Armitage, and J. C. Tash, Physicochemical limnology of Lake Bonney, Antarctica, *Limnol. Oceanogr., 9*, 207–217, 1964.

Armitage, K. B., and H. B. House, A limnological reconnaissance in the area of McMurdo Sound, Antarctica, *Limnol. Oceanogr., 7*, 36–41, 1962.

Bell, R. A. I., Lake Miers, South Victoria Land, Antarctica, *N.Z. J. Geol. Geophys., 10*(2), 540–556, 1967.

Black, R. F., Saline discharges from Taylor Glacier, Victoria Land, Antarctica, *Antarctic J. U.S., 4*, 89–90, 1969.

Black, R. F., and C. J. Bowser, Salts and associated phenomena of the termini of the Hobbs and Taylor Glaciers, Victoria Land, Antarctica, *IUGG, Commission on Snow and Ice, Pub. 79*, 226–238, 1968.

Black, R. F., M. L. Jackson, and T. E. Berg, Saline discharge from Taylor Glacier, Victoria Land, Antarctica, *J. Geology, 73*, 175–181, 1965.

Burton, H. R., Chemistry, physics and evolution of Antarctic saline lakes, *Hydrobiologia, 82*, 339–362, 1981.

Bydder, E. L., and R. Holdsworth, Lake Vanda (Antarctica) revisited, *N.Z. J. Geol. Geophys., 20*, 1027–1032, 1977.

Caldwell, D. R., The maximum density points of pure and saline water, *Deep–Sea Res., 25*, 175– 181, 1978.

Caldwell, D. R., T. M. Dillon, J. M. Brubaker, P. A. Newberger, and C. A. Paulson, The scaling of vertical temperature gradient spectra, *J. Geophys. Res., 85*(C4), 1917–1924, 1980.

Caldwell, D. R., T. M. Chriss, P. A. Newberger, and T. M. Dillon, The thinness of oceanic temperature gradients, *J. Geophys. Res., 86*(C5), 4290–4292, 1981.

Carlson, C. A., F. M. Phillips, D. Elmore, and H. W. Bentley, Chlorine-36 tracing of salinity sources in the Dry Valleys of Victoria Land, Antarctica, *Geochimica et Cosmo-chimica Acta, 54*, 311–318, 1990.

Chen, C.-T. A., and Millero, F. J., Precise thermodynamic properties for natural waters covering only the limnological range, *Limnol. Oceanogr. 31*(3), 657–662, 1986.

Chinn, T., Hydrology and climate in the Ross Sea area, *J. Royal Soc. N.Z., 11*, 373–386, 1981.

Chinn, T. J., Physical hydrology of the Dry Valleys lakes, in *Physical and Biogeochemical Processes in Antarctic Lakes, Antarctic Research Series 59*, edited by W. J. Green and E. I. Freeman, pp. 1–51, AGU, Washington D. C., 1993.

Chinn, T. J. H., and A. D. H. Woods, Hydrology and Glaciology, Dry Valleys, Antarctica, Annual Report for 1981–82, Ministry of Works and Development, Christchurch, Rept. No. WS 1017, 63 pages, 1984.

Clow, G. D., C. P. McKay, G. M. Simmons, Jr., and R. A. Wharton, Jr., Climatological observations and predicted sublimation rates at Lake Hoare, Antarctica, *J. Climate, 1*(7), 715–728, 1988.

Convitz, P. A., D. M. McKnight, L. M. MacDonald, and A. Fountain, Hydrological processes influencing streamflow variation in Fryxell Basin, Antarctica, This volume.

Craig, J. R., R. D. Fortner, and B. L. Weand, Halite and hydrohalite from Lake Bonney, Taylor Valley, Antarctica, *Geology, 2*(8), 389–390, 1974.

Culkin, F., and N. D. Smith, Determination of the concentration of potassium chloride solution having the same conductivity, at $15°C$ and infinite frequency, as standard seawater of salinity 35.0000 ppt (chlorinity 19.37394 ppt), *IEEE J. Oceanic Eng., OE-5*(1), 22–23, 1980.

Dana, G. L., R. A. Wharton, and R. Dubayah, Solar radiation in the McMurdo Dry Valleys, This volume.

Decker, E. R., and G. J. Bucher, Geothermal studies in Antarctica, *Antarctic J. U.S.,12*(4), 102– 104, 1977.

Dillon, T. M., and D. R. Caldwell, The Batchelor spectrum and dissipation in the upper ocean, *J. Geophys. Res., 85*(C4), 1910-1916, 1980.

Farmer, D. M., and L. Armi, Maximal two-layer exchange over a sill and through the combination of a sill and contraction with barotropic flow, *J. Fluid Mech., 164*, 53–76, 1986.

Fofonoff, N. P., and R. C. Millard, Jr., Algorithms for computation of fundamental properties of seawater, *UNESCO Technical Papers in Marine Science 44,* Division of Marine Sciences, UNESCO, Paris, 53 pages, 1983.

Fritsen, C. H., E. E. Adams, C. M. McKay, and J. C. Priscu, Permanent ice covers of the McMurdo Dry Valleys lakes: liquid water content, This volume.

Gibson, C. H., and W. H. Schwarz, The universal equilibrium spectra of turbulent velocity and scalar fields, *J. Fluid Mech., 16,* 365–384, 1963.

Gill, A. E., *Atmosphere-Ocean Dynamics,* Academic Press, New York, 662 pages, 1982.

Gregg, M. C., The microstructure of the ocean, *Scientific American, 228*(2), 65–77, 1973.

Gregg, M., T. Meagher, A. Pederson, and E. Aagaard, Low noise temperature microstructure measurements with thermistors, *Deep-Sea Res., 25,* 843–856, 1978.

Gregg, M. C., and W. C. Hess, Dynamic response calibration of Sea-Bird temperature and conductivity probes, *J. Atmospheric and Oceanic Technology,* 304–313, 1985.

Gumbley, J. W., The sedimentology of three Antarctic lakes. Unpublished M.Sc. Thesis lodged in the Library, University of Waikato, Hamilton, New Zealand, 1975.

Hawes, I., Turbulence and its consequences for phytoplankton development in two ice covered Antarctic lakes, *Br. Antarct. Surv. Bull.,* No. 60, 69–81, 1983.

Hawes, I., J. Hall, C. Howard-Williams, M. James, and A.-M. Schwarz, Seasonal and long-term changes in the physical, chemical and biological features of the Lake Vanda water column, in *Proc. Intl. Workshop on Polar Desert Ecosystems, 1–4 July 1996, Christchurch, N.Z.,* 1996.

Hendy, C. H., A. T. Wilson, K. B. Popplewell, and D. A. House, Dating of geochemical events in Lake Bonney, Antarctica, and their relation to glacial and climate changes, *N.Z. J. Geol. Geophys., 20*(6), 1103–1122, 1977.

Hewitt, G. F., Tables of the resistivity of aqueous sodium chloride solutions, U.K. Atomic Energy Authority Research Group Report, Chemical Engineering Division, Atomic Energy Research Establishment, Harwell, Berkshire, H. M. S. O., 16 pages, 1960.

Hoare, R. A., Problems of heat transfer in Lake Vanda, a density stratified Antarctic lake, *Nature, 210*(5038), 787–789, 1966.

Hoare, R. A., Thermohaline convection in Lake Vanda, Antarctica, *J. Geophys. Res., 73*(2), 607–612, 1968.

Hoare, R. A., K. B. Popplewell, D. A. House, R. A. Henderson, W. M. Prebble, and A. T. Wilson, Lake Bonney, Taylor Valley, Antarctica: a natural solar energy trap, *Nature, 202*(4935), 886–888, 1964.

Hoare, R. A., K. B. Popplewell, D. A. House, R. A. Henderson, W. M. Prebble, and A. T. Wilson, Solar heating of Lake Fryxell, a permanently ice-covered Antarctic lake, *J. Geophys. Res. 70*(6), 1555–1558, 1965.

Huppert, H. E., and E. G. Josberger, The melting of ice in cold stratified water, *J. Physical Oceanogr., 10,* 953–960, 1980.

Huppert, H. E., and J. S. Turner, Double diffusive convection and its implications for temperature and salinity structure of the ocean and Lake Vanda, *J. Physical Oceanogr., 2,* 456–461, 1972.

Huppert, H. E., and J. S. Turner, On melting icebergs, *Nature, 271,* 46–48, 1978.

Huppert, H. E., and J. S. Turner, Iceblocks melting into a salinity gradient, *J. Fluid Mech., 100,* 367–384, 1980.

Hutchinson, G. E., *A Treatise on Limnology, Volume 1, Geography, Physics, and Chemistry,* John Wiley and Sons, Inc., New York, 1015 pages, 1957.

Imberger, J., Mixing in reservoirs, in *Mixing in Inland and Coastal Waters,* by H. B. Fischer, J. Imberger, E. J. List, R. C. Y. Koh, and N. H. Brooks, pp. 150–222, Academic Press, New York, 483 pages, 1979.

Imberger, J., Thermal characteristics of standing waters: an illustration of dynamic processes, *Hydrobiologia, 125,* 7–29, 1985a.

Imberger, J., The diurnal mixed layer, *Limnol. Oceanogr., 30*(4), 737–770, 1985b.

Imberger, J., and J. C. Patterson, Physical limnology, *Advances in Applied Mechanics,* pp. 303–475, Academic Press, New York, 1990.

Imberger, J., and G. N. Ivey, On the nature of turbulence in a stratified fluid. Part II: Application to lakes, *J. Physical Oceanogr., 21*(5), 659–680, 1991.

Jeevaraj, C. G., and J. Imberger, Experimental study of double diffusive instability in sidewall heating, *J. Fluid Mech., 222,* 565–586, 1991.

Josberger, E. G., and S. Martin, A laboratory and theoretical study of the boundary layer adjacent to a vertical melting ice wall in salt water, *J. Fluid Mech., 111,* 439–473, 1981.

Kestin, J., and W. A. Wakeham, *Transport Properties of Fluids: Thermal Conductivity, Viscosity, and Diffusion Coefficient,* Hemisphere Publishing Corp., New York, 344 pages, 1988.

Keys, J. R., The saline discharge at the terminus of the Taylor Glacier, *Antarctic J. U.S., 14,* 82–85, 1979.

Lawrence, M. J. F., and C. H. Hendy, Water column and sediment characteristics of Lake Fryxell, Taylor Valley, Antarctica, *N.Z. J. Geol. Geophys., 28,* 543–552, 1985.

Lewis, E. L., The Practical Salinity Scale 1978 and its antecedents, *IEEE J. Oceanic Engineering, OE-5*(1), 3–8, 1980.

Lewis, M. R., J. J. Cullen, and T. Platt, Phytoplankton and thermal structure in the upper ocean: consequences of nonuniformity in the chlorophyll profile, *J. Geophys. Res., 88*(C4), 2565–2570, 1983.

Lizotte, M. P., and J. C. Priscu, Spectral irradiance and bio-optical properties in perennially ice-covered lakes of the Dry Valleys (McMurdo Sound, Antarctica), in *Contributions to Antarctic Research III, Antarctic Research Series 57,* edited by D. H. Elliott, pp. 1–14, AGU, Washington D. C., 1992.

Lizotte, M. P., and J. C. Priscu, Distribution, succession, and fate of phytoplankton in the dry valleys lakes of Antarctica, based on pigment analysis, This volume.

Lyons, W. B., L. R. Bartek, and P.A. Mayewski, A climate history of the McMurdo Dry Valleys since the last glacial maximum: a synthesis, in *Ecosystem Processes in Antarctic Ice-Free Landscapes,* edited by W. B. Lyons, C. Howard-Williams, and I. Hawes, A. A. Balkema, Rotterdam, In press.

Lyons, W. B., K. A. Welch, K. Neumann, J. K. Toxey, R. McArthur, and C. Williams, Geochemistry linkages among glaciers, streams, and lakes within the Taylor Valley, Antarctica, This volume.

McKay, C. P., G. D. Clow, R. A. Wharton, Jr., and S. W. Squyres, Thickness of ice on perennially frozen lakes, *Nature, 313*(6003), 561–562, 1985.

McKay, C. P., G. D. Clow, D. T. Andersen, and R. A. Wharton, Jr., Light transmission and reflection in perennially ice-covred Lake Hoare, Antarctica, *J. Geophys. Res., 99*(C10), 20427–20444, 1994.

Matsubaya, O., H. Sakai, T. Torii, H. Burton, K. Kerry, Antarctic saline lake stable isotopic ratios, chemical compositions and evolution, *Geochimica et Cosmochimica Acta, 43,*7– 25, 1979.

Matsumoto, G. I., Geochemical features of the McMurdo Dry Valleys lakes, Antarctica, in *Physical and Biogeochemical Processes in Antarctic Lakes, Antarctic Research Series 59,* edited by W. J. Green and E. I. Freeman, pp. 95–118, AGU, Washington D. C., 1993.

Matthews, P. C., and S. I. Heaney, Solar heating and its influence on mixing in ice-covered lakes, *Freshwater Biology, 18,* 135–149, 1987.

Meagher, T. B., A. M. Pederson, and M. C. Gregg, A low-noise conductivity micro-structure instrument, *Oceans 82: Conference Record: Industry, Government, Education–Partners in Progress,* Conference sponsored by Marine Technology Society, IEEE Council on Oceanic Engineering, Sept., 20–22, 1982, Washington D.C., 283–290, 1982.

Miller, L. G., and G. R. Aiken, Effects of glacial meltwater inflows and moat freezing on mixing in an ice-covered Antarctic lake as interpreted from stable isotope and tritium distributions, *Limnol. Oceanogr,* 41, 966–976, 1996.

Millero, F. J., and W. H. Leung, The thermodynamics of sea-water at one atmosphere, *Amer. J. Sci., 276,* 1035–1077, 1976.

Millero, F. J., and A. Poisson, International one-atmosphere equation of state of seawater, *Deep-Sea Res., 28A,* 625–629, 1981.

Millero, F. J., C.-T. Chen, A. Bradshaw, and K. Schleicher, A new high pressure equation of state for seawater, *Deep-Sea Res., 27A,* 255–264, 1980.

Millero, F. J., D. Dawson, and A. Gozalez, The density of artificial river and estuarine waters, *J. Geophys. Res., 81,* 1177–1179, 1976.

Millero, F. J., G. Perron, and J. F. Desnoyers, Heat capacity of seawater solutions from 5°C to 35°C and .05 to 22 ppt chlorinity, *J. Geophys. Res., 78*(21), 4499–4506, 1973.

Montgomery, R. B., Oceanographic data, in *American Institute of Physics Handbook,* Sec. 2, Mechanics, 115–134, 1957.

Mortimer, C. H., and F. J. H. Mackereth, Convection and its consequences in ice-covered lakes, *Verh. Int. Verein. Limnol., 13,* 923–932, 1958.

Newman, F. C., Temperature steps in Lake Kivu: a bottom-heated saline lake, *J. Phys. Oceanogr., 6,* 157–163, 1976.

Osborn, T. R., Temperature microstructure in Powell Lake, *J. Phys. Oceanogr., 3,* 302–307, 1973.

Palmisano, A. C., and G. M. Simmons, Jr., Spectral down-welling irradiance in an Antarctic lake, *Polar Biol., 7,* 145–151, 1987.

Parker, B. C., G. M. Simmons, Jr., K. G. Seaburg, D. D. Cathey, and F. C. T. Allnutt, Comparative ecology of plankton communities in seven Antarctic oasis lakes, *J. Plankton Res., 4*(2), 271–285, 1982.

Patterson, J. C., and Hamblin, P. F., Thermal simulation of a lake with winter ice-cover, *Limnol. Oceanogr., 33*(3), 323– 338, 1988.

Pederson, A. M., A small *in situ* conductivity instrument, *Proceedings OCEAN 73 - IEEE Conference on Engineering in the Ocean,* 68–75, 1973.

Pederson, A. M., and M. C. Gregg, Development of a small in-situ conductivity instrument, *IEEE J. Oceanic Eng., OE-4*(3). 69–75, 1979.

Polzin, K. L., K. G. Speer, and J. M. Toole, Intense mixing of Antarctic Bottom Water in the equatorial Atlantic Ocean, *Nature, 380,* 54–55, 1996.

Pond, S., and G. L. Pickard, *Introductory Physical Oceanography, 2nd ed.,* Pergamon Press, Oxford, 329 pages, 1983.

Priscu, J. C., Phytoplankton nutrient deficiency in lakes of the McMurdo Dry Valleys, Antarctica, *Freshwater Biol., 34,* 215–227, 1995.

Priscu, J. C., The biogeochemistry of nitrous oxide in permanently ice-covered lakes of the McMurdo Dry Valleys, Antarctica, *Microbially Mediated Atmospheric Change, Special Issue of Global Change Biology,* In press.

Priscu, J. C., M. T. Downes, and C. P. McKay, Extreme super-saturation of nitrous oxide in a poorly ventilated Antarctic lake, *Limnol. Oceanogr., 41*(7), 1544–1551, 1996.

Ragotzkie, R. A., and G. E. Likens, The heat balance of two Antarctic lakes, *Limnol. Oceanogr., 9,* 412–425, 1964.

Reid, R. C., J. M. Prausnitz, and T. K. Sherwood, *The Properties of Gases and Liquids, Third Edition,* McGraw-Hill, New York, 688 pages, 1977.

Richman, J., and C. Garrett, The transfer of energy and momentum by the wind to the surface mixed layer, *J. Physical Oceanogr., 7,* 876–881, 1977.

Riley, J. P., and G. Skirrow, editors, *Chemical Ocean-ography, 2nd Edition, Vol. 1,* Academic Press, London, 1975.

Robinson, R. A., and R. H. Stokes, *Electrolyte Solutions,* Butterworths Scientific Publications, London, 512 pages, 1955.

Schladow, S. G., and J. Imberger, Sidewall effects in a double diffusive system, *J. Geophys. Res., 92*(C6), 6501–6514, 1987.

Schladow, S. G., E. Thomas, and J. R. Koseff, The dynamics of intrusions into a thermohaline stratification, *J. Fluid Mech., 236,* 127–165, 1992.

Scott, R. F., *The Voyage of the Discovery, Vol. 2,* Smith Elder, London, 1905.

Sea-Bird Electronics, SBE 25 Sealogger CTD Operating Manual, SeaBird Electronics, Inc., Bellevue, Washington, 1989a.

Sea-Bird Electronics, The temperature and conductivity duct: installation, use, and data processing steps to minimize salinity spiking error, SeaBird Electronics, Inc., Bellevue, Washington, 1989b.

Shirtcliffe, T. G. L., Lake Bonney, Antarctica: cause of the elevated temperatures, *J. Geophys. Res., 69*(24), 5257–5268, 1964.

Shirtcliffe, T. G. L., Thermosolutal convection: observation of an overstable mode, *Nature, 213,* 489–490, 1967.

Shirtcliffe, T. G. L., and R. F. Benseman, A sun-heated Antarctic lake, *J. Geophys. Res., 69*(16), 3355–3359, 1964.

Spigel, R. H., and J. C. Priscu, Evolution of temperature and salt structure of Lake Bonney, a chemically stratified Antarctic lake, *Hydrobiologia, 321,* 177–190, 1996.

Spigel, R. H., I. Forne, I. Sheppard, and J. C. Priscu, Differences in temperature and conductivity between the east and west lobes of Lake Bonney: evidence for circulation within and between lobes, *Antarctic J. of the U.S., 26,* 221–222, 1991. Tabor, H., Solar ponds, *Solar Energy, 27*(3), 181–194, 1981.

Turner, J. S., The coupled transports of salt and heat across a sharp density interface, *Int. J. Heat and Mass Trans., 8,* 759–767, 1965.

Turner, J. S., *Buoyancy Effects in Fluids,* Cambridge University Press, Cambridge, 367 pages, 1973.

Veronis, G., Effect of a stabilizing gradient of solute on thermal convection, *J. Fluid Mech., 34*(2), 315–336, 1968.

Vincent, W. F., Production strategies in Antarctic inland waters: phytoplankton eco-physiology in a permanently ice-covered lake, *Ecology, 62*(5), 1215–1224, 1981.

Vincent, W. F., Antarctic limnology, in *Inland Waters of New Zealand,* edited by A. B. Viner, pp. 379–412, DSIR Bulletin 241, Science and Information Publishing Centre, Department of Scientific and Industrial Research, Wellington, 494 pages, 1987.

Welch, H. E., and M. A. Bergmann, Water circulation in small arctic lakes in winter, *Can. J. Fish. Aquat. Sci., 42,* 506–520, 1985.

Welch, K. A., and W. B. Lyons, Comparative limnology of the Taylor Valley lakes, *Antarctic J. U.S.,* in press.

Wharton, R. A., Jr., C. P. McKay, G. M. Simmons, Jr., and B. C. Parker, Oxygen budget of a perennially ice-covered Antarctic Dry-Valley lake, *Limnol. Oceanogr., 31,* 437–443, 1986.

Wharton, R. A., Jr., C. P. McKay, G. D. Clow, R. C. Mancinelli, and G. M. Simmons, Jr., Perennial N_2 supersaturation in an Antarctic lake. *Nature, 325,* 343–345, 1987.

Wharton, R. A., Jr., C. P. McKay, G. D. Clow, and D. T. Andersen, Perennial ice covers and their influence on Antarctic lake ecosystems, in *Physical and Biogeochemical Processes in Antarctic Lakes, Antarctic Research Series 59,* edited by W. J. Green and E. I. Freeman, pp. 53–70, AGU, Washington D. C., 1993.

White, F. M., *Fluid Mechanics,* McGraw-Hill, New York, 701 pages, 1979.

Williams, D. L., and R. P. von Herzen, Heat loss from the earth: new estimate, *Geology, 2*(7), 327–328, 1974.

Wilson, A. T., Evidence from chemical diffusion of a climatic change in the McMurdo Dry Valleys 1,200 years ago, *Nature, 201*(4915), 176–177, 1964.

Wilson, A. T., A review of the geochemistry and lake physics of the Antarctic dry areas, in *Dry Valley Drilling Project, Antarctic Research Series 33,* edited by L. D. McGinnis, pp.185–192, AGU, Washington D. C., 1981.

Wilson, A.T., and H. W. Wellman, Lake Vanda: an Antarctic lake, *Nature, 196*(4860), 1171- 1173, 1962.

Wilson, A. T., C. H. Hendy, T. R. Healy, J. W. Gumbley, A. B. Field, and C. P. Reynolds, Dry Valley lake sediments: a record of Cenozoic climatic events, *Antarctic J. U.S., 9,* 134–135, 1974a.

Wilson, A. T., R. Holdsworth, and C. H. Hendy, Lake Vanda: source of heating, *Antarctic J. U.S., 9,* 137–138, 1974b.

Wood, I. R., A lock exchange flow, *J. Fluid Mech., 42,* 671-687, 1970.

Wu, J., Wind induced turbulent entrainment across a stable interface, *J. Fluid Mech., 61,* 275–297, 1973.

Yoshida, Y., T. Torii, Y. Yusa, S. Nakaya, and K. Moriwaki, A limnological study of some lakes in the Antarctic, in *Quaternary Studies,* edited by R. P. Suggate and M. M. Cresswell, pp. 311–320, The Royal Society of New Zealand, Wellington, 1975.

Yusa, Y., On the water temperature in Lake Vanda, Victoria Land, Antarctica, *Memoirs of the National Institute of Polar Research, Tokyo, Special Issue 4,* 75–89, 1975.

Yusa, Y., A study of thermosolutal convection in saline lakes, *Memoirs of the Faculty of Science, Kyoto Univ., Series A of Physics, Astrophysics, Geophysics and Chemistry, 35*(1), 149–183, 1977.

Robert H. Spigel, Department of Civil Engineering, University of Canterbury, Private Bag 4800, Christchurch, New Zealand

John C. Priscu, Department of Biological Sciences, Montana State University, Bozeman, Montana, 59717, U.S.A.

(Received January 23,1997;
accepted May 31, 1997)

OPTICAL PROPERTIES OF THE MCMURDO DRY VALLEY LAKES, ANTARCTICA

Clive Howard-Williams, Anne-Maree Schwarz, Ian Hawes

National Institute of Water and Atmospheric Research Ltd., Christchurch, New Zealand

John C. Priscu

Department of Biological Sciences, Montana State University, Bozeman, Montana,

The optical properties of the ice and water columns of lakes of the McMurdo Dry Valleys are described. Attenuation of light is dominated by the effects of the permanent ice cover, which reduces incident irradiance between 78 and 99%. The ice cover also imparts a strong blue to blue-green bias to its spectral distribution. Attenuation by ice can be highly variable over short time and distance scales. This is related to the nature of incident light (direct or diffuse), ice temperature (which affects crystal structure), snow cover, solar angle, and the amount of sediment and air spaces within the ice. Transmission is highest in ice at low temperature, with diffuse incident irradiance, in the absence of snow and at low sediment and bubble contents. Within the water columns, most attenuation is due to water itself. The lakes typically have extremely low concentrations of dissolved yellow substances. In some strata, phytoplankton and suspended sediments can make significant impacts on water clarity. This is particularly evident in the deep chlorophyll-*a* layers in some lakes. Overall, the lakes of the McMurdo Dry Valleys can be characterized as being extreme shade environments, with what light there is being in the blue or blue-green portion of the spectrum. The demands that this environment imposes on phototrophs is briefly discussed.

INTRODUCTION

The optical properties of inland waters are highly variable and reflect properties of both the lake itself and, often more importantly, its catchment. Lakes of the McMurdo Dry Valleys are end members of the limnological spectrum for several reasons, many of which are discussed in other chapters of this volume. First, they are permanently covered with 3 to 4 m of ice. This feature alone sets them apart from most other lakes on the planet, even those at the same latitude in the arctic [*Adams et al.*, this volume]. Second, the inflows are generated only by melting glacier ice, rather than snow-melt or liquid precipitation [*Conovitz et al.*, this volume]. Third, the catchments are essentially devoid of vegetation. Fourth, many are in closed drainage basins (endorheic drainage), and with small inflow volumes exhibit long hydraulic residence times.

Finally, lack of exposure to wind induced mixing and the presence of strong salinity gradients result in many lakes having highly stratified water columns [*Spigel and Priscu*, this volume].

In general, nutrient concentrations in the inflowing streams are low at the point of entry to the lakes [*Howard-Williams et al.*, 1986]. Low trophogenic zone nutrient concentrations support low phytoplankton population densities [*Priscu*, 1995], and biological attenuation of light is weak. Because glacier melt in the McMurdo Dry Valleys is a relatively slow process [*McKnight et al.*, this volume], suspended sediment loads in most of the streams are usually low. There are, however, a few inflows with seasonally high suspended sediment loads [*Howard-Williams et al.*, 1986; *Webster et al.*, 1996] but these streams are the exception rather than the rule. Streams in glacier-fed catchments with no vegetation may be expected to have low concentrations

TABLE 1. General Characteristics of the McMurdo Dry Valley Lakes Considered in this Chapter.
m.a.s.l. = Meters Above Mean Sea Level

Lake	Elevation (m.a.s.l.)	Depth (m)	Drainage	Notes
Fryxell	17	18	Endorheic	Meromictic
Bonney west	60	40	Endorheic	Meromictic
Bonney east	60	40	Endorheic	Meromictic
Hoare	73	34	Endorheic	Proglacial, some mixing
Vanda	123	75	Endorheic	Meromictic
Miers	240	20	Exorheic	Meromictic
Wilson	100?	>100	Endorheic	Meromictic, proglacial

of dissolved organic matter [*McKnight et al.*, 1991], a major attenuating component of light in aquatic ecosystems [*Hutchinson*, 1957; *Kirk*, 1994]. The combination of low suspended solids and poorly developed terrestrial dissolved organic carbon (DOC) sources might be expected to result in highly transparent lakes.

Perennial ice cover has a profound effect on both the quality and quantity of the light available for photosynthesis [*Priscu*, 1991; *Neale and Priscu*, this volume; *McKay et al.*, 1994,]. The ice cover can be clear or contain wind-blown sediments, gas bubbles, and crystal structures within it which attenuate light by absorption and scattering. Lake ice is therefore highly variable in appearance: clear, white, blue, or even brown [*Adams et al.*, this volume]. Ice rapidly attenuates light at the red end of the spectrum, thus shifting the wavelengths of the light which enters the water column below [*Palmisano and Simmons*, 1987].

Because of its influence on preventing mixing of the water column of the lakes, a second effect of the ice cover is to enhance water column stability to the point of exhibiting meromixis because of low turbulence which is insufficient to mix old salt layers (Table 1) [*Spigel and Priscu*, this volume]. This allows a series of "attenuating layers" to develop as discrete vertically stratified zones of organisms or particulate matter. For instance, Lakes Vanda, Fryxell, and Bonney have layers of phytoplankton in relatively high concentrations at some depth below the underside of the ice, associated with discrete nutrient supply near the oxycline [*Vincent*, 1988; *Priscu*, 1995]. These Deep Chlorophyll Maxima (DCM) should selectively attenuate light more rapidly than the water column above and below and also alter the downwelling spectrum producing a distinct change in transmittance profiles. In temperate lakes, such deep blooms are

sensitive to minor changes in water column turbidity and irradiance [*Vincent*, 1983]. The locations of the DCM in dry valley lakes must also be partly dependent on the clarity of the overlying water column [see *Vincent*, 1981].

This paper provides a review of existing data and compilations of both new and existing data in which the optical properties of the lakes of the McMurdo Dry Valleys are compared. The macro- and micro-scale structure of light attenuation in each lake is examined to determine the influence of suspensoid layers on the light regime at depth. The implications of this for photosynthetic organisms is discussed.

INFLUENCE OF THE PERENNIAL ICE COVER

By far the most important attenuating layer in the lakes is the ice cover. Attenuation of light by this layer has been the subject of detailed study on sea ice in the McMurdo Sound region [*Buckley and Trodahl*, 1987; *Trodahl et al.*, 1989] and on some of the lakes [e.g., Lake Hoare, *Palmisano and Simmons*, 1987; *McKay et al.*, 1994; Lake Bonney, *Priscu*, 1991; *Adams et al.*, this volume; *Fritsen et al.*, this volume].

Attenuation by the ice cover results from absorbance within the ice and reflection at the surface and within the ice, and varies markedly between lakes (Table 2). Lake Vanda's ice cover is the thinnest (Table 2) and most transparent (K_{ice} = 0.6 m^{-1}) resulting in approximately 13% of incident Photosynthetically Active Radiation (PAR: 400–700 nm) reaching the water column below the ice. In contrast, the ice cover of Lake Fryxell (K_{ice} = 1.1 m^{-1}) transmits only 1% of incident PAR to the water column. Total attenuation coefficients of the ice cover provide a limited insight into the processes which influence the amount of light reaching the water column, since there can be considerable

TABLE 2. Summary Table of Attenuation of Incident Irradiance by the Ice Cover (K_{ice} is the Extinction Coefficient for the Ice Cover), Mean Ice Thickness at the Time of Transparency Measurements, and Ice Transparency (% PAR Transmitted) in the Dry Valley Lakes in Mid Summer (November to January).

Lake		K_{ice}	Thickness (m)	n	% incident of PAR under ice	n	References
Vanda	mean	0.60	3.4	9	13.17	11	*Goldman et al., 1967*
	range	0.49–0.67	3.1–4.5		5.2–20		*Seaburg et al., 1983*
							Vincent and Vincent, 1982
							Parker et al., 1982
							Priscu, 1989
							This study
							Vincent, 1988
Bonney	mean	0.85	4.3	3	2.73	3	*Goldman et al., 1967*
(East lobe)	range	0.76–0.98	4.0–4.5		1.7–3.3		*Seaburg et al., 1983*
							Parker et al., 1983
							Spigel, unpublished
							Priscu, 1991
Hoare	mean	1.08	3.95	6	1.59	6	*Seaburg et al., 1983*
	range	0.96–1.33	2.6–5.5		0.5–2.8		*Palmisano and Simmons, 1987*
							Parker et al., 1982
							Hawes and Schwarz, unpublished
Fryxell	mean	1.08	4.3	5	1.34	5	*Parker et al., 1982*
	range	1.0–1.21	3.8–4.6		0.5–3.2		*Vincent, 1981*
							Priscu, 1989
							Vincent, 1988

variation in surface albedo and attenuating properties within the ice itself.

The structure of the ice cover of Lake Hoare, and its influence on light transmission, was studied in detail by *McKay et al.* [1994]. They found that the two major factors influencing light transmission were the amount of sediment (technically, sand and gravel) [*Adams et al.*, this volume] in the ice, and gas bubble density and alignment. The amount of sediment in the ice cover of Lake Hoare ranged from 0.2–2.0 g cm^{-2}, and most was concentrated in the upper 1 m of the ice cover. Variability in sediment concentration resulted in a three-fold variability in under-ice PAR on a spatial scale of meters [*Wharton et al.*, 1989]. In a recent study using a diver operated spectroradiometer we found that penetration of PAR through Lake Hoare ice varied almost ten-fold from 0.6 to 5% also over distance scales of meters (authors' unpublished data). Below the sediment layer *McKay et al.* [1994] found the ice was very clear. Here the vertical alignment of bubbles coincides with the occurrence of vertically oriented [c-axis, sensu *Wilson*, 1981; *Adams et al.*, this volume] ice crystals which are several centimeters in diameter and extend for meters down through the ice.

Both absorption and scattering processes are important in the attenuation of PAR in lake ice. Absorption takes place within ice itself and in the entrapped wind-blown sand and gravel particles in the ice. The red end of the PAR spectrum is absorbed to the greatest extent by ice, while the specific wavelength

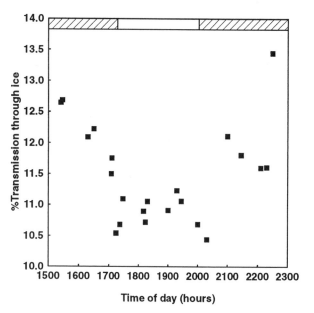

Fig. 1. Percentage transmission through 3.5 m of ice on Lake Vanda over an 8 hour period on 10 January 1996. The sky was cloudless between approximately 1700 and 2000 hrs. Shaded section of upper bar shows cloud cover.

absorbed by particles will depend on the type of particle. Organic particles may be expected to strongly absorb blue wavelengths. Scattering by particles, ice bubbles, and ice crystal structure also depends on the relative size of the scattering object and the wavelength of light [Kirk, 1994]. By increasing the path length for light, scattering can increase the probability of light absorption as well as that of reflectance [e.g., Trodhal et al., 1989]. In the ice covers of similar thickness (Table 2) differences in light scattering will be due to particle concentrations, fractures, and bubble density rather than ice thickness alone. The relative importance of the absorption and scattering components imparted by the sediment particles in the ice versus those of other attenuating substances has not been determined and will vary greatly with ice type.

DIEL SHIFTS IN TRANSMISSION THROUGH ICE COVER

Lakes that are shaded by mountains for a period each day (e.g., Vanda, Bonney, Hoare) experience an abrupt daily change in total solar radiation [Dana et al., this volume] and also a change from direct to diffuse radiation. Data from Lake Bonney show that the relative transmission through the ice almost doubles under diffuse light when compared to direct light

[Priscu, 1991]. Changes in percent transmission over time scales of hours due to variations in cloud cover can also be seen which are comparable to the diel shifts from direct to diffuse light. Data for an 11 hour period on Lake Vanda (Figure 1) show an increase in transmission of the order of 1.3 times when conditions changed from bright sun to overcast. The net effect of this may be to decrease the temporal variance in PAR beneath the ice on a given day, since during periods of reduced incident radiation due to clouds, transmission will be highest. The mechanism for this is not yet clear as comparisons of spectral reflectance under diffuse and direct light for two different ice types show few differences (Figure 2).

SEASONAL CHANGES IN ICE TRANSPARENCY

Buckley and Trodahl [1987] demonstrated major shifts in the transparency of sea ice following an air temperature rise from −15°C to −5°C and the associated draining of surface brine. In the fresh water ice of the dry valley lakes, such changes in brine content will not occur, although seasonal changes in optical transparency have also been noted for Lakes Bonney, Vanda, and Hoare [*Priscu*, 1991; *McKay et al.*, 1994; *Wharton et al.*, 1989; authors' unpublished data]. For example,

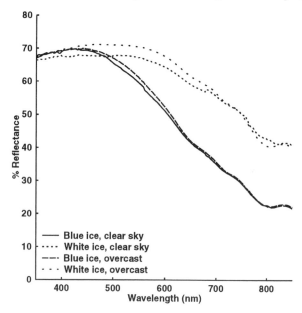

Fig. 2. Wavelength dependence of surface reflectance at the ice cover of Lake Hoare, December 1996. Data were collected with a LiCOR LI 1800 spectroradiometer suspended 1m above the ice to measure downward and upward irradiance in clear and overcast conditions.

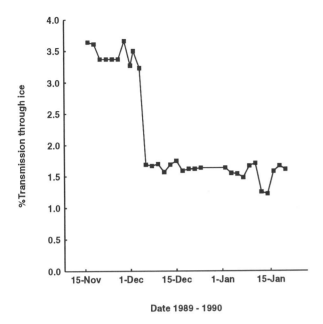

Fig. 3. Percent transmission of PAR through ice on Lake Bonney between November 1989 and January 1990 [redrawn from *Priscu*, 1991]. Transmission was computed from daily averages in PAR.

percent transmission dropped from 3.2 to 1.7% and attenuation increased from 0.8–0.98 m^{-1} in a week, coincident with visual observations of ice surface fracturing on lake Bonney (Figure 3) [*Priscu*, 1991]. Transmission apparently declined because of hoarfrost formation within near surface bubbles and fracturing along the grain boundaries of individual ice crystals with increasing ice temperature [*Adams et al.*, this volume].

Our data for Lake Vanda also show a distinct seasonal change with transmission declining from 21% in September to 13% by mid summer and increasing again in January. As in the case of Lake Bonney, visual observations showed a change in ice appearance from blue in October/November to white in December, suggesting seasonal changes in light scattering.

Detailed analysis of the ice cover of Lake Hoare suggests that the decreased transparency is also due to deep warming of the ice and to the formation of light scattering Tyndall figures [*Mae*, 1975; *Walker*, 1986; *McKay et al.*, 1994]. Visual changes to the ice may be the "whitening" or apparent fine fracturing which occurs at this time as noted above for Lakes Bonney and Vanda.

Seasonal changes in transparency may follow a predictable sequence [*McKay et al.*, 1994]. Early in the season, ice is cold, completely frozen and percent transmission is at its highest. With increased radiative heating and Tyndall figure formation, scattering and internal absorption increase thereby reducing transmission. As summer progresses, ice melts and destroys the fine cracks and other scattering "structures" [see *Adams et al.*, this volume] and transmission increases. As winter approaches, ice temperature declines, cracks and fissures reappear and transmission decreases. Transmission varied over the summer in Lake Hoare by a factor of three.

INFLUENCE OF SNOW COVER

Snow cover on ice can significantly increase light attenuation. Thin snow cover (18 mm) on sea ice at McMurdo Sound has been shown to reduce transmission of all wavelengths in the PAR spectrum to 30% [*Trodahl and Buckley*, 1990]. Snow cover, with its high albedo, has been shown to be a major contributor to attenuation of light in ice covered maritime Antarctic lakes [*Hawes*, 1985] and in sub-Arctic lakes [*Adams*, 1978; *Roulet and Adams*, 1984; *Bolsenga et al.*, 1996]. However persistent snow cover is rare on most dry valley lakes and mainly affects underwater light regimes by increasing spatial and temporal variability (patchiness). Spatial variability in snow cover can result in an overestimation of average PAR transmission to the underlying water column if not corrected for [*Roulet and Adams*, 1984].

SPECTRAL INFLUENCE OF THE ICE COVER

Spectral Reflectance

Spectral reflectance for clear and white ice varies little over the PAR wavelengths [*Grenfell and Maykut*, 1977; *McKay et al.*, 1994], although for blue ice we have recorded a significant reduction in spectral reflectance between 500 and 700 nm (Figure 2). Albedo depends mostly on the amount of sediment in the ice [*McKay et al.*, 1994], the degree of fracturing [*Adams et al.*, this volume] and on whether the ice is smooth or irregular [*Goldman et al.*, 1967]. High sediment content reduces albedo, while fracturing and irregular surfaces lead to high values.

Spectral Attenuation and Transmission

Pure ice transmits in the blue region of the spectrum and strongly absorbs at wavelengths greater than 600

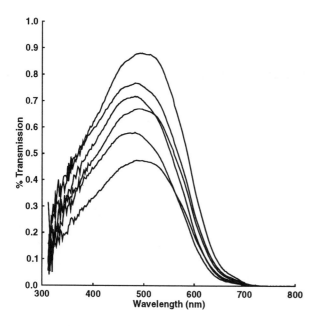

Fig. 4. Percent transmission through ice on Lake Hoare in December 1996 for the waveband 300–800 nm. Spectra were collected in clear sky conditions at six different positions by divers operating an LI 1800 spectroradiometer on the under-surface of the ice, well away from the dive hole.

nm. Maximum light transmission through the ice covers of lakes Hoare and Joyce occurred between 450 and 490 nm (Figure 4), consistent with earlier studies based on instruments with broad bandpass filters [*Palmisano and Simmons*, 1987; *Lizotte and Priscu*, 1992a; *Vincent*, 1988]. There is considerable variability in downward irradiance in the 450–550 waveband at different positions on Lake Hoare (Figure 4), further illustrating the patchiness in light transmission through an ice cover.

Seasonal changes in ice transparency described above have an influence on the spectral properties of the transmitted light [*Wharton et al.*, 1989]. Between November and January transmitted light in the waveband 400–600 nm declined relative to that at 700 nm probably due to increased scattering of blue light brought about by fracturing. Although the specific absorbance of red light by ice is much greater than that of blue light, short wavelengths are more subject to scattering.

It is interesting to note that seasonal changes in UV transparency of sea ice may be more marked than for PAR [*Trodahl and Buckley*, 1990]. Early season ice (when, incidentally, the ozone hole is at its maximum) is usually highly transparent to UV. High transparency at this time is due to low scattering, and short wave-

length UV light is more prone to scattering than longer wavelengths. Although such studies are yet to done on dry valley lakes, we anticipate that scattering may also significantly influence UV transmittance through lake ice with a high early season UV transparency that reduces as ice fracturing increases.

THE WATER COLUMN

Bulk Attenuation

Gross differences in attenuation between the lakes can be summarized by the integrated water column attenuation coefficients for downwelling irradiance (K_d) and for scalar irradiance (K_0). Despite changes in instrumentation which affect estimations of irradiance [*Kirk*, 1994], comparisons of the relatively long term records for the lakes (1963-1996) suggest that attenuation coefficients vary more as a function of time and lake than as a result of instrumentation. Table 3 provides a compilation of attenuation coefficients for the lakes at different times and depths.

The lakes can be arranged on a gradient of attenuation coefficient from lowest to highest as follows: Vanda, Miers, Bonney, Hoare, Fryxell, Wilson. The data from Lake Wilson are based on one series of measurements taken in January [*Webster et al.*, 1996] and the extent of seasonal variability for this lake is not known. Coefficients span an order of magnitude among lakes from extreme clarity at Lake Vanda (K_d <0.05 m^{-1}) to relatively turbid Lake Wilson (K_d = 0.8 m^{-1}). Considerable variation in attenuation with depth is recorded (Table 3) because of discrete layers of attenuating materials. Attenuation coefficients measured over small depth intervals are particularly unreliable as indicators of integral attenuation in vertically structured lakes such as these. For example, although Hoare is considered to be more transparent than Fryxell [*Lizotte and Priscu*, 1992a], there have been higher values of K_0 recorded in some layers in the former (Table 3).

Temporal Changes in Attenuation

There is little consistent evidence of a seasonal change in integral water column attenuation in dry valley lakes (Table 3). Attenuation in Lake Vanda, from just below the ice to 40 m, varied more between years than within a summer season (October-February). Data obtained with the same instrumentation show that K_0 in 1993-1994 averaged 0.053 m^{-1} with a range of

0.046–0.060 (Table 3), as compared to 1995-1996 when it averaged 0.036 m^{-1} (range 0.033–0.037). Although K_0 increased to a maximum in late December 1993 in Lake Vanda this pattern was not evident in data from other years, nor could it be explained by changes in biotic attenuation as evidenced by chlorophyll-*a* concentration. In the upper layers of Lake Fryxell (5–12m) there was a gradual seasonal increase in attenuation in the water column from 0.528 to 0.621 between November 15, 1994 and January 18, 1995. Over the same period attenuation increased in the west lobe of Lake Bonney from 0.170 on November 6, 1994 to 0.214 on December 21, 1994 (Priscu, unpublished data). These increases are consistent with the period of glacial stream flow into the lakes.

The differences between years in Lake Vanda (Table 3) may be associated with small differences in chlorophyll-*a* and inorganic turbidity. Turbidity may be caused by the inflow from the Onyx River, which has a suspended solids concentration ranging from 2.7 to 130 mg m^{-3} [*Howard-Williams et al.,* 1986]. The Onyx River has a considerable effect on the water in the Lake Vanda moat at the east end of the lake. This area becomes markedly turbid following the onset of flow of the river. For instance, in 1993 at the time of initial river flow on January 1 there was no significant moat and the water beneath the seasonal moat ice had a K_d of 0.04 m^{-1}. By January 15 this had risen by a factor of six to 0.24 m^{-1}. However the river water remains largely confined to the moat and consequently has little effect on light attenuation within the main water body. The tendency for the inflowing river water to follow the moat is consistent with similar observations in Lake Fryxell [*McKnight and Andrews,* 1993]. The influence of meltwater streams is likely to be most marked in Lake Wilson. Here, the single large inflow stream in 1993-1994 was highly turbid, with a suspended solids concentration of glacial flour reaching 40 g m^{-3}; the lake had no moat during this period [*Webster et al.* 1996].

Reflectance

Reflectance [E_u (PAR)/E_d (PAR)] in Lakes Vanda, Bonney and Wilson ranges from 0.062 to 0.17. These values are all high and even in highly transparent Lake Vanda values have been recorded ranging from 0.062 to 0.092. Reflectance is largely from suspended inorganic material, and is usually higher in waters containing glacial flour than in other natural waters [*Howard-Williams and Vincent,* 1984]. It is likely to be a seasonally varying attribute, governed by the extent of river inflows. Of the dry valley lakes, Lake Wilson is the only one where a high suspended sediment (mostly glacial flour) concentration of up to 5.6 g m^{-3} has been recorded in the lake at a period of low river inflows [*Webster et al.,* 1996]. This compares with a concentration of <0.1 g m^{-3} in Lake Vanda throughout summer (authors' unpublished data). Not suprisingly Lake Wilson is the most turbid site, with a reflectance of 0.17.

Spectral Attenuation

Comparisons of spectral data obtained with a Biospherical Instruments MER 1000 spectroradiometer for Lakes Bonney, Hoare, and Fryxell [*Palmisano and Simmons,* 1987; *Lizotte and Priscu,* 1992a] with our data and those of *Vincent and Vincent* [1982] and *Goldman et al.* [1967] for Lake Vanda are presented as spectral attenuation coefficients (K_d (λ)) for three wavelengths in Table 4. Spectral attenuation coefficients across the PAR waveband for Lakes Hoare and Joyce are shown in Figure 5. These data were obtained using a diver operated LiCOR LI 1800 scanning spectroradiometer suspended 0.5 and 4.5 m below the ice cover. The most penetrating wavebands for Lakes Hoare and Joyce were for green light (490–500 nm) (Figure 5). A noticeable difference between the lakes was the relatively low attenuation at wavelengths below 470 nm in Lake Joyce compared to that in Lake Hoare. In Lake Fryxell, maximum penetration was between 520 and 580 nm [*Palmisano and Simmons,* 1987] whereas Lake Bonney passes blue-green light with the most penetrating waveband between 480 and 520 nm. Although full spectral data are not available for Lake Vanda the analyses with optical filters suggests that blue wavelengths (midpoint 420 nm) penetrate deeply (Table 4). With an attenuation coefficient at 420 nm of 0.066 m^{-1}, Lake Vanda is considerably more transparent to blue light than either Lakes Hoare or Joyce (cf. Figure 5).

Red light is strongly attenuated by the ice cover (see above) and the spectral composition of the PAR immediately beneath the ice is blue and green light (Figure 4). *Lizotte and Priscu* [1992a] suggested that the marked reduction of far red light (*ca.* 680 nm) under the ice means that natural fluorescence by phytoplankton at 683 nm will become an important source of long wavelengths at depth. This effect can be seen in a comparison of spectral scans for upward and downward irradiance flux in Lake Hoare (Figure 6). There is a

TABLE 3. Attenuation Coefficients for Discrete Depths in the Water Column.

Lake	Year	Depth (m)	K_o	K_d	Method*	Reference
Vanda	Jan-63			0.049	p	*Goldman et al., 1967*
	Jan-63			0.042	p	*Goldman et al., 1967*
	Feb-63			0.041	p	*Goldman et al., 1967*
	Dec-80			0.050	c	*Vincent and Vincent, 1982*
	1980-1981			0.055	c	*Kasper et al., 1982*
	Nov-93	3.5–40	0.055	0.052	s,c	This study
	Dec-93	3.5–40	0.060	0.05	s,c	This study
	Jan-94	3.5–40	0.051	0.055	s,c	This study
	Nov-94	3.5–40	0.033	0.04	s,c	This study
	Dec-94	3.5–40	0.045	0.05	s,c	This study
	Sep-95	3.5–40	0.035		s	This study
	Oct-95	3.5–40	0.036		s	This study
	Jan-96	3.5–40	0.037	0.034	s,c	This study
Hoare	Dec-82			0.164	p	*Parker et al., 1982*
	Nov-94	4.5–22	0.127		s	This study
	Dec-94	5–22	0.219		s	This study
	Dec-94	4.5–22	0.173		s	This study
	Dec-94	4.5	0.420		PNF	This study
	Dec-94	11	0.120		PNF	This study
	Jan-95	4.5–22	0.216		s	This study
	Dec-96	10–18		0.17	c	This study
Fryxell	Nov-79	5–7.5		0.074	c	*Vincent, 1981*
	Nov-79	7.5–9.5		0.250	c	*Vincent, 1981*
	Dec-90	7	0.275		s	*Lizotte and Priscu, 1992*
	Dec-90	8.5	0.352		s	*Lizotte and Priscu, 1992*
	Nov-94	5–11.5	0.528		s	This study
	Dec-94	5–12	0.549		s	This study
	Jan-95	5–12	0.621		s	This study
Bonney	Jan-63			0.141	p	*Goldman et al., 1967*
(west lobe)	Nov-82			0.158	c	*Parker et al., 1982*
	Nov-94	4.5–20		0.102	c	This study
	Dec-94	4.5–20		0.095	c	This study
	Jan-95	4.5–20		0.127	c	This study
Bonney	Nov–82			0.138	p	*Parker et al., 1982*
(east lobe)	Dec–89	5–12	0.120		s	*Lizotte and Priscu, 1992b*
	Dec–89	12–20	0.106		s	*Lizotte and Priscu, 1992b*

* The method of collection is indicated by p = irradiance photometer; s =scalar sensor; c = cosine corrected PNF = Biosphereical Instruments Corp. Profiling natural fluorometer.

TABLE 3. Attenuation Coefficients for Discrete Depths in the Water Column.

Lake	Year	Depth (m)	K_o	K_d	Method*	Reference
	Nov-90	6	0.143		s	*Lizotte and Priscu*, 1992
	Dec-90	6	0.138		s	*Lizotte and Priscu*, 1992
	Jan-91	6	0.246		s	*Lizotte and Priscu*, 1992
	Nov-90	10	0.094		s	*Lizotte and Priscu*, 1992
	Dec-90	10	0.107		s	*Lizotte and Priscu*, 1992
	Jan-91	10	0.195		s	*Lizotte and Priscu*, 1992
	Nov-90	17	0.118		s	*Lizotte and Priscu*, 1992
	Dec-90	17	0.109		s	*Lizotte and Priscu*, 1992
	Jan-91	17	0.125		s	*Lizotte and Priscu*, 1992
	Nov-94	4.5–22	0.095		s	This study
	Dec-94	4.5–22	0.095		s	This study
	Jan-95	4.5–22	0.127		s	This study
Wilson	Jan-93		0.800		c	*Webster et al.*, 1996
Miers	Jan-95		0.110		PNF	This study

TABLE 4. Spectral Values for Downward Attenuation.

Lake	Date	Depth (m)	Spectral K_d Values			Wavelength of most penetrating waveband	Reference
			440	520	680		
Vanda	Jan-63		0.038	0.055			*Goldman et al.*, 1967
	Jan-63		0.035	0.06			*Goldman et al.*, 1967
	Feb-63		0.031	0.058			*Goldman et al.*, 1967
	Dec-80		0.04	0.06	0.46		*Vincent and Vincent*, 1982
	Dec-93		0.056	0.085	0.712		This study
	Dec-94		0.051	0.039			This study
	Dec-94		0.066	0.07			This study
Hoare	Dec-82	6–14	0.418	0.282	0.715	515–542	*Palmisano and Simmons*, 1987
	Dec-82	3.7–6	0.372	0.271	0.625	515–542	*Palmisano and Simmons*, 1987
	Dec-82	2.7–9.8	0.264	0.241	0.486		*Palmisano and Simmons*, 1987
Fryxell	Dec-90	6–9	0.72	0.4	0.59	520–580	*Lizotte and Priscu*, 1992

Data from *Goldman et al.* (1967) were obtained using optical filters of 420 nm and 540 nm. *Vincent and Vincent* (1982) used optical filters with mid-points at 440, 520, and 680 nm. Remaining data were obtained using a MER 1000 Biospherical Instruments spectroradiometer with a band width of 10 nm.

* The method of collection is indicated by p = irradiance photometer; s =scalar sensor; c = cosine corrected PNF = Biosphereical Instruments Corp. Profiling natural fluorometer.

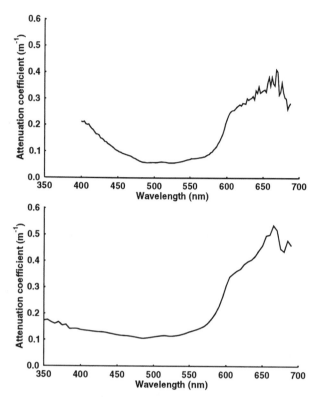

Fig. 5. Spectral attenuation coefficients for the upper 5m of the water columns of (A) Lake Hoare, and (B) Lake Joyce in December 1996. Data were collected as for Figure 4, with the instrument deployed at 4.5 and 9.5 m.

distinct peak in upward irradiance above 650 nm and centered on 680 nm. Reflectance at 680 nm was 0.35, consistent with a high proportion of red light due to chlorophyll-*a* fluorescence. The effect may be compounded by Raman emission, which occurs when predominantly blue-green downward irradiance undergoes a change to longer wavelengths when scattered [*Marshall and Smith,* 1990]. Raman emission shows up as an emission band roughly 100 nm on the long wavelength side of the exciting wavelength [*Kirk,* 1994]. However the position of the peak situated at 680 nm suggests that chlorophyll-*a,* rather than Raman emission dominates. In situ "production" of red light in the water column may be measurable in many other ice capped polar lakes and seas when ice covers are thick and the availability of red light from sunlight is negligible. Spectral reflectance data for Lake Hoare calculated from the data in Figure 6 shows a distinct peak centered around 680 nm, consistent with phytoplankton fluorescence and not Raman emission. Red light emission processes may explain the apparent

decrease in spectral attenuation above 650 nm (Figure 5a) in Lake Hoare.

Scattering and Absorption

The relative importance of scattering and absorption in lake water can be assessed by comparing their respective coefficients [*Kirk,* 1994]. Scattering coefficients (b) for the upper waters of three dry valley lakes are generally higher than the absorption coefficients (a) (Table 5). The ratio of scattering to absorption was between 6.4 and 7.6 in Lakes Vanda and Fryxell, and between 2 and 4.8 in Lake Bonney (Table 5). In general, absorption coefficients of Antarctic lake waters are lower than those recorded for even alpine or high Arctic sites, and the small particle size of glacial flour will impart a high specific scattering coefficient [*Kirk,* 1994]

Factors Affecting Attenuation

Constituents of the water column such as phytoplankton, dissolved organic matter (yellow color), non-algal particulates as well as the water itself, influence absorption and scattering coefficients, which in turn determine the apparent optical properties such as K_d [*Kirk,* 1994]. In Lakes Bonney and Fryxell absorption was typically dominated by water (38–75%) with phytoplankton usually secondary (11–47%) and a

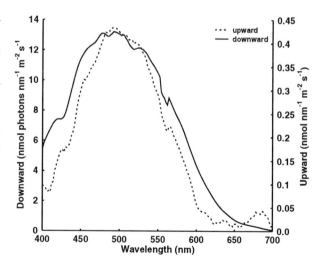

Fig. 6. Spectral scans of upward and downward irradiance (E_u (λ) and $E_d(\lambda)$) at a depth of 5.5 m in Lake Hoare (December 1996). Slight irregularities in the original scans have been smoothed by plotting the running averages over 10 nm band widths. Data were collected as for Figure 4.

TABLE 5. Absorption (a) and Scattering (b) Coefficients for the Upper Water Layers of Three Dry Valley Lakes.

Lake	Date	Depth	Coefficient			
			a	b	b/a	
Vanda	1994-1995	5	0.014	0.103	7.45	This study
	1994-1995	12	0.023	0.173	7.45	This study
	1994-1995	25	0.021	0.162	7.6	This study
	1994-1995	30	0.022	0.167	7.45	This study
	1994-1995	40	0.02	0.13	6.5	This study
Fryxell	Dec. 1990		0.151	0.971	6.4	*Lizotte and Priscu*, 1992
	Dec. 1990		0.184	1.313	7.1	*Lizotte and Priscu*, 1992
Bonney	Nov. 1990	6	0.096	0.365	3.8	*Lizotte and Priscu*, 1992
(east lobe)	Dec.1990	6	0.098	0.316	3.2	*Lizotte and Priscu*, 1992
	Jan. 1991	6	0.205	0.402	1.9	*Lizotte and Priscu*, 1992
	Nov.1990	10	0.065	0.227	3.8	*Lizotte and Priscu*, 1992
	Dec. 1990	10	0.066	0.32	4.8	*Lizotte and Priscu*, 1992
	Jan. 1991	10	0.124	0.554	4.46	*Lizotte and Priscu*, 1992
	Nov. 1990	17	0.081	0.288	3.55	*Lizotte and Priscu*, 1992
	Dec. 1990	17	0.08	0.224	2.8	*Lizotte and Priscu*, 1992
	Jan. 1991	17	0.086	0.308	3.6	*Lizotte and Priscu*, 1992

variable contribution from dissolved organic matter (0.46%) [*Lizotte and Priscu*, 1992a]. In Lake Vanda the contributions to the absorption coefficient at 440 nm in the upper 20m of the water column showed that absorption was also dominated by water (54%) although particulate matter was more important (44%) than chlorophyll-*a* (1.2%). Although chlorophyll-*a* is the dominant pigment [*Lizotte and Priscu*, this volume], it is only one in an assemblage of different pigments in the phytoplankton [*Hoepffner and Sathyendranath*, 1992] and so may slightly underestimate the proportional contribution by phytoplankton. An analysis of absorbance due to dissolved organic matter (g_{440}) was carried out on samples collected in January 1996 in Lake Vanda and December 1996 in Lakes Hoare and Joyce. Dissolved organic matter (g_{440}) was not detectable in Lake Vanda at 10 and 38 m, and was 0.06 m^{-1} at 63 m depth. Values were slightly higher in Lakes Joyce (0.10–0.12 m^{-1} to 10 m depth) and Lake Hoare (0.15–0.25 m^{-1} to 16 m). Analysis of the 63 m data in Lake Vanda showed that dissolved color estimated by g_{440} contributed less than 1% to total absorption at 440 nm and will be less than this higher in the water column. This is consistent with a vegetation free catchment and little internal generation of DOC. The situation in Lake Vanda contrasts with Lake Fryxell and Bonney, where internal generation of DOC can be high despite a vegetation free catchment [*McKnight et al.*, 1991; *Lizotte and Priscu*, 1992a].

Optical Structure of the Water Column

Chemically stratified lakes are often characterized by layers of particulate and dissolved material associated with density layers. In the ice covered lakes of the dry valleys discrete density layers are particularly marked [*Spigel and Priscu*, 1996 and this volume]. Of the factors known to affect water clarity in such layers, chlorophyll-*a* is perhaps the most dynamic. Within Lake Vanda two well defined convection cells are found, one from 5–25 m depth and the other from 31–45 m depth [*Spigel and Priscu*, this volume]. The lower of these tends to have a slightly higher chlorophyll-*a* concentration than the upper. Well defined Deep Chlorophyll Maxima (DCM) are particularly well developed in Lakes Bonney, Fryxell, and

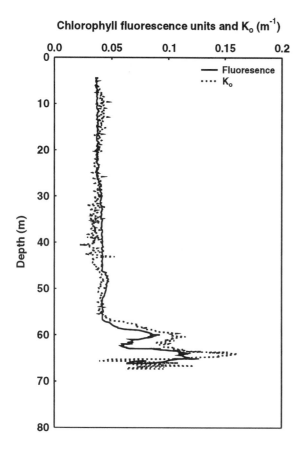

Fig. 7. Attenuation (K_0, dotted line) and upwelling natural chlorophyll-*a* fluorescence (relative units, solid line) over the depth profile in Lake Vanda. Data were obtained with a Biospherical Instruments Corp. PNF 300 profiler.

Vanda [*Vincent, 1982; Vincent and Vincent, 1981; Lizotte and Priscu,* 1992b; *Lizotte et al.,* 1996] usually associated with stable non-convecting layers where gradients of nutrients are sharp. The increased chlorophyll-*a* and associated particulate matter in the DCM layers have a marked effect on attenuation of PAR.

The influence of the DCM on fine scale (30 cm) structure of K_0 in Lake Vanda is well illustrated in Figure 7, which shows chlorophyll-*a* fluorescence, measured using a Chelsea Instruments in situ fluorometer, and K_0, measured using a Biospherical Instruments PNF300 profiling fluorometer, with depth. The break between the upper and lower convecting cells at 25–30 m is evident in the fluorescence trace. Below 45 m vertical structure is more marked. The small fluorescence peak at 50 m, and the double peak in fluorescence at 60–65 m, was closely mirrored by K_0, which increased from 0.055 to 0.12 between 60 and 65 m. Not

suprisingly, beam transmission in Lake Vanda (Figure 8) followed a similar but reciprocal pattern. In all cases vertical structure apparently recorded an increase in phytoplankton biomass in the DCM at 60–65 m depth. Sharp reductions in transmittance below 67 m depth may reflect the high biomass of non-chlorophyll containing bacteria [*Vincent,* 1988] or mineral precipitates (e.g., Fe and Mn-oxyhydroxides) at the oxycline [*Webster,* 1993; *Green et al.,* this volume].

In both lobes of Lake Bonney, chlorophyll-*a* appears to structure the K_0 profile within the trophogenic zone (Figure 9a, b). The general pattern down to 12 m follows that of the chlorophyll-*a* profile reported by *Lizotte and Priscu* [1994] and *Priscu* [1995] with maxima at 5 m and 12 m. The K_0 maximum near 25 m in the east lobe of Bonney (Figure 9a) reflects a non-photosynthetically active chlorophyll layer located below the chemocline, whereas the K_0 maximum at 24 m in the west lobe (Figure 9b) is caused by non-chlorophyllous matter (Priscu, unpublished). Not all dry valley lake exhibits a vertical zonation of optical properties. For example, Lake Miers shows relatively little optical structure in the water column

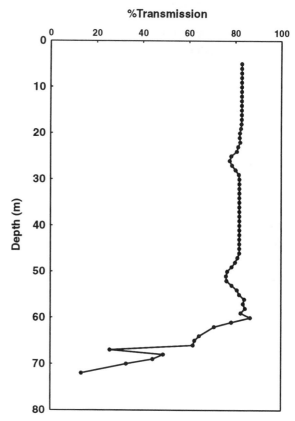

Fig. 8. Vertical profile of beam transmission in Lake Vanda measured at 1 m intervals.

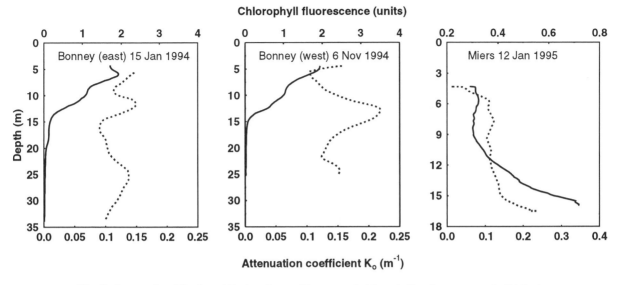

Fig. 9. Attenuation (K_0, dotted line) and upwelling natural chlorophyll-*a* fluorescence (solid line) in Lake Bonney (east lobe), Lake Bonney (west lobe) and Lake Miers. K_0 is in ln units m-1; fluorescence is in relative units. Data were collected as for Figure 7.

(Figure 9c) with K_0 values remaining constant from 5 to (15 m, increasing gradually below this depth. Lake Miers also has little structure to its chlorophyll-a to a depth of 10 m, but shows a gradual increase in concentration with depth below this (Figure 9c).

IMPLICATIONS FOR PHOTOTROPHS

The optical properties of the ice cover and water columns of the McMurdo Dry Valley lakes regulate the quantity and quality of light which penetrates to a given depth. Ice cover has been shown to have an overwhelming effect in the amount of light which is present in these lakes, reducing incident PAR to 1–13% leading to the development of shaded aquatic ecosystems. The ice cover effects the spectral distribution of the PAR reaching the water column by effectively eliminating red light. The success of phototrophs under such conditions will depend on the efficiency with which they can capture and utilize available quanta [*Neale and Priscu*, 1995; *Neale and Priscu*, this volume; *Seaburg et al.*, 1983].

Low intensity of light available under ice selects for shade adapted phytoplankton and benthic algae [*Priscu*, 1989]. *Hawes and Schwarz* [in press] described the photosynthesis-light relationships of benthic algal/cyanobacterial mats and reported saturation light intensities in the region of 2–5% of incident irradiance. *Vincent* [1981] showed that photosynthesis of phytoplankton from Lake Fryxell (9 m) saturated at

less than 1% incident irradiance and those from Lake Bonney require only 15–45 µmol photons m-2 s-1, equivalent to approximately 2–8% of average daily incident PAR [*Lizotte and Priscu,* 1992b]. Despite the high degree of shade adaptation seen in these organisms, algal photosynthesis is rarely light saturated [*Vincent,* 1981; *Lizotte and Priscu,* 1992b].

Primary producers in the dry valley lakes appear to be specially adapted to utilize the under-ice spectrum [*Neale and Priscu,* 1995; *Neale and Priscu,* this volume]. The benthic communities are characterized by a high proportion of cyanobacterial taxa [*Hawes and Schwarz,* in press], which typically contain high concentrations of phycobilin pigments. High phycobilin contents have been observed in planktonic cyanobacteria in Lake Vanda, evident as a red-pink coloration seen under microscopic examination of fresh material. Phycobilins absorb maximally at 400–500 nm [*Rowan,* 1989], and are well adapted for harvesting the green light which penetrates the ice. *Vincent and Vincent* [1982] also commented on the possible significance of light quality on the relative distribution of the yellow pigmented alga *Ochromonas* immediately beneath the ice in Lake Vanda and the red pigmented *Chroomonas* at depth.

Despite the physical characteristics of the dry valley lakes that position them as end members of a limnological spectrum, there is still considerable variability among the lakes. A characteristic is the vertical structure in optical properties imposed by a series of attenuating layers. The topmost is the thick ice cover, which removes between 87 and 99% of incident PAR depending on the lake. Scattering is the dominant attenuating process within the ice. Selective removal of red light results in a spectral shift in the light transmitted to the water columns to predominantly shorter wavelengths. Attenuation by ice is variable and dependent on ice thickness, structure, and sediment content. Stable water columns, due to a lack of wind mixing under the ice, combined with meromixis confine phytoplankton populations to discrete depth layers each with its characteristic light spectrum. Phytoplankton and benthic autotrophs in dry valley lakes may be expected to be highly sensitive to light quality. Water column attenuation can vary by an order of magnitude due to differences in phytoplankton biomass and suspended solids concentration. The ratio of light scattering to absorption is high due to fine glacial flours which effectively scatter light, rather than to high concentrations of suspended solids per se. Relative to temperate and Arctic lakes, absorption is likely to play a smaller role in overall light attenuation because of the low concentrations of DOC in the inflowing waters. However the evidence shows that internally generated DOC from lake communities is significant at certain depths in some of the lakes (Lakes Fryxell and Bonney) further imposing a layered structure to the optical properties.

It is clear that light penetration is a key influence on primary production and consequently carbon production in these cold desert ecosystems. Owing to the tight association between light availability and primary production, these aquatic ecosystems are, perhaps more than most others, driven by their optical properties.

Acknowledgments. We acknowledge the assistance with data collection of Dale Anderson, Mark James, and Julie Hall. We also thank Christian Fritsen, J.C. Ellis-Evans, and an anonymous reviewer for very helpful comments on the text. The original data in this Chapter were funded through the New Zealand Foundation for Research Science and Technology under contract Numbers CO1406 and CO1601, and the US National Science Foundation, Office of Polar Programs under grants DPP 88-20591, OPP 91-17907 and OPP 92-11773 to JCP.

REFERENCES

Adams, W. A., Effects of ice cover on the solar radiation regime in Canadian lakes. *Verh. Int. Ver. Limnol. 20,* 142–149, 1978.

Adams, E. E., J. C. Priscu, C. F. Fritsen, S. R. Smith and S. L. Brackman, Permanent ice covers of the McMurdo Dry Valley lakes, Antarctica: Bubble formation and metamorphism, this volume.

Bolsenga, S.J., M. Evans, H.A. Vanderploeg, and D.G. Norton, PAR transmittance through thick, clear freshwater ice. *Hydrobiologia 330*, 227–230, 1996.

Buckley, R. G. and H. J. Trodahl, Scattering and absorption of visible light by sea ice, *Nature, 326,* 867–869, 1987.

Dana, G.L., R.A. Wharton Jr., and T. Dubayah, Solar radiation in the McMurdo Dry Valleys, Antarctica, this volume.

Conovitz, P.A., D.M. McKnight, L.M. McDonald, A. Fountain, and H.R. House, Hydrologic processes influencing streamflow variation in Fryxell Basin, Antarctica, this volume.

Fritsen, C. H., E. E. Adams, C. P. McKay, and J. C. Priscu, Permanent ice covers of the McMurdo Dry Valley lakes, Antarctica: Liquid water content, this volume.

Goldman, C. R., D. T. Mason, and J. E. Hobbie, Two antarctic desert lakes, *Limnol. and Oceanogr., 12,* 295–310, 1967.

Grenfell, T. C., and G. A. Maykut, The optical properties of ice and snow in the Arctic Basin. *J Glaciol 18,* 445–463, 1977.

Hawes, I., Light climate and phytoplankton photosynthesis in maritime Antarctic lakes, *Hydrobiologia 123,* 69–79, 1985.

Hawes, I., and A-M. Schwarz, Photosynthesis in benthic cyanobacterial mats from Lake Hoare, Antarctica, *Antarctic J. of the US,* in press.

Hoeppfner, N., and S. Sathyendranath, Bio-optical characteristics of coastal waters: absorption spectra of phytoplankton and pigment distribution in the Western North Atlantic. *Limnol. Oceanogr. 37,* 1660–1679, 1992.

Howard-Williams, C., and W.F. Vincent, Optical properties of New Zealand lakes I: attenuation, scattering, and a comparison between downwelling and scalar irradiances, *Arch Hydrobiol. 19,* 318–330, 1984.

Howard-Williams, C., C. L. Vincent, P. A. Broady, W. F. Vincent, Antarctic stream ecosystems: variability in environmental properties and algal community structure, *Int. Revue ges. Hydrobiol., 71,* 511–544, 1986.

Hutchinson, G. E., *A Treatise on Limnology. I. Geography, Physics and Chemistry.* New York, John Wiley and Sons Inc. 1015 pp, 1957.

Kirk, J. T. O., *Light and Photosynthesis in Aquatic Ecosystems.* 2nd Edition, Cambridge University Press, Cambridge, 509 pp, 1994.

Lizotte, M. P., and J. C. Priscu, Spectral irradiance and bio-optical properties in perennially ice-covered lakes of the dry valleys (McMurdo Sound, Antarctica). Contributions to Antarctic Research III, *Antarctic Research Series, 57,* 1–14, American Geophysical Union, Washington, D.C., 1992a.

Lizotte, M. P., and J. C. Priscu, Photosynthesis-irradiance relationships in phytoplankton from the physically stable water column of a perennially ice-covered lake (Lake Bonney, Antarctica), *J. Phycol. 28,* 179–185, 1992b.

Lizotte, M. P., and J. C. Priscu, Natural fluorescence and quantum yields in vertically stationary phytoplankton from perennially ice-covered lakes. *Limnol. Oceanogr. 39,* 1399–1410, 1994.

Lizotte, M. P. and J. C. Priscu, Pigment analysis of the McMurdo Dry Valley lakes of Antarctica, this volume.

Lizotte, M. P., T. R. Sharp, and J. C. Priscu, Phytoplankton dynamics in the stratified water column of Lake Bonney, Antarctica. I: biomass and productivity during the winter-spring transition, *Polar Biol. 16,* 155–162, 1996.

Mae, S., Tyndall figures at grain boundaries of pure ice, *Nature 257,* 382–383, 1975.

Marshall, B. R., and R. C. Smith, Raman scattering and in-water ocean optical properties, *Appl. Opt. 29*, 71–84, 1990.

McKay, C. P., G. D. Clow, D. T. Anderson, and R. A. Wharton Jr., Light transmission and reflection in perennially ice-covered Lake Hoare, Antarctica, *J. Geophys. Res. 199*, 20,427–20,444, 1994.

McKnight, D. M. and E. D. Andrews, Hydrologic and geochemical processes at the stream-lake interface in a permanently ice-covered lake in the McMurdo dry valleys, Antarctic, *Verh. Internat. Verein. Limnol. 25*, 957–959, 1993.

McKnight, D. M., G. R. Aiken, and R. L. Smith, Aquatic fulvic acids in microbially based ecosystems - results from two desert lakes in Antarctica, *Limnol. Oceanogr. 36*, 998–1006, 1991.

McKnight, D. M., A. Alger, C. Tate, G. Shupe, and S. Spaulding, Longitudinal patterns in algal abundance and species distribution in meltwater streams in Taylor Valley, Southern Victoria Land, Antarctica, this volume.

Neale, P. J., and J. C. Priscu, The photosynthetic apparatus of phytoplankton from a perennially ice-covered Antarctic lake: acclimation to an extreme shade environment, *Plant Cell Physiol. 36*, 253–263, 1995.

Neale, P.J. and J.C. Priscu, Fluorescence quenching in phytoplankton of the McMurdo Dry Valley Lakes (Antarctica): implications for the structure and function of the photosynthetic apparatus, this volume.

Palmisano, A., and G. M. Simmons Jr., Spectral downwelling irradiance in an Antarctic lake, *Polar Biol. 7*, 145–151, 1987.

Priscu, J. C., Photon dependence of inorganic nitrogen transport by phytoplankton on perennially ice-covered antarctic lakes, *Hydrobiologia, 172*, 173–182, 1989.

Priscu, J. C., Variation in light attenuation by the permanent ice cap of Lake Bonney during spring and summer, *Ant. J. of the U.S.*, 223–224, 1991.

Priscu, J. C., Phytoplankton nutrient deficiency in lakes of the McMurdo Dry Valleys, Antarctica. *Freshwat. Biol. 34*, 215–227, 1995.

Roulet, N. T., and W. P. Adams, Illustrations of the spatial variability of light entering a lake using an empirical model, *Hydrobiologia 109*, 64–67. 1984.

Rowan, K. S., *Photosynthetic pigments of algae*. Cambridge University Press, New York, 334 p, 1989

Seaburg, K. G., M. Kasper, and B. C. Parker, Photosynthetic quantum efficiencies of phytoplankton from perennially ice covered antarctic lakes, *J. Phycol. 19*, 446–452, 1983.

Spigel, R. H., and J. C. Priscu, Evolution of temperature and salt structure of Lake Bonney, a chemically stratified antarctic lake, *Hydrobiologia 321*, 177–190, 1996.

Spigel, R.H., and J.C. Priscu, Physical limnology of the McMurdo Dry Valley Lakes, this volume

Trodahl, H. J., and R. G. Buckley, Enhanced ultraviolet transmission of Antarctic sea ice during the austral spring, *Geophys. Res. Lett. 17*, 2177–2179, 1990.

Trodahl, H. J., R. G. Buckley, and M. Vignaux, Anisotropic light radiance in and under sea ice, *Cold Regions Sci. and Techn. 16*, 305–308, 1989.

Vincent, W. F., Production strategies in Antarctic inland waters: phytoplankton eco-physiology in a permanent ice-covered lake, *Ecology 62*, 1215–1224, 1981.

Vincent, W.F. Phytoplankton production and winter mixing: contrasting effects in two oligotrophic lakes, *J. Ecol., 71*, 1–20, 1983.

Vincent, W. F., *Microbial ecosystems of Antarctica*, Cambridge University Press, 304 pp, 1988.

Vincent, W. F., and C. L. Vincent, Factors controlling phytoplankton production in Lake Vanda (77°S), *Can. J. of Fish and Aquat. Sci. 39*, 1602–1609, 1982.

Walker, J., The amateur scientist: exotic patterns appear in water when it is freezing or melting, *Scientific American 255*, 114–119, 1986.

Webster, J. G., Trace metal behavior in oxic and anoxic Ca-Cl brines of the Wright Valley drainage, Antarctica. *Chem. Geol. 112*, 255–274, 1993.

Webster J., I. Hawes, M. Downes, M. Timperley, and C. Howard-Williams. Evidence for regional climate change in the recent evolution of a high latitude pro-glacial lake, *Antarctic Science, 8*, 49–59 1996.

Wharton, R. A. Jr., G. M. Simmons Jr., and C. P. McKay, Perennially ice-covered Lake Hoare, Antarctica: physical environment, biology and sedimentation. *Hydrobiologia 172*, 305–320, 1989.

Wilson, A. T., A review of the geochemistry and lake physics of the Antarctic dry areas. *In* L. D. McGinnis (ed.), *Dry Valley Drilling Project, Antarctic Research Series 33*, 185–192. American Geophysical Union, Washington, D.C., 465 pp, 1981.

I. Hawes National Institute for Water and Atmospheric Research, Kyle Street, P.O. Box 8602, Christchurch, New Zealand

C. Howard-Williams, National Institute for Water and Atmospheric Research, Kyle Street, P.O. Box 8602, Christchurch, New Zealand

A.-M. Schwarz, National Institute for Water and Atmospheric Research, Kyle Street, P.O. Box 8602, Christchurch, New Zealand

J. C. Priscu, Dept. of Biology, Lewis Hall, Montana State University, Bozeman, Montana, 59717, USA

(Received September 12, 1996;
accepted February 25, 1997)

COBALT CYCLING AND FATE IN LAKE VANDA

William J. Green

School of Interdisciplinary Studies, Miami University, Oxford, Ohio

Donald E. Canfield

Institute of Biology, Odense University, Odense M, Denmark

Philip Nixon

School of Interdisciplinary Studies, Miami University, Oxford, Ohio

Lake Vanda is a closed-basin, permanently ice-covered lake located in the Wright Valley, Antarctica. Among its most important and geochemically significant features is the fact that its 68-m watercolumn has been stratified for some 1200 years. This has resulted in the evolution of distinct layers, which range from cool, fresh, oxygen-rich, moderately basic surface waters beneath the ice to warm, saline, sulfide-bearing, acidic waters at depth. Here we present the first detailed vertical profiles for dissolved and total cobalt in the Vanda watercolumn. In the well-oxygenated upper waters, dissolved cobalt concentrations are found to be less than 0.2 nM. Significant concentrations begin to appear at 55 m, with the onset of manganese oxide dissolution. A cobalt maximum (13 nM) is observed at 61 m, near the top of the anoxic zone. Beneath this depth, cobalt concentrations begin to decrease in response to increasing sulfide levels. Cobalt distributions and residence times are discussed in terms of the apparent role played by manganese oxide phases in both the transport and release of cobalt.

INTRODUCTION

The weathering products of continental rocks include virtually every element in the periodic table. Among these are the trace metals, elements such as Co, Ni, Cu, Zn, Cd, Hg, and Pb, which are generally present in unpolluted fresh and marine waters at the parts per billion or parts per trillion levels. The dissolution of these elements from host minerals, their transport via streams and rivers, and their eventual removal to the sediments have been the subjects of a large and growing geochemical literature [*Solomons and Forstner*, 1984; *Chester*, 1990].

Since modern industrial societies are, in some sense, built upon the use of metals, human activities have significantly altered the geochemical cycles of these elements. Some aquatic environments receive large quantities of metals from industrial waste waters,

sewage discharges, urban runoff, and effluents from working and abandoned mines. The movement of metals through the atmosphere has also been greatly augmented by the increasing worldwide use of fossil fuels and by such activities as cement production and extractive metallurgy. For example it has been estimated that more than 50% of all trace metals entering the Great Lakes are now transported via the atmosphere, and the anthropogenic contribution to global lead emissions is roughly thirty times higher than natural sources [*Nriagu and Pacyna*, 1988].

Scientific interest in metals derives, in part, from the fact that they are closely linked to biological processes. Many trace metals serve as essential micronutrients for phytoplankton. Copper, for example, is involved in photosynthetic electron transport, nickel in certain hydrolysis reactions, and cobalt in transfers involving hydrogen and carbon [*Stumm and Morgan*, 1996]. And

205

yet, at sufficiently elevated levels, metals can be chronically or acutely toxic [*Morel and Morel-Laurens*, 1983]. Toxicity itself depends upon a number of factors, but it appears that the toxicity of the free metal ion in solution is a crucial variable [*Sunda et al.*, 1978]. *Sunda and Ferguson* [1983] have tentatively shown that organisms dwelling in low productivity, oligotrophic waters, such as the open ocean, may be more sensitive to excursions in metal concentrations than those dwelling in waters of higher productivity.

Cobalt, the subject of this study, occurs in more than 200 known ores, only a few of which are of commercial value. Although certain of these ores have been used for thousands of years to impart a blue color to glass and pottery, it has only been in this century that the element has received extensive use. Ceramics, high temperature alloys, cobalt-containing steels used in permanent magnets, and catalysts in hydrogenation and rehydrogenation reactions all employ cobalt in one form or another. The element also serves as the centerpiece of the vitamin B_{12} molecule. Thus, as is the case with many trace metals, the environmental behavior of cobalt is of interest both because of the element's biochemical significance and because of its increasing use in a variety of industrial applications.

A number of geochemical studies have charted the behavior of cobalt in soils and in natural waters. The early work of *Jenne* [1968] proposed that manganese and iron oxides and hydrous oxides exert important controls on the availability of Co and other heavy metals in soils and fresh water sediments. *Johnson et al.* [1988] contrasted the behavior of cobalt and copper in the Santa Monica Basin and presented evidence that suggested that, while copper was transported largely by organic matter, cobalt was scavenged by manganese oxide particles. This observation is consistent with *Li*'s [1981] analysis of pelagic red clays, which shows that cobalt is associated with manganese rather than with iron in these widespread marine sedimentary deposits.

Murray's [1975] laboratory study of the interaction of metal ions with manganese dioxide showed that cobalt was adsorbed to a greater extent than other ions. *Murray* obtained the following sequence for the affinity of metal ions for manganese dioxide surface:

$$Mg < Ca < Sr < Ba < Ni < Zn < Mn < Co.$$

A two step mechanism involving de-protonation of the oxide surface followed by metal uptake was proposed. This is shown below:

$$Mn - O - H + Co^{2+} = Mn - O - Co^+ + H^+.$$

Burns [1976] has shown that this adsorbed cobalt may eventually be incorporated into manganese oxide lattices as an octahedrally coordinated, low-spin, Co^{3+} ion. This energetically favorable inclusion may account for the fact that cobalt is so strongly partitioned into manganese crusts and other marine manganese minerals.

The work of *Balistrieri et al.* [1994] on meromictic Hall Lake indicates that, in natural waters, the fate of many trace metals is linked to that of manganese and iron oxide phases. In this system, *Balistrieri et al.* [1994] reported excellent correlations between dissolved concentrations of Mn and Co, suggesting that these elements are co-cycled across oxic-suboxic and suboxic-anoxic redox boundaries. Similar correlations have also been observed in other lakes by *Green et al.* [1989] and by *Balistrieri et al.* [1992].

The objective of the present study is to examine the behavior of cobalt across the permanent redox boundary of Lake Vanda, with a view toward gaining a better understanding of how this element is cycled in natural waters. The unique watercolumn structure of Lake Vanda provides an excellent opportunity to carry out investigations of element behavior and fate across a broad range of geochemical environments.

SITE DESCRIPTION

Lake Vanda is located in Wright Valley, a stark, glacier-carved valley bordered by the high rock walls of the Asgard and Olympus ranges and covered by dark, undulating drift deposits. The lake is capped by a smooth, permanent, 4-m thick ice cover, is approximately 5.6 km long, 1.5 km wide, and, at the time of this study (1986–1987 austral summer), 68.8 m deep in the western depression. Water is supplied by the Onyx River during a six-week period from about mid-December to early February. The Onyx, which has its source at the Wright Lower Glacier 27 km to the east, has an annual discharge rate of about 2 billion liters. Figure 1 shows the location of the lake and its principal inflow.

Lake Vanda can be viewed as a meromictic, two layer system [*Spigel and Priscu*, this volume]. The waters above about 45 m are cool, well-mixed, fresh, and rich in dissolved oxygen. Below 45 m the lake is strongly stratified, with temperatures increasing to 25°C and salinity rising to 120 parts per thousand at 68 m. The bottom waters are anoxic. According to *Wilson*'s [1964] hypothesis, present stratification is the result of flooding (1200 years ago) of a shallow, saline brine by fresh waters from the Onyx River. Chemical profiles

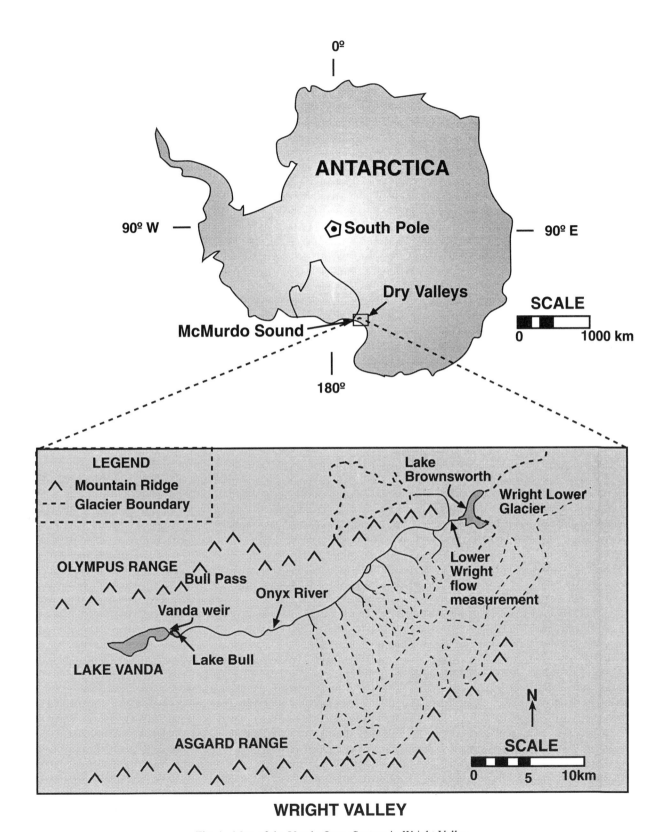

Fig. 1. Map of the Vanda-Onyx System in Wright Valley.

have developed since that time, largely through molecular diffusion [*Canfield and Green,* 1985]. *Canfield et al.* [1995] have argued that the horizontal transport of solutes at depth by advection is highly unlikely.

An unusual feature of Lake Vanda's chemistry is the presence of calcium and chloride as the major ionic species. Based on a study of the Onyx River and on the lake's major ion chemistry, *Green and Canfield* [1984] concluded that the salts are derived from two principal sources, namely the deep groundwater reservoir of the Don Juan Basin and the Onyx River itself. The mixing model developed by *Green and Canfield* [1984] shows that the groundwater once contributed most of the calcium and chloride found in the lake, while the Onyx River was the dominant source of the magnesium, potassium, sulfate, and bicarbonate. Sodium appears to have come equally from both sources.

The biology of Lake Vanda, which is recognized as one of the clearest and least productive lakes in the world, has been described by *Vincent and Vincent* [1982]. Low algal standing crops and low rates of primary productivity are found at all depths. However a narrow band of more productive water is present just above the anoxic zone, where an active population of low-light adapted phytoplankton has been observed [*Goldman et al.,* 1967; *Vincent and Vincent,* 1982].

Studies of selected metal distributions in Lake Vanda have been reported by *Green et al.* [1986], *Green et al.* [1989], *Green et al.* [1993], *Webster* [1994], and *Canfield et al.* [1995]. However only the work of *Canfield et al.* [1995] on Pb and Pb-210 presents data at closely spaced intervals (1 m) through the stable redox boundary. All of the investigations to date indicate low metal concentrations in the upper, oxygen-rich waters of the lake and metal enrichment in waters just above the anoxic zone. This study provided an opportunity to look at cobalt profiles throughout the lake, especially at close intervals through the redox zone. One of our objectives is to understand the way in which cobalt concentrations are controlled in natural waters.

FIELD AND LABORATORY METHODS

Lake samples were collected on December 12, 1986, a time before the rapid rise in lake level that has been recorded over the past nine years. A 6 liter, all plastic Kemmerer bottle was lowered by a nylon line to obtain samples for dissolved oxygen, sulfide, and pH analysis. Dissolved oxygen concentrations were measured on

first aliquots using the Winkler titration, as described by *Grasshoff* [1976]. The methylene blue procedure, developed by *Cline* [1969], was used for sulfide determination. pH was measured with an Orion Ionalyzer (model 339A) connected to an Orion gel-filled electrode. Samples for metals analysis were collected using a peristaltic pump and Tygon tubing which had been thoroughly cleaned at the Eklund Biological Laboratory, McMurdo Station, by circulating 10% nitric acid, 1% ultrapure nitric acid, and, finally, deionized water for several hours. In the field, tubing was further cleaned by passing through it several liters of water from depth.

Polyethylene bottles used for the collection and storage of filtered and unfiltered trace metal samples were cleaned at the Eklund Biological Laboratory before transport to the field. Bottles were soaked in 10% nitric acid for 5 days, in 1% ultrapure nitric acid for 1 day, and then rinsed three times with deionized water and again with sample before filling. In-line Nucleopore (0.2 micron pore size) filters were soaked for 24 hours in deionized water, for 2 days in 10% nitric acid, and for 24 hours in 1% ultrapure nitric acid. Each filter was transported to the field in acid cleaned plastic containers which had been rinsed with deionized water. Finally, filters were rinsed with 5 to 10 ml of sample. Filtered and unfiltered samples were acidified with 2 ml of ultrapure nitric acid to lower the pH to less than 2.

Dissolved iron concentrations were measured at the Eklund Laboratory using the Ferrozine colorimetric technique developed by *Stookey* [1970]. Dissolved manganese was analyzed using the formaldoxime method of *Brewer and Spencer* [1971]. Replicate determinations were reproducible to better than 0.1 μM up to 10 μM and reproducible to better than 0.1 μM at higher concentrations. (It should be noted that these data have been reported graphically in *Canfield et al.* [1995]).

Dissolved cobalt and "total acid soluble" cobalt concentrations were determined using a Perkin-Elmer 3030 Graphite Furnace Atomic Absorption Spectropho-tometer. Samples (300 ml), buffered to pH 4.5 by ultrapure acetic acid and ultrapure ammonium hydrox-ide, were doubly extracted into Freon TF using the chelating agents APDC and DDDC, and then back-extracted into 3 ml of ultrapure nitric acid. The method is similar to that discussed by *Bruland et al.* [1979]. The detection limit for cobalt was the concentration that, upon 100-fold concentration, produced a 1% absorption. This value was 0.17 nM.

Samples were also collected from the Onyx River, the sole inflow to Lake Vanda. These were analyzed in a similar manner for dissolved and "total acid soluble" cobalt.

RESULTS

Chemical Characteristics

Data for pH, dissolved oxygen, and sulfide are presented in Table 1 and are plotted in Figure 2. In the upper waters of the lake, pH values gradually increase with depth down to 35 m, where pH equals 8.2. Below this depth, values decrease rapidly down to 56 m, where pH equals 6.3. Between 56 and 58 m the pH remains nearly constant at 6.4, but below this depth it begins to fall again. *Canfield et al.* [1995] noted that the "pH plateau" between 56 and 58 m occurs in the vicinity of the dissolved oxygen submaximum at 58 m. The oxygen submaximum, itself, has been attributed by *Vincent et al.* [1981] to high rates of photosynthesis at this depth in the watercolumn. In the anoxic brine, we record a low value of 5.9 pH units at 64 m. While no

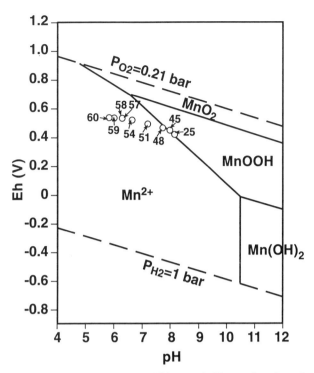

Fig. 2. Dissolved oxygen, sulfides, and pH as a function of depth in Lake Vanda.

TABLE 1. Dissolved Oxygen, pH, and Sulfides in Lake Vanda.

Depth (m)	Oxygen (μM)	pH	Sulfide (μM)
5	536	7.82	NA
15	560	8.00	NA
25	721	8.12	NA
35	718	8.20	NA
45	731	7.95	NA
48	689	7.70	NA
49	692	7.61	NA
50	685	7.50	NA
51	592	7.18	NA
52	559	6.93	NA
53	554	6.75	NA
54	524	6.60	NA
55	480	6.43	NA
56	457	6.30	NA
57	460	6.30	NA
58	589	6.29	NA
59	570	6.02	NA
60	311	5.90	NA
61	18	5.90	0
62	0	5.79	105.0
63	0	5.70	245.8
64	0	5.60	469.5
64.5	0	NA	NA
65	0	5.70	726.8
66	0	5.65	1012.3

attempt has yet been made to model pH in Lake Vanda, this parameter is most likely set by the H_2CO_3 / HCO_3^- / CO_3^{-2} system, with decomposition of organic matter and carbon dioxide production in bottom waters driving the pH to more acidic values.

Dissolved oxygen is high throughout the upper lake, especially between 35 and 45 m, where concentrations are in excess of 700 μM. Between 50 and 56 m there is a strong decline in oxygen levels down to 457 μM. Two meters below this depth, at 58 m, lies the oxygen submaximum, with concentrations of 589 μM. (This submaximum has been observed over many years, both by ourselves and others, and appears to be a relatively permanent feature of the stable Vanda watercolumn.) Between 59 and 60 m there is a steeply negative oxygen gradient, and the 61-m depth of the lake is anoxic. Below this depth, down to the sediments, sulfide levels increase linearly to a high of 1 mM at 66 m.

Iron and Manganese Profiles

Iron concentrations in the upper waters of the lake are known to be quite low [*Green et al.*, 1989], ranging from only 9 to 73 nM. Between 60 and 61 m dissolved iron increases dramatically, roughly a 100 fold in-

TABLE 2. Iron, Manganese, and Cobalt in Lake Vanda.

Depth (m)	Fe (μM)	Mn (μM)	Co (nM)	Co* (nM)
5	†0.016	†0.012	<0.17	<0.17
10	†0.023	†0.015	<0.17	0.53
25	†0.041	†0.034	<0.17	0.24
35	†0.009	†0.042	<0.17	0.25
45	†0.018	†0.072	<0.17	NA
45.7	NA	NA	<0.17	0.26
46.6	NA	NA	<0.17	0.29
47.5	NA	NA	<0.17	0.42
48	†0.019	†0.13	<0.17	NA
48.5	NA	NA	<0.17	NA
49	NA	NA	<0.17	NA
49.4	NA	NA	<0.17	0.44
50	NA	NA	<0.17	NA
50.3	NA	NA	<0.17	0.74
51	NA	0.546	NA	NA
51.3	NA	NA	0.034	1.1
52	NA	1.44	NA	NA
52.2	NA	NA	NA	1.4
53	NA	2.77	NA	NA
53.2	NA	NA	NA	NA
54	†0.016	5.08	NA	NA
54.1	NA	NA	NA	NA
55	NA	7.18	0.41	2.2
56	NA	7.40	1.0	4.4
57	NA	6.52	3.4	4.1
58	†0.073	4.86	3.1	3.4
59	NA	5.63	3.0	1.6
60	0.251	13.60	6.5	5.7
61	22.1	68.30	13.0	10.0
62	29.2	68.30	1.2	6.6
63	28.5	54.40	0.43	1.5
64	19.1	53.70	<0.17	1.5
65	13.5	54.40	<0.17	1.3
66	9.36	56.40	NA	2.2

NA = Not analyzed.

* = Dissolved + particulate metal.

† = Concentrations taken from *Canfield et al.* [1995].

crease, as shown in Table 2 and in Figure 3. Since it is between these depths that waters become anoxic, we attribute the concentration change to the reductive dissolution of iron oxides and hydrous oxides, according to the following half-reaction:

$$Fe(OH)_3 + 3H^+ + e^- = Fe^{2+} + 3H_2O.$$

This reaction is usually driven in natural waters by organic carbon, which serves as the principal source of electrons [*Elderfield*, 1985]. In Lake Vanda, it is also driven to the right by the increasing concentrations of hydrogen ion at depth. Below the dissolved iron maximum at 62 m, concentrations decrease with depth in response to increasing sulfide levels. The vertical profile, therefore, shows a band of high iron concentrations between 61 and 63 m, where dissolved oxygen is

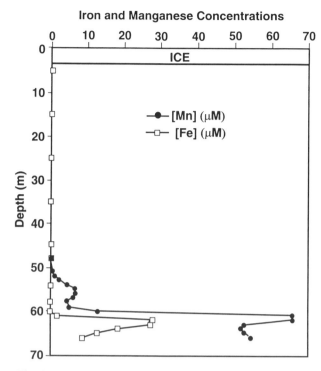

Fig. 3. Dissolved Mn and Fe as a function of depth in Lake Vanda.

depleted and where sulfide concentrations are still relatively low.

Manganese data are given in Table 2 and plotted in Figure 3. Dissolved manganese concentrations are very low in the upper lake, but begin to increase just below 50 m. Between 51 and 56 m there is a 14 fold elevation in manganese concentrations. This suggests that in this region of falling pH, but still ample dissolved oxygen, manganese oxides are being reductively dissolved as shown below:

$$MnO_2 + 4H^+ + 2e^- = Mn^{2+} + 2H_2O.$$

Thus, in Lake Vanda, there is nearly a 10-m interval between the development of the Mn-cline and the Fe-cline, that is, between the onset of significant manganese oxide reduction and significant iron oxide reduction. This large spatial interval will be important in the interpretation of cobalt behavior.

Among the most interesting features of the manganese profile is the occurrence, at 56 m, of a manganese submaximum. Figure 4 shows that this coincides with the oxygen subminimum at the same depth. This may suggest a very close, near-to-equilibrium coupling between the oxygen and manganese systems, as represented below:

$$2Mn^{2+} + O_2 + 2H_2O = 2MnO_2 + 4H^+.$$

Oxygen increases at 57 and 58 m result in dissolved manganese depletion at these depths (the above reaction is shifted to the right), with the subminimum occurring at 58 m. Below 58 m, levels of dissolved manganese again begin to rise, with a five-fold increase occurring at the oxic/anoxic interface between 60 and 61 m.

Dissolved and Total Acid Soluble Cobalt in Lake Vanda

The data for cobalt, as obtained on filtered and unfiltered samples from the same depths, are presented in Table 2 and graphed in Figure 5. Dissolved concentrations in the upper lake are all low, less than 0.017 nM. It is not until 51.3 m that significant cobalt appears in the watercolumn. A dissolved Co gradient appears between 55 and 57 m, with a submaximum at the latter depth. Values decrease at 57, 58, and 59 m, to a subminimum at this depth of 3.0 nM. Between 59 and 61 m—the oxic/anoxic interface—there is a strong

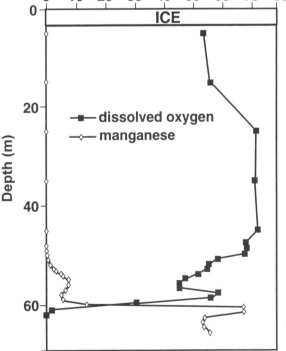

Fig.4. Manganese and oxygen as a function of depth in Lake Vanda.

Co gradient, with values at 61 m reaching a high of 13.0 nM. Below 62 m, concentrations decrease down to the sediments.

Comparison of the dissolved and total acid soluble cobalt profiles in Figure 6 shows that there is evidence for particulate cobalt throughout the lake. We consider the difference between dissolved and total acid soluble cobalt to be a measure of the particulate metal. Particulate Co appears to be concentrated between about 50 and 57 m and, below, between 62 and 66 m.

Cobalt in the Onyx River

Table 3 presents data for dissolved and total acid soluble cobalt in the Onyx River. We note at this point that dissolved concentrations are very low and that most of the cobalt delivered to Lake Vanda is delivered in particulate form. The data will be used to determine cobalt residence times in the lake.

DISCUSSION

Watercolumn Cobalt

The behavior of cobalt in Lake Vanda appears to be intimately linked to the pH and redox conditions of the watercolumn and especially to the solution chemistry of

TABLE 3. Cobalt in the Onyx River.

Date	Dissolved Cobalt	Total Cobalt
12/11/86	NA	8.6
12/13/86	NA	4.5
12/14/86	<0.17	2.5
12/15/86	NA	2.6
12/16/86	NA	2.9
12/17/86	NA	1.4
12/20/86	NA	1.2
12/21/86	<0.17	1.2
12/22/86	<0.17	0.92
12/24/86	<0.17	0.70
12/25/86	0.34	0.71
12/26/86	0.53	1.5
12/26/86	<0.17	NA
12/28/86	<0.17	1.0
12/29/86	0.17	0.98
12/30/86	0.22	0.46
12/31/86	<0.17	1.1
1/1/87	<0.17	1.2
1/2/87	<0.17	3.0
1/4/87	NA	3.3
1/5/87	0.60	3.8
1/7/87	NA	3.4

Concentrations expressed as nM.
NA = Not analyzed.

manganese. In their recent study of Pb-210 in the lake, *Canfield et al.* [1995] identified a zone between about 50 and 60 m where reductive dissolution of manganese oxides appears to be occurring. They noted that while this region was well oxygenated, eH-pH conditions favored higher levels of dissolved manganese (see Figure 6). This zone was called the "aerobic manganese reduction" zone, or AMR, and it was noted that the top of the AMR lies 11 m above the zone of iron reduction. It is within the AMR that similarities in the manganese and cobalt profiles can be recognized.

Past studies of the Vanda watercolumn have called attention to the apparent importance of manganese oxide phases in the transport and cycling of trace metals. Thus *Green et al.* [1993] argued that in waters between 50 and 60 m, the vertical profiles of Ni, Cu, Cd, and Co (although in that early study only two points were presented for dissolved Co) suggested that these metals were being transported and released largely by manganese oxide phases. A similar conclusion was reached by *Canfield et al.* [1995] in their study of Pb and Pb-210.

We believe that the cobalt profile presented here is consistent with the manganese oxide transport and cycling theory. We envision the following: Dissolved manganese is transported to Lake Vanda by the Onyx River. This manganese is eventually oxidized to form a solid oxide or oxyhydroxide, possibly associated with the surfaces of clay or organic particles. As these particles sink, trace metals are scavenged onto their oxide-bearing surfaces. However in the chemically aggressive waters of the AMR zone, these manganese oxide phases are reduced and, in the process of reduction, adsorbed, and lattice-bound metals are released. The fate of Co is thus tied to that of manganese. This is represented in the cartoon in Figure 7.

The relatively high concentration of particulate cobalt between about 50 and 55 m and the apparent delayed onset of cobalt dissolution relative to manganese (the dissolved Mn gradient appears 5 m higher in the watercolumn than the dissolved Co gradient) suggest that upwardly diffusing manganese may be re-oxidizing above 55 m and forming fresh particle surfaces for cobalt sorption. Such a mechanism has been proposed by *Canfield et al.* [1995] to explain the high levels of particulate lead at these depths.

Since iron and manganese oxides and oxyhydroxides have often been considered important carrier phases for trace metals, it is noteworthy that in Lake Vanda there is little evidence for iron participation in the cycling of cobalt. At the depth where significant iron dissolution does begin to occur and where evidence of cobalt release

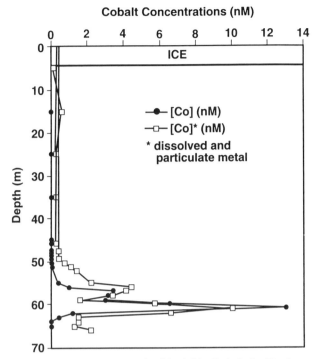

Fig. 5. Dissolved and total acid soluble Co in Lake Vanda.

might be expected, namely near 61 m, cobalt concentrations begin to decline, a response to the presence of hydrogen sulfide.

Residence Times

Using the average concentration of total acid soluble Co in the Onyx River (2.2 nM), *Chinn's* [1981] estimate of annual river discharge (2.0×10^9 l yr^{-1}), our total acid soluble Co profile in the lake, and the lake volume estimates of *Green and Canfield* [1984], we calculate a residence time of Co in the mixed upper layer of Lake Vanda (down to 50 m) of 5.9 years.

In an earlier study [*Green et al.*, 1986] we determined the residence times of Mn, Fe, Cu, and Cd to be 9.4, 1.4, 172, and 82 years, respectively. In terms of its duration in the watercolumn, Co behaves in a manner similar to that of Fe and Mn. Cobalt's rapid removal is consistent with its tendency to be strongly adsorbed onto manganese oxide surfaces [*Murray*, 1975].

CONCLUSIONS

The long stratification history of Lake Vanda and its stable zonation into well-defined oxic, suboxic, and

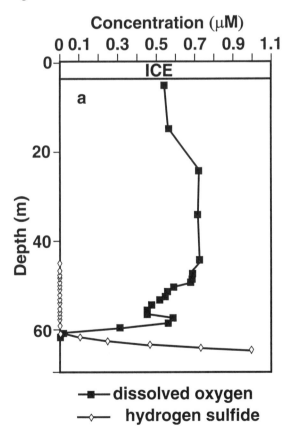

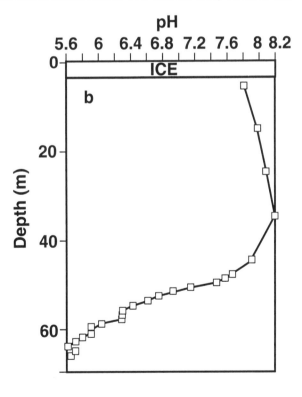

Fig. 6. Eh-pH diagram for Mn, with Eh-pH values for selected depths in the Vanda watercolumn. For details see *Canfield et al.* [1995].

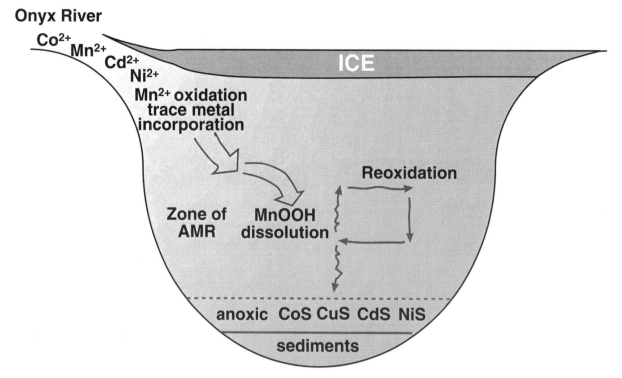

Fig. 7. A sketch of important processes involving trace metals in the Vanda-Onyx system.

anoxic regions make it an excellent system in which to study the fate of trace elements. We view the lake as a natural geochemical laboratory, a flask, in which nature has constructed and maintained a watercolumn that from top to bottom is representative of a wide range of biogeochemical conditions. Cold, fresh, oxygen-rich surface waters gradually give way, with depth, to warm, saline, sulfide-bearing layers that have remained relatively constant over decades and possibly over centuries. In many of its features the Vanda watercolumn mimics marine pore waters, summer stratified lakes, anoxic fjords, and ancient stratified oceans, and many of the chemical profiles observed in the lake are profiles in extremis.

This study of cobalt, along with supporting data on oxygen, sulfides, pH, manganese, and iron, suggests that the geochemistry of this element is closely linked to that of manganese. Low cobalt concentrations and short residence times in the oxic, high pH waters above 50 m indicate that the element is quickly removed from this environment. However, in the AMR zone, Co concentrations increase dramatically, suggesting release from dissolving manganese oxide phases. In the deep anoxic zone, cobalt is most likely precipitated as a sulfide.

The vertical profile for Co, and especially the tendency for this element to accumulate in the AMR zone, suggests that other aquatic environments, especially those which tend to favor aerobic manganese oxide reduction (acidified lakes, estuarine sediments, hypolimnetic waters during summer stratification) may also tend to favor Co accumulation.

REFERENCES

Balistrieri, L. S., J. W. Murray, and B. Paul. The biogeochemical cycling of trace metals in the water column of Lake Sammamish, Washington: response to seasonally anoxic conditions. *Limnol. and Oceanogr.*, 31, 529–547, 1992.

Balistrieri, L. S., J. W. Murray, and B. Paul. The geochemical cycling of trace elements in a biogenic meromictic lake. *Geochim. Cosmochim. Acta*, 58, 3993–4008, 1994.

Brewer, P. G. and D. W. Spencer. Colorimetric determination of manganese in anoxic waters. *Limnol. Oceanogr.*, 16, 107–110, 1971.

Bruland, K. W., R. P. Frank, G. A., Knauer, and J. H. Martin. 1979. Sampling and analytical methods for the determination of copper, cadmium, zinc and nickel at the nanogram per liter level in seawater. *Anal. Chim. Acta*, 105, 233–245., 1979.

Burns, R. G. The uptake of cobalt in ferro-manganese nodules, soils and synthetic manganese (IV) oxides. Geochim. *Cosmochim Acta*, 40, 95–100, 1976.

Canfield, D. E. and W. J. Green. The cycling of nutrients in a closed-basin antarctic lake: Lake Vanda. *Biogeochemistry*, 1, 233–256, 1985.

Canfield, D. E., W. J. Green, and P. Nixon. Pb-210 and stable lead through the redox transition zone of an antarctic lake. Geochim. *Cosmochim. Acta*, 59, 2459–2468, 1995.

Chester, R. *Marine Geochemistry*. Nunwin-Hyman, London. 698 pp., 1990.

Chinn, T. J. H. Hydrology and climate in the Ross Sea area. *J. Royal Soc. New Zeal.*, 11, 373–386, 1981.

Cline, J. D. Spectrophotometric determination of hydrogen sulfide in natural waters. *Limnol. Oceanogr.*, 14, 454–458.

Elderfield, H. 1985. Element cycling in bottom sediments. *Phil. Trans. Soc. Lon. A.*, 315, 19–23, 1969.

Goldman, C. R., D. T. Mason, and J. E. Hobbie. Two antarctic desert lakes. *Limnol. Oceanogr.*, 12, 295–310, 1967.

Grasshoff, K. Methods of seawater analysis. *Verlag-chemie*, 1976.

Green, W. J. and D. E. Canfield. Geochemistry of the Onyx River (Wright Valley, Antarctica) and its role in the chemical evolution of Lake Vanda. *Geochim. Cosmochim. Acta*, 48, 2457–2467, 1984.

Green, W. J., D. E. Canfield, G. F. Lee, and R. A. Jones. Mn, Fe, Cu, and Cd distributions and residence times in closed-basin Lake Vanda. *Hydrobiologia*, 134, 237–248, 1986.

Green, W. J., T. G. Ferdelman, and D. E. Canfield. Metal dynamics in Lake Vanda (Wright Valley, Antarctica). *Chem. Geol.*, 76, 85–94, 1989.

Green, W. J., D. E. Canfield, S. Yu, K. E. Shave, T. G. Ferdelman, and G. Delanois. Metal transport and release processes in Lake Vanda: the role of oxide phases. In: Green, W. J. and E. I. Friedman (eds.), *Physical and Biogeochemical Processes in Antarctic Lakes*. American Geophysical Union, Washington, DC, 1993.

Jenne, E. A. Controls on Mn, Fe, Co, Ni, Cu, and Zn concentrations in soils and water: the significant role of hydrous Mn- and Fe-oxides. *Amer. Chem. Soc. Adv. Chem. Ser.*, 337–387, 1968.

Johnson, K. S., P. M. Stout, W. M. Berelson, and C. M. Sakamoto-Arnold. Cobalt and copper distributions in the waters of the Santa Monica Basin, California. *Nature*, 332, 527–530, 1988.

Li, Y.-H. Ultimate removal mechanisms of elements from the ocean. Geochim. Cosmochim. Acta, 45, 1659–1664, 1981.

Morel, F. M. M. and N. M. L. Morel-Laurens. Trace metals and plankton in the oceans: facts and speculations. In C. S. Wong et al. (eds.), *Trace Metals in Seawater*. Plenum Press, NY. 441–869, 1983.

Murray, J. W. The interaction of metal ions at the manganese dioxide-solution interface. *Geochim. Cosmochim. Acta*, 39, 505–519, 1975.

Nriagu, J. O. and J. M. Pacyna. Quantitative assessment of worldwide contamination of air, water and soils with trace metals. *Nature*, 333, 134–139, 1988.

Solomons, W. and U. Forstner. *Metals in the hydrocycle*. Springer-Verlag, New York, 1984.

Spigel, R. H. and J. C. Priscu. Physical limnology of the McMurdo Dry Valley Lakes, this volume.

Stookey, L. Ferrozine-a: new spectrophotometric reagent for iron. *Anal. Chem.*, 42, 779–786, 1970.

Stumm, W. and J. J. Morgan. *Aquatic Chemistry*. Wiley, New York. 1022 pp., 1996.

Sunda, W. G. and R. L. Ferguson. Sensitivity of natural bacterial communities to the addition of copper and cupric ion activity: a bioassay of copper complexation in seawater. In C. S. Wong et al. (eds.), *Trace Metals in Seawater*. Plenum Press, NY. 871–891, 1983.

Sunda, W. G., D. W. Engel, and R. M. Thuotte. Effect of chemical speciation on the toxicity of Cd to grass shrimp. Palaemonetes pugio: importance of free cadmium ion. *Environ. Sci. Technol.*, 12, 409–413, 1978.

Vincent, W. F. and C. L. Vincent. Factors controlling phytoplankton production in Lake Vanda (77 S). *Can. J. Fish. Aquat. Sci.*, 39, 1602–1609, 1982.

Vincent, W. F., M. T. Downes, and C. L. Vincent. Nitrous oxide cycling in Lake Vanda, Antarctica. *Nature*, 292, 618–620, 1981.

Webster, J. G. Trace metal behavior in oxic and anoxic Ca-Cl brine of the Wright Valley, Antarctica. *Chem. Geol.*, 112, 255–274, 1994.

Wilson, A. T. Evidence from chemical diffusion of a climate change in the McMurdo Dry Valleys 1200 years ago. *Nature*, 201, 176–177, 1964.

Donald E. Canfield, Institute of Biology, Odense University, Campusvej 55, 5230 Odense M, Denmark.

William J. Green, School of Interdisciplinary Studies, Miami University, Oxford, OH 45056.

Philip Nixon, School of Interdisciplinary Studies, Miami University, Oxford, OH 45056.

(Received September 5 , 1996;
Accepted April 17, 1997)

THE ABUNDANCE OF AMMONIUM-OXIDIZING BACTERIA IN LAKE BONNEY, ANTARCTICA DETERMINED BY IMMUNOFLUORESCENCE, PCR AND IN SITU HYBRIDIZATION

Mary A. Voytek

Institute of Marine and Coastal Sciences, Rutgers University, New Brunswick, New Jersey

Bess B. Ward

Marine Sciences Program, University of California, Santa Cruz, Santa Cruz, California

John C. Priscu

Department of Biological Sciences, Montana State University, Bozeman, Montana

Previous studies of biogeochemical cycling in Lake Bonney, Antarctica, suggest that nitrification plays a central role in controlling the depth distributions of oxidized and reduced inorganic nitrogen in both east and west lobes. For example, there is a mid-depth N_2O-N maximum of 41.6 µg at l^{-1} (> 500,000% saturation) in the east lobe, the highest level reported for a natural system. The source of this N_2O peak is thought to be nitrification under conditions of low oxygen tension. Although nitrifying bacteria have been detected, attempts to isolate and culture them have been unsuccessful. This study examines three techniques for the determination of abundance of nitrifying bacteria in this lake. Applying a polymerase chain reaction (PCR) assay developed for the detection of ammonium-oxidizing bacteria belonging to the beta and gamma subclasses of the Proteobacteria, immunofluorescent antibody assay (IFA) and fluorescent probe *in situ* hybridization (FISH) techniques, the distribution and relative abundance of ammonium-oxidizers was examined. In general, nitrifiers were detected at depths above the pycnocline and usually associated with decreasing concentrations of NH_4^+ and increasing concentrations of NO_3^- or NO_2^-. These data are consistent with the chemical distributions and the role of nitrifying bacteria in determining the distribution of nitrogen compounds in this lake.

INTRODUCTION

Lake Bonney, a perennially ice-covered, glacial-fed lake within the Taylor Valley, provides an unusual system in which to study the microbially mediated processes of nitrification and the bacteria responsible. Physical processes, such as turbulent mixing, are severely limited due to the ice cover and in situ biogeochemical reactions and diffusion are the dominate processes controlling chemical distributions. The lake is divided into two lobes by a narrow, shallow strait which results in the hydrographic isolation of the deeper layers of each basin. As a result of this isolation and important differences in the recent history (1000–2000 years) of the two lobes, their deep water chemistry differs markedly [*Spigel and Priscu*, 1996; *Lyons et al.*, 1997]. Both lobes are chemically stratified; the chemocline in each basin is located below the sill depth and separates the upper trophogenic zone which is fresh, well-oxygenated, and low in nitrogenous compounds from the hypersaline, anoxic lower zone where photoautotrophic activity is absent and inorganic nitrogen is regenerated [*Priscu et al.*, 1995]. In the west lobe, the distribution of inorganic nitrogen species below the chemocline is consistent with a typical stratified system in which the surface layer is nitrogen depleted, and the deep anoxic layer has high NH_4^+ concentrations but the oxidized forms of nitrogen are depleted presumably due to denitrification. Denitrification has been measured in the deep layer of this lobe [*Priscu et al.*, 1995]. In contrast the deep anoxic layer of the east lobe has high concentrations of NO_2^-, NO_3^-, and NH_4^+. Denitrification was not detectable below the chemocline in the east lobe [*Priscu et al.*, 1995]. Additionally N_2O appears at very high levels (exceeding 580,000% over air saturation) at the oxic/anoxic interface of the east lobe, while trace levels exist in the west lobe [*Priscu et al.*, 1996; *Priscu*, 1997]. These observations suggest that denitrification dominates chemical distributions in the west lobe and that nitrification, rather than denitrification, dominates the chemical distributions in the east lobe.

It is widely accepted that microorganisms are the primary mediators in most biogeochemical processes of the nitrogen cycle in aquatic systems. Nitrification is the two-step process of oxidation of NH_4^+ through NO_2^- to NO_3^- and is carried out by two separate groups of chemolithotrophic bacteria (NH_4^+ oxidizers and NO_2^- oxidizers). Under low oxygen tensions, nitrification can also result in significant production of trace gases such as N_2O and NO in aquatic environments and their subsequent release into the atmosphere [*Poth and Focht*, 1985; *Downes*, 1988; *Goreau et al.*, 1980]. Nitrification has been studied for its potential importance as an oxygen sink, a mechanism for loss of nitrogen (via coupling to denitrification), and as a source of NO_2^- and NO_3^- supplying new nitrogen to primary producers in surface waters. The presence of nitrifying bacteria and the process of nitrification can be inferred by chemical profiles of the substrates and products of the reactions. However in environments containing oxygen gradients, nitrification is often tightly coupled to the heterotrophic microbial process, denitrification (dissimilatory NO_3^- reduction), sharing many of the same substrates and intermediates. This spatial and chemical overlap confounds direct interpretation. To determine the importance of each process in controlling the biogeochemistry in these complex systems, information on the abundance and distribution of the organism responsible is needed in conjunction with rate measurements and nutrient profiles.

Although nitrifiers have been isolated from diverse environments and are generally ubiquitous in soils, freshwater and marine environments [*Koops and Möller*, 1992], they account for a very small proportion of the total bacterial population in natural environments. Moreover little is known about the diversity or genetic composition of lacustrine nitrifiers. The number of species that have been isolated and described from lakes is small, even compared with the accepted low diversity of nitrifiers in terrestrial and marine environments [*Hall*, 1986]. Sensitive and specific methods of detection and identification are necessary in order to study their ecology and their role in natural systems.

Early efforts to enumerate viable nitrifiers in lakes relied on the MPN (most probable number) technique. This technique is problematic in that it is inherently imprecise and grossly underestimates the population [*Hall*, 1986]. Moreover comparison between studies is difficult due to the sensitivity of bacteria to enrichment conditions. Another method, fluorescent polyclonal antibodies, has been used to study the serological diversity and distribution of nitrifying bacteria in soils [*Besler and Schmidt*, 1978; *Schmidt*, 1974], sewage [*Yoshioka et al.*, 1982], lake sediments [*Smorczewski and Schmidt*, 1991], and marine environments [*Ward and Perry*, 1980; *Ward and Carlucci*, 1985]. This method detects nitrifiers based on cross-reactivity to antibodies directed against cell wall components of known isolates; thus it requires the isolation and culture of bacterial strains. It is widely accepted that, although many bacterial cells present in a natural population appear viable, they cannot be isolated due to the limitation of classical cultivation techniques such as selective media. Moreover in the case of chemoautotrophs, which grow very slowly, and therefore require months or even years for isolation and purification, development of detection methods which do not require culturing would be advantageous.

More recently, studies based on the analysis of 16S rRNA and the use of the polymerase chain reaction have investigated the spatial distribution in the natural environment of microbial taxa which have not been grown in culture [*Giovanni and Cary*, 1993]. This technique has been applied to the detection and analysis of NH_4^+ oxidizers of the beta subclass in natural samples [*McCaig et al.*, 1994; *Nejdat and Abeliovich*, 1994; *Teske et al.*, 1994; *Voytek and Ward*, 1995; *Hiorns et al.*, 1995]. The technique provides mainly qualitative or semiqualitative information and cannot differentiate between dead, dormant, and active cells.

Fluorescent-probe in situ hybridization (FISH) extends the use of comparative 16S rRNA sequencing to rapid determinative and autocological studies of specific microorganisms in the natural environment [*Giovanni et al.*, 1988; *De Long et al.*, 1989; *De Long*, 1993; *Amann et al.*, 1991, 1995]. The technique relies on the permeability of most bacterial cells, when fixed, to short fluorescently labeled oligonucleotide probes. These probes can be designed to be specific on different taxonomic levels ranging from domains to subspecies [*Giovanni et al.*, 1998; *Amann et al.*, 1995]. The detection of individual bacteria is dependent on the presence of sufficient ribosomes per cell (on the order of 10^3 per cell). Therefore not only can bacteria be differentiated on many phylogenetic levels, but the technique also gives information on physiological state of the bacteria on the basis of the number of ribosomes per cell.

A first step in discerning the underlying mechanisms controlling nutrient distributions is to determine the distribution of the bacteria responsible for the processes of nitrification and denitrification. Information on the distribution of denitrifying bacteria and measurements and estimate of rates of transformation in Lake Bonney are presented elsewhere [*Priscu et al.*, 1996; *Priscu*, in press; *Ward and Priscu*, in press]. In this study, we used three techniques (IFA, PCR, and FISH) to determine the abundance of nitrifying bacteria in the east and west lobes of Lake Bonney.

METHODS AND MATERIALS

Study Site

Lake Bonney (77°43′S, 162°20′E) is a large permanently ice-covered (4–5m thick) lake within the Taylor Valley. The lake is separated into two lobes by a narrow sill (approximately 50 m wide and 13 m deep). The area of the east lobe of Lake Bonney is 3.5 km² and the maximum depth is 40 m. The temperature increases from 0°C just beneath the ice to 6.1°C at 15 m and decreases to –2°C near the bottom [*Priscu et al.*, 1993; *Priscu et al.*, 1996]. Above 20 m, dissolved O_2 is supersaturated, 45.8 mg l⁻¹, and the water is fresh above 13 m. Below 20 m, the water becomes anoxic and hypersaline, reaching salinities more than 5 times seawater. A strong chemocline occurs at about 17 m (Figure 1A). The concentrations of oxidized forms of nitrogen increase in the deeper layers and remain high to the bottom. A broad N_2O-N peak occurs between 20–24 m, reaching values exceeding 41 µM. The west lobe of Lake Bonney is smaller than the east lobe with a total surface area of 1.3 km² and a maximum depth of 40 m. In the fresh layer above 15m, the dissolved

O_2 is supersaturated [*Priscu et al.*, 1993; *Priscu et al.*, 1996]. The deep water is anaerobic and salinities reach 6 times seawater. The temperature range is from –2 to 3°C. The chemocline and oxycline are shallower in the west lobe, beginning at 13 m and 15 m respectively (Figure 1B). Nitrogenous compounds are low in the surface waters. NH_4^+ is regenerated beneath the trophogenic zone and steadily increases to 300 µM at the bottom (40m). NO_2^-, NO_3^-, and N_2O-N peak between 13–17 m, with concentrations reaching 0.6 µM, 25 µM, and 1.1 µM respectively, and then decrease rapidly.

Sample Collection

Sampling was done through a 25 cm diameter hole in the surface ice (4 m thick) at a central site in each lobe during the 1993-1994 and 1994-1995 austral spring and summer (October–December). Bacterioplankton samples were collected in Niskin bottles from ten depths. Depths are reported relative to the free water surface, i.e., the level to which water rose in the sampling hole. Samples (200 ml) for IFA and total bacterial counts were preserved in buffered formalin, 2% final concentration, and stored at 4°C. Subsamples (1–2 ml) were fixed in 4% freshly prepared formaldehyde for in situ hybridization probing (see below). Samples for DNA analysis (approximately 4 liters) were concentrated 100 fold by ultrafiltration using a Filtron (Northborough, MA) open channel Ultrasette with a 300 kd nominal molecular weight cutoff membrane or a Pellicon tangential flow filtration system using GVLP, 0.22 µm pore size ultrafilters (Millipore). The concentrate was filtered onto a 47 mm, 0.2 µm pore size Gelman Supor filter. Filters were stored frozen in EDTA (0.5 M, 0.5 ml) until total DNA was extracted.

Staining and Enumeration of Total Bacteria and Nitrifiers Serotypes

Replicate 5 and 10 ml aliquots of the formalin preserved samples were filter concentrated to 1 ml onto prestained Poretics 0.2 µm polycarbonate filters and stained with DAPI (final concentration 0.05 µg ml⁻¹; Sigma Chemical Corp.) following the protocol of *Porter and Feig* [1980]. Total bacterial counts were obtained by enumeration with a Zeiss epifluorescent scope at x1000, using a 100W HBO mercury lamp and Zeiss filter set: BP365/10 excitation, LP395 barrier, and FT510 dichromatic beam splitting filters.

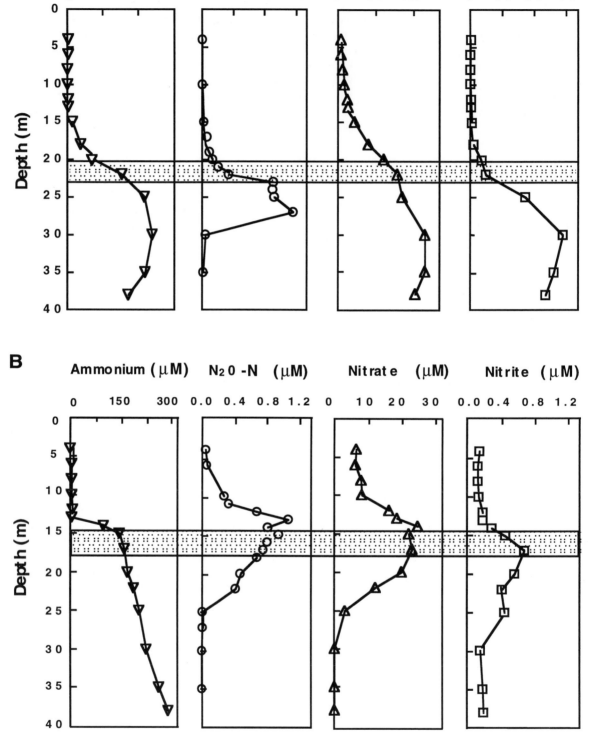

Fig. 1. Dissolved inorganic nitrogen (NH_4^+, N_2O, NO_3^-, NO_2^-) in the east lobe (A) and west lobe (B) of Lake Bonney, November 1993. Stipled bars demarcate the oxic/anoxic transition zone.

Immunofluorescent enumeration of NH_4^+ oxidizing bacteria was done by the method described by *Ward and Carlucci* [1985] using antisera raised against *Nitrosomonas marina* and *Nitrosococcus oceanus*. The antisera used are fully described by *Ward and Carlucci* [1985]. Two sets of duplicate 20 ml aliquots were filtered onto prestained Poretics 0.2 μm polycarbonate filters. Separate filters were stained with each antiserum and duplicates of each serotype were counted. Stained samples were enumerated with a Zeiss epifluorescence scope at x1000, using a 100W HBO mercury lamp and Zeiss filter set: 450DF55 excitation, 505DF35 barrier, and 505drlext02 dichromatic filters. A minimum of 200 cells or 300 fields were counted to calculate abundances.

DNA Extraction and PCR Amplification

High molecular weight DNA was extracted and purified from the frozen filters of concentrated bacteria following a standard protocol with slight modification [*Ausubel et al.*, 1989; *Kerkhof and Ward*, 1993, *Voytek and Ward*, 1995]. The quality of the extracted DNA template was confirmed using the universal 16S rRNA primers EUB 1 and EUB 2 [27 forward and 1525 reverse; *Liesack et al.*, 1991]. PCR prime corresponding to conserved sequences within the 5' and 3' regions of the 16S rDNAs of beta subclass [NITA and NITB; *Voytek and Ward*, 1995] and gamma-subclass [NOC1 and NOC2; *Voytek and Ward*, 1997] NH_4^+ oxidizers were used to detect these groups. Amplification with these primers yields a 1080 bp NIT product and a 1130 bp NOC product. Two stage PCR amplifications were performed following the protocol described in *Voytek and Ward* [1995; 1997] with slight modification. To reduce amplification of non-specific products the total number of cycles was reduced to 30.

Semiquantification of Ammonium Oxidizers

Direct quantification of the number of organisms in a natural sample based on specific PCR products amplified from DNA extracted from these samples is difficult. In all steps of the collection, extraction, and amplification procedures, error and uncertainty are introduced. Some cells are more efficiently collected by ultrafiltration and not all cells readily lyse in standard DNA extraction protocols. Additionally the template copy number per cell may vary almost ten-fold and some templates may amplify more efficiently than others [*Suzuki and Giovanonni*, 1996]. Furthermore *Farrelly et al.* [1995] have shown that without

foreknowledge of the genome size and gene copy number in all members, accurate estimations of the abundance of a particular species is impossible. The two-step amplification protocol used here reinforces these errors. For these reasons, an absolute determination of the abundance of NH_4^+ oxidizers with PCR was not possible. In order to determine the relative abundance of NH_4^+ oxidizers in each natural sample, the EUB amplified product was diluted in TE before amplification with the NOC or NIT primers. The dilution series include 5 dilutions (undiluted, 1:2, 1:5, 1:10, and 1:50). The greater the dilution factor, the higher the number of the original template molecules, and thus the higher the number of NH_4^+ oxidizers present in the original sample. A number was assigned to a sample based on the highest dilution in the series that yielded the correct product (e.g., amplification of undiluted samples was assigned 1; 1–50 was assigned 5).

In Situ Hybridization

Cells collected by centrifugation from subsamples (1–2 ml) were washed in PBS and resuspended in 4% formaldehyde, followed by storage at 4°C for 4–16 h. Preserved cells were spotted on gelatin subbed slides and treated with serial ethanol rinses (50, 75, and 100%) as described by *DeLong* [1993]. The following labeled probes were used: (i) EUB 338 complementary to a region of the 16S rRNA conserved in the domain Bacteria [*Amann et al.*, 1995]; (ii) NIT A, B, and C, oligonucleotides complementary to and specific for selected regions of the 16S rRNA molecules of the beta subclass NH_4^+ oxidizers of the Proteobacteria [*Voytek and Ward*, 1995]. The hybridization properties of the FITC labeled oligonucleotides were determined empirically using fixed whole cells from homologous (*Nitrosomonas europaea* cultures) and heterologous species (*Escherichia coli* and *Nitrosococcus oceanus*, a gamma subclass nitrifier, cultures). The NIT probe was also tested with the pure culture of *Nitrosomonas europaea* diluted approximately 10 to 1 with *E. coli*. The optimal conditions for hybridization are reported in *Voytek* [1996]. DAPI staining was performed following the protocol of *Hicks et al.* [1992] for dual staining with DAPI and fluorescent rRNA probes. Slides were stored desiccated at −20°C in the dark. Epifluorescence microscopy was used to visualize cells using the filter sets described above. A minimum of 200 cells or 300 fields were counted to calculate abundances of hybridized cells.

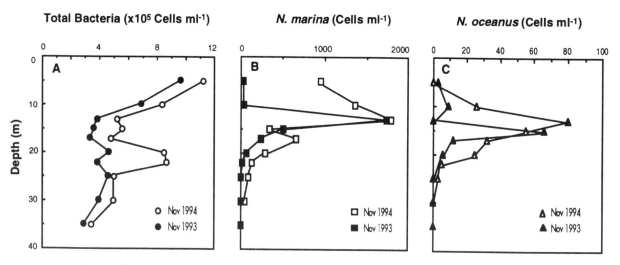

Fig. 2. The depth distribution of total bacteria (A), *Nitrosomonas* serotyped ammonium oxidizers (B) and *Nitrosococcus oceanus* serotyped ammonium oxidizers (C) in the east lobe of Lake Bonney, November 1993 and November 1994. Abundances determined by epifluorescence microscopy using DAPI or FITC. See text for details.

RESULTS

Total Bacterial Community and DAPI Counts

The bacterioplankton population in all samples counted contained a variety of morphotypes but was dominated by rod-shaped cells less than 1 μm in length. The second most abundant cell type was long and filamentous (10 μm). In both lobes and in both years the total abundance of bacteria was between 10^5 and 10^6 cells ml^{-1} (Figures 2A and 3A). Although the

absolute numbers of cells are slightly higher in 1994, the depth profiles from year to year did not change substantially. In both lobes, the highest concentrations of cells were observed in the surface and cell number decreased with depth. In the east lobe, there was a small subsurface peak of bacteria at 20 m.

Abundances of N. europaea and N. oceanus Serotypes by IFA

The cell concentrations of *N. marina* serotype were

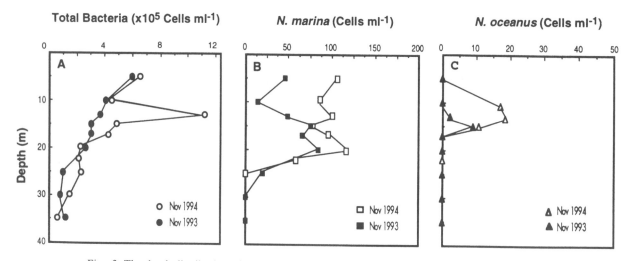

Fig. 3. The depth distribution of total bacteria (A), *Nitrosomonas* serotyped ammonium oxidizers (B) and *Nitrosococcus oceanus* serotyped ammonium oxidizers (C) in the west lobe of Lake Bonney, November 1993 and November 1994. Abundances determined by epifluorescence microscopy using DAPI or FITC. See text for details.

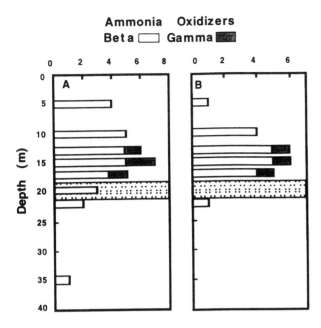

Fig. 4. Relative abundance of beta and gamma subclass ammonium oxidizers in the east lobe of Lake Bonney, November 1993 (A) and November 1994 (B) determined by PCR. Stipled bars demarcate the oxic/anoxic transition zone.

on the order of 10^2 to 10^3 ml^{-1} and were one to two orders of magnitude higher than the *N. oceanus* serotype in the west and east lobes, respectively (Figures 2B, C and 3B, C). In general, cell concentrations of both serotypes were much higher in the east lobe. *N. marina* had a broad surface maximum in the west lobe (Figure 3B) and a large peak at 13m in the east lobe (approximately 2000 cells ml^{-1}; Figure 2B). The distribution of *N. oceanus* was much more limited; cell densities were low in the surface and peaked at 13 m in both lobes. At depths below the peak abundances, concentrations fell off steeply. Virtually no NH$_4^+$ oxidizers of either serotype were detected deeper than 22 m.

Semiquantification of Ammonia Oxidizers by PCR

In both 1993 and 1994, nitrifying bacteria were detected by the PCR assay in samples taken from seven depths above 25 m in the east lobe of Lake Bonney (Figure 4). In 1993, the relative abundances of both groups were somewhat higher and beta nitrifiers were also detected at 35 m (Figure 5B). In both years, the beta NH$_4^+$ oxidizers were far more abundant than the gamma and had a broader distribution pattern. The highest abundances of beta nitrifiers were observed in the aerobic waters above the chemocline while gamma

nitrifiers had a small peak around 15 m, just above the oxygen transition zone (Figure 4). The overall abundance of nitrifiers in the west lobe was considerably lower (Figure 5). Beta nitrifiers were most abundant above the oxycline and the distribution of the less abundant gamma nitrifiers was limited to the oxygen transition zone. During the 1993 sampling season, NH$_4^+$ oxidizing bacteria were detected throughout the water column (Figure 5A). The distribution of nitrifiers was more limited in samples taken at the same time the following year (Figure 5B).

In Situ Hybridization

In probe hybridizations of pure cultures, close to 100% of the cells were detectable using the EUB 338 for cells in both *E. coli* and *N. europaea*. The stained *N. europaea* cells were much dimmer than the *E. coli* cells. The NIT A, B, and C probes hybridized to >98% of the *N. europaea* cells in culture. In the mixed sample of 10% of the *N. europaea* culture diluted in *E. coli*, the NIT probes hybridized to >96% of the *N. europaea* added (approximately 10% of the total bacterial cells present). None of the nitrifier specific probes hybridized to the heterologous strains. Samples collected from the east lobe, November 1993, were probed with fluorescently labeled oligonucleotides. Cell concentrations were estimated using DAPI counts for

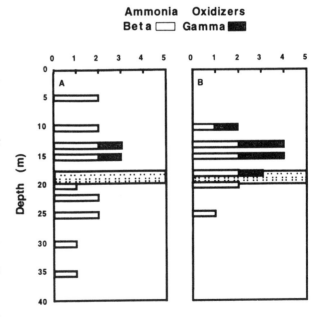

Fig. 5. Relative abundance of beta and gamma subclass ammonium oxidizers in the west lobe of Lake Bonney, November 1993 (A) and November 1994 (B) determined by PCR. Stipled bars demarcate the oxic/anoxic transition zone.

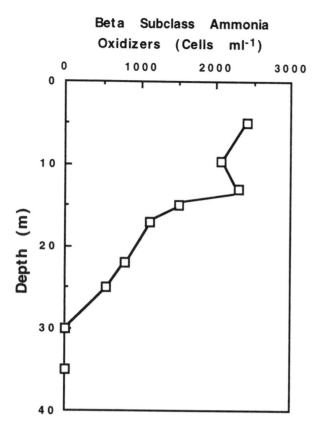

Beta Subclass Ammonia Oxidizers (Cells ml^{-1})

Fig. 6. The distribution of beta subclass ammonium oxidizers in the east lobe of Lake Bonney, November 1993. Abundances determined by fluorescent in situ hybridization (FISH).

total bacteria to correct for any loss of cell material during processing. Beta-subclass NH_4^+ oxidizers were observed in 7 of the 10 samples taken at depths in the east lobe (Figure 6). The highest cell density observed was approximately 2300 cells ml^{-1} at 5 m. This peak was a part of a broad maximum of FISH labeled cells between 5–13 m. Cell densities decreased sharply below 13 m. The overall brightness (i.e., signal strength) of the individual cells appeared to decrease with depth and increasing salinity. Background fluorescence also increased in the deeper samples.

DISCUSSION

The overall patterns of nitrifier abundances found in the two lobes of Lake Bonney were very similar with both the IFA and PCR detection techniques. In general, there were more beta subclass than gamma subclass NH_4^+ oxidizers present in this lake, and the beta subclass was more widely distributed. One reason for

this difference is simply that there are many genera of NH_4^+ oxidizers in the beta subclass and they have been found in soils, marine, and freshwater environments. The gamma subclass has one genetically confirmed member, thus far, and it is the marine species *Nitrosococcus oceanus*. It is not surprising that one or two species from a limited environmental niche may be less abundant and have a restricted distribution compared to a group with environmentally diverse members.

The IFA and PCR techniques are expected to detect similar subsets of the total NH_4^+ oxidizer population. Overall the pattern of nitrifiers was similar, but there were several samples in which PCR detected nitrifiers where IFA did not. Most of the observed discrepancy was in the distribution of beta-nitrifiers. The PCR assay was based on primers derived from the sequences of many genera of beta nitrifiers while the antiserum used in the IFA assay was raised to a single species of beta subclass NH_4^+ oxidizer. Therefore we might expect the IFA assay to pick up only a narrow subset of the total beta nitrifier population present while the PCR technique could detect a broader species assemblage. Additionally the efficacy of the IFA assay is dependent on the specificity of the antisera used. As mentioned above, IFA has been used to study nitrifying bacteria in many environments [e.g. *Schmidt*, 1974; *Yoshioka et al.*, 1982; *Smorczewski and Schmidt*, 1991; *Ward and Carlucci*, 1985]. Generally in these studies, antiserum is raised to a strain isolated originally from a similar environment to the one being investigated. No NH4+ oxidizers have been isolated from Lake Bonney despite three years of cultivation efforts. The antisera we used in this study were raised against two marine isolates from temperate waters: *N. marina* and *N. oceanus*. The *N. marina* antiserum cross reacts with the soil nitrifier, *N. europaea* and a subset of the nitrifier strains isolated from a variety of marine environments [*Ward and Carlucci*, 1985; *Voytek and Ward*, 1995]. Neither of the antisera cross react with most of the known soil strains or the estuarine nitrifier isolate (*Nitrosococcus mobilis*) available in culture [*Voytek and Ward*, 1995]. In contrast, all of these NH_4^+ oxidizers can be amplified with the NITA-B primers [*Voytek and Ward*, 1995]. The biological habitats of Lake Bonney are quite different from temperate freshwater or marine environments and the nitrifier community probably reflects those environmental differences. Therefore we suspect that our results from the IFA may have been more closely correlated to the PCR results if we had been able to use antisera from nitrifiers isolated from this system.

Less frequently, IFA detected nitrifiers and the PCR technique did not (e.g., gamma subclass nitrifiers at 20 and 22 m in the east lobe; Figures 2C and 4). This may be due to interference of the PCR amplification from competing non-target DNA. PCR primers amplify target DNA less efficiently when the target is rare relative to the total DNA present. Interference would be most severe in samples containing a potentially large quantity of competing template. A peak in total bacterial cells was observed at these depths (Figure 2A), therefore interference might be greatest there. Alternatively, immunofluorescence may be over-estimating the number of "live" cells in the surface. Both detection techniques (PCR and immuno-fluorescence) may overestimate to some extent since they rely on features of the cell that persist after the cell has lost viability. Survival studies have suggested that a metabolic shutdown occurs when bacteria are severely nutrient limited [*Moyer and Morita*, 1989a and 1989b]. In these studies bacterial cells remained physically intact and maintained their cellular protein levels but cellular DNA content decreased to less than 5% and viability dropped to <0.3% [*Moyer and Morita*, 1989b]. Additionally *Zweifel and Hagström* [1995] found that only a small fraction (2–32%) of intact bacteria enumerated by traditional fluorescent staining techniques actually contained nucleoids. It is possible that the immunofluorescence technique picks up intact cells that contain no or substantially degraded genomic DNA, which would not be picked up by PCR. If this is the case, the estimate of abundance obtained by quantitative PCR assays would be expected to correlate more closely with the activity of nitrifiers than the estimates obtained by immunofluorescence.

The depth profile of fluorescently labeled (FISH) beta subclass nitrifiers generally followed the shape of the profile generated by the less quantitative PCR method. The highest numbers of fluorescently stained cells were found in the surface waters above 15m (Figure 6). In the east lobe, cell densities were on the order of 2000 cells ml^{-1} in the surface and decreased sharply with depth. No cells were detected in samples from 30 m or deeper. The high density of fluorescently labeled cells at 13 m corresponds to the peak observed of *N. marina* serotyped cells (Figures 2b and 6); however much lower densities of *N. marina* serotype cells were observed at shallower depths. This suggests that the peak of beta nitrifiers at 13 m is dominated by a nitrifier species of the *N. marina* serotype. Above that depth however, the nitrifiers present were detected by PCR and FISH and are, therefore, genotypically related

but are not phenotypically similar to the *N. marina* serotype. Like the PCR technique, FISH is more inclusive and can detect a wider variety of beta-subclass NH$_4^+$ oxidizers.

The fluorescence signal strength was higher in hybridized cells examined at depths above 20 m. This may be an indication of higher metabolic activity of the cells at these depths. Indeed the profiles of nitrogenous compounds (Figure 1a and 1b) [*Voytek,* 1996] and the rates of nitrification measured in this lake [*Priscu et al.*, 1996] support this hypothesis. Below 20 m, fluorescently labeled cells were present at concentrations of 100–1000 cells ml^{-1}, but the signal strength was significantly attenuated. In recent papers by *Priscu* [1996] and *Priscu et al.* [1996], the lower N$_2$O peak observed in the east lobe of Lake Bonney was attributed to relict nitrifying activity. No current activity has been detected at this depth; therefore it was postulated that nitrifier cells active years ago were responsible. Presumably cells that are not meta-bolically active would rapidly lose rRNA and not be detected. Low-temperature, highly-saline waters may serve as an excellent medium for preservation of non-active, intact cells and would explain persistent detection, albeit at a reduced signal, where no activity is found. An alternative explanation of the diminished signal in samples taken from the deeper depths is the effect of the hypersaline waters from which they were harvested. The prehybridization washing steps in the in situ protocol may not have been adequate to remove the excess salts and may have changed the hybridization conditions in these samples. Higher salt concentrations reduces the stringency of the hybridization solution and may cause more non-specific binding of the probes. The higher background fluorescence observed in these samples may be an indication of this.

Small subunit rRNA-based oligonucleotide probes have become increasingly useful as a tool for characterizing microbial cells in environmental samples [*Amann et al.*, 1995]. The in situ identification and enumeration of individual bacteria by FISH is dependent on the presence of sufficient ribosomes per cell. The detection limit is around 10^3 to 10^4 ribosomes per cell [*DeLong et al.*, 1989; *Amann et al.*, 1995], and therefore sensitivity depends in part on the physiological state of the target cells. Faster growing or more physiologically active cells tend to have more ribosomes and hence bind proportionately more probe molecules. One limitation to FISH is the potential inability to detect slow growing cells (e.g., some bacterial symbionts), which may have too few ribosomes per cell to provide a signal. To gain greater sensitivity, researchers have attempted to label oligo-nucleotides with multiple fluorescent molecules, but this resulted in poor or non-specific binding [*DeLong*, 1993]. Greater sensitivity may be achieved by employing indirect labeling methods (e.g., biotin/avidin systems [*Singer et al.*, 1986]), coupled with more sensitive detection systems (e.g., chemi-luminescent detection). Cell detection and enumeration by flow cytometry [*Amann et al.,* 1991] is another potential technique for improving the efficiency of counting cells that are in low abundance or rare in natural samples.

Overall we found little variation between samples taken in 1993 and 1994 for both total bacterial counts and for the specific subsets of nitrifying bacteria. This is consistent with the observation of little seasonal or year to year variation in total bacterial counts in samples taken over the past ten years of sampling in Lake Bonney (Priscu, unpublished data). The total bacteria counts measured here were on the order of 10^5–10^6 cells ml^{-1}, which is typical for polar lakes [*Vincent*, 1988; *Simmons et al.*, 1993]. The beta subclass nitrifiers represented up to 1% of the total bacterial population in the east lobe of Lake Bonney determined by FISH and IFA. The gamma subclass was a much less significant fraction of the total population in both lobes. These abundances are quite high for aquatic environments. Using IFA, studies in marine systems have estimated the abundance of nitrifying bacteria to be on the order of 1–10 cells ml^{-1}, and, unlike in our studies, the cell densities were relatively constant throughout the water column [*Ward and Carlucci*, 1985; *Ward*, 1986]. Based on MPN or calculations from measuring nitrification activity, the estimated range of nitrifier cell densities in lakes is between 1 and 10^4 cells ml^{-1} [*Hall*, 1986].

The depth distributions of NH_4^+ oxidizers were very similar no matter which of the three methods were used. NH_4^+ oxidizing bacteria were detected throughout the water column in both lobes but were less abundant in the west than in the east lobe. In general the highest cell numbers of beta nitrifiers were found in the surface waters above the oxycline or in association with the nutricline. Despite the abundance of NH_4^+ oxidizers in the surface waters, NO_2^- and NO_3^- do not accumulate, presumably because these nutrients are readily used by phytoplankton and bacteria in the euphotic zone [*Priscu et al.*, 1995]. The distribution of the less abundant gamma nitrifiers was generally limited to the oxygen transition zone. The distribution pattern of nitrifiers

suggests that nitrification was probably responsible for the shallow portion of the peaks of N_2O, NO_2^-, and NO_3^- observed below 20 and 13 m in the east and west lobes respectively. Both beta and gamma nitrifiers were low in number or absent in the deep waters. Based on the low FISH signal strength of cells from deeper depths and undetectable nitrification rates for these depths, nitrifiers detected in the anoxic bottom waters were probably inactive.

Single cell analyses have proven extremely useful for determining the abundance, activities, and variability of microbial species in their natural environments. Unfortunately very few microbes are amenable to cultivation and so they remain difficult to differentiate, identify, and characterize. PCR, with its capacity to amplify specific sequences of DNA from a few copies of the target molecule, is a tool which resolves some of the limitations of traditional culturing techniques and immunofluorescence. PCR eliminates the need to isolate individual nitrifiers or to subject them to often inconclusive taxonomic tests. The advantages of PCR have been demonstrated in the detection and analysis of NH_4^+ oxidizers of the beta subclass [*McCaig et al.*, 1994; *Nejdat and Abeliovich*, 1994; *Voytek and Ward*, 1995; *Hiorns et al.*, 1995] and gamma subclass nitrifiers [*Voytek and Ward*, 1997; *Voytek*, 1996] in natural samples. FISH extends our ability to understand the structure and dynamics of natural microbial communities by not only providing quantitative information on community structure but also qualitative information of the physiological state of the bacteria present.

Acknowledgements. We are indebted to John Priscu for inviting us to work in Lake Bonney and R. D. Bartlett, M. T. Geissler, and C. Woolston for logistical and technical help in Antarctica. This work was supported by the National Science Foundation (OCE-9115040 and DPP-9117907).

REFERENCES

Amann, R., N. Springer, W. Ludwig, H.-D. Görtz, and K.-H. Schleifer, Identification *in situ* and phylogeny of uncultured bacterial endosymbionts, *Nature*, 351, 161–164, 1991.

Amann, R.I. W. Ludwig, and K-H Schleifer, Phylogenetic identification and *in situ* detection of individual microbial cells without cultivation, *Microbiol. Rev.* 59, 143–169, 1995.

Ausubel, F.M., R. Brent, R.E. Kingston, D.D. Moore, J.G. Seidman, J.A. Smith, and K. Struhl (eds.), *Current protocols in molecular biology.* Green Publishing Associates and Wiley-Interscience, New York., 1989

Besler, L.W. and E.L. Schmidt, Serological diversity within a terrestrial ammonia-oxidizing population, *Appl. Environ. Microbiol*, 36, 584–593, 1978.

DeLong, E.F., Single cell identification using fluorescently labeled, ribosomal RNA-specific probes. in *Handbook of Methods in Aquatic Microbial Ecology*, edited by P.F. Kemp, B.F. Sherr, E.B. Sherr, and J.J. Cole, pp. 285–294, Lewis Publishers, Boca Raton, FL, 1993.

DeLong, E.F., G.S. Wickam, and N.R. Pace, Phyolgenetic stains: ribosomal RNA based probes for the identification of single cells, *Science*, 243, 1360–1363, 1989.

Downes, M.T., Aquatic nitrogen transformations at low oxygen concentrations,*Appl. Environ. Microbiol.*, 54, 172–175, 1988.

Farrelly, V., F.A. Rainey, and E. Stackebrandt, Effect of genome size and *rrn* gene copy number of PCR amplification of 16S rRNA genes from a mixture of bacterial species, *Appl. Environ. Microbiol.*, 61, 2798–2801, 1995.

Giovannoni, S.J., E.F. DeLong, T.M. Schmidt, and N.R. Pace, Phylogenetic group-specific oligodeoxynucleotide probes for identification of single microbial cells, *J. Bacteriol.*, 170, 720–726, 1988.

Goreau, T.J., W.A. Kaplan, S.C. Wofsy, M.B. McElroy, F.W. Valois, and S.W. Watson, Production of NO_2 and N_2O by nitrifying bacteria at reduced concentrations of oxygen,*Appl. Environ. Microbiol.*, 40, 526–532, 1980.

Hall, G.H., Nitrification in lakes, in *Nitrification*, edited by J.I. Prosser, pp 127–156, IRL Press, Washington, D.C., 1986.

Head, I.M, W.D. Hiorns, T. Martin, A.J. McCarthy, and J.R. Saunders, The phylogeny of autotrophic ammonia-oxidizing bacteria as determined by analysis of 16S ribosomal RNA gene sequences, *J. Gen. Microbiol.*, 139, 1147–1153, 1993.

Hicks, R.E., R.I. Amann, and D.A. Stahl, Dual staining of natural bacterioplankton with 4', 6-diamidino-2-phenylindole and fluorescent oligonucleotide probes targeting kingdom-level 16S rRNA sequences, *Appl. Environ. Microbiol.*, 58,2158–2163, 1992.

Hiorns, W.D., R.C. Hastings, I.M. Head, A.J. McCarthy, J R. Saunders, R.W. Pickup, and G.H. Hall, Amplification of 16S ribosomal RNA genes of autotrophic ammonia-oxidizing bacteria demonstrates the ubiquity of *Nitrosospiras* in the environment, *Microbiol.*, 141, 2793–2800, 1995.

Kaplan, W.A. and S.C. Wofsy, The biogeochemistry of nitrous oxide: a review. in *Advances in Aquatic Microbiology*, vol. 3, pp. 181–206, Academic Press Inc., London, 1985.

Kerkhof, L. and Ward, B.B., Comparison of nucleic Acid hybridization and fluorometry for measurement of the relationship between RNA/DNA ratio and growth rate in a marine bacterium, *Appl. Environ. Microbiol*, 59, 1303–1309, 1993.

Koops, H.-P. and U.C. Möller, The lithotrophic ammonia-oxidizing bacteria, in *The Prokaryotes: a handbook on the biology of bacteria: ecophysiology, isolation, identification, applications*, 2nd ed., edited by A. Balows, H.G. Truper, M. Dworkin, W. Harder and K.-H. Schleifer, pp. 2625–2637, Springer-Verlag, New York, 1992.

Liesack, W., H. Weyland, and E. Stackenbrandt, Potential risks of gene amplification by PCR as determined by 16S rDNA analysis of a mixed-culture of strict barophilic bacteria, *Microb. Ecol.*, 21, 191–198, 1991.

Lyons, W.B., S.W. Tyler, R.A. Wharton, Jr., and D.M. McKnight, The late Holocene/Paleoclimate history of the McMurdo Dry Valleys Antarctic as derived from lacustrine iotope data, *Paleogeography, paleoclimatology and paleoecology*, in press.

McCaig, A.E., T.M. Embley, and J.I. Prosser, Molecular analysis of enrichment cultures of marine ammonia oxidisers, *FEMS Microb. Let.*, 120, 363–367, 1994.

Moyer, C.L. and R.Y Morita, Effect of growth rate and starvation-survival on cellular DNA, RNA and protein of a psychrophilic marine bacterium, *Appli. Environ. Microbiol.*, 55, 2710–2716, 1989a.

Moyer, C.L. and R.Y. Morita, Effect of growth rate and starvation-survival on the viability and stability of a psychrophilic marine bacterium, *Appl. Environ. Microbiol.*, 55, 1122–1127, 1989b.

Nejdat, A. and A. Abeliovich, Detection of *Nitrosomonas* spp. by the polymerase chain reaction, *FEMS Microb. Let.*, 120, 191–194, 1994.

Porter, K.G. and Y.S. Feig, The use of DAPI for identifying and counting aquatic microflora, *Limnol. Oceanogr.*, 25, 943–948, 1980.

Poth, M. and D.D. Focht, ^{15}N kinetic analysis of N_2O production by *Nitrosomonas europaea*: an examination of nitrifier dentrification, *Appl. Environ. Microbiol.*, 49, 1134–1141, 1985.

Poulsen, L.K, G. Ballard, and D.A. Stahl, Use of rRNA fluorescence *in situ* hybridization for measuring the activity of single cells in young and established biofilms, *Appl. Environ. Microbiol.*, 59, 13554–1360, 1993.

Priscu, J.C., The biogeochemistry of nitrous oxide in permanently ice-covered lakes of the McMurdo Dry Valleys, Antarctica, in *Microbially Mediated Atmospheric Change*, Special Issue of Global Change Biology, in press.

Priscu, J. C., Phytoplankton nutrient deficientcy in lakes of the McMurdo Dry Valleys, Anarctica, *Freshwater Biol.*, 34, 215–227, 1995.

Priscu, J.C., M.T. Downes, and C.P. McKay, Extreme supersaturation of nitrous oxide in a non-ventilated antarctic lake, *Limnol. Oceanogr.*, 41,1544–1551, 1996.

Priscu, J.C. and M.T. Downes, Relationship among nitrogen uptake, ammonium oxidation and nitrous oxide concentration in the coastal waters of western Cook Strait, New Zealand, *Estuar. Coast. Shelf Sci.*, 20, 529–542, 1985.

Priscu, J.C., B. B. Ward, and M. T. Downes, Water column transformations of nitrogen in Lake Bonney, a perennially ice-covered Antarctic lake, *Ant. J. U.S.* 28 (5), 237–239, 1993.

Priscu, J.C., W.F. Vincent, and C. Howard-Williams, Inorganic nitrogen uptake and regeneration in perennially ice-covered Lakes Fryxell and Vanda, Antarctica, *J. Plankton Res.*, 11, 335–351, 1989.

Schmidt, E.L., Quantitative autecological study of micro-organisms in soil by immunofluorescence, *Soil Sci.* 118, 141–149, 1974.

Simmons, G.M., J.R. Vestal, and R.A. Wharton,

Environmental regulators of microbial activity in continental Antarctic lakes, in *Antarctic Microbiology*, edited by W. Vincent, pp. 491–541, Wiley-Liss, Inc., 1993.

Singer, R.H., J.B. Lawrence, and C. Villnave, Optimization of *in situ* hybridization using isotopic and non-isotopic detection methods, *Biotechniques*, 4, 230, 1986.

Smorczewski, W.T. and Schmidt, E.L., Numbers, activities, and diversity of autotrophic ammonia-oxidizing bacteria in a freshwater, eutrophic lake sediment, *Can. J. Microbiol.,* 37, 828–833, 1991.

Spigel, R.H. and J.C. Priscu, Evolution of temperature and salt structure of Lake Bonney, a chemically stratified Antarctic lake, *Hydrobiologia*, 1996.

Suzuki, M.T. and S.J Giovannoni, Bias caused by template annealing in the amplification of mixtures of 16S rRNA genes in PCR, *Appl. Environ. Microbiol.,* 62: 625–630, 1996.

Teske, A., E. Alm, J.M. Regan, S. Toze, B.E. Rittmann, and D. A. Stahl, Evolutionary elationships among ammonia- and nitrite oxidizing bacteria. *J. Bacteriol*, 176 (21), 6623–6630, 1994..

Vincent, W. 1988. *Microbial Ecosystems of Antarctica*. Cambride University Press, Cambridge, England. 304pp.

Voytek, M.A., Relative abundance and species diversity of autotrophic ammonium-oxidizing bacteria in aquatic systems, Ph.D. Dissertation, University of California, Santa Cruz, 1996.

Voytek, M.A., J.C. Priscu, and Ward, B.B, Detection of the beta and gamma subclass ammonia-oxidizing bacterium in six lakes in Antarctica using the polymerase chain reaction, in press, 1997.

Voytek, M.A. and Ward, B.B., Detection of ammonium-oxidizing bacteria of the beta-subclass proteobacteria in aquatic samples with the PCR. *Appl. Environ. Microbiol.,* 56, 2430–2435, 1995.

Voytek, M.A. and Ward, B.B., Detection of the ammonia-oxidizing bacterium, *Nitrosococcus oceanus,* in seawater using the polymerase chain reaction. *FEMS Microbiol. Lett.,* submitted, 1997.

Ward, B.B., Nitrification in marine environments. in *Nitrification,* edited by J. I. Prosser, pp 157–184, IRL Press, Washington, D.C., 1986.

Ward, B.B., and A.F., Carlucci Marine ammonia- and nitrite-oxidizing bacteria: Serological diversity determined by immunofluorescence in culture and in the environment. *Appl. Environ. Microbiol.* 50: 194–201, 1985..

Ward, B.B., and M.J. Perry,. Immunofluorescent assay for the marine ammonium-oxidizing bacterium, *Nitrosococcus oceanus. Appl. Environ. Microbiol.,* 39, 913–918, 1980.

Ward, B.B., and J.C. Priscu, Detection and characterization of denitrifying bacteria from a permanently ice-covered Antarctic lake, *Hydrobiologia*, in press, 1997.

Ward, D.M., R. Weller, and M.M. Bateson, 16S rRNA sequences reveal numerous uncultured microorganisms in a natural community, Nature, 345:63–65, 1990.

Watson, S.W., Characteristics of a marine nitrifying bacterium, *Nitrosocystis oceanus* sp. n. *L i m n o l. Oceanogr.,* 10, R274–289, 1965.

Watson, S.W., E. Bock, E. Harms, H. -P. Koops, and A.B. Hooper, Nitrifying bacteria in *Bergey's Manual of Systematic Bacteriology* vol. 3, edited by J.T. Staley, M.P. Bryant, N. Pfennig and J.G. Holt, pp. 1808–1834, The Williams & Wilkins Co., Baltimore, 1989.

Yoshioka, T., H. Hisayoshi, and Y. Saijo, Growth kinetic studies of nitrifying bacteria by the immunofluorescent counting method, *J. Gen. Appl. Microbiol.,* 28, 169–180, 1982.

Zweifel, U.L. and A. Hagström, Total counts of marine bacteria include a large fraction of non-nucleoid-containing bacteria (ghosts), *Appl. Environ. Microbiol.,* 61, 2180–2185, 1995.

J. C. Priscu, Department of Biological Sciences, Lewis Hall, Montana State University, Bozeman, MT 59717

M. A. Voytek, Institute of Marine and Coastal Sciences, Rutgers University, New Brunswick, NJ 08903

B.B. Ward, Marine Sciences Program, University of California, Santa Cruz, Santa Cruz, CA 95654

(Received October 1 1996;
Accepted May 1, 1997)

PIGMENT ANALYSIS OF THE DISTRIBUTION, SUCCESSION, AND FATE OF PHYTOPLANKTON IN THE MCMURDO DRY VALLEY LAKES OF ANTARCTICA

Michael P. Lizotte

Department of Biology and Microbiology, University of Wisconsin Oshkosh, Oshkosh, Wisconsin

John C. Priscu

Department of Biological Sciences, Montana State University, Bozeman, Montana

Phytoplankton populations in lakes of the McMurdo Dry Valleys have been the subject of taxonomic and ecologic study since the early 1960s. Populations in the major lakes studied (Lakes Bonney, Fryxell, Hoare, and Vanda) include various species of chlorophytes, chrysophytes, cryptophytes, and cyanobacteria. Earlier reports were based primarily on microscopic analyses of preserved water samples. We sampled suspended particulate matter from the four lakes listed above for analysis of algal pigments by high-performance liquid chromatographic (HPLC) methods. Fresh waters beneath ice cover in all lakes were dominated by cryptophyte algae, based on alloxanthin-dominated pigment signatures. Deeper, more saline waters in Lake Bonney were dominated by chrysophytes (fucoxanthin-containing algae) and chlorophytes (chlorophyll-*b*-containing algae). Comparisons with cell counts from Lake Bonney and with published reports of species composition from all the lakes imply that cryptophytes and chrysophytes may have been underestimated by previous microscopic cell counts of preserved water samples. Temporal trends in Lake Bonney showed all three chlorophyll maxima (5 m, 12 m, and 18 m) contained significant quantities of pigments at the onset of light in September and sequential development of deeper phytoplankton populations through the spring growth season. Particles collected early in spring may include significant amounts of detritus that contain pigments. These pigmented particles could include remnants from algal blooms of the previous year's growth season, which may overwinter without significant breakdown because of (1) the slow rate of photo-oxidation under dark, cold conditions, and (2) a paucity of grazing. This explanation is supported by a trend of decreasing chlorophyll breakdown products during the spring, which appear to be photo-oxidized as light intensity increases in Lake Bonney.

INTRODUCTION

The lakes located in the dry valleys near McMurdo Sound, Antarctica contain highly stratified phytoplankton populations under unique conditions. Perennial ice-cover, low advective stream inflow, and strong vertical gradients in salinity [e.g., *Spigel and Priscu*, this volume] make these lakes among the most hydrodynamically stable aquatic systems known. The plankton of these lakes are entirely microbial, primarily algae and bacteria, though protozoans and rotifers may be abundant in certain lakes [*James et al.*, this volume]. These microbial populations exhibit growth under a broad range of temperature, salinity, and nutrient conditions. Low irradiance influences the phytoplankton of these lakes, which show extreme physiological acclimation [e.g., *Lizotte and Priscu*, 1992a, 1992b, 1994; *Neale and Priscu*, 1995, this volume; *Lizotte et al.*, 1996].

Phytoplankton pigment composition was recognized as a measure that could address the following objectives of our dry valley lakes studies: 1) characterize photoacclimation with depth and over the growth season; 2) distinguish whether changes in photosynthetic parameters over the season are due to shifts in cellular processes or to taxonomic changes; and 3) determine the degree of vertical separation between taxonomic groups of phytoplankton in response to physical and chemical gradients. In addition, pigment signatures allowed us to determine the relative contribution of major phytoplankton groups to total biomass being monitored as chlorophyll-*a* concentration, based on the abundance of specific accessory pigments [e.g., *Everitt et al.* 1990].

Comparisons between this chemotaxonomic approach and microscopic phytoplankton cell counts would also allow us to verify and make comparisons with past studies of phytoplankton species composition in the McMurdo Dry Valley lakes. Studies of phytoplankton composition in the dry valley lakes since the early 1960s have been based on microscopy [*Armitage and House*, 1962; *Goldman et al.*, 1967; *Koob and Leister*, 1972; *Vincent*, 1981; *Parker et al.*, 1982], but only the most recent reports have been quantitative [e.g., *Spaulding et al.*, 1994]. Herein we review the published taxonomic work, and, in the context of introducing a chemotaxonomy approach, we discuss the possibility of methodological biases. For example, different groups of phytoplankton have been reported to dominate Lake Bonney, depending on whether results were based on fresh samples [*Koob and Leister,* 1972] or on preserved water samples [e.g., *Parker et al.,* 1982]. We also determined whether some phytoplankton groups may be underestimated by microscopy due to insufficient preservation methods, as demonstrated in marine systems [*Gieskes and Kraay*, 1983; *Buma et al.*, 1990].

We proposed previously [*Lizotte and Priscu,* 1992a] that shade-adapted phytoplankton populations from perennially ice-covered lakes in the McMurdo Dry Valleys present a system analogous to multiple deep chlorophyll maxima. Herein we report on the vertical distribution, seasonal growth, and overwinter fate of distinct phytoplankton communities that make up the chlorophyll maxima in four dry valley lakes.

METHODS

Site Description

Four lakes in the McMurdo Dry Valleys were sampled: Bonney, Fryxell, Hoare, and Vanda. Ice covers of 3 to 5 m thickness are present year-round [*Adams et al.*, this volume], with some ice thinning and the melting of a moat along the shore during austral summer. Streams fed by meltwater from glaciers deliver fresh water if and when spring-summer temperatures permit [*Lyons et al.*, this volume; *Conovitz et al.*, this volume]. None of these lakes have surface drainage, but the lakes lose water at the surface due to evaporation and ice ablation. Shallow depths in these lakes are freshwater and are supersaturated in O_2 and N_2. The deepest waters of these lakes are depleted in oxygen, with H_2S present in the deepest waters of Lakes Fryxell, Hoare, and Vanda.

Lake Bonney is located at the head of the Taylor Valley (77°43'S, 162°20'E). This lake has two basins separated by a narrow sill approximately 40 m wide and 12 m deep. Maximum water column depths are about 40 m in each basin. The west lobe is 1.3 km^2 and abuts the face of the Taylor Glacier, which feeds several major streams draining into the basin. The east lobe is 3.5 km^2 and receives major stream input from glaciers east of the lake. The depth of the steepest chemocline, below which oxygen is depleted, was 13 m in the west lobe and 20 m in the east lobe at the time of this study.

Lake Hoare (77°38'S, 163°07'E) is a 1.9 km^2 freshwater reservoir that is dammed by the Canada Glacier. Maximum depth is 34 m. Temperature and sodium concentrations are relatively uniform to a depth of approximately 28 m, below which is an anaerobic layer.

Lake Fryxell is at the mouth of the Taylor Valley (77°37'S, 163°07'E) about 7 km from McMurdo Sound. Surface area is 7.1 km^2 and maximum depth is 21 m. The main chemocline and oxycline is at ca. 9 m.

Lake Vanda is located in the Wright Valley (77°32'S, 161°33'E) and has an area of 6.7 km^2. Maximum depth is about 80 m and the anaerobic hypolimnion is below a depth of approximately 60 m. Lake Vanda receives significant amounts of fresh water in some years from the Onyx River.

Sample Collection

Water samples were collected over three field seasons from 1989 to 1991 between late winter (September 9) and summer (January 9). Collections were made from stations at the center of each basin through ice holes of 0.25 to 1.0 m diameter. Most samples of Lake Bonney were from the east lobe. Depths were measured relative to the piezometric level in ice holes (approximately 0.3 m below the top of the ice cover). Water was collected with 10-liter Niskin bottles and placed in HDPE jars stored (< 6 h) in a dark cooler for transport to the laboratory.

Water subsamples (1.5 to 7.0 liters) were passed through glass-fiber filters (Whatman GF/C) to concentrate particulate material for pigment analysis. Filters were placed in cryogenic vials and stored in liquid nitrogen. For microscopy cell counts, 100-ml subsamples were placed in glass bottles and preserved with acid-Lugol's solution [APHA, 1985] at 1% final concentration.

Pigment Analyses

Pigments were extracted from filtered material by maceration in 5 ml of acetone on ice. The grinding apparatus was rinsed twice with 1 ml of acetone. Macerated and rinsed materials were placed in the dark at 4°C for 2 h to 3 h. Glass fibers and particulates were then removed by filtration through a Whatman GF/F filter. The volume of acetone extract was recorded and the sample was stored in the dark on ice. Acetone extract was mixed 1:1 (v:v) with an ion-pairing agent (IPA) solution containing tetrabutylammonium acetate [Mantoura and Llewellyn, 1983]. Immediately after mixing with IPA, 1.0 ml of sample was injected into a Waters HPLC system consisting of a 15-cm C-8 reverse-phase column, two pumps (model 510), a gradient controller (model 680), a photodiode array detector (PDA, model 991) and a fluorometer (model 370) containing cutoff filters for detecting the fluorescence of chlorophylls. The elution gradient was linear from 60%:40% (v:v) acetone:IPA to 100% acetone in 20 min at a flow rate of 2 ml min^{-1}. Pigments were detected by absorption at either 440 nm (for carotenoids, chlorophylls, and chlorophyllides) or 410 nm (for phaeophytins and phaeophorbides).

Pigment standards were acquired from commercial sources (Sigma Chemical) or made from algal cultures or fresh spinach. Standards for chlorophyll-a, chlorophyll-b, α-carotene, β-carotene, lutein, and bacteriochlorophyll-a were available commercially. Phaeophytin-a was made from chlorophyll-a by acidification. Lutein, violaxanthin, and neoxanthin were isolated from spinach leaf extracts using an HPLC preparatory column. Alloxanthin, diadinoxanthin, fucoxanthin, chlorophyllides, and phaeophorbides were identified in samples based on absorption spectra measured by the

PDA detector (from 380 nm to 750 nm, every 1.3 nm, at 2.3 s intervals). Peaks for chlorophylls, chlorophyllides, phaeophorbides, and phaeophytins were corroborated by detection with the fluorometer.

Pigments were quantified by applying specific extinction coefficients [E; Rowan, 1989] to absorbances measured by the PDA detector. For each pigment, E was weighted for the ratio between absorbance at the peak wavelength for E (A_{max}) and absorbance at the detection wavelength (A_{det} at either 440 nm or 410 nm). $A_{max} : A_{det}$ ratios were estimated from the slopes of least-square regressions of A_{max} and A_{det} values from PDA spectra collected during HPLC runs of standards and samples. Pigment concentrations (C) were calculated as:

$$C = Area * (A_{max} : A_{det}) * F * V_e / (E * V_f * V_i)$$

where Area is the integrated area of a peak on the chromatogram, F is the flow rate, V_e is the volume of extract, V_f is the volume of water filtered, and V_i is the volume of extract injected into the HPLC.

Phytoplankton Cell Counts

Water samples preserved in acid-Lugol's solution were placed in 100-ml Utermöhl settling chambers and allowed to settle for 5 days. Phytoplankton cells were counted for the entire sample using an inverted compound microscope [Lund et al., 1958]. Species designations are based on the key of Seaberg et al. [1979]. Cell dimensions were measured for a subset of each taxonomic group and biovolumes were calculated using appropriate geometric shapes.

RESULTS

Chromatograms of samples from three depths in Lake Bonney (Figure 1) show the efficacy of the HPLC method for separating the major phytoplankton pigments present. The population immediately beneath the ice cover had a pigment signature with three major contributors, chlorophyll-a (chl-a), alloxanthin, and chl-c (= chl-c_1 + chl-c_2). This chromatogram indicates dominance by cryptophyte algae. The 13-m sample was more complex because of the addition of a significant fucoxanthin peak, which indicates the presence of chromophyte algae. An even greater diversity of pigments was observed at 18 m, with chl-b, lutein, violaxanthin, and neoxanthin indicating a substantial chlorophyte population. Microscopic examinations of

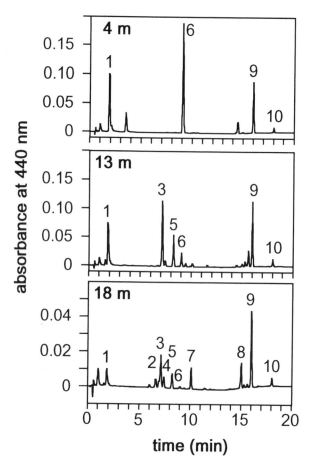

Fig. 1. Chromatograms for samples collected from the east lobe of Lake Bonney on December 5, 1991. Peaks identified by numbers: 1) chlorophyll-*c*, 2) neoxanthin, 3) fucoxanthin, 4) violaxanthin, 5) diadinoxanthin, 6) alloxanthin, 7) lutein + zeaxanthin, 8) chlorophyll-*b*, 9) chlorophyll-*a*, 10) β-carotene.

chlorophyll-c_3 in our samples (which can be separated by our method), a single chl-*c* pigment class was quantified. Our HPLC method was also unable to differentiate between α-carotene and β-carotene, thus the peak identified as β-carotene may include various forms of related carotenes. Though we did not test for the ability to distinguish lutein from zeaxanthin, similar protocols are unable to make that separation [e.g., *Mantoura and Llewellyn*, 1983]; current protocols are capable [e.g., *Wright et al.*, 1991]. For our samples, we identified a peak as lutein + zeaxanthin, based on 1) an absorbance spectra resembling lutein and 2) the presence of other chlorophyte accessory pigments that were co-detected (e.g., chl-*b*, violaxanthin, and neoxanthin) and our microscopy observations of much larger biovolumes of chlorophytes (which typically contain more lutein than zeaxanthin) than cyanobacteria (which contain zeaxanthin but no lutein). This method is also incapable of separating diadinoxanthin and 19'-hexanoyloxyfucoxanthin (a pigment found in some prymnesiophytes). However our cell counts showed that chrysophytes were abundant when the peak identified as diadinoxanthin was present, and we are unaware of any reports of prymnesiophyte species from these lakes. Finally our samples were not expected to show evidence for conversion of xanthophylls by de-epoxidation (e.g., violaxanthin to zeaxanthin in chlorophytes, or diadinoxanthin to diatoxanthin in chrysophytes) because of the low-light environment and the relatively long, dark periods of sample storage during transport after collection (1 to 4 h).

Vertical profiles of pigment concentrations collected over the spring-summer phytoplankton growth season in Lake Bonney showed that the three distinct populations outlined above with diagnostic pigments (alloxanthin, fucoxanthin, and chl-*b*) develop sequentially (Figure 2). From 4 m to 8 m, the cryptophyte marker alloxanthin is dominant on all dates. The 10 m and 12 m populations are alloxanthin-dominated early in the season, but a significant fucoxanthin peak was observed in December and January. Fucoxanthin is the dominant marker pigment from 13 m to 20 m as late as November, with a deep population of chl-*b* containing algae growing at 17 m to 20 m in December and January.

Ratios of pigments diagnostic for certain taxa (such as alloxanthin, fucoxanthin, and chl-*b*) to chl-*a* can be used as a relative measure of the contribution of certain taxa (cryptophytes, chrysophytes, and chlorophytes, respectively) to phytoplankton biomass. When pigment ratios measured over two field seasons are plotted

these samples showed that significant populations of corresponding algae were coincident with the accessory pigments identified as markers: the cryptophyte *Chroomonas lacustris* with alloxanthin; the chrysophyte *Ochromonas* sp. with fucoxanthin; and the chlorophyte *Chlamydomonas subcaudata* with chl-*b*.

We developed our HPLC method to be as simple as possible with the capability of separating the major pigments present in the study lakes. More complex methods can separate a greater variety of algal pigments [e.g., *Wright et al.*, 1991]. Like most general HPLC methods developed for phytoplankton pigment analysis, this method is not capable of distinguishing among chl-c_1, chl-c_2, and Mg 2,4-divinylpheaoporphyrin-a_5 monomethyl ester. Because of this limitation, and the lack of significant amounts of

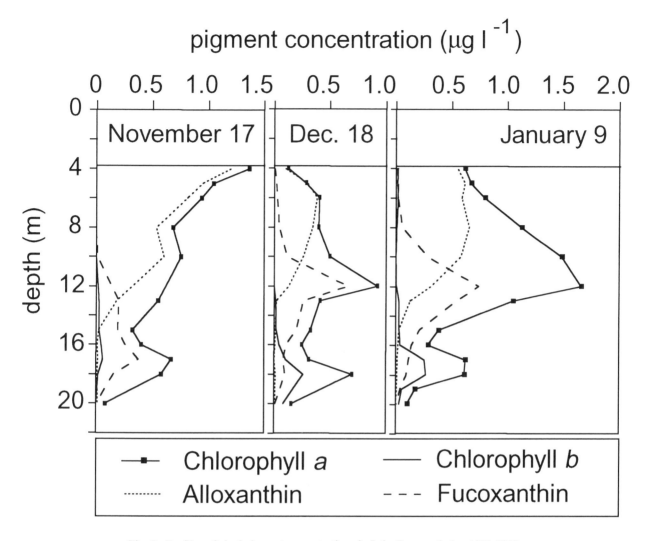

Fig. 2. Profiles of algal pigment concentrations in Lake Bonney during 1990-1991.

against date for the three chlorophyll maxima in Lake Bonney, we see a pattern of sequential blooms by different taxa at each depth (Figure 3). Early in the season, the chl-*a* signal appears to be associated with significant contributions of all three marker pigments. Later in the season, each chlorophyll maxima has a chl-*a* signal associated primarily with one marker pigment: alloxanthin beneath the ice, fucoxanthin at mid-depth, and chl-*b* in the deepest population. These shifts may represent a decrease in taxonomic diversity as dominant taxa bloom.

For Lake Bonney, we can compare pigment ratios to the biovolume of taxonomic groups (determined from cell counts on preserved water samples) for three dates in 1990-1991 (Table 1). Alloxanthin:chl-*a* molar ratios ranged from 1.2 to 1.4 for under-ice populations,

with negligible ratios for fucoxanthin:chl-*a* and chl-*b*:chl-*a*. These results demonstrate the dominance of cryptophyte algae under-ice, in contrast to biovolume results showing that chlorophytes (42 to 53%) were equal to or greater than cryptophytes (37 to 40%). At the middle chlorophyll maxima (13 m), fucoxanthin:chl-*a* ratios imply dominance by chrysophytes, (0.49 to 0.87), whereas biovolume results show that chrysophytes (56 to 63%) were joined by significant contributions of chlorophytes (21 to 32%) despite low chl-*b*:chl-*a* ratios (0.03 to 0.04). Only for the late season samples from the deep chlorophyll maximum were the ratios of chl-*b*:chl-*a* high enough (0.33 to 0.40) to corroborate the substantial contribution by chlorophytes to biovolume (86 to 92%).

Overall we measured very low ratios of chl-*b*:chl-*a*

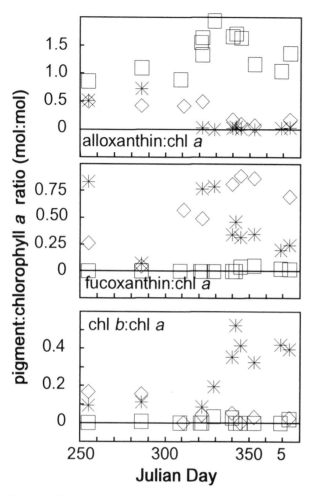

Fig. 3. Seasonal trends in the molar ratios of accessory pigments to chlorophyll-*a* in Lake Bonney. Samples were collected from 4 to 5 m (squares), 13 m (diamonds), and 17 m (stars) during field seasons in 1990–1991 and late 1991.

the basins and throughout the epilimnia of Lakes Fryxell, Hoare, and the west lobe of Bonney. Chrysophyte algae, detected by the presence of fucoxanthin, were dominant only in the mid-depths of the epilimnion in the east lobe of Lake Bonney, and were significant at all depths in Lake Vanda. Chlorophyte algae, detected by the presence of chl-*b*, were dominant only in the bottom of the epilimnion in the east lobe of Lake Bonney, but were significant immediately beneath the ice in Lake Hoare. The only samples that showed evidence of carotenoids from cyanobacteria (tentatively identified as myxoxantho-phyll, but not quantified) were from Lake Vanda. The xanthophyll typically used to detect cyanobacteria is zeaxanthin; however spectral scans of the "lutein + zea-xanthin" peak resembled lutein. The dominant pigment in the hypoliminion of Lake Fryxell (10 to 11 m) was bacteriochlorophyll-*a* (not quantified), indicating a large population of photosynthetic bacteria. The next most prevalent pigments at these depths were alloxanthin, chl-*a*, and chl-*c*, in ratios similar to the shallower phytoplankton populations of Lake Fryxell.

Pigment signatures in the east lobe of Lake Bonney become most complex near the interface between the super-oxygenated epilimnion and the oxygen-poor hypolimnion (ca. 20 m). This part of the water column, 18 m to 21 m, was always the chl-*a* minimum, and had relatively high concentrations of chlorophyll breakdown products compared to shallower or deeper waters (Table 3). The main breakdown products present were chlorophyllides and phaeophorbides. Ratios of phaeophytin-*a* : chl *a* were low at all depths (< 0.02). Above 15 m, the relative amounts of chlorophyll breakdown products were much lower.

Seasonal trends for concentrations of phaeo-phorbides, chlorophyllides, and chl-*a* in the oxycline waters of Lake Bonney were reconstructed from late season collections in 1990–1991 and early season collections in 1991 (Figure 4). The late season bloom of phytoplankton just above the oxycline was observed in the data from 17 m. Smaller increases in chl-*a* were noted at 20 m and 23 m late in the season. Both of the chlorophyll breakdown products showed decreasing trends from the winter-spring transition through the austral spring. A late spring sample collected from 18 m on December 5, 1991 matched the December 1990 increase in chl-*a*, as well as the decreases in phaeophorbide and chlorophyllide concentrations to negligible amounts.

In the oxygen-poor waters deeper than 21 m in Lake Bonney, the ratio of chlorophyll breakdown products to

in samples containing <60% biovolume as chloro-phytes. One possibility is that the cells counted as chlorophytes were atypical in chl-*b* : chl-*a* or in pigment per biovolume. A significant amount of the chlorophyte biovolume was represented by cells that resembled chlamydomonad zygospores [*Sharp*, 1993]. Early in the growth season, these lakes also show poor correlations between chl-*a* concentrations and photosynthetic production, implying that a large fraction of the biomass is associated with inactive cells [*Lizotte et al.*, 1996].

Pigment composition in the oxygenated epilimnia of various lakes in the dry valleys shows that alloxanthin-containing cryptophytes dominated most phytoplankton populations (Table 2). This cryptophyte signature was clearly seen in shallow populations of all

TABLE 1. Relative Abundance of Phytoplankton Groups in Lake Bonney during 1990–1991, as Determined by Microscopy (% of algal biovolume) and by Pigment Analysis (molar ratios).

Date	Depth (m)	Cryptophytes (%)	Cryptophytes (allox:chl-a)	Chrysophytes (%)	Chrysophytes (fucox:chl-a)	Chlorophytes (%)	Chlorophytes (chl-b:chl-a)
Nov. 17	4	40	1.38	n.d.	n.d.	42	n.d.
	13	15	0.52	56	0.49	29	0.04
	17	n.d.	0.03	42	0.77	58	0.09
Dec. 18	4	37	1.21	n.d.	0.05	53	n.d.
	13	11	0.07	63	0.87	21	0.03
	17	n.d.	n.d.	14	0.35	86	0.33
Jan. 9	4	39	1.41	n.d.	0.02	50	0.02
	13	6	0.18	59	0.70	32	0.03
	17	n.d.	0.03	8	0.24	92	0.40

Abbreviations: allox = alloxanthin; fucox = fucoxanthin; chl = chlorophyll; n.d. = not detected.

TABLE 2. Concentrations of chl-a and Molar Ratios of Accessory Pigments to Chlorophyll-a for Phytoplankton Maxima in the Dry Valley Lakes in 1990.

Lake	Date	Depth (m)	chl-a (μg l^{-1})	allox:chl-a	fucox:chl-a	chl-b:chl
Bonney (east lobe)	Dec. 18	4	0.13	1.21	0.05	n.d.
		13	0.41	0.07	0.87	0.03
		17	0.31	n.d.	0.35	0.33
Bonney (west lobe)	Dec. 1	4	0.70	1.58	0.02	n.d.
		13	0.89	1.11	0.12	n.d.
Hoare	Dec. 8	5	0.41	0.72	n.d.	0.16
		10	2.1	0.87	n.d.	n.d.
		12.5	3.2	0.89	n.d.	n.d.
Fryxell	Dec. 7	5	2.3	0.99	0.01	0.06
		7	2.6	1.03	0.01	0.05
		8.5	4.1	0.97	0.03	0.05
Vanda	Nov. 28	3	0.09	0.46	0.41	n.d.
		20	0.14	0.49	0.41	n.d.
		57.5	0.37	0.11	0.27	n.d.

Abbreviations: allox = alloxanthin; fucox = fucoxanthin; chl = chlorophyll; n.d. = not detected.

TABLE 3. Mean (± standard deviation) of chl-*a* Concentration ($\mu g \ l^{-1}$) and the Molar ratio of Chlorophyll Degradation Products to chl-*a* in the East Lobe of Lake Bonney.

Depth (m)	n	chl-*a*	chlorophyllide:chl-*a*	phaeophorbide:chl-*a*	phaeophytin:chl-*a*
4	10	1.11 (0.58)	n.d.	0.03 (0.04)	0.001 (0.002)
5	8	0.80 (0.40)	n.d.	0.05 (0.05)	0.008 (0.015)
6	8	0.73 (0.30)	0.03 (0.06)	0.07 (0.09)	0.005 (0.004)
10	7	0.78 (0.41)	0.01 (0.02)	0.03 (0.04)	0.002 (0.002)
12	6	1.02 (0.39)	0.04 (0.04)	0.06 (0.07)	0.003 (0.006)
13	9	0.54 (0.34)	0.04 (0.06)	0.07 (0.10)	0.003 (0.007)
15	5	0.27 (0.10)	0.10 (0.13)	0.13 (0.16)	0.013 (0.026)
17	9	0.37 (0.23)	0.21 (0.39)	0.17 (0.29)	0.010 (0.029)
18	6	0.47 (0.31)	0.44 (0.60)	0.31 (0.45)	0.001 (0.002)
19	2	0.11 (0.06)	0.68 (0.64)	0.43 (0.37)	0.006 (0.006)
20	5	0.08 (0.04)	0.75 (0.55)	0.60 (0.34)	n.d.
21	2	0.08 (0.02)	0.85 (0.50)	0.67 (0.38)	n.d.
23	6	0.24 (0.05)	0.31 (0.11)	0.20 (0.11)	0.009 (0.013)
25	4	0.40 (0.06)	0.18 (0.08)	0.12 (0.06)	0.019 (0.004)
30	4	0.43 (0.06)	0.14 (0.02)	0.09 (0.04)	0.011 (0.002)
35	5	0.13 (0.03)	0.44 (0.10)	0.36 (0.10)	n.d.

Abbreviations: chl = chlorophyll; n.d. = not detected; n = sample number.

chl-*a* were lower than the peak depth (18 to 21 m), but still relatively high (Table 3). The peak concentrations for chlorophyll breakdown products were found at 23 to 25 m in conjunction with a peak in chl-*a*. Pigment signatures below 20 m most closely resembled phytoplankton from 17 to 20 m, with chl-*b* and lutein + zeaxanthin as the main accessory pigments. The hypolimnetic chl-*a* concentrations were relatively consistent on all sampling dates compared to the upper water column where the phytoplankton bloom in the spring. This consistency is demonstrated by the lower variability statistics in Table 3, and by the examples of reconstructed seasonal trends in Figure 4.

DISCUSSION

Analyses of algal pigments in suspended particulates from the dry valley lakes produced new insights to the composition of phytoplankton communities and the efficacy of the most common method for preserving phytoplankton for microscopy. For Lake Bonney, our seasonal studies also illuminated temporal trends in algal biomass, its taxonomic composition, and the fate of this algal material.

Late in spring, the chlorophyll maxima were typically dominated by a single algal group based on pigment signatures. The most common pigment signature encountered in these four lakes was for cryptophyte algae. Cryptophytes appear to dominate the entire oxygenated water column in Lakes Fryxell, Hoare, and the west lobe of Bonney, as well as the shallowest waters in the east lobe of Bonney. Chrysophytes were dominant only in the mid-depth maximum of the east lobe of Lake Bonney, but were co-dominant with cryptophytes in Lake Vanda. Chlorophytes were dominant only in the deepest

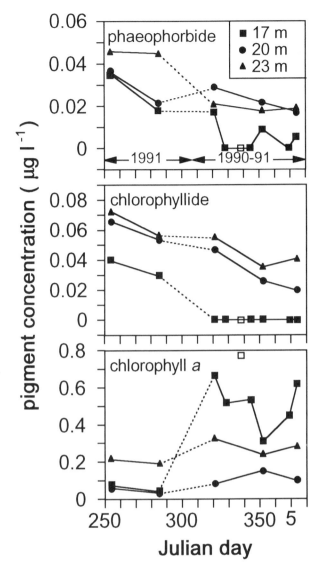

Fig. 4. Reconstructed seasonal trend of concentrations of phaeophorbides, chlorophyllides, and chlorophyll-*a* at the oxycline (20 m), 3 m above (17 m) and 3 m below (23 m) in the east lobe of Lake Bonney. Note that samples collected during the winter-spring transition are from 1991, the year after samples were collected for spring-summer. A summer 1991 sample from 18 m is shown as open squares.

maximum of the east lobe of Lake Bonney.

These results can be contrasted with published reports of phytoplankton populations in these four lakes (Table 4). If we assume our pigment analysis as a baseline, past surveys based on cell counts have overestimated the importance of chlorophytes and cyanobacteria while cryptophytes and chrysophytes were underestimated. It is important to note that most taxonomic surveys conducted in the past were not

quantitative, but they did make qualitative statements about relative prevalence of species. Part of the discrepancy could be because past rankings were based on cell concentration rather than biochemical, biomass, or biovolume concentrations; the former tend to overestimate the contribution of small-celled species (i.e., for some of the species of *Oscillatoria* and *Chlorella* reported in these lakes). Our comparison between pigment analysis and cell counts supports the possibility that certain taxa are lost from water samples preserved for microscopy [*Gieskes and Kraay*, 1983; *Buma et al.*, 1990].

Another possibility is that the differences observed among studies over the past three decades (Table 4) reflect actual changes in species composition from year to year [*Spaulding et al.*, 1994]. The most obvious trend in the literature we reviewed is the increasing prominence (or recognition) of cryptophytes in these dry valley lakes. Recent studies in coastal waters off the Antarctic peninsula have observed that increases in cryptophyte populations are correlated with the timing of freshwater input from glaciers [*Moline and Prezelin*, 1996]. If cryptophyte populations are responsive to increased freshwater flux, particularly from glacier meltwaters, then they could be an important indicator organism for ecosystem-level shifts in the dry valley lakes. For the long-term ecological research projects currently underway on these lakes, it would seem prudent to use both the specific taxonomic methods of microscopy and the coarse taxonomic methods of pigment analysis to monitor phytoplankton populations until the possibility of methodological biases can be eliminated.

Phytoplankton populations in the perennially ice-covered lakes were typically separated into taxonomically distinct vertical strata. Seasonal trends in Lake Bonney show that the cryptophyte community immediately under the ice peaked during November, while deeper populations of chrysophytes, and chlorophytes continued to grow into the summer. Profiles for fucoxanthin indicate that flagellated chrysophytes, may have moved up the water column before forming a distinct peak at 12 m depth. This depth is just above the nutricline [e.g., *Priscu*, 1995], which may present a limit to further upward migration.

The peak for ratios of chlorophyll degradation products in Lake Bonney was found near the bottom of the euphotic zone, in the region of the chemocline where sinking particles may accumulate after achieving neutral buoyancy in the dense brine. The fact that the concentrations of chlorophyll breakdown products are much higher at the beginning of spring than at the end

TABLE 4. Phytoplankton Taxa (in order of abundance) Reported for
McMurdo Dry Valley Lakes.

Reference (Methods*)	Bonney	Fryxell	Hoare	Vanda
Armitage and House [1962] (70 mm mesh net)	C			C
Goldman [1964] (not reported)				D
Goldman et al. [1967] (Lugol's, settled)	B C			B D C
Koob and Leister [1972] (Craf's, filtered)	C D			
Vincent [1981] (Lugol's, settled)		B C A		
Parker et al. [1982] (Lugol's, settled)	A C	A C	A C D	B D
Seaburg et al. [1983] (Lugol's, settled)	A C D	A C	A D C	B C A D
Spaulding et al. [1994] (Lugol's, settled)		A C D		
this study (Lugol's, settled)	C A B			
this study (HPLC)	A B C	A C	A C	A B

*The analysis methods reported are given in parentheses.

A = Cryptophyte; B = Chrysophyte; C = Chlorophyte; and D = Cyanobacteria.

implies that their accumulation may be based on the previous year's phytoplankton production in the upper water column. Seasonal increases in chl-*a* below 20 m were not associated with measurable primary production [*Sharp*, 1993], and the pigment signatures imply that this chl-*a* was due to cells settling from phytoplankton populations at 17 to 20 m. The seasonal trend of decreasing concentrations for degradation products in conjunction with the onset of light reaching the water column in spring implies that photooxidation may be a major loss mechanism for pigments. Another possible mechanism is bacterial degradation: there is a peak in bacterial populations at the oxycline, and an annual cycle of activity linked to primary production [e.g., *Goldman et al.*, 1964; *Ellis-Evans*, 1985; *Priscu*, 1992] may lead to increased decomposition during the spring season.

Acknowledgments. We thank P.J. Neale, R.H. Spigel, I. Forne, B. T. Hatcher, R. Nugent, J. Rudek, and B. Kelly for their assistance in the field. We also thank S.T. Kottmeier, Ant. Support Assoc., and the U.S. Navy for logistical support.

This work was supported by the National Science Foundation Division of Polar Programs grants DPP-8820591, DPP-9211773, and OPP-9419423 to J.C.P.

REFERENCES

APHA, *Standard Methods for the Examination of Water and Wastewater*, 16th ed., American Public Health Association, Washington, DC, 1985.

Armitage, K. B. and H. B. House, A limnological reconnaissance in the area of McMurdo Sound, Antarctica, *Limnol. Oceanogr.*, 7, 36–41, 1962.

Buma, A. G. J., P. Treguer, G. W. Kraay, and J. Morvan, Algal pigment patterns in different watermasses of the Atlantic sector of the Southern Ocean during fall 1987, *Polar Biol.*, 11, 55–62, 1990.

Ellis-Evans, J. C., Decomposition processes in maritime Antarctic lakes, in *Antarctic Nutrient Cycles and Food Webs*, edited by W.R. Siegfried, P.R. Condy, and R.M. Laws, pp. 253–260, Springer-Verlag, Berlin, 1985.

Everitt, D. A., S. W. Wright, J. K. Volkman, D. P. Thomas, and E. J. Lindstrom, Phytoplankton community compositions in the western equatorial Pacific determined from chlorophyll and carotenoid pigment distributions, *Deep-Sea Res.*, 37, 975–997, 1990.

Gieskes, W. W. C., and G. W. Kraay, Dominance of Cryptophyceae during the phytoplankton spring bloom in the central North Sea detected by HPLC analysis of pigments, *Mar. Biol.*, *75*, 179–185, 1983.

Goldman, C. R., Primary productivity studies in Antarctic lakes, in *Biologie Antarctique*, edited by R. Carrick, M. Holdgate, and J. Prevost, pp. 291–299, Hermann, Paris, 1964.

Goldman, C.R., D.T. Mason, and J.E. Hobbie,Two Antarctic desert lakes, *Limnol. Oceanogr.*, *12*, 295–310, 1967.

Koob, D. D., and G. L. Leister, Primary productivity and associated physical, chemical, and biological characteristics of Lake Bonney: a perennially ice-covered lake in Antarctica, *Antarct. Res. Ser.*, *20*, 51–68, 1972.

Lizotte, M. P., and J. C. Priscu, Photosynthesis-irradiance relationships in phytoplankton from the physically stable water column of a perennially ice-covered lake (Lake Bonney, Antarctica), *J. Phycol.*, *28*, 179–185, 1992a.

Lizotte, M. P., and J. C. Priscu.. Spectral irradiance and bio-optical properties in perennially ice-covered lakes of the dry valleys (McMurdo Sound, Antarctica), *Antarct. Res. Ser.*, *57*, 1–14, 1992b.

Lizotte, M. P. and J. C. Priscu, Natural fluorescence and quantum yields in vertically stationary phytoplankton from perennially ice-covered lakes, *Limnol. Oceanogr., 39*, 1399–1410, 1994.

Lizotte, M. P., T. R. Sharp, and J. C. Priscu, Phytoplankton dynamics in the stratified water column of Lake Bonney, Antarctica: I. Biomass and productivity during the winter-spring transition, *Polar Biol., 16,* 155–162, 1996.

Lund, J. W. G., C. Kipling, and E. D. LeCren, The inverted microscope method of estimating algal numbers and the statistical basis for estimations by counting, *Hydrobiologia, 11,* 143–170, 1958.

Mantoura, R. F. C., and C. A. Llewellyn, The rapid determination of algal chlorophyll and carotenoid pigments and their breakdown products in natural waters by reverse-phase high-performance liquid chromatography, *Anal. Chim. Acta, 151,* 297–314, 1983.

Moline, M. A., and B. B. Prezelin, Long-term monitoring and analyses of physical factors regulating variability in coastal Antarctic phytoplankton biomass, in situ productivity and taxonomic composition over subseasonal, seasonal and interannual time scales, *Mar. Ecol. Prog. Ser., 145,* 143–160, 1996.

Neale, P. J. and J. C. Priscu, The photosynthetic apparatus of phytoplankton from a perennially ice-covered Antarctic lake: Acclimation to an extreme shade environment, *Plant Cell Physiol.*, *36*, 253–263, 1995.

Neale, P. J. and J. C. Priscu, Fluorescence quenching in phytoplankton of the McMurdo Dry Valley lakes (Antarctica): Implications for the structure and function of the photosynthetic apparatus, this volume

Parker, B. C., G. M. Simmons, Jr., K. G. Seaburg, D. D. Cathey, and F. C. T. Allnutt, Comparative ecology of plankton communities in seven Antarctic oasis lakes, *J. Plankton Res.*, *4*, 271–286, 1982.

Priscu, J. C., Particulate organic matter decomposition in the water column of Lake Bonney, Taylor Valley, Antarctica. *Antarct. J. U.S.*, *27*, 260–262. 1992

Priscu, J. C., Phytoplankton nutrient deficiency in lakes of the McMurdo dry valleys, Antarctica, *Freshwater Biol., 34*, 215–227, 1995.

Rowan, K. S., *Photosynthetic Pigments of Algae*, Cambridge University Press, New York, 334 pp., 1989.

Seaburg, K. G., B. C. Parker, G. W. Prescott, and L. A. Whitford, The algae of southern Victorialand, Antarctica, *Bibliothecia Phycologica*, *46*, 1–169, 1979.

Seaburg, K. G., M. Kaspar, and B. C. Parker, Photosynthetic quantum efficiencies of phytoplankton from perennially ice covered Antarctic lakes, *J. Phycol.*, *19*, 446–52, 1983.

Sharp, T. R., *Temporal and Spatial Variation of Light, Nutrients and Phytoplankton Production in Lake Bonney, Antarctica*, M.S. Thesis, Montana State University, 166 pp., 1993.

Spaulding, S. A., D. M. McKnight, R. L. Smith, and R. Dufford, Phytoplankton population dynamics in perennially ice-covered Lake Fryxell, Antarctica, *J. Plankton Res., 16*, 527–541, 1994.

Spigel, R. H., and J. C. Priscu, Physical limnology of the McMurdo Dry Valley Lakes, this volume.

Vincent, W. F., Production strategies in Antarctic inland waters: phytoplankton eco-physiology in a permanently ice-covered lake, *Ecology, 62*, 1215–1224, 1981.

Vincent, W. F., and C. L. Vincent, Factors controlling phytoplankton production in Lake Vanda (77°S), *Can. J. Fish. Aq. Sci., 39*, 1602–1609, 1982.

Wright, S. W., S. W. Jeffrey, R. F. C. Mantoura, C. A. Llewellyn, T. Bjornland, D. Repeta, and N. Welschmeyer, Improved HPLC method for the analysis of chlorophylls and carotenoids from marine phytoplankton, *Mar. Ecol. Prog. Ser.*, *77*, 183–196, 1991.

M. P. Lizotte, Department of Biology and Microbiology, University of Wisconsin, Oshkosh, Oshkosh ,WI 54901-8640

J. C. Priscu, Department of Biological Sciences, Montana State University, Bozeman, MT 59717

(Received October 2, 1996;
accepted April 30, 1997)

FLUORESCENCE QUENCHING IN PHYTOPLANKTON OF THE MCMURDO DRY VALLEY LAKES (ANTARCTICA): IMPLICATIONS FOR THE STRUCTURE AND FUNCTION OF THE PHOTOSYNTHETIC APPARATUS

Patrick J. Neale

Smithsonian Environmental Research Center, Edgewater, Maryland

John C. Priscu

Department of Biological Sciences, Montana State University, Bozeman, Montana

Phytoplankton in perennially ice-covered lakes of the McMurdo Dry Valleys experience a light environment that is unusually stable due to constant shade (< 1–3% of incident) and a narrow spectral distribution (blue-green). This relative constancy is due to optical attenuation and spectral filtering through the ice cover and an absence of vertical mixing. We have studied the structure and function of the photosynthetic apparatus of phytoplankton in Lakes Bonney, Fryxell, and Hoare (Taylor Valley) using a variety of methods. Some photosynthetic characteristics of phytoplankton in the dry valley lakes indicate low-light acclimation, including a low irradiance for the onset of light saturation of photosynthesis (I_k) and high sensitivity to photoinhibition. Other characteristics seem contrary to expectations for an extreme shade environment. Antenna sizes are not large (average Chl:P_{700} = 743 mol mol^{-1}) and the maximum quantum yield of photosynthesis is low. We obtained further information on the structure and function of the photosynthetic apparatus in these phytoplankton through analysis of the slow (minutes time-scale) fluorescence transients. Our approach used a sensor for upwelling radiance at 683 nm (natural fluorescence) mounted over a transparent container and illuminated with fixed intensity blue-green irradiance. For samples from Lake Bonney and Lake Hoare, the steady-state fluorescence yield (F_s) after about five minutes of illumination was lower (quenched) for irradiances greater than 10 μmol photons m^{-2} s^{-1}. This indicated induction of protective mechanisms to dissipate excess excitation irradiance (non-photochemical quenching) at an unprecedented low irradiance. In situ, these assemblages appear to be at, or just below, the threshold for the induction of non-photochemical quenching. Conversely, the Lake Fryxell assemblage had a high maximum quantum yield of photosystem II photochemistry and did not show F_s quenching at irradiances up to 100 μmol photons m^{-2} s^{-1}. Overall the results are consistent with photosynthetic acclimation to minimize excitation pressure, i.e., an energetic imbalance between photochemical sources and metabolic sinks. Acclimation of the dry valley phytoplankton assemblages is constrained by a strikingly narrow irradiance range from low to high excitation pressure. We hypothesize that optimum light harvesting subject to maintenance of low excitation pressure is possible because of the constant shade environment.

INTRODUCTION

Phytoplankton in the lakes of the McMurdo Dry Valleys, Antarctic grow in a unique environment due to a perennial ice-covers 3–5 m thick. In Lake Bonney, irradiance beneath the ice is less than 1 to 3% of incident, with maximum irradiance less than 50 μmol photons m^{-2} s^{-1} [*Priscu*, 1991; *Lizotte and Priscu*, 1992b]. In addition, the ice cover selectively transmits blue-green irradiance, with maximum flux near 500 nm. Irradiances at wavelengths shorter than 440 nm have intensities of <50% relative to the spectral peak, and irradiances at wavelengths longer than 600 nm have intensities lower than 10% of the spectral peak

241

TABLE 1. Symbols and Abbreviations Used in the Text

Symbol	Description	Units
Chl	Chlorophyll-*a*	mg m^{-3}
F'_0	Minimum fluorescence yield after actinic illumination	m^{-1} (see footnote)
F'_m	Maximum fluorescence yield under actinic illumination	m^{-1}
F_0	Minimum fluorescence yield after dark adaptation	(see footnote)
F_m	Maximum fluorescence yield after dark adaptation	(see footnote)
ϕ_m	Maximum quantum yield of PSII photochemistry (electrons generated per photons absorbed)	dimensionless
ϕ_p	Quantum yield of PSII photochemistry (electrons generated per photons absorbed)	dimensionless
F_s	Steady state fluorescence yield under actinic illumination	m^{-1}
F_s(max)	Maximum F_s	m^{-1}
I	Irradiance	μmol photons m^{-2} s^{-1}
Lu_{683}	Upwelling radiance at 683 nm	nmol photons m^{-2} s^{-1} steradian^{-1}
P_{700}	Photosystem I reaction center	
PAM	Pulse Amplitude Modulation fluorometry	
PAR	Photosynthetically Available Radiation (400–700 nm)	μmol photons m^{-2} s^{-1}
p^B_m	Light saturated rate of photosynthesis normalized to biomass	mg C mg Chl^{-1} h^{-1}
PSII	Photosystem II	

The fluorescence yields measured with the PNF have not been adjusted for absorption, thus units are given as m^{-1}, this is equivalent to the term "fluorescence coefficient" used by *Chamberlin et al.* [1990]. Fluorescence measured on the Turner Designs (F_0 and F_m) are on a arbitrary output scale (volts).

[*Lizotte and Priscu*, 1992b; *Howard-Williams et al.*, *this volume*].

The ice cover also prevents stirring of lake waters by surface winds. The McMurdo Dry Valley lakes are highly stratified [*Angino et al.*, 1964], and turbulence is very low or non-existent [*Spigel and Priscu, this volume*]. Owing to the lack of vertical mixing and the continuous daylight characteristic of the Antarctic summer, the phytoplankton in the dry valley lakes grow in an unusually stable shade environment [*Vincent*, 1981b]. This setting offers a unique opportunity to examine the acclimation of natural phytoplankton populations to their irradiance environment.

Light availability appears to be a primary factor regulating phytoplankton growth in the dry valley lakes. Photosynthesis-irradiance response curves imply that phytoplankton photosynthesis is always light limited [*Lizotte and Priscu*, 1992a]. Likewise, seasonal variation in primary production by different biomass maxima is linearly related to available irradiance [*Lizotte et al.*, 1996]. Survival of a given population depends on the efficiency with which light is absorbed and used for photosynthesis, both processes are dependent on the structure and function of the photosynthetic apparatus.

In addition to low intensity, irradiance spectral distribution affects light utilization by phytoplankton of the dry valleys. Most non-phycobilin containing eukaryotic microalgae (and in particular chlorophytes) have a lower efficiency of photosynthesis in blue-green light compared to other wavebands (review *Larkum and Barrett* [1983]). Better utilization of blue-green light could be achieved through synthesis of light-harvesting pigment-protein complexes which absorb in the predominant spectral bands and efficiently transfer energy to the reaction center, increasing the maximum quantum yield for photosynthesis. Such chromatically selective pigmentation has been reported for several algal groups, but has been rarely reported for cryptophytes and chlorophytes [*Larkum and Barrett*, 1983], flagellate groups commonly found in the dry valley lakes [*Lizotte and Priscu, this volume*]. Nevertheless our studies provide two lines of evidence that pigmentation of Lake Bonney populations responds to spectral irradiance. Firstly, phytoplankton have an abundance of light-harvesting carotenoids such as alloxanthin (in cryptophytes) and violaxanthin (in chlorophytes) that are particularly efficient in absorption of blue-green light [*Neale and Priscu*, 1995; *Lizotte and Priscu, this volume*]. Secondly, blue-green and red light are utilized with equal efficiency for light-dependent electron transport by photosystem II (PSII) in *Chlamydomonas subcaudata* isolated from Lake Bonney [*Neale and Priscu*, 1995]. This contrasts with the common strain of *C. reinhardtii* that is much less efficient at utilization of blue-green *versus* red light in PSII photochemistry [*Neale and Priscu*, 1995].

With the low average light intensity below the ice cover, low-light acclimation of photosynthesis [*Falkowski and LaRoche*, 1991] would be expected in phytoplankton populations of the dry valley lakes. Observations on the relationship of photosynthesis to irradiance (P-I) curves are consistent with this expectation [*Lizotte and Priscu*, 1992a; *Lizotte and Priscu*, 1994]. Low-light acclimation is also indicated by high sensitivity of photosynthesis to inhibition by photosynthetically available radiation (PAR) and ultraviolet radiation (UV) [*Neale et al.*, 1994]. This sensitivity implies that phytoplankton in the dry valley lakes lack defense mechanisms required to minimize damage resulting from exposure to the full solar spectrum at near surface intensities. However not all characteristics of the phytoplankton in dry valley lakes match those expected of low-light acclimated populations. Maximum quantum yields for photosynthesis are low relative to the theoretical maximum, especially in shallow populations situated away from the nutricline [*Lizotte and Priscu*, 1992a; *Lizotte and Priscu*, 1994]. Also, it was expected that the overall antenna size of the light-harvesting chlorophyll in photosynthesis, indicated by the ratio of total chlorophyll to the photosystem I reaction center [Chl:P_{700} ratio] would be large in Lake Bonney phytoplankton. This expectation has precedent in another microalgal population living under low irradiance conditions, sea-ice algal diatoms that have very large ratios of Chl:P_{700} [*Barlow et al.*, 1988]. In contrast average Chl:P_{700} in Lake Bonney phytoplankton (743 mol mol^{-1}) is similar to many

measurements with high-light grown microalgae [*Neale and Priscu*, 1995]. However the Chl:P_{700} ratio may significantly underestimate the actual photosynthetic antenna size given the high concentration of light-harvesting carotenoids.

In this report, we present data on the fluorescence characteristics of phytoplankton from Lake Bonney and other perennially ice-covered lakes in the Taylor Valley. Our objective is to better elucidate responses of the photosynthetic apparatus to the interacting effects of low light intensity, light spectral composition, nutrient supply, and temperature. In particular, we focus on the dynamic aspects of the variation in fluorescence yield in these populations. Studies of the dynamic response of fluorescence to variations in incident irradiance on time scales of seconds to minutes is a powerful tool to diagnose the regulation of light harvesting and photosynthetic electron transport [*Büchel and Wilhelm*, 1993]. In general these studies have been aided by developments of instruments to measure fluorescence yield during on-going illumination of photosynthetic tissue. One approach that has been widely used in higher plants and dense cultures of algae is Pulse Amplitude Modulation (PAM) fluorometry [*Schreiber et al.*, 1986]. Studies using this or similar techniques have shown that chlorophyll fluorescence yield can either increase or decrease as illumination increases. The direction of this change in fluorescence yield depends on the balance between the ability to absorb light and store energy in chemical form (light reactions) and the enzymatic capacity to utilize that energy in metabolic processes, in particular carbon fixation. Here we use such fluorescence trend analysis as a tool to better understand the structure and function of the photosynthetic apparatus of phytoplankton in the lakes of the McMurdo Dry Valleys.

Changes in fluorescence yield are linked to changes in the dissipation of the excitation energy absorbed by PSII, the primary source of chlorophyll fluorescence at ambient temperature [see reveiw by *Büchel and Wilhelm*, 1993]. Light absorbed by the photosynthetic apparatus has three possible fates: 1) photochemical production of reduced energy carriers, ultimately leading to photosynthetic electron transport; 2) re-emission as fluorescence; or 3) non-radiative decay (heat production). When light is low and limiting to photosynthesis, a maximum proportion of absorbed photons is used for photochemistry. Under these conditions, fluorescence emission and non-radiative dissipation are low, and fluorescence is said to be quenched photochemically. When light intensity increases, the inherent turnover rate of PSII starts to limit the efficiency of photochemistry; some of photons find occupied or "closed" reaction centers. Fluorescence yield rises and remains elevated as long as the increased photochemical production is balanced by downstream dissipation. However continued increases in irradiance can result in a dangerous imbalance, which is little tolerated by photosynthetic organisms. In particular, electrons from reduced chromophores and electron carriers can be diverted to formation of possibly damaging radicals. There are mechanisms by which the overall activity of PSII is down regulated to better match downstream demand for the products of photosynthetic electron transport, primarily carbon fixation (for a more detailed discussion see review by *Büchel and Wilhelm,* [1993]). The mechanisms deactivate the excitation energy absorbed by PSII and increase dissipation to heat, so that neither photochemistry or fluorescence emission can occur. As a consequence of this increase in the so-called "non-photochemical quenching," fluorescence yield is lowered. Detection of non-photochemical quenching of fluorescence is therefore indicative of limitation of photosynthesis by downstream processes.

Because of feedback to PSII photochemistry, variations in the yield of in vivo fluorescence can be caused by many factors besides the absolute light intensity, for example nutrient supply and temperature. The basic principle is that photosynthetic systems appear to be highly regulated to keep photochemistry and coupled metabolic processes in balance. This principle has been described as "excitation pressure" regulation [*Maxwell et al.,* 1995a]. *Maxwell et al.* showed that *Chlorella vulgaris* acclimates to minimize excitation pressure under growth conditions. As a result, *C. vulgaris* grown under high light and temperature are comparable to those grown at lower light and low temperature in terms of the structure and function of the photosynthetic apparatus, e.g., pigment content photosynthesis-irradiance relationship, and resistance to photoinhibition. In both cases acclimation counteracted a growth environment induced imbalance between upstream generation and downstream utilization of reductant, whether because of overexcitation of photochemistry (due to high light) or thermodynamic limits on metabolism (due to low temperature).

In this report, we describe a simple method for measuring fluorescence transients over slow time scales of minutes to seconds in natural populations of phytoplankton. The method uses a sensor designed to measure upwelling irradiance at 683 nm, Lu_{683} or "natural fluorescence." Instead of using the sensor as commonly employed to measure profiles of in situ fluorescence stimulated by solar irradiance, we measured fluorescence in the laboratory with the instrument mounted over a sample in a transparent container illuminated by a blue-green light source with similar spectral composition as in situ irradiance. We then measured relative fluorescence yield as the ratio of measured Lu_{683} to measured incident PAR in the container. We have used this method to measure slow fluorescence transients as a function of irradiance for phytoplankton from perennially ice-covered lakes of the Taylor Valley (Lake Bonney, Lake Fryxell, and Lake Hoare). The transients are interpreted in terms of excitation pressure theory to infer factors controlling the structure and function of the photosynthetic apparatus of phytoplankton in the dry valley lakes.

METHODS

Study Sites

Phytoplankton populations were sampled from the depths of peak biomass in the east lobe of Lake Bonney (4.5 and 17 m), Lake Hoare (10 m), and Lake Fryxell (8.5 m). Biomass ranged from 0.7 to 6.6 mg Chl m^{-3} [*Lizotte and Priscu,* 1994]. More extensive physical, chemical and biological descriptions of these lakes are given elsewhere in this volume [*Lizotte and Priscu, Spigel and Priscu, Howard-Williams et al.,* this volume]. Phytoplankton are typically dominated by phytoflagellates in all lakes. In Lake Bonney, the assemblage is dominated by a cryptophyte (*Chroomonas lacustrus*) under-ice (4.5 to 8 m) and a chlorophyte (*Chlamydomonas subcaudata*) in the deep layer (17 to 20 m) [*Koob and Leister, 1972; Parker et al., 1982; Sharp 1993*]. Based on HPLC pigment studies, cryptophytes also dominated the depths sampled at Lake Hoare (10 m) and Lake Fryxell (8.5 m) [*Lizotte and Priscu,* this volume; see also *Vincent,* 1981b]. All samples were collected from October through December, 1990.

Fluorescence Measurements

A 10 liter sample was taken using a Niskin bottle through a drill hole in the ice cover and stored in the dark at ambient temperature. After at least 30 min of dark-adaptation, a 1 liter aliquot was placed in a clear polycarbonate container and the Lu_{683} (Chl fluorescence) sensor of a Biospherical Instruments PNF-300 profiling natural fluorometer was positioned directly over the uncovered top of the container. The container was then illuminated by a halogen projector (500W) source filtered through two blue-green (Corning 4-97) filters and one or more nickel screen neutral density filters to vary overall light intensity. The spectral composition of the actinic irradiance was measured with a MER-1000 spectroradiometer as described by [*Lizotte and Priscu*, 1992b] and was similar to underwater irradiance in Lake Bonney [*Lizotte and Priscu*, 1992b] (Figure 1). The intensity of actinic irradiance was measured in the center of container using a scalar quantum sensor (Biospherical Instruments QSL-100) for PAR (400-700 nm). The PNF-300 output was sampled once per second; instrument response time (90% of step change) was 3.5 seconds. Chlorophyll fluorescence was corrected for a background measured on filtered lake water and then scaled by the factor of 4π divided by scalar irradiance to estimate a relative fluorescence yield (m^{-1}), uncorrected for the presence of the air-water interface.

Maximum quantum yield of PSII photochemistry (ϕ_m) was also measured on samples after 30 min dark adaptation using a Turner Designs Model 10 fluorometer. Sample (2–3 ml) was placed in a cuvette and an initial reading was taken as soon as a stable reading was attained (e.g., within 10 s). The cuvette was then removed from the fluorometer and the inhibitor dichlorophenyl-dimethyl urea (DCMU), which blocks PSII electron transport, was added from a stock solution of 325 mM in 95% ethanol. The final concentration was 10 μM as previously described [*Vincent, 1981a; 1981b*]. The sample was returned to the fluorometer and a second reading (F_m) was taken at a higher stable emission level attained within 30 s. These two measurements were used to calculate ϕ_m according to the formula $(F_m - F_0)/F_m$ ([cf., *Vincent* [1981a]). All field measurements were performed in dim light (< 1 μmol photons m^{-2} s^{-1}) and at ambient air temperature (about 0°C).

Additional laboratory measurements were made with a PAM fluorometer (Walz, Effeltrich, Germany) using the basic fiber optic setup as described [*Schreiber et al.*, 1986]. The tip of the fiber optic bundle of the PAM was placed next to a Pyrex flask containing a culture of *Thalassiosira pseudonana* (Clone 3H, CCMP) with a biomass of 200 mg Chl m^{-3}. The flask was illuminated with irradiance from a halogen light source filtered through a neutral density screen to obtain a range of intensities. Scalar quantum irradiance in the flask was measured with a Biospherical Instruments QSL-100 light meter.

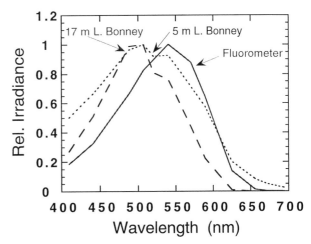

Fig. 1. Spectral downwelling irradiance normalized to maximum irradiance for two depths in Lake Bonney (December 10, 1990 measurements redrawn from *Lizotte and Priscu* [1992b]), and for the actinic illumination used for laboratory measurements of fluorescence transients using the PNF-300. The profile of spectral irradiance at 17 m is plotted with long dash, at 5 m with short dashes, and of the actinic irradiance with a solid line. The actinic illumination consists of a halogen (500W) source filtered through two blue-green (Corning 4-97) filters. In all cases peak irradiance was in the 500 to 550 nm range.

RESULTS

Maximum Quantum Yield of PSII

The vertical stratification and differentiation characteristic of the phytoplankton community in the dry valley lakes also applies to differences in the maximum quantum yield of PSII photochemistry (ϕ_m). In the east lobe of Lake Bonney during October through early December 1990, ϕ_m was significantly lower in the assemblage immediately below the ice (4.5 m) than in the deep chlorophyll (Chl) maximum (15–18 m) (Figure 2). In contrast, ϕ_m was highest near the surface in Lake Hoare and was mostly lower at depth including the biomass peak near 10 m (Figure 2). In Lake Fryxell, only the depth of the biomass peak (8.5 m) was sampled for fluorescence assay. At this depth, ϕ_m was 0.64. The Lake Fryxell assemblage was the only assemblage with a ϕ_m in the range expected for "healthy" populations or cultures (0.6 to 0.7) [*Büchel and Wilhelm*, 1993; *Geider et al.*, 1993].

Time Course of Fluorescence Yield

An example of the time-dependent variation in fluorescence yield for shallow (4.5 m) phytoplankton

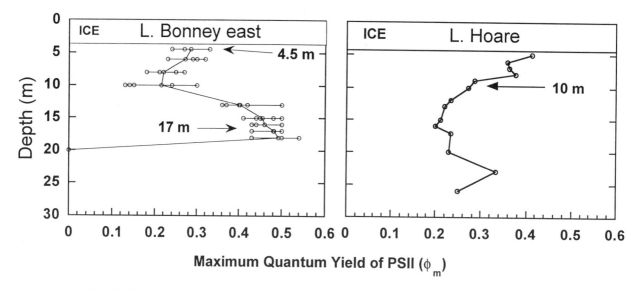

Fig. 2. Profiles of the maximum quantum yield of PSII (ϕ_m) as estimated from the ratio ($F_m -$ F_0)/F_m of fluorescence measured in a Turner Designs fluorometer in absence (F_0) and presence (F_m) of the photosynthetic inhibitor dichlorophenyl-dimethyl urea (DCMU). Composite of five profiles sampled in the east lobe of Lake Bonney during October through December 1990 (line connects means) and a profile from Lake Hoare on December 9, 1990. Arrows indicate position of chlorophyll maxima sampled for fluorescence transient analysis. All samples were dark adapted at least 30 min before measurement. Horizontal lines in the upper portion of each panel denote ice thickness.

from Lake Bonney is presented in Figure 3. At steady, low irradiance (<6 μmol photons m^{-2} s^{-1}), light limiting for photosynthesis [*Lizotte and Priscu*, 1992a], fluorescence yield was constant. However at irradiances near or above the threshold of light saturation of photosynthesis (I_k, 15–40 μmol photons m^{-2} s^{-1}, [*Lizotte and Priscu*, 1992a]), fluorescence rose to a peak and then rapidly declined to a new steady state (F_s) after about five minutes of illumination. The initial peak is called the Kautsky induction and reflects PSII dynamics on time scales of seconds [*Büchel and Wilhelm*, 1993]. Of particular interest is the steady state F_s level attained after the induction and whether this is lower (more quenched) compared to the F_s level in low irradiance. For the 4.5 m samples from Lake Bonney, F_s was progressively lower as irradiance increased above 10 μmol photons m^{-2} s^{-1} (Figure 3).

Also shown in Figure 3 is the fluorescence transient that occurs after addition of DCMU. There is an immediate fluorescence yield increase from F_s to a maximum yield under illuminated conditions (F'_m). The latter yield continues to rise as illumination continues in the presence of DCMU, indicating a steady reversal in non-photochemical quenching when

photosynthetic electron transport is blocked. At increasing irradiance, there is a decline in both F_s and F'_m as well as in the difference F'_m- F_s, the steady state variable fluorescence.

A different type of time course was exhibited by the deep (17 m) assemblage in Lake Bonney (Figure 4). The reduction in F_s is not as rapid as at 4.5 m, and addition of DCMU leads to a rapid reversal of quenching. This progressive rise in fluorescence after addition of DCMU is so rapid that it was not clear what F'_m would have been immediately post addition, and so a steady-state variable fluorescence is difficult to estimate. The 17 m assemblage was also unusual in that, after DCMU addition, fluorescence first increased and then decreased again. This implies that some portion of the quenching was inducible even in the presence of DCMU. This decrease was not observed in the Lake Hoare and Lake Fryxell samples, even though they also had rapid increases in F' with continued exposure in the presence of DCMU (data not shown).

The fluorescence transients defined by the PNF for phytoplankton from the dry valley lakes allow for a systematic examination of the dependence of fluorescence yield on incident irradiance (F_s versus I). This

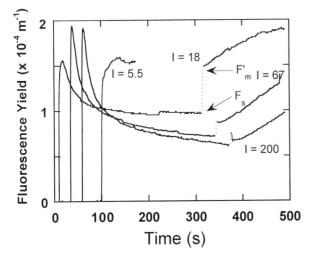

Fig. 3. Time course of relative fluorescence yield for Lake Bonney phytoplankton sampled at 4.5 m on November 20, 1990. Measurements were made with four separate 1 1 aliquots exposed to scalar irradiances of 5.5, 18, 67, and 200 μmol photons m^{-2} s^{-1}; the corresponding trace is labeled with the actinic intensity. For clarity start times are offset 25 seconds. The discontinuity (dotted line) at the elapsed time of 300 seconds corresponds to the addition of DCMU at a final concentration of 10 μM. As intensity increases, the steady state yield (F_s) and DCMU enhanced yield (F'_m) is progressively lower. There is a 10 s pause in data acquisition during the addition of the DCMU which is not shown on the plot.

type of study has rarely been done for any natural population of phytoplankton, much less in the dry valleys, despite "natural fluorescence" being advocated as an indicator of phytoplankton photosynthesis [*Kiefer et al.*, 1989]. A summary of the irradiance-dependent variation of F_s at 4.5 m in Lake Bonney is shown in Figure 5. Experiments from different dates were combined by normalizing to average yield for irradiance less than 10 μmol photons m^{-2} s^{-1}. A consistent decrease in F_s relative to low-light levels (quenching) occurred for all irradiances greater than 10 μmol photons m^{-2} s^{-1}. Based on the fit to a rectangular hyperbola, F_s reached one-half of the average low light value at approximately 80 μmol photons m^{-2} s^{-1} (Figure 5). While there are few other observations to compare with this, the induction of quenching at such a low irradiance appears to be unprecedented. *Chamberlin et al.* [1990] combined data from various open ocean and coastal marine environments and, though there is a large amount of scatter in their analysis, the data indicate that quenching of F_s only occurs at "high" irradiance (e.g., > 500 μmol photons m^{-2} s^{-1}).

Other published F_s versus I curves, which are mostly for cultures, also show no F_s quenching until irradiance is 10 or 100 times higher than the threshold in Lake Bonney. An example result is shown in Figure 5, the F_s versus I curve determined by PAM fluorometry for a culture of *Thalassiosira pseudonana* grown at 170 μmol photons m^{-2} s^{-1}. In this culture, F_s increased until irradiance exceeded 165 μmol photons m^{-2} s^{-1}. Quenching below the average F_s in low light did not occur until irradiance exceeded 300 μmol photons m^{-2} s^{-1}. The overall pattern of level or increasing yield at lower irradiance, followed by decreased yield at higher irradiance, is similar to the F_s versus I curve of the Lake Bonney sample, however the pattern occurs at irradiances about 30 times higher in the *T. pseudonana* culture (Figure 5). Similar results have been reported for a related diatom species, *T. weisflogii* [*Falkowski et al.*, 1986]. *Ting and Owens* [1993] measured the steady-state fluorescence parameters of *Phaeodactylum tricornutum* grown at 100 μmol photons m^{-2} s^{-1}. They did not report F_s per se, but reported that significant increases in non-photochemical quenching only occurred at irradiances exceeding 400 μmol photons m^{-2} s^{-1}.

Such high thresholds for fluorescence quenching are not only a feature of diatom cultures. In measurements

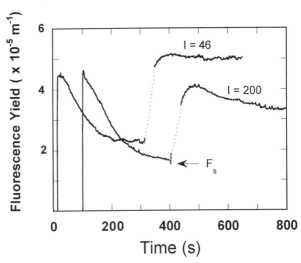

Fig. 4. Time course of relative fluorescence yield for Lake Bonney phytoplankton sampled at 17 m on November 22, 1990. Measurement conditions as in Figure 3, except only scalar irradiances of 46 and 200 μmol photons m^{-2} s^{-1} are shown. The discontinuity corresponding to DCMU addition is shown with a dotted line. Fluorescence yield increased rapidly with continued illumination after DCMU addition, therefore no attempt was made to back extrapolate to a F'_m yield.

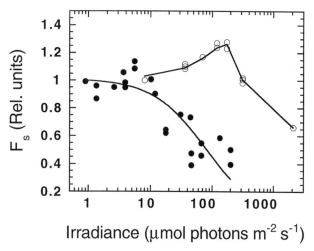

Fig. 5. Relative fluorescence yield (F_s) as a function of scalar irradiance for Lake Bonney phytoplankton (4.5 m) sampled during November and December, 1990 (closed circles), and of *Thalassiosira pseudonana* (Clone 3H) grown at 170 μmol photons m^{-2} s^{-1} (open circles) as measured with a PAM fluorometer. Note log scale for irradiance, fluorescence yield is normalized to the average low light (I < 10 μmol photons m^{-2} s^{-1}) for each sample. The data from Lake Bonney was fit to the function $F_s = 1/(1 + kI)$ using nonlinear regression, the estimate of k was 0.012 with a standard error of 0.003 ($R^2 = 0.70$), implying a light intensity of 80 μmol photons m^{-2} s^{-1} (with approximate standard error of ± 25) for 50% quenching of F_s.

with cultures of the chlorophyte alga *Ankistrodesmus braunii* growing at 80 μmol photons m^{-2} s^{-1}, *Schreiber et al.* [1995b] found that F_s yield increased as irradiance was increased to 600 μmol photons m^{-2} s^{-1}. Using blue-light illumination similar to that used in the present study, *Falkowski et al.* [1988] found that F_s increased up to an intensity of 60 μmol photons m^{-2} s^{-1} in cultures of *Chlorella vulgaris* that were low light acclimated (growth at 40 μmol photons m^{-2} s^{-1}). The increase in F_s was eliminated when irradiance was supplemented with far-red (photosystem I absorbed) irradiance, and in both cases quenching only occurred when irradiance exceeded 165 μmol photons m^{-2} s^{-1}.

A low threshold for fluorescence quenching was also observed in other dry valley phytoplankton assemblages. The 17 m assemblage from Lake Bonney and the 10 m assemblage from Lake Hoare exhibited similar F_s versus I curves, with significant quenching at irradiance greater than 10 μmol photons

m^{-2} s^{-1} (Figure 6). At irradiances between approximately 1 and 6 μmol photons m^{-2} s^{-1}, fluorescence appeared to increase. Variability in the data precludes a definitive assignment of curve shape, however all curves from Lake Bonney and Lake Hoare have higher yields around 5 μmol photons m^{-2} s^{-1}. F_s decreases to one-half the low light level at 61 μmol photons m^{-2} s^{-1} (Lake Hoare) and 138 μmol photons m^{-2} s^{-1} (17 m Lake Bonney).

The Lake Fryxell assemblage is the one exception to the general pattern of low quenching thresholds. These phytoplankton exhibited a general increase in F_s over the range 1 to 100 μmol photons m^{-2} s^{-1} (Figure 6), a response that resembles the previously cited culture studies.

The large decreases in F_s in Lake Bonney and Lake Hoare suggest induction of non-photochemical quenching at extremely low irradiances. This conclusion can be confirmed directly for the 4.5 m assemblage in Lake Bonney through analysis of F'_m. In the

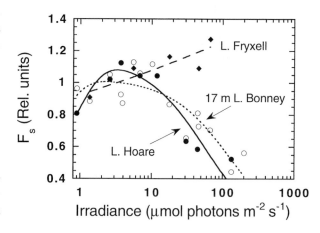

Fig. 6. Relative fluorescence yield (F_s) as a function of scalar irradiance for Lake Bonney phytoplankton (17 m) sampled during November and December, 1990 (open circles), 10 m Lake Hoare (closed circles), and 8.5 m Lake Fryxell (closed diamonds), both sampled in December, 1990. Fluorescence yield is normalized to the average F_s in low light (I < 10 μmol photons m^{-2} s^{-1}), The Lake Hoare and 17 m Lake Bonney data were fit to the function $F_s = f(I)/(1+kI)$, where f(I) is a saturating function F_s (max)$(1-\exp(-\alpha I))$ and F_s(max) is peak F_s, and α is a saturation parameter. Based on the estimated k's, the corresponding estimated irradiances for 50% quenching from F_s(max) are 61 μmol photons m^{-2} s^{-1} (approximate standard error ± 20) for Lake Hoare ($R^2 = 0.85$) and 138 μmol photons m^{-2} s^{-1} (approximate standard error ± 40) for 17 m, Lake Bonney ($R^2 = 0.73$).

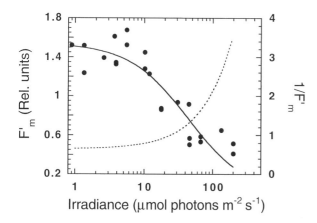

Fig. 7. Maximum fluorescence yield (F'_m) as a function of scalar irradiance for Lake Bonney phytoplankton (4.5 m) sampled during November and December, 1990 (solid line, left hand axis). Fluorescence yield is normalized to the average F_s in low light ($I < 10$ μmol photons m^{-2} s^{-1}) in each case. The data were fit to the function $F'_m = F'_m(max)/(1 + kI)$. The corresponding estimate irradiances for 50% quenching from $F'_m(max)$ are 44 μmol photons m^{-2} s^{-1} (approximate standard error ± 9, $R^2 = 0.83$). The fitted curve was then used to estimate $1/F'_m$ (dotted line, right hand axis), a measure of non-photochemical quenching.

case of an active fluorometer, such as the PAM, F'_m is measured concurrently with F_s by applying a short duration, high intensity flash [*Schreiber et al.*, 1995a]. However pulsed illumination of the large volume used for the PNF measurements was not possible; therefore F'_m was obtained by blocking the PSII acceptor side with DCMU. In samples from 4.5 m in Lake Bonney, addition of DCMU led to a rapid rise to F'_m followed by a slower secondary increase. The F'_m yield is readily distinguished since the secondary increase was comparatively slow (Figure 3). The decrease in F'_m with irradiance for the 4.5 m Lake Bonney assemblage was stronger than the decrease in F_s (Figure 7). A decrease to 50% of the low light F_m occurred at approximately 50 μmol photons m^{-2} s^{-1}. The relative increase in non-photochemical quenching is estimated as $1/F'_m$ [*Havaux et al.*, 1991], which shows a sharp increase at irradiances exceeding 10 μmol photons m^{-2} s^{-1}. Non-photochemical quenching is more conventionally estimated by the ratio $1 - (F'_m - F'_0)/(F_m - F_0)$ [*Büchel and Wilhelm*, 1993], however this could not be implemented since F'_0 (F_0 immediately after cessation of actinic illumination) cannot be measured after the addition of DCMU.

The post-illumination level of DCMU-induced fluorescence was also measured in the 17 m Lake Bonney (Figure 4), Lake Hoare, and Lake Fryxell assemblages (data not shown). In all cases the fluorescence immediately after addition was lower than the F_m attained under low light (data not shown). However the secondary increase was much more rapid in these samples. Thus F'_m could not be quantitatively determined in these cases. Qualitatively, quenching of F'_m was apparent for irradiance greater than 10 μmol photons m^{-2} s^{-1} for 17 m in Lake Bonney and Lake Hoare, and greater than 20 μmol photons m^{-2} s^{-1} in Lake Fryxell.

A primary motivation for simultaneous measurement of F'_m and F_s is the estimation of ϕ_p, where $\phi_p = (F'_m - F_s)/F'_m$. The parameter ϕ_p is interpreted as the quantum yield of PSII photochemistry, also known as the Genty yield [*Genty et al.*, 1989]. PAM measurements of ϕ_p been shown to be highly correlated with the quantum yield of photosynthesis in leaves [*Genty et al.*, 1989] and algal cultures [*Kroon et al.*, 1993]. Photosynthesis calculated from ϕ_p measured by a related type of instrument, the "pump and probe" fluorometer, was correlated with ^{14}C based estimates of primary production in the northwest Atlantic [*Kiefer and Reynolds*, 1992; *Kolber and Falkowski*, 1993]. Apart from these latter measurements, little is known about how closely ϕ_p is correlated with the quantum yield of photosynthesis in natural assemblages of phytoplankton. The ϕ_p versus irradiance relationship for phytoplankton at 4.5 m in Lake Bonney is shown in Figure 8. A single curve was fit to the composite data based on the assumption that quantum yield of PSII would be proportional to

$$(1 - e^{(-I/I_k)})/I$$

where I_k is irradiance for the onset of light saturation of photosynthesis. This analysis suggested an overall I_k of 30 μmol photons m^{-2} s^{-1} for phytoplankton from 4.5 m in Lake Bonney. An independent estimate of I_k was made using photosynthesis (carbon assimilation) versus irradiance curves [*Lizotte and Priscu*, 1992a; *Lizotte and Priscu*, 1994]. The I_k of the phytoplankton from 4.5 m in the east lobe of Lake Bonney ranged from 23 to 33 (average 28) during November and December 1990. Thus there is good agreement between the light-saturation characteristics of PSII photochemistry and photosynthesis in this assemblage.

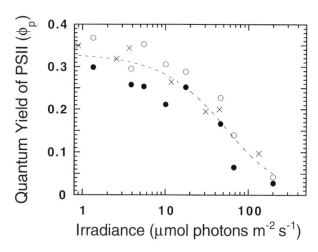

Fig. 8. Quantum yield of PSII photochemistry (ϕ_p) in relation to incident irradiance for phytoplankton at 4.5 m in Lake Bonney for three sample dates: November 20 (open circles), November 21 (solid circles), and December 9 (x), 1990. Data from all three dates were used to fit the function M(1–exp $(-I/I_k))/$ I, where M is an arbitrary scaling constant (effectively the molar ratio of incident to absorbed photons). The fitted value for M was 10, and I_k was 30.5 µmol photons m^{-2} s^{-1} (approximate standard error ± 5.1, R^2 = 0.87).

DISCUSSION

Quenching Analysis Using Lu_{683}

We report here a novel application of a natural fluorescence (Lu_{683}) sensor for determination of fluorescence quenching parameters in natural populations of phytoplankton. Much of the debate about the relationship between Lu_{683} and rates of photosynthesis centers on how to properly adjust for fluorescence quenching [*Chamberlin et al.*, 1990; *Cullen and Lewis*, 1995; *Kiefer and Reynolds*, 1992; *Lizotte and Priscu*, 1994]. Published results on the relationship of fluorescence yield (F_s) to ambient irradiance show much scatter, possibly due to physiological variations between the different populations included in the analysis. Improved precision might be obtained using field measurements of F_s versus I curves for specific assemblages of interest. Recently this has been facilitated by the development of active fluorometers using low measuring irradiance, i.e., the PAM with a high-sensitivity cuvette system [*Schreiber*, 1994] or the Fast Repetition Rate (FRR) fluorometer [*Falkowski and Kolber*, 1995]. However if this type of sophisticated instrumentation is lacking, useful laboratory

measurements can be with a Lu_{683} sensor as described in this report. At biomass levels below 0.5 mg Chl m^{-3}, a larger volume of sample water may needed to obtain sufficient signal. Other types of active excitation field fluorometers, such the Sea-Tech fluorometer, generally use too high excitation irradiance to measure a true F_s yield [*Neale et al.*, 1989].

Implications for Acclimation of the Photosynthetic Apparatus

Viewed strictly from the standpoint of growth in a low-light environment, the phytoplankton of dry valley lakes present a paradoxical picture. This is best seen with the relatively well-studied assemblages from the east lobe of Lake Bonney. In some respects, these assemblages appear to be low-light acclimated, for example in their extreme sensitivity to UV inhibition [*Neale et al.*, 1994] and low I_k [*Lizotte and Priscu*, 1992a]. However I_k is comparable to that measured in other Antarctic habitats (see review in *Cabrera and Montecino* [1990]), and lower I_k values have been measured [*Priscu et al.* 1987; *Lizotte and Priscu*, 1992a]. Indeed there are several aspects that seem contrary to extreme low light adaptation. Antenna sizes, based on Chl:P_{700} ratios, are not particularly high [*Neale and Priscu*, 1995]. Also, α, the initial slope of P^B versus I curve, and the quantum yield of photosynthesis are well below the highest values measured for phytoplankton in low-light environments [*Lizotte and Priscu*, 1994].

This seemingly paradoxical picture is perhaps better understood if acclimation of the photosynthetic apparatus is viewed more as a response to excitation pressure [*Maxwell et al.*, 1995a] than to the light environment, per se. The results of the fluorescence quenching analysis on dry valley lakes phytoplankton show that irradiance even slightly above the normal range of in situ irradiance results in a large drop in the quantum yield of PSII photochemistry (ϕ_p) and an increase in non-photochemical quenching. Though excitation pressure (= 1 – q_p, sensu *Maxwell et al.*, [1995a]) was not quantitatively estimated (F'$_0$ could not be measured), the drop in ϕ_p implies that excitation pressure was high. *Maxwell et al.* [1995a] interpreted induction of non-photochemical quenching at low irradiance as indicative of acclimation to low excitation pressure. Characteristics of acclimation to low excitation pressure also include higher sensitivity to photoinhibition and lower accumulation of xanthophyll quenching pigments, which are also properties of the

Lake Bonney assemblage [*Neale et al.*, 1994; *Neale and Priscu*, 1995; *Lizotte and Priscu*, this volume]

These characteristics were especially evident in the 4.5 m assemblage. Midday irradiance just below the ice is sufficient to nearly saturate rates of photosynthesis and induce non-photochemical quenching. Interestingly, this assemblage is acclimated to low excitation pressure, even though it is apparently often on the verge of experiencing high excitation pressure. Apparently, the maximum sustainable rate of electron transport only slightly exceeds the rate occurring under mean in situ irradiance, a condition consistent with the extremely low maximum rates of photosynthesis (PB_m) in this assemblage [*Lizotte and Priscu*, 1992a]. At least two factors appear to be limiting overall photosynthetic capacity: 1) chronic nutrient stress (at least during the November–December period [*Priscu*, 1995]); and 2) low temperature, i.e., 0°C just below the ice cover. These factors may also explain the low maximum quantum yield of PSII observed in this assemblage, though it unknown whether this arises because of a direct effect on PSII structure [*Kolber et al.*, 1988], a limitation of the PSII repair cycle with accumulation of inactive units, [*Smith et al.*, 1990], or some other mechanism.

The threshold light intensities for fluorescence quenching in the 17 m assemblage from Lake Bonney and the 10 m assemblage in Lake Hoare are similar to the 4.5 m assemblage. However these assemblages experience generally lower light intensities than the 4.5 m assemblage. Thus there is a greater "safety margin" between in situ irradiance and "high" irradiances in these assemblages. The margin is probably made possible by a more ample nutrient supply to these phytoplankton, owing to their proximity to the lake nutricline [*Lizotte and Priscu*, 1994; *Priscu*, 1995] and higher water temperatures. Both PB_m and maximum quantum yield of photosynthesis are higher in these assemblages compared to 4.5 m assemblage [*Lizotte and Priscu*, 1994]. Nevertheless the maximum quantum yield of PSII (ϕ_m) was still below the level observed in "healthy" phytoplankton. This may be due to low flux rates of nutrients to these populations since supply is by molecular diffusion [*Priscu*, 1995; *Spigel and Priscu*, this volume].

The one exception to this general paradigm of photosynthetic acclimation in the dry valley lakes is the 8.5 m assemblage from Lake Fryxell. In contrast to phytoplankton in Lake Bonney and Lake Hoare, this assemblage had a high ϕ_m and a relatively high maximum quantum yield of photosynthesis [*Lizotte and Priscu*, 1994]. The biomass peak in Lake Fryxell

is also larger than in the other lakes, suggesting more robust growth. The F_s was not quenched at low intensities signifying a low excitation pressure even at irradiances above the reported I_k for this assemblage of 10 μmol photons m^{-2} s^{-1} [*Lizotte and Priscu*, 1994]. This assemblage appears to be responding like a nutrient sufficient culture, suggesting higher nutrient flux to this environment despite low ambient concentrations. A slight inconsistency with this interpretation is the reported low PB_m as determined by ^{14}C incorporation over several hour incubations [*Priscu et al.*, 1989]. One possible explanation is that these phytoplankton are able to sustain high rates of electron transport over the short exposures used for the fluorescence measurements, yet rates decrease over longer exposures (cf. [*Marra*, 1978]).

The remarkable feature of the Lake Bonney and Lake Hoare phytoplankton is that they exhibit low excitation-pressure acclimation despite the fact that a strikingly small increase in growth irradiance would apparently shift them into a high excitation-pressure state. The success of this growth strategy is undoubtedly aided by the relatively stable shade regime experienced by phytoplankton in the lakes of the McMurdo Dry Valleys. Thus the photosynthetic apparatus acclimates so as to have the maximum light harvesting capacity consistent with the requirement of low excitation pressure under growth conditions. These phytoplankton may be physiologically capable of increasing light harvesting capacity, but such an increase would quickly overwhelm cellular metabolic capacity to utilize that energy as manifested by induction of non-photochemical quenching. Overall these results lead to the hypothesis that growth in an environment of low nutrient supply, low temperature, and stable shade entails an extremely fine-tuned allocation of resources between the light harvesting (e.g., PSII) and energy utilization (e.g., Calvin cycle enzymes) components of the photosynthetic apparatus. Indeed there is evidence that excitation pressure, perhaps as manifested as a change in the chloroplast ATP/NADPH ratio [*Melis et al.*, 1985], regulates chloroplast gene expression in *Dunaliella salina* [*Maxwell et al.*, 1995b]. Further experimentation would be needed to show whether such a fine-tuned allocation is actually operating within the phytoplankton assemblages in the lakes of the McMurdo Dry Valleys.

Acknowledgments. We thank Michael Lizotte, Thomas Sharp, Robert Spigel, and Ian Forne for field assistance. P.J.N. thanks Anastasios Melis for use of laboratory facilities in the Dept. of Plant Biology, University of California at Berkeley. This work was supported by NSF grants DPP-88-20591, OPP-92-11773 and OPP-94-19423 to J.C.P.

REFERENCES

Angino, E.E., K.B. Armitage, and J.C. Tash, Physiochemical limnology of Lake Bonney, Antarctica, *Limnol. Oceanogr.*, 9, 207–217, 1964.

Barlow, R.G., M. Gosselin, L. Legendre, J.-C. Therriault, S. Demers, R.F.C. Mantoura, and C.A. Llewellyn, Photoadaptive strategies in sea-ice microalgae, *Mar. Ecol.*, 45, 145–152, 1988.

Büchel, C., and C. Wilhelm, *In vivo* analysis of slow chlorophyll fluorescence induction kinetics in algae: Progress, problems and perspectives, *Photochem. Photobio.*, 58, 137–148, 1993.

Cabrera, S., and V. Montecino, Photosynthetic parameters of the entire euphotic phytoplankton of the Bransfield Strait, summer 1985, *Polar Biol.*, 10, 507–513, 1990.

Chamberlin, W.S., C.R. Booth, D.A. Kiefer, J.R. Morrow, and R.C. Murphy, Evidence for a simple relationship between natural fluorescence, photosynthesis and chlorophyll in the sea, *Deep-Sea Res.*, 37, 951–973, 1990.

Cullen, J.J., and M.R. Lewis, Biological processes and optical measurements near the sea-surface: some issues relevant to remote sensing, *J. Geophys. Res.*, 100 (C7), 13,255–13,266, 1995.

Falkowski, P.G., and Z. Kolber, Variations in chlorophyll fluorescence yields in phytoplankton in the world oceans, *Aust. J. Plant Physiol.*, 22, 341–55, 1995.

Falkowski, P.G., Z. Kolber, and Y. Fujita, Effect of redox state on the dynamics of photosystem II during steady-state photosynthesis in eucaryotic algae, *Biochim. Biophys. Acta*, 933, 432–443, 1988.

Falkowski, P.G., and J. LaRoche, Acclimation to spectral irradiance in algae, *J. Phycol.*, 27, 8–14, 1991.

Falkowski, P.G., K. Wyman, A. Ley, and D. Mauzerall, Relationship of steady state photosynthesis to fluorescence in eucaryotic algae, *Biochim. Biophys. Acta*, 849, 183–192, 1986.

Geider, R.J., R.M. Greene, Z. Kolber, H.L. MacIntyre, and P.G. Falkowski, Fluorescence assessment of the maximum quantum efficiency of photosynthesis in the western North Atlantic, *Deep-Sea Res.*, 40, 1204–1224, 1993.

Genty, B., J.M. Briantais, and N. Baker, The relationship between the quantum yield of photosynthetic electron transport and quenching of chlorophyll fluorescence, *Biochim. Biophys. Acta*, 990, 87–92, 1989.

Havaux, M., R.J. Strasser, and H. Greppin, A theoretical and experimental analysis of the qP and qN coefficients of chlorophyll fluorescence quenching and their relation to photochemical and nonphotochemical events, *Photosyn. Res.*, 27, 41–55, 1991.

Howard-Williams, C., A.-M. Schwarz, and I. Hawes, Optical properties of dry valley lakes, *Antarct. Res. Ser.*, this volume.

Kiefer, D.A., W.S. Chamberlin, and C.R. Booth, Natural fluorescence of chlorophyll-*a*: relationship to photosynthe-

sis and chlorophyll concentration in the western South Pacific gyre, *Limnol. Oceanogr.*, 34, 868–881, 1989.

Kiefer, D.A., and R.A. Reynolds, Advances in understanding phytoplankton fluorescence and photosynthesis, in *Primary productivity and biogeochemical cycles in the sea*, edited by P.G. Falkowski, and A.D. Woodhead, pp. 155–174, Plenum Press, New York, 1992.

Kolber, Z., and P.G. Falkowski, Use of active fluorescence to estimate phytoplankton photosynthesis *in situ*, *Limnol. Oceanogr.*, 38, 1646–1665, 1993.

Kolber, Z., J.R. Zehr, and P.G. Falkowski, Effects of growth irradiance and nitrogen limitation on photosynthetic energy conversion in photosystem II, *Plant Physiol.*, 88, 923–929, 1988.

Koob, D.D., G.L. Leister, Primary productivity and associated physical, chemical and biological characteristics of Lake Bonney, a prennially ice-covered lake in Antarctica, *Antarct. Res. Ser.*, 20, 51–68, 1972

Kroon, B., B.B. Prézelin, and O. Schofield, Chromatic regulation of quantum yields for photosystem II charge separation, oxygen evolution, and carbon fixation in *Heterocapsa pygmaea* (Pyrrophyta), *J. Phycol.*, 29, 453–462, 1993.

Larkum, A., and J. Barrett, Light-harvesting processes in algae, *Adv. Bot. Res.*, 10, 1–219, 1983.

Lizotte, M.P., and J.C. Priscu, Pigment analysis of the distribution, succession, and fate of the phytoplankton in the McMurdo Dry Valley lakes of Antarctica, *Antarct. Res. Ser.*, this volume.

Lizotte, M.P., and J.C. Priscu, Photosynthesis-irradiance relationships in phytoplankton from the physically stable water column of a perennially ice-covered lake (Lake Bonney, Antarctica), *J. Phycol.*, 28, 179–185, 1992a.

Lizotte, M.P., and J.C. Priscu, Spectral irradiance and bio-optical properties in perennnially ice-covered lakes of the dry valleys (McMurdo Sound, Antarctica), *Antarct. Res. Ser.*, 57, 1–14, 1992b.

Lizotte, M.P., and J.C. Priscu, Natural fluorescence and quantum yields in vertically stationary phytoplankton from perennially ice-covered lakes, *Limnol. Oceanogr.*, 39 (6), 1399–1410, 1994.

Lizotte, M.P., T.R. Sharp, J.C. Priscu, Phytoplankton dynamics in the stratified water column of Lake Bonney, Antarctica 1. Biomass and productivity during the winter-spring transition, *Polar Biol.*, 16, 155–162, 1996.

Marra, J., Effect of short-term variation in light intensity on photosynthesis of a marine phytoplankter: a laboratory simulation study, *Mar. Biol.*, 46, 191–202, 1978.

Maxwell, D.P., D.E. Laudenbach, and N.P.A. Huner, Redox regulation of light-harvesting complex II and cab mRNA abundance in *Dunaliella salina*, *Plant Physiol.*, 109, 787–795, 1995b.

Maxwell, D.P., S. Falk, and N.P.A. Huner, Photosystem II excitation pressure and development of resistance to photoinhibition. I. Light-harvesting complex II abundance and zeaxanthin content in *Chorella vulgaris*, *Plant Physiol.*, 107, 687–694, 1995a.

Melis, A., A. Manodori, R.E. Glick, M.L. Ghirardi, S.W. McCauley, and P.J. Neale, The mechanism of photosynthetic membrane adaptation to environmental stress conditions: a hypothesis on the role of electron-transport capacity and of ATP/NADPH pool in the regulation of thylakoid membrane organization and function, *Physiol. Veg.*, 23 (5), 757–765, 1985.

Neale, P.J., J.J. Cullen, and C.M. Yentsch, Bio-optical inferences from chlorophyll a fluorescence: What kind of fluorescence is measured in flow cytometry?, *Limnol. Oceanogr.* 34, 1739–1748, 1989.

Neale, P.J., M.P. Lesser, and J.J. Cullen, Effects of ultraviolet radiation on the photosynthesis of phytoplankton in the vicinity of McMurdo Station (78°S), in *Ultraviolet radiation and biological research in Antarctica*, edited by C.S. Weiler and P.A. Penhale., pp. 125–142, 1994.

Neale, P.J., and J.C. Priscu, The photosynthetic apparatus of phytoplankton from a perennially ice-covered Antarctic lake: Acclimation to an extreme shade environment, *Plant Cell Physiol.*, 36, 253–263, 1995.

Parker, B.C., G.M. Simmons, K.G. Seaburg, D.D. Cathey and F.C.T. Allnutt, Comparative ecology of plankton communities in seven Antarctic oasis lakes, *J. Plankton Res.*, 4, 271–285.

Priscu, J.C., Variation in light attenuation by the permanent ice cap of Lake Bonney during spring and summer, *Antarct. J. U.S.*, 26, 223–224, 1992

Priscu, J.C., Phytoplankton nutrient deficiency in lakes of the McMurdo dry valleys, Antarctica, *Freshwater Biol.*, 34, 215–227, 1995

Priscu, J.C., L.R. Priscu, W.F. Vincent, and C. Howard-Williams, Photosynthate distribution by microplankton in permanently ice-covered Antarctic desert lakes, *Limnol. Oceanogr.*, 32, 260–270, 1987.

Schreiber, U., New emitter-detector-cuvette assembly for measuring modulated chlorophyll fluorescence of highly diluted suspensions in conjunction with the standard PAM fluorometer, *Zeitschrift für Naturforschung*, 49c, 646–656, 1994.

Schreiber, U., H. Hormann, C. Neubauer, and C. Klughammer, Assessment of photosystem II photochemical quantum yield by chlorophyll fluorescence quenching analysis, *Aust. J. Plant Physiol.*, 22, 209–20, 1995b.

Schreiber, U., T. Endo, M. Hualing, and K. Asada, Quenching analysis of chlorophyll fluorescence by the saturation pulse method: Particular aspects relating to the study of eukaryotic algae and cyanobacteria, *Plant Cell Physiol.*, 36 (5), 873–882, 1995a.

Schreiber, U., U. Schliwa, and B. Bilger, Continuous recording of photochemical and nonphotochemical chlorophyll fluorescence quenching with a new type of modulation fluorometer, *Photosyn. Res.*, 10, 51–62, 1986.

Sharp, T.R. *Temporal and spatial variation of light, nutrients and phytoplankton production in Lake Bonney, Antarctica*, Msc. thesis, Montana State University, p. 166, 1993.

Smith, B.M., P.J. Morrissey, J.E. Guenther, J.A. Nemson, M.A. Harrison, J.F. Allen, and A. Melis, Response of the photosynthetic apparatus in *Dunaliella salina* (Green Algae) to irradiance stress, *Plant Physiol.*, 93, 1433–1440, 1990.

Spigel, R.H., and J.C. Priscu, Physical limnology of the McMurdo Dry Valley Lakes, *Antarct. Res. Ser*, this volume.

Ting, C.S., and T.G. Owens, Photochemical and nonphotochemical fluorescence quenching processes in the diatom *Phaeodactylum tricornutum*, *Plant Physiol.*, 101, 1323–1330, 1993.

Vincent, W.F., Photosynthetic capacity measured by DCMU-induced chlorophyll fluorescence in an oligotrophic lake., *Freshwater Biol.*, 11, 61–78, 1981a.

Vincent, W.F., Production strategies in Antarctic inland waters: phytoplankton eco-physiology in a permanently ice-covered lake, *Ecology*, 62, 1215–1224, 1981b.

P. J. Neale, Smithsonian Environmental Research Center, P O Box 28, Edgewater, MD 21037

J. C. Priscu, Department of Biological Sciences, Montana State University, Bozeman, MT, 59717

(Received October 1, 1996;
accepted April 24, 1997)

PROTOZOOPLANKTON AND MICROZOOPLANKTON ECOLOGY IN LAKES OF THE DRY VALLEYS, SOUTHERN VICTORIA LAND

Mark R. James

National Institute of Water and Atmospheric Research Ltd., Christchurch, New Zealand

Julie A. Hall

National Institute of Water and Atmospheric Research Ltd., Hamilton, New Zealand

Johanna Laybourn-Parry

Department of Physiology and Environmental Science, University of Nottingham, Loughborough, UK

Antarctic lakes are extreme environments with unique characteristics including lack of wind-driven mixing, low planktonic abundance and species diversity, and dominance by microbial communities. There have been several studies of the phytoplankton community structure and distribution in McMurdo Dry Valley lakes, but very little is known of the protozoo- and microzooplankton communities in these lakes. This chapter combines published data from limited studies to date with new distribution and experimental data on Lakes Vanda and Bonney describing the physical and chemical features which influence their distribution and trophic interactions. Diversity and abundance of ciliated protozoa was greatest in the more productive lakes (Bonney and Fryxell). New data on seasonal distribution of protozoa in Lake Vanda showed there was a small overwintering population dominated by *Euplotes* and *Askenasia*. *Askenasia* and other members of the Didinidae dominated the spring increase in abundance with *Euplotes* increasing in abundance in the austral summer, potentially in response to increased bacterial abundance. Four distinct communities were found in the permanently stratified Lake Vanda: the community just below the ice in the upper euphotic zone consisted of *Askenasia* and *Urotricha*; the lower isohaline cell contained several taxa including *Askenasia*, *Monodinium*, and a spirotrich; heliozoa dominated the transition zone to saline water; and the community at 60 m was almost exclusively the bacterivorous *Euplotes*. Ciliated protozoa and rotifers were absent from depths >64 m and transplant experiments demonstrated these groups could not survive in the anoxic bottom waters despite their presence in similar waters in temperate lakes. Their absence is probably because of toxic chemicals and permanent anoxia. Observations in Lake Vanda suggest the structure and vertical stratification of protozoan communities are principally determined by food resources, but low abundance and grazing rates suggest they do not control phytoplankton or bacterial populations.

INTRODUCTION

The importance of the microbial plankton component in carbon flow and nutrient regeneration was first elaborated by *Pomeroy* [1974] and later expanded by *Williams* [1981] and *Azam et al.* [1983] for marine environments. More recently studies on freshwater lakes have demonstrated that microbial plankton play an equally important role in freshwater environments [e.g., *Weisse et*

al., 1990; Simek and Straskrabová, 1992; Laybourn-Parry et al., 1994; Stone et al. 1993]. The so-called microbial loop depends upon dissolved organic carbon (DOC) derived mostly from exuded photosynthate from the phytoplankton (Figure 1). In lakes, macrophytes and allochthonous inputs of DOC may also be important. This DOC provides an energy source for the bacterioplankton which are in turn grazed by heterotrophic nanoflagellates (HNAN) and some ciliates. The protists are subsequently

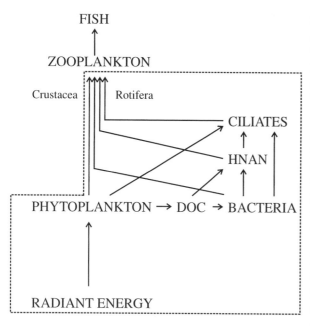

Fig. 1. Plankton community structure in lakes. Components enclosed in dashed box show the structure of McMurdo Dry Valley lake communities. DOC = dissolved organic carbon, HNAN = heterotrophic nanoflagellates.

exploited by the traditional zooplankton (crustaceans and rotifers). In some freshwater lakes, rotifers and cladocerans also exploit bacteria and compete with the protozoa for that food resource (Figure 1). Thus, in most freshwater lakes, the microbial loop is in effect a series of additional trophic levels in a complex food web. However, in extreme environments, the metazoan plankton are sparse or absent and the plankton are dominated by the microbial component [*Laybourn-Parry et al.*, 1994; 1995]. Such environments, which lack top-down control, offer a unique opportunity to study trophic interactions within the microbial plankton in the absence of predation pressure.

Less than five percent of the Antarctic continent is ice-free. These ice-free regions often contain streams and lakes formed by glacial melt. The oldest lakes (hundreds of thousands of years) are perhaps found amongst the permanently ice-covered lakes of the McMurdo Dry Valleys in southern Victoria Land. More recent fillings of some of these lakes, for example Lake Vanda, may have occurred over the last 1200 years [*Chinn,* 1993].

The extreme cold, brief growing season and isolation of the Antarctic continent have resulted in low planktonic biomass and species diversity in Antarctic lakes and a dominance by microbial foodwebs. Crustaceans, which dominate the zooplankton in temperate and tropical fresh waters, have been recorded in lakes of the Vestfold Hills

region of eastern Antarctica [*Laybourn-Parry and Marchant,* 1992; *Laybourn-Parry and Bayliss,* 1996] but they are generally rare. The few limnological studies of McMurdo Dry Valley lakes have confirmed the absence of crustacean zooplankton and dominance by protozoans. The environmental extremes experienced would be expected to accentuate adaptations which enable planktonic groups to flourish. Strong gradients have been found for chemical and physical parameters [*Vincent,* 1987] within McMurdo Dry Valley lakes which suggests a similar pattern may exist for protozoo- and microzooplankton. The potential wide spectrum of ecological niches could be expected to result in higher species diversity than observed elsewhere in Antarctica.

The chemical and physical diversity of the McMurdo Dry Valley lakes suggest the plankton community should also differ between lakes. A preliminary comparison of seven lakes in this Antarctic oasis by *Parker et al.* [1982] suggested greater plankton diversity in the more oligotrophic lakes (Vanda and Bonney-East lobe) but protozoo- and microzooplankton data was limited and no vertical profiles were reported. Lake Fryxell is one of the few lakes in the McMurdo Dry Valleys where the plankton community has been intensively studied during the austral summer and while the phytoplankton species diversity is low [*Spaulding et al.,* 1994] the protozoan community is relatively abundant and more diverse than other systems in Antarctica yet studied [*Laybourn-Parry et al., 1997*]. Lake Fryxell, however, is at the more eutrophic end of the trophic spectrum for McMurdo Dry Valley lakes (Table 1) and whether these observations apply to other lakes in the region is not known.

Unlike the lakes which have been intensively studied elsewhere in continental Antarctica (e.g., Vestfold Hills), the lakes of the McMurdo Dry Valleys are permanently ice-covered although some have temporary moats which develop for brief periods in summer. Because of the ice cover the lakes lack wind-driven turbulence but may have inflow or thermahaline convection driven turbulence. This lack of wind-driven turbulence has important implications for vertical zonation. The distribution of phytoplankton and protozooplankton plankton showed marked zonation in Lake Fryxell, probably in response to light, nutrient, salinity, and temperature gradients, which influence the food resources and niche separation [*Spaulding et al.,* 1994; *Laybourn-Parry et al., 1997*]. Whether this zonation is a feature of all McMurdo Dry Valley lakes is unknown.

This chapter presents an overview of protozoo- and microzooplankton ecology in the McMurdo Dry Valley lakes based on limited published data and also includes new distribution and experimental data collected between

TABLE 1. Physical, Chemical, and Biological Features of Selected McMurdo Dry Valley Lakes.

Lake	Area km²	Depth m	Temperature °C	Conductivity μmho cm⁻¹	Chlorophyll-a mg m⁻³ (max.)
Vanda	5.0	75	4.0–22.7	870–94 000	1.7
Fryxell	7.0	20	0–3.5	590–8600	13.2
Bonney - west lobe	1.0	40	−2.5–4.0	960–>50 000	3.3
Bonney - east lobe	2.9	40	−1.0–7.0	1000–>50 000	3.6

1992 and 1995 from Lake Vanda, Lake Bonney and Lake Fryxell. First we briefly describe the physical and chemical features of these lakes. We then describe the species diversity and abundance of protozoo- and microzooplankton and examine whether physical and chemical variables might have affected the protozoo- and microzooplankton community structure and vertical distribution. Finally we consider trophic interactions within the plankton community and potential carbon flows in this unique environment.

DESCRIPTION OF LAKES

Lake Vanda

Lake Vanda, in the Wright Valley, is probably the best known of continental Antarctic lakes. The lake exhibits unique physical characteristics with temperatures close to 4°C at the surface and 23°C at the bottom and a gradient from fresh water just below the ice to three times the conductivity of seawater in bottom waters. This chemical gradient influences the distribution of biological components. Surface waters are among the clearest in the world and the deep light penetration (extinction coefficients 0.04 m⁻¹, 0.06 m⁻¹ and 0.46 m⁻¹ for blue, green, and red light respectively) and increased availability of dissolved nutrients through diffusion from the deep anoxic layer result in deep chlorophyll maximum (DCM) at 62.5 m. Despite several reports on phytoplankton [Parker et al., 1982; Vincent and Vincent, 1982; Vincent and Howard-Williams, 1985] the protozooplankton of Lake Vanda have received little attention. Parker et al. [1982] recorded two species of protozoan ciliates, while Cathey et al. [1981], who collected samples from limnetic and littoral habitats, recorded four taxa of ciliates and one rotifer. The rotifers were recorded only in the littoral region.

Lake Bonney

Lake Bonney, in the Taylor Valley, has a much narrower range of temperatures with +2°C from just below the ice down to a depth of 10 m after which the temperature drops

exponentially to –5°C at 30 m. Conductivity increases from fresh water just below the ice (<1 mho cm⁻¹) to 85 mho cm⁻¹ at 10 m and, like Lake Vanda, has three times the conductivity of seawater at the bottom [Spigel and Priscu, this issue]. The east and west lobes are separated by a sill about 8 m below the ice surface. A chlorophyll-a maximum occurs at 13 m. Parker et al. [1982] recorded only two taxa of ciliated protozoa, but, in a more intensive study of the littoral and planktonic region, Cathey et al. [1981] recorded six ciliate taxa and three rotifer taxa, with the latter again confined to the littoral region.

Lake Fryxell

Lake Fryxell is strongly stratified with an oxycline at 7.5–9.5 m, below which the waters are anoxic. Water temperatures are close to 0°C below the ice, rising to 3.5°C at 12 m. Conductivity also increases with depth from fresh water just below the ice to brackish water at the bottom (19 m). Fryxell is the most productive lake on a volumetric basis in the McMurdo Dry Valleys with chlorophyll-a concentrations up to 10.4 mg m⁻³. Eleven ciliated protozoan taxa have been recorded by Laybourn-Parry et al. [1997]. Most taxa were confined to particular strata and changes in both the community composition and vertical distribution of taxa could occur interannually. Laybourn-Parry et al. also established that two species of Philodina (Rotifera) were also in the plankton.

SAMPLING AND ANALYTICAL PROCEDURES

Water samples were collected from Lake Fryxell (January 1992, 1994), from Lake Vanda (November/December 1994, September 1995) and Lake Bonney (December 1994). The collections in September 1995 were the first attempt to study the protozoo- and microzooplankton community in winter and determine whether the community develops in spring when light first reappears.

Samples were collected using either a 3 liter Van Dorn bottle (Lake Fryxell) or a 1.5 liter sampler (Lakes Vanda and Bonney) lowered through a hole drilled in the ice cover at a central station in Lakes Fryxell and Vanda and in each lobe of Lake Bonney. One liter water samples were fixed in either Lugol's iodine (1% final concentration) or buffered glutaraldehyde/paraformaldehyde (2% final concentration) for protozoan identification and enumeration. Identifications were also made by observations of live material. Vertical net hauls (20 μm mesh) were made in Lake Vanda to collect animals for experiments. Enumeration and identification of samples followed procedures outlined in *Hall et al.*[1993], *James* [1991], and *Laybourn-Parry et al.* [1997]. Because of the time required to count samples replicate counts were limited to only a few depths. Coefficient of variation was <20% for major taxa and total ciliate abundance.

Experimental Studies

Trophic interactions were determined in Lake Vanda by dilution experiments based on the method of *Landry and Hassett* [1982]. Preliminary observations indicated zooplankton were not found below 64 m in Lake Vanda. To test potential toxicity of the deeper waters, experiments were conducted with different ratios of deep and shallow water. These experiments were conducted on *Philodina* in the laboratory. The ratios tested were 0, 6.25, 12.5, 25, 50, and 100% dilution of 55 m water with water from 67 m giving conductivities of 4, 8.2, 12.4, 20.8, 37.5, and 71 mS cm^{-1}. Experiments were run at 4°C for 48 h and survivorship assessed by light microscopy. Field experiments for toxicity were also carried out in September 1995 using 250 ml perspex chambers with 20 μm permeable membranes to allow the exchange of water while retaining the animals. The chambers were lowered to depths of 40, 50, 55, 59 m, and at 1 m intervals between 60 and 67 m. Survivorship of rotifers was determined after three days by light microscopy. Survival was classified as living and mobile, contracted and immobile, and not contracted and immobile.

RESULTS AND DISCUSSION

Species Diversity

The lakes of the McMurdo Dry Valley region of southern Victoria Land have been noted for their lack of crustacean zooplankton and low diversity of ciliated protozoans. Our findings suggest that bdelloid rotifers belonging to the genus *Philodina* are present in the plankton. Even though bdelloid rotifers are morphologically adapted to a benthic habitat, they were recorded in Lakes Vanda, Bonney, and Fryxell (Table 2). They were not found in the anoxic bottom layer of Lake Vanda (65–75m) or Lake Fryxell (>10 m) which suggests they were either migrating from the edge of the lake, were an established part of the plankton community, or were entrained from benthos in thermohaline convection cells. There is evidence that greater inflows of water are responsible for littoral species becoming established in the plankton. This has been noted for phytoplankton by *Spaulding et al.* [1994] and may also be responsible for the establishment of rotifers in the plankton. Two species of *Philodina* have been recorded along with other rotifers in the littoral zones of the Larsemann Hills oasis lakes in eastern Antarctica [*Dartnall*, 1995].

The diversity of ciliated protozoa was greatest in the more productive lakes (Bonney (7–9 taxa) and Fryxell (10 taxa)) than the ultra-oligotrophic Lake Vanda where only 4–5 taxa were recorded. *Cathey et al.* [1981] also recorded four taxa in Lake Vanda. Such diversity is low by comparison with lower latitudes but is higher than other Antarctic freshwater lakes, e.g., Crooked Lake in the Vestfold Hills, and one of Antarctica's largest freshwater lakes, has only three species of ciliate [*Laybourn-Parry et al.*, 1991]. Most lakes in the Vestfold Hills have fewer than six taxa and are dominated by either haptorids or oligotrichs including *Strombidium*. Lakes in eastern Antarctica however are very different environments and systems comparable with McMurdo Dry Valley lakes are scarce. The community in the McMurdo Dry Valley lakes is more diverse and includes hypotrichs, choreotrichs, litostomes, and peritrichs. *Laybourn-Parry et al.* [1997] attributed the greater diversity in Lake Fryxell, compared with lakes in eastern Antarctica to the lake's age, relatively high temperatures, and a range of physicochemical conditions down the water column which offered a variety of niches and food resources. Earlier observations in McMurdo Dry Valley lakes implied plankton communities were more diverse in Lake Vanda than Lake Fryxell, but more detailed examination of new data from the present study indicates the reverse applies to ciliated protozoa, i.e., greater diversity occurs in the more productive lake. The greater diversity observed in lower latitude temperate lakes is a result of seasonal changes in physical structure (e.g., stratification and deoxygenation), the greater productivity, and episodic changes in food resources. It is interesting that *Euplotes*, a bacterivorous ciliate, was found in all McMurdo Dry Valley lakes but other bacterivores (*Vorticella*, hymenostomes) were only found in more productive Lake Bonney and Lake Fryxell. Vorticellids in lower latitude systems are common in eutrophic and mesotrophic lakes but rare or absent in oligotrophic waters [*Laybourn-Parry*, 1994].

TABLE 2. Microzooplankton Taxa Identified in Lakes of the McMurdo Dry Valleys, southern Victoria Land.

Location	L. Vanda				L. Bonney		L. Fryxell
Date	25 Nov 94	7 Dec 94	26 Dec 94	Sep 95	E	W	
Taxa							
Ciliated protozoa							
Euplotes	*	*	*	*	*	*	*
Oligotrichs	*	*			*	*	
Strombidium		*					*
Halteria					*	*	*
Askenasia	*	*	*	*	*	*	*
Monodinium	*		*		*	*	*(2 spp.)
Urotricha	*		*		*	*	*
Hymostome					*		
Bursellopsis		*(?)			*		
Vorticella					*	*	*
Bursaria							*
Chilodonella							*
Nassula							*
No. of ciliate taxa	(5)	(5)	(4)	(2)	(9)	(7)	(11)
Heliozoa	*	*	*	*	*(scarce)	*(scarce)	*
Rotifers	*	*			*(scarce)	*(scarce)	*

Fryxell data are from *Laybourn-Parry et al.*, 1995. E and W are East and West lobes of Lake Bonney.
* Denotes taxa were present.   *(?) Denotes identification preliminary

Abundance

The abundance of ciliates in the McMurdo Dry Valley lakes is comparable with Char Lake in the Arctic [*Rigler et al.*, 1974] and oligotrophic lakes in lower latitudes of the southern and northern hemispheres [*Laybourn-Parry et al.*, 1991; *James et al.*, 1995]. The higher abundance in Lake Fryxell (Table 3) is consistent with the lake being more productive than Lake Bonney or Vanda. Abundance in Lake Vanda was lower in early December than in November but increased again by late December to a maximum of 1270 l⁻¹. However peaks may have occurred at depths which were not sampled. This first record of the protozoo- and microzooplankton community for late winter in Lake Vanda showed there was a small over-wintering population of ciliates and Heliozoa (Table 3).

Planktonic Heliozoa have largely been overlooked in limnological studies but they can be an important component of lower latitude lakes [*Arndt, 1993*] and have been recorded in low concentrations in lakes of the Vestfold Hills [*Laybourn-Parry and Marchant, 1992*]. Arndt concluded that the importance of Sarcodines was independent of trophic state or latitude of lakes, but this is not consistent with relatively high numbers found in oligotrophic Lake Taylor and Crooked Lake in the Vestfold Hills compared with eutrophic lakes at lower latitudes

[*James, 1995*]. *Laybourn-Parry et al.* [1994] consider that Sarcodines are characteristic of oligotrophic lakes but only sporadic components of eutrophic systems. It is interesting that heliozoans were a major component of the protozooplankton community in ultra-oligotrophic Lake Vanda (over 50% of abundance) but were not found or were scarce in Lake Bonney and Lake Fryxell. Suctorians (Figure 2) have been recorded in Lakes Fryxell [*Laybourn-Parry et al., 1997*], Hoare, Miers, and Brownsworth [*Cathey et al., 1981*] but were not found during the present study. These are ambush predators and require a reasonably high prey concentration of ciliates to survive, hence their absence from ultra-oligotrophic Antarctic systems. The lack of data however, for some of these groups, illustrates the poor state of knowledge of protozoa in Antarctic lakes. Rotifers, mostly the bdelloid *Philodina*, were found in samples from Lakes Vanda and Fryxell but were not recorded in Lake Bonney.

Heterotrophic nanoflagellates (HNAN, Figure 2) were recorded in all lakes. Abundances were at the lower end of those reported for a range of trophic spectra [*Pick and Caron, 1987; Nagata, 1988; Bloem and Bar-Gilissen, 1989; Berninger et al., 1991, Laybourn-Parry et al., 1994*] reflecting the oligotrophic nature of most of the McMurdo Dry Valley lakes. HNAN were most abundant in the west lobe of Lake Bonney followed by Lake Fryxell and Lake

TABLE 3. Abundance of Microzooplankton Groups in McMurdo Dry Valley Lakes.

Lake	Date	Total ciliates Nos. per liter	Heliozoa Nos. per liter	Rotifers Nos. per liter (x10³l⁻¹)	HNAN
Vanda	18 Dec 93	1000–6540	0–3050	ND	68–790
	25 Nov 94	13–812	0–67	0–13	10–185
	7 Dec 94	7–~170	0–100	7–33	52–645
	25 Dec 94	28–1273	99–459	ND	23–61
	28 Sep 95	0–200	12	ND	41–128
Bonney east lobe	9 Dec 94	8–1612	0	0	27–161
west lobe	10 Dec 94	0–1780	0	0	162–1158
Fryxell	12 Jan 92	0–7720	0	4–34	28–739

HNAN = heterotrophic nano-flagellates.   ND = not detected.

Vanda (Table 3). Data from different sampling times for Lake Vanda demonstrate that seasonal changes in abundance do occur with numbers ranging from 1.28 x 10^5 cells l^{-1} in September to 6.5 x 10^5 cells l^{-1} in the DCM in mid-summer. Seasonal changes in abundance of flagellates have also been demonstrated in lower latitude and other Antarctic systems [*Laybourn-Parry and Bayliss,* 1995; *Laybourn-Parry et al., 1995*].

Stratification and Seasonal Development

A unique feature of McMurdo Dry Valley lakes is their permanent physical and chemical stratification [*Spigel and Priscu,* this volume]. This stratification is reflected in the vertical profiles for microzooplankton. Most taxa are confined to particular zones which are defined by homogenous layers separated from each other by density gradients. This is particularly evident in Lake Vanda where a large isohaline and isothermal band extends from 17 to 37 m. Below this depth there is an exponential increase in conductivity and temperature with 120 mS cm⁻¹ and 23°C at the bottom (75 m). The lake is anoxic below 66 m which appears to exclude all protozoans from depths greater than 64 m. Protozoans are found in anoxic layers in temperate lakes however [*Esteban et al., 1993*], which implies other chemical or physical factors are excluding protozoans.

Four distinct communities of microzooplankton were observed in Lake Vanda (Figure 3a). The community in the upper euphotic zone just below the ice consisted of *Askenasia* (Figure 2) and *Urotricha* at the beginning of the season (November 1994 , Figure 3a, also present under 'others' in September 1995, Figure 3b) and an unidentified Didinidae by late December. Ciliate abundance was highest in this zone. In the lower isohaline cell between 17 and 37 m, the ciliate community was dominated by either

Didinidae which includes *Askenasia* and *Monodinium* (25 November and 26 December) or a spirotrich (7 December). Heliozoa dominated the community in the third zone where there was a transition to saline water. The fourth community consisted almost exclusively of the bacterivorous *Euplotes* with a large peak at 60 m coinciding with the peak in bacterial abundance in oxygenated waters (M. R. James, unpublished data). This peak appeared to develop late in 1994 between 25 November and 26 December, potentially in response to increased bacterial abundance. *Laybourn-Parry et al.* [1996] observed an increase in abundance of a small scuticociliate in early December in Heywood Lake, Signy Island followed by increased abundance of the oligotrich *Halteria grandinella*. Oligotrichs were never very abundant in Lake Vanda. *Spaulding et al.* [1994] and *Laybourn-Parry et al.* [1997] however noted that there can be significant interannual variability in both flagellate and ciliate species dominance. These observations on ciliate community structure are based on very limited data and emphasize the need for more detailed long-term studies. Rotifers were generally a minor component of the microzooplankton community and were most abundant in the middle zone (17–55 m).

The conditions that produce the zonation in the protozoo- and microzooplankton community are likely to be related directly to the physical and chemical structure which causes zonation of protozoan food sources (bacteria, phytoplankton). Potential biological interactions are discussed below.

Lizotte et al. [1996] found phytoplankton photosynthesis in Lake Bonney began in late winter (before 9 September) but until this study we have only been able to speculate about what happens to the protozoan community in McMurdo Dry Valley lakes in winter. Collections made in September 1995 revealed that there

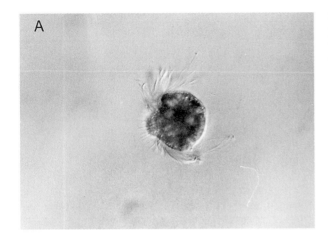

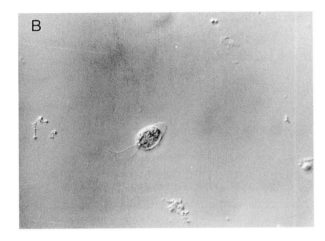

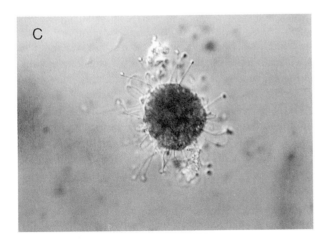

Fig. 2. Photograph of some common porotozoan taxa (A) a haptorid, *Askenasia* sp., (B) a heterotrophic flagellate and (C) a suctorian from Lake Fryxell.

is probably a reduced protozoan community over the winter period (Figure 3b). The same pattern of vertical zonation was evident during late winter in Lake Vanda as summer with an upper community dominated by *Askenasia* (Figure 2a*)*, a middle zone of Heliozoa and spirotrichs, and a deep community composed almost exclusively of *Euplotes*. Unfortunately there was no sample immediately below the ice (~3 m) on 11 September, but the peak observed on 28 September was probably present on both occasions. These data imply there is a small resident population during winter which develops during the short period of daylight each summer.

A pattern similar to that in Lake Vanda was observed in Lake Bonney (Table 4) with highest numbers of ciliates just below the ice in both lobes. Abundance in this zone was comparable with that recorded just under the ice in Lake Vanda. In Lake Bonney, ciliate abundance was low at the chlorophyll-*a* maximum at 10 m in the east lobe and they were absent at the chlorophyll-*a* maximum slightly deeper (12 m) in the west lobe. The haptorid *Askenasia* (Figure 2a) dominated the community just below the ice, but like Vanda the middle strata was dominated by spirotrichs.

Laybourn-Parry et al. [1997] also found protozoa were confined to particular strata in Lake Fryxell. An unidentified prostomatid was only found in the deeper, more saline layers, and high numbers of *Monodinium* were found immediately below the ice. The occurrence of high numbers of Didinidae (*Askenasia, Monodinium*) immediately below the ice appears to be a feature common to the larger McMurdo Dry Valley lakes.

Heterotrophic flagellates, dominated by *Ochromonas,* which is mixotrophic, were most abundant between 60 and 64 m in Lake Vanda and at 8 m in the west lobe of Bonney. These deep maxima (Figure 4, Table 4) coincided with peaks in bacterial biomass (M. R. James, unpublished data). Photosynthetic uptake experiments (I. Hawes, unpublished data) have also demonstrated that this zone receives sufficient light for net photosynthesis and phytoplankton have access to nutrients diffusing out of the deep anoxic layers. Previous studies on photosynthesis revealed three distinct algae communities down the water column of Lake Vanda; the lowest at 55–57.5 m was comprised of shade-adapted algae [*Vincent*, 1981; *Vincent and Vincent*, 1982]. These algae probably provide the DOC which supports bacteria and, in turn, HNAN production at this depth. These conditions favor motile phytoplankton which can maintain their vertical position to maximize light and nutrient availability, although there is some evidence that density gradients may also promote accumulation of particulate material in this zone.

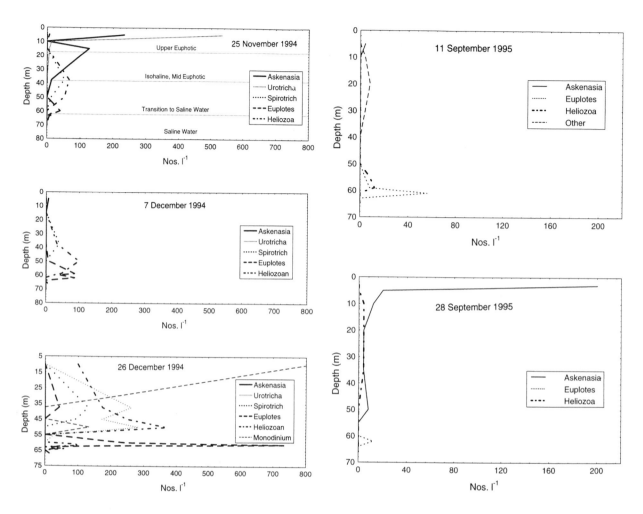

Fig. 3a and b. Vertical profiles of major ciliate taxa and Heliozoa in Lake Vanda. No sample was taken immediately below the ice on 11 September.

The absence of ciliated protozoa and rotifers from depths greater than 64 m is of particular interest; the lack of predation/grazing on flagellates may contribute to the existence of a DCM at this depth. Previous studies in the melt pond region of the McMurdo Ice Shelf have shown that most of the protozoo- and microzooplankton taxa, including the rotifer *Philodina,* can survive in water of conductivities up to 10 mS cm^{-1} but were not found in very saline ponds (Salt Pond, 54 mS cm^{-1}) [*James et al., 1995*]. The conductivity of water at the DCM in Lake Vanda (63 m) was 50 mS cm^{-1}.

The salinity and temperature tolerances for *Philodina gregaria* have been examined in laboratory experiments [*Spurr, 1975*]. *P. gregaria* survived ionic concentrations up to 250 kg m^{-3} Na$^+$ + Cl$^-$ and water temperatures up to 32°C but survival was <20% at concentrations over 15 kg m^{-3} Na$^+$ + Cl$^-$. This latter ionic concentration corresponds

to a conductivity of 30 mS cm^{-1} found at 60-61 m in Lake Vanda. In the transplant experiments no rotifers survived when incubated at depths below 55 m and those below 60 m were not contracted suggesting they probably died instantly. The laboratory experiments showed *Philodina* survived only in 55 m water which had not been diluted with water from 67 m. Even a 6.25% dilution, which only increased the conductivity to 8.2 mS cm^{-1}, was sufficient to cause 100% mortality. The results from these experiments in Lake Vanda suggest that, while high conductivity might be important, there must be other chemicals present below 60 m to which rotifers are even more sensitive. These chemicals have not yet been identified.

While high conductivity may partially explain the absence of rotifers, some ciliated protozoans were found in highly saline Salt Pond (54 mS cm^{-1}) on the McMurdo

TABLE 4. Microzooplankton Abundance in Lake Bonney

	East Lobe Depth(m)				West Lobe Depth (m)			
Taxa	4.5	8	14	18	4.5	8	13	15
Ciliates (nos. l^{-1})								
Euplotes	76	0	0	0	4	0	0	0
Askenasia	1032	0	0	0	628	4	0	0
Urotricha	256	4	4	0	100	44	0	0
Spirotrichs	232	4	0	0	208	284	0	0
Total ciliates	1612	12	8	0	1344	340	0	0
Heterotrophic flagellates (nos. ml^{-1})	161	36	27	66	80	1158	*	347

* no sample taken

Ice Shelf [*James et al.*, 1995]. Ciliates are ubiquitous in most aquatic habitats including the extreme environments of geothermal lakes and anoxic bottom waters of eutrophic lakes at lower latitudes worldwide [*Goulder,* 1974; *Bark,* 1981; *Laybourn-Parry et al.*, 1990; *James et al., 1995*]. It is therefore surprising that ciliates are absent from the anoxic waters of all McMurdo Dry Valley lakes, although, unlike lower latitude lakes, they do not have a period of complete water column mixing.

Trophic Interactions and Nutrition

The observations on vertical distribution and results from standard grazing experiments give insight into the relationships between the protozooplankton and their food resources. The clear zonation of different taxa suggests strong niche separation, which is probably based on available food resources and physical and chemical conditions. The high abundance of Didinidae (e.g., *Askenasia*) just below the ice suggests they rely either on an algal community associated with the ice interface or are mixotrophic. No pigmented cells were obvious in live observations and unfortunately cells were not examined using fluorescence microscopy. They certainly feed on algae at lower latitudes. Observations of the food vacuoles of *Askenasia volvox* suggest they can consume algae, bacteria, non-pigmented flagellates, and other ciliates [*Tamar, 1973*]. Feeding experiments on *Askenasia* from an oligotrophic lake in New Zealand however, found no evidence for ingestion of bacteria, small phytoplankton or similar (1.0 μm) sized particles [*James,* 1995]. Although "blooms" of flagellates have been found in Antarctic lakes immediately under the ice (e.g., Sombre Lake, Signy Island [*Hawes,* 1983]), aggregations in this zone have not been observed in McMurdo Dry Valley lakes, though there is some evidence for a peak in primary production in this zone of Lake Fryxell and Lake Hoare and secondary peaks in Lake Bonney [*Priscu,* 1995]. These observations provide evidence for potential grazer control of phytoplankton biomass in this zone. Tamar's study showed flagellates may be an important food source for *Askenasia;* small chrysophytes often dominate the community just below the ice in Antarctic lakes [*Hawes,* 1983; *Lizotte et al.* 1996].

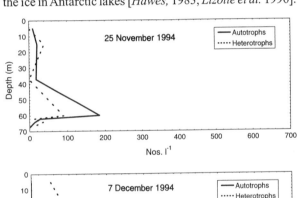

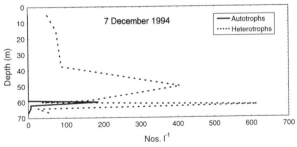

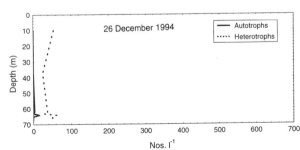

Fig. 4. Vertical profiles of autotrophic and heterotrophic flagellates in Lake Vanda.

However, planktonic organisms must be nutritional opportunists, and it is likely that *Askenasia* has a wide range of potential prey or food. The late winter/spring sampling of Lake Vanda is the first record for McMurdo Dry Valley lakes of a protozoan peak immediately under the ice. Chlorophyll fluorescence peaks have been recorded immediately under the ice in Lake Bonney as early as 9 September [*Lizotte et al.* 1996]. Further work is required with more detailed sampling in this zone to determine whether this peak is a consistent feature that may decrease through grazing pressure by the ciliate community which develops in summer. There is some evidence however, from studies in Lake Bonney that the upper phytoplantion population becomes nutrient limited in late spring/summer [*Lizotte et al.* 1996].

The community in the isohaline mid-euphotic zone of Lake Vanda (17-37 m) is largely dominated either by *Askenasia* or spirotrichs. Most of the spirotrichs were ca. 25 μm in length and previous studies in New Zealand lakes would indicate they feed mostly on particles <5 μm diameter. Based on clearance rates for spirotrichs in temperate latitudes (1.85 μl ind^{-1} h^{-1} at 10-15°C, *James,* 1995) and a maximum abundance of 260 l^{-1} in Lake Vanda, this group would potentially remove <2% of the algal population per day. Assuming a similar clearance rate for *Askenasia* and a maximum density of 1270 l^{-1} for Lake Vanda and 1000 l^{-1} for Lake Bonney, *Askenasia* could potentially remove up to 5.6 and 4.4% of standing phytoplankton crop per day, respectively. Such grazing pressure could be significant because of the slow growth rates for phytoplankton. Doubling times of HNAN can be up to 41 days in Antarctic lakes [*Laybourn-Parry et al.,* 1995] although *Priscu et al.* [1987] estimated average doublings d^{-1} of 0.308 for microplankton in Lake Vanda. The results from experiments with HNAN by *Laybourn-Parry et al.* [1995] suggested that Antarctic protists show no adaptation to low temperatures, so clearance rates for ciliates could be considerably lower than those used above. There has been no assessment of the potential grazing pressure on picophytoplankton populations which occur at concentrations up to 3.8 x 10^6 cells l^{-1} in Lake Vanda. Based, however, on maximum ciliate densities in Lake Vanda (1270 l^{-1}) and clearance rates measured for ciliates in melt ponds on the McMurdo Ice Shelf of 1.9 μl ind^{-1} h^{-1} [*James et al.* 1995], the impact will be <10% of standing stock per day.

At least two species of Heliozoa were recorded in the transition zone between the lower isohaline cell and the deeper high salinity zone in Lake Vanda. Feeding studies of heliozoans are rare, but they appear to be able to exploit both phytoplankton and HNAN [*Patterson and Hausmann,*

1981]. During peak abundance they may have a significant impact on flagellates and ciliates [*Arndt, 1993*].

Two ciliates known to feed heavily on bacteria are *Vorticella* in Lake Fryxell and *Euplotes* in Lake Vanda. Although *Vorticella* was only found in low numbers in Lake Fryxell (up to 80 l^{-1}), in Lake Vanda, *Euplotes* developed a relatively abundant population (730 l^{-1}) at 60-61 m, which is immediately above the bacterial peak. *Euplotes* are capable of clearing bacteria at rates of 0.1 μl ind^{-1} h^{-1} [*James et al.,* 1995]. This corresponds to <0.2% of the bacterial population cleared per day in Lake Vanda indicating they are unlikely to be a controlling factor for bacterial production in this zone. However, *Laybourn-Parry et al.* [1995] found HNAN removed up to 9.7% of bacterial production per day in Crooked Lake, in the Vestfold Hills. Bacterivorous ciliates were not an important component of the zooplankton community in Lake Bonney.

Dilution experiments were also carried out in Lake Vanda to determine whether there was a grazing impact on phytoplankton or bacteria. Unfortunately no experiments were conducted with water immediately below the ice, but there was no detectable grazing impact at 15, 37.5, 55 or 63 m. There was also very little phytoplankton or bacterial growth during the experiments, but experimental nutrient addition stimulated phytoplankton growth in water from 15, 37.5 and 55 m. Results from these experiments are consistent with the calculations above and suggest that nutrient limitation is likely to be the most important variable controlling phytoplankton populations. Although *Priscu* [1995] found no response to nutrient additions in bioassay experiments in Lake Vanda, the dissolved inorganic nitrogen to soluble reactive phosphorus ratio implied phosphorus deficiency.

Mixotrophy

The combination of heterotrophy and autotrophy is common among planktonic protists, particularly in oligotrophic environments [*Laybourn-Parry,* 1992]. In the "feast and famine" environment common to the plankton, mixotrophy confers a distinct competitive advantage. Among the ciliates, mixotrophy can take two forms: organellar, which involves the sequestration of the plastids from ingested phytoflagellates and algae, and cellular, where endosymbiotic zoochlorellae are involved. Three groups of mixotrophs have been identified in the McMurdo Dry Valley lakes. A plastidic species of *Strombidium* (probably *S. viride*) has been recorded in Lake Fryxell as well as a species of *Bursaria* containing zoochlorellae [*Laybourn-Parry et al., 1997*]. *Strombidium* is common in the McMurdo Dry Valley lakes (Table 2) and though most records are derived from Lugol's fixed material which precludes determination of mixotrophy, it is likely to be plastidic. Other ciliates recorded may also be plastidic. For example *Halteria*, which occurs commonly in Lakes Bonney and Fryxell (Table 2), may harbor zoochlorellae [*Finlay et al.*, 1988]. This aspect of protozooplankton biology in the McMurdo Dry Valley lakes needs further study.

Heliozoa in Lake Vanda contained zoochlorellae. About 20 chlorophyll containing structures, each 2-3 μm in diameter, were observed inside the heliozoan cells. These are likely to be zoochlorellae. Facultative mixotrophy involving the sequestration of plastids by heliozoans has been observed at lower latitudes [*Patterson and Durrschmidt*, 1987] and zoochlorellae are common among other planktonic sarcodines, for example foraminiferans and radiolarians.

Among the phytoflagellates recorded in Lake Fryxell by *Spaulding et al.* [1994] are *Dinobryon* and *Ochromonas*, both of which are known to be phagotrophic on bacteria [*Sanders and Porter*, 1988]. In Lake Vanda, *Ochromonas* is very common at a depth close to the peak abundance of bacteria (I. Hawes, unpublished data). *Ochromonas* ingest bacteria potentially to obtain nutrients and vitamins for photosynthesis and to supplement their carbon budgets. It is likely that these flagellates are operating in the mixotrophic mode in the McMurdo Dry Valley lakes because *Ochromonas*, in lakes of both maritime Antarctica and the Vestfold Hills, have been observed ingesting bacteria and microspheres (J. Laybourn-Parry, unpublished data).

SUMMARY

The protozooplankton in McMurdo Dry Valley lakes is diverse relative to lakes elsewhere in Antarctica. The permanent zonation and simple foodwebs characteristic of these lakes offer a unique opportunity to study microbial plankton dynamics in the absence of top-down control. New data and observations in Lake Vanda indicate that the structure and vertical stratification of the communities are principally determined by food resources. The inability of protozoo- and microzooplankton to survive in anoxic bottom waters in McMurdo Dry Valley lakes cannot be explained simply by lack of oxygen or high salinity and suggests some other chemical factor is limiting their distribution. Because of the diversity and range of feeding strategies, protozooplankton and microzooplankton can exploit the limited resources available in the extreme conditions of these oligotrophic lakes. The first records for late winter in Lake Vanda showed there is a small overwintering population of ciliates and Heliozoa which develops in spring. The lack of data from these unique systems emphasizes the need for more detailed long-term studies.

Acknowledgements. We thank the New Zealand Antarctic Programme (NZAP) and the US Navy for logistic support in the field, I. Hawes, C. Howard-Williams, A-M. Schwarz, and M. Weatherhead for their collaboration with experiments and helpful discussions. The research was carried out with funding from the Foundation for Research, Science and Technology (New Zealand).

REFERENCES

Arndt, H. A., Critical review of the importance of rhizopods (naked and testate amoebae) and actinopods (heliozoa) in lake plankton, *Mar. Microbial Food Webs 7,* 3–30, 1993.

Azam, F, T. Fenchel, J. G. Field, J. S. Gray, L. A. Meyer-Reil, and F. Thingstad, The ecological role of water-column microbes in the sea, *Mar. Ecol. Prog. Ser. 10,* 257–263, 1983.

Bark, A. W., The spatial and temporal distribution of planktonic and benthic protozoan communities in a small eutrophic lake, *Hydrobiologia 85,* 239–255, 1981.

Berninger, U-G., B. J. Finlay, and P. Kuuppo-Leinikki, Protozoan control of bacterial abundances in freshwater, *Limnol. Oceanogr., 36,* 139–147, 1991.

Bloem, J. and M-J. B. Bar-Gilissen, Bacterial activity and protozoan grazing potential in a stratified lake, *Limnol. Oceanogr., 34,* 279–309, 1989.

Cathey, D. D., B. C. Parker, G. M. J. Simmons, W. H. J. Yongue, and M. R. Van Brunt, The microfauna of algal mats and artificial substrates in Southern Victoria Land lakes of Antarctica, *Hydrobiologia 85,* 3–15, 1981.

Chinn, T., Physical hydrology of Dry Valley lakes, *Antarctic Research Series, 59,* 1–52, 1993.

Dartnall, H. J. G., Rotifers and other aquatic invertebrates, from the Larsemann Hills, Antarctica, *Papers and Proceedings Roy. Soc. Tasmania, 129,* 17–23, 1995.

Esteban, G., B. J. Finlay, and T. M. Embley, New species double the diversity of anaerobic ciliates in a Spanish lake, *FEMS Microbiology Letters 109,* 93–100, 1993.

Finlay, B. J., K. J. Clarke, A. J. Cowling, R. M. Hindle, A. Rogerson, and U-G. Berninger, On the abundance and distribution of protozoa and their food in a productive freshwater pond, *Eur. J. Protistol., 23,* 205–217, 1988.

Goulder, R., The seasonal and spatial distribution of some benthic ciliated protozoa in Esthwaite Water, *Freshwater Biol., 4,* 127–147, 1974.

Hall, J. A., D. P. Barrett, and M. R. James, The importance of phytoflagellate, heterotrophic flagellate and ciliate grazing on bacteria and picophytoplankton sized prey in a coastal marine environment, *J. Plank. Res. 15,* 1075–1086, 1993.

James, M. R., Sampling and preservation methods for the quantitative enumeration of microzooplankton, *N. Z. J. Mar. Freshwat. Res., 25,* 305–310, 1991.

James, M. R., A comparison of microzooplankton in aquatic food webs with special emphasis on ciliates. Ph.D. Dissertation, University of Otago, Dunedin, New Zealand, 1995.

James, M. R., R. D. Pridmore, and V. J. Cummings, Planktonic communities of melt ponds on the McMurdo Ice Shelf, Antarctica, *J. Plank. Res. 15,* 555–567, 1995.

Landry, M. R. and R. P. Hassett, Estimating the grazing impact of marine microzooplankton, *Mar. Biol. 67,* 283–288, 1982.

Laybourn-Parry, J., Protozoan Plankton Ecology, Chapman and Hall, London, 1992.

Laybourn-Parry, J., Seasonal successions of protozooplankton in freshwater ecosystems of different latitudes, *Mar. Microbial Food Webs, 8,* 145–162, 1994.

Laybourn-Parry, J., and P. Bayliss, Seasonal dynamics of the plankton community in Lake Druzhby, Princess Elizabeth Land, Eastern Antarctica, *Freshwater Biol. 35,* 57–68, 1996.

Laybourn-Parry, J., and H. J. Marchant, The microbial plankton of freshwater lakes in the Vestfold Hills, Antarctica, *Polar Biol. 12,* 405–410, 1992.

Laybourn-Parry, J., P. Bayliss, and J. C. Ellis-Evans, The dynamics of heterotrophic nanoflagellates and bacterioplankton in a large ultra-oligotrophic Antarctic lake, *J. Plank. Res. 17,* 1835–1850, 1995.

Laybourn-Parry, J., J. C. Ellis-Evans, and H. Butler, Microbial dynamics during the summer ice-loss phase in maritime Antarctic lakes, *J. Plank. Res. 18(4),* 495–511, 1996.

Laybourn-Parry, J., M. R. James, D. McKnight, J. Priscu, S. Spaulding, and R. Shiel, The microbial plankton of Lake Fryxell, Southern Victoria Land, Antarctica, *Polar Biol.17,* 54–61, *1997.*

Laybourn-Parry, J., J. Olver, and S. Rees, The hypolimnetic protozoan plankton of a eutrophic lake, *Hydrobiologia 203,* 111–119, 1990.

Laybourn-Parry, J., H. J. Marchant, and P. Brown, The plankton of a large oligotrophic freshwater Antarctic lake, *J. Plank. Res. 13,* 1137–1149, 1991.

Laybourn-Parry, J., J. Walton, J. Young, R. I. Jones, and A. Shine, Protozooplankton and bacterioplankton in a large oligotrophic lake - Loch Ness, Scotland, *J. Plank. Res. 16,* 1655–1670, 1994.

Lizotte, M. P., T. R. Sharp, and J. C. Priscu, Phytoplankton dynamics in the stratified water column of Lake Bonney, Antarctica. 1. Biomass and productivity during the winter-spring transition, *Polar Biol. 16,* 155–162, 1996.

Nagata, T., The microflagellate-picoplankton food linkage in the water column of Lake Biwa, *Limnol. Oceanogr., 33,* 504–517, 1988.

Parker, B. C., G M J. Simmons, K. G. Seaburg, D. D. Cathey, and F. C. T. Allnutt, Comparative ecology of plankton communities in seven Antarctic oasis lakes, *J. Plank. Res. 4,* 271–286, 1982.

Patterson, D. J., and M. Durrschmidte, Selective retention of chloroplasts by algivorous heliozoa: fortuitous chloroplast symbiosis? *Eur. J. Protistology 23,* 51–55, 1987.

Patterson, D. J., and K. Hausmann, Feeding in *Actinophrys sol* (Protozoa: Heliozoa), I., Light Microscopy, *Microbios, 31,* 39–55, 1981.

Pick, F. R., and D. A. Caron, Picoplankton and nanoplankton biomass in Lake Ontario: relative contribution of phototrophic and heterotrophic communities, *Can. J. Fish. Aquat. Sci., 44,* 2164–2172, 1987.

Priscu, J. C., Phytoplankton nutrient deficiency in lakes of the McMurdo dry valleys, Antarctica, *Freshwater Biol. 34,* 215–227, 1995.

Priscu, J. C., L. R. Priscu, W. F. Vincent, and C. Howard-Williams, Photosynthate distribution by microplankton in permanently ice-covered Antarctic desert lakes, *Limnol. Oceanogr., 32(1),* 260–270, 1987.

Pomeroy, L. R., The ocean's food web, a changing paradigm, *Bioscience 24,* 499–504, 1974.

Rigler, F. H., M. E. MacCallum, and J. C. Roff, Production of zooplankton in Char Lake, *J. Fish. Res. Bd. Canada 31,* 637–646, 1974.

Sanders, R. W., and K. G. Porter, Phagotrophic phytoflagellates, *Adv. Microbial Ecol., 10,* 167–192, 1988.

Simek, K., and V. Straskrabová, Bacterioplankton production and protozoan bacterivory in a mesotrophic reservoir, *J. Plank. Res. 14,* 773–787, 1992.

Spaulding, S. A., D. M. McKnight, R. L. Smith, and R. Dufford, Phytoplankton population dynamics in perennially ice-covered Lake Fryxell, Antarctica, *J. Plank. Res. 16,* 527–541, 1994.

Spurr, B., Limnology of Bird Pond, Ross Island, Antarctica, *N.Z. J. Mar. Freshwat. Res. 9,* 547—562, 1975.

Stone, L., T. Berman, R. Bonner, S. Barry, and S. W. Weeks, Lake Inneret: a seasonal model for carbon flux through the planktonic biota, *Limnol. Oeanogr. 38,* 1680–1695, 1993.

Tamar, H., Observations on *Askenasia volvox* (Claparede and Lachmann, 1859), *J. Protozool. 20,* 46–50, 1973.

Vincent, W. F., Production strategies in Antarctic inland waters: phytoplankton eco-physiology in a permanent ice-covered lake, *Ecology 62,* 1215–1224, 1981.

Vincent, W. F., Antarctic limnology, *in:* Inland waters of New Zealand, edited by A. B. Viner, pp. 379–412, DSIR Information Publishing Centre, Wellington, 1987.

Vincent, W. F., and C. Howard-Williams, Ecosystem properties of Dry Valley lakes, *N.Z. Antarctic Record 6* (supplement), 11–20, 1985.

Vincent, W. F., and C. L. Vincent, Factors controlling phytoplankton production in Lake Vanda, *Can. J. Fish. Aquat. Sci., 39,* 1602–1609, 1982.

Weisse, T., H. Müller, R. M. Pinto-Coelho, A. Schweizer, D. Springmann, and G. Baldringer, Response of the microbial loop to the phytoplankton spring bloom in a large pre-alpine lake, *Limnol. Oceanogr. 35,* 781–794, 1990.

Williams, P. J. Le-B., Incorporation of microheterotrophic processes into the classical paradigm of the plankton food web, *Kieler Meeresforsch, Songerh., 5:L,* 1–28, 1981.

J. A. Hall National Institute of Water and Atmospheric Research Kyle Street, P.O. Box 11 115, Hamilton, New Zealand

M. R. James National Institute of Water and Atmospheric Research Kyle Street, P.O. Box 8602,Christchurch, New Zealand

J. Laybourn-Parry Department of Physiology and Environmental Science, University of Nottingham, Loughborough, United Kingdom

(Received July 11, 1996;
accepted February 25, 1997)

PERMANENT ICE COVERS OF THE MCMURDO DRY VALLEYS LAKES, ANTARCTICA: LIQUID WATER CONTENTS

Christian H. Fritsen[1], Edward E. Adams[2], Christopher P. McKay[3], and John C. Priscu[1]

A novel method of analyzing ice temperature records is applied to several years of data from Lake Hoare and Lake Bonney (Antarctica) to estimate vertical distributions of liquid water in the perennial ice covers at the end of summer melting seasons. Three years of ice temperature data at Lake Bonney (1993-1995) show that the ice contained 20% liquid water located at 1 to 2.5 m below the ice surface near the end of all three melting seasons. Liquid water fractions at Lake Hoare were low (<15%) throughout the ice column in 1986. In 1987 and 1988 liquid water fractions increased to a maxima of 70% at depths between 1.5 and 2.5 m. Maxima in liquid water content for both lakes were coincident with layers of bubbles having arching morphologies which were predominantly associated with pockets of sedimentary material (silts, sand, and gravel). Interpreting these bubble morphologies in context of the ice energy budgets indicates that the majority (>90%) of the liquid water in the ice is generated when visible radiation is absorbed by lithogenic matter within the ice during the austral summer. Our work also demonstrates the potential use of this energy budget analysis for monitoring ecosystem-level responses to climate variability.

INTRODUCTION

Ice covers on lakes influence the exchange of momentum, thermal energy, and materials between the water column and the atmosphere. Thus the presence of an ice cover as well as the physical attributes of the ice cover influences a lake's ecosystem processes. The McMurdo Dry Valleys contain numerous lakes with thick (3 to 20 m) perennial ice covers. Because of their permanence, the ice covers control the lakes sedimentation processes [e.g., *Neddell et al.,* 1987; *Squyres et al.,* 1991; *Anderson et al.,* 1993; *Doran et al.,* 1994], the quantity and quality of light penetrating into the water column for photosynthesis [*Palmisano and Simmons,* 1987; *Lizotte and Priscu,* 1992; *McKay et al.,* 1994], as well as the transfer of gases into and out of the lakes [*Craig et al.,* 1992; *Wharton et al.,* 1993; *Priscu,* 1996].

These ice covers exist as solid physical barriers during winter months. Each austral summer, however, the combination of relatively warm air temperatures (~0°C) and continual solar radiation leads to partial melting of the ice covers. At this time, it is presumed that the ice covers become permeable to materials [*Neddell et al.,* 1987; *Squyres et al.,* 1991; *Anderson et al.,* 1993] and gases [*Anderson et al.,* 1993].

Permeability is a physical characteristic of ice covers which is difficult to quantify directly because its complex nature involves several factors including porosity, connectedness, and tortuosity. However the permeability of liquid-containing composite materials is generally related to the amount of the liquid phase within the solid matrix. Therefore developing methods for determining the liquid water fractions of the permanent ice covers improves our ability to monitor the physical properties of the ice covers that directly relate to permeability and indirectly relate to the exchanges of matter and energy between the lakes and their surroundings.

An ice cover's liquid water fraction also is a quantity that can be used derive energy budgets over seasonal time scales. The ice covers on the McMurdo

[1]Department of Biological Sciences, Montana State University, Bozeman, Montana
[2]Department of Civil Engineering, Montana State University, Bozeman, Montana
[3]Solar Systems Exploration Branch, NASA Ames Research Center, Moffett Field, California

Dry Valley lakes are believed to be sensitive indicators of climate change [*Wharton et al.* 1993]. Hence it is logical to assume that monitoring both annual and seasonal energy budgets will provide additional indices of climate variability that can be related to global climate change.

Internal melting in permanent ice covers has not been studied or quantified previously, despite its apparent importance in ecosystem processes and its potential for monitoring climate variability. However we have recently devised a novel method to quantify the liquid water contents of the ice covers at the end of the austral summer. The method consists of analyzing energy budgets of known depth intervals in the ice during the austral autumn when freezing fronts are propagating from the surface to the bottom of the ice covers. We apply this method to ice temperature records collected from Lakes Bonney and Hoare and provide the first quantitative estimates of the liquid water contents of the ice covers which are interpreted in context of their morphological features and seasonal climatological parameters. These analyses yield direct insights into the overriding mechanisms that control internal ice melting, ice permeability to materials, as well as ice-energy transfer processes.

METHODS

Site Description and Instrumentation

Lakes Hoare and Bonney are located in the Taylor Valley, Antarctica. General hydrology of the lakes and the physicochemical and biological characteristics of their water columns are described in detail by others [e.g. *Green et al.*, 1993; *Spigel and Priscu,* this volume]. The ice covers have surface topographies consisting of alternating ridges and troughs that are generally oriented parallel to the valley floor and prevailing winds. Previous evaluation of the annual mass balance of the Lake Hoare ice cover in the late 1980s indicated that about 35 cm yr[-1] of ice sublimated or evaporated from the ice surface, while an additional 60 to 70 cm of the ice melted at the base during the austral summer [*Clow et al.*, 1988]. This annual loss was roughly balanced by new ice growth at the base of the ice in the austral autumn and winters. From these analyses it is apparent that the ice covers are physically dynamic features with a net upward movement of the ice at approximately 30 to 40 cm yr[-1].

Thermocouples (Campbell, Type T) were placed in the 4 m thick ice cover near the center of the east lobe of Lake Bonney in December of 1992 at 0.5, 1.0, 2.0, and 3.5 m from the ice surface. Thermocouples at Lake Hoare were placed in the ice (original depths 0.5, 1.0, 1.5, 2.0, 2.5, and 3.0 m) in December of 1985 and retrieved in the austral summer of 1988. Data were collected at hourly intervals in Lake Bonney using a Campbell 21X data logger powered by gel-cell batteries. This data logger concurrently recorded downwelling and upwelling photosynthetically active radiation (PAR) measured with LiCor 192S sensors set 1 m above the ice surface. Instantaneous readings, daily averages, maximum, and minimum daily temperatures in the Lake Hoare ice cover were recorded at local noon. Because ablation at the ice surface occurs throughout the year, the thermocouples experienced net movement relative to the ice surface throughout the seasons. However the array of thermocouples is presumed to have moved as one unit and the depths of the thermocouples relative to each other are assumed to have remained constant.

Liquid Water Contents Estimated from Autumnal Energy Budgets

The ice temperature records were analyzed during the periods when freezing fronts were propagating through the ice in the austral autumn. The energy budget for this period (conceptually illustrated in Figure 1) was used to calculate the changes in latent heat for depth intervals defined by the thermocouples. The change in latent heat was then used to estimate the amount of liquid water generated during the proceeding summer melt season.

The change of total internal energy within a layer of ice, E_T, is controlled by the fluxes of energy in and out of the layer. Within the ice column, energy is transferred primarily by short-wave radiation, E_r, and heat conduction, E_c, such that,

$$E_T = E_r + E_c.$$

Temperature gradients ($\delta T/\delta z$) above defined layers were used to calculate the time-integrated conductive heat loss out of the layers according to

$$E_c = \int k_i \frac{\partial T}{\partial z} \, dt$$

where k_i is the bulk thermal conductivity coefficient for the ice. Temperature gradients below the freezing fronts

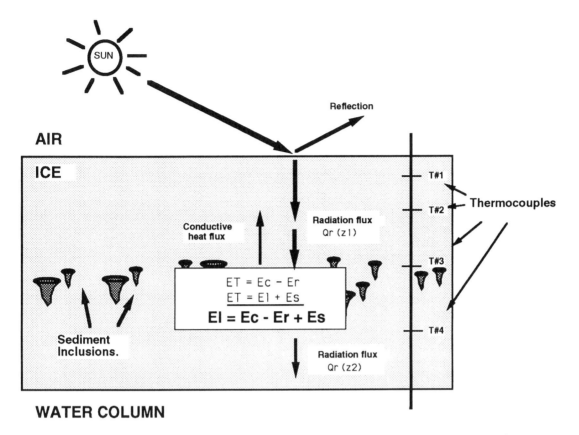

Fig. 1. Conceptual model of the energy budget used to derive the time and depth integrated flux of latent heat from layers within the permanent ice covers. The thermocouples and the sediment inclusions are illustrated schematically.

are isothermal at 0°C when freezing fronts are progressing through the defined layers. Therefore, no heat is transferred by conduction from below.

The coefficient for the bulk thermal conductivity of the ice is estimated as a volume-weighted function of the individual conductivity coefficients of the materials comprising the total ice volume such as pure ice (k_{pi}), air (k_a), and gravel (k_g), such that

$$k_i = k_{pi} V_{pi} + k_a V_a + k_g V_g$$

where V_{pi}, V_a, and V_g are the fractional volumes of the materials within a given ice volume.

The absorption of short-wave (penetrative) radiation in a layer of ice was estimated from a simple radiative-transfer model where down-welling irradiances above the ice surface, $I_d(0+,PAR)$, were propagated into the ice, $I_d(0-,PAR)$, after allowing for reflection, R_o, from the ice surface by the following:

$$I_d(0-,PAR) = I_d(0+,PAR) \, (1-R_o).$$

Downwelling irradiances at depth, $I_d(z,PAR)$, were then modeled assuming an exponential decrease with increasing depth in the ice, such that

$$\frac{\partial I_d (z, PAR)}{\partial z} = -K_d(z) \, I_d(0-, PAR)$$

where $K_d(z)$ is the diffuse downwelling attenuation coefficient of the medium, a composite apparent optical property [sensu Kirk, 1994]. During September 1995, spherical radiometers (LiCor) were frozen in the ice of the east lobe of Lake Bonney at 1 m intervals to monitor in-ice scalar irradiances, $I_o(z, PAR)$. These in-ice irradiances were then used to compute attenuation coefficients, $K_o(z, PAR)$. Values for $K_o(z, PAR)$ during the first several months of their deployment ranged from 1.2 to 1.8 m^{-1} in the upper 2.5 m. Assuming attenuation coefficients for scalar irradiances approximate those for downwelling irradiances, we utilize a value of 1.5 m^{-1} for K_d in our initial model. Analysis of the energy budget's sensitivity to a range of K_d values are addressed in a later section.

$I_d(z, PAR)$ is a flux of photons over a broad band (400 to 700 nm) of electromagnetic radiation and is not an exact measurement of the energy flux. The energy flux at various depths in the ice was estimated by converting the modeled PAR values (μmol photons m^{-2} s^{-1}) to radiative energy fluxes, Q_r (W m-2), using a factor of 4.2 μmol photons J^{-1} [*Morel and Smith*, 1974]. Because this factor was derived from measurements in oceanic waters, an ice radiative transfer model [*Fritsen et al.*, 1992] was used to check the applicability of this relationship to in-ice irradiances of PAR. This comparison showed that the factor varied from 4.1 to 4.5 depending on optical depths and constituents within the ice.

The depth and time-integrated absorption of short-wave energy, E_r, was then estimated as

$$Er = \int \int \frac{\partial Q_r(z)}{\partial z} dz\, dt$$

where the integrals are over the depth of the thermocouple spacing and the period of active freezing. Because the attenuation of downwelling irradiance is due to scattering plus absorption, the radiative transfer model is likely to yield an upper estimate of E_r. Again, sensitivity analysis, presented below, addresses uncertainties in the energy budget calculations associated with the assumptions of this radiative transfer model.

Net changes in total internal energy within a layer resulting from radiation and conductive heat exchanges are realized as changes in latent heat energy, E_l, and specific heat energy, E_s;

$$E_T = E_l + E_s$$

The change in specific heat was calculated as the product of the bulk specific heat capacity (c_i), bulk density (ρ_i), and the depth averaged change in ice temperature (T_i) within the layer. The bulk specific heat capacity and the bulk density were assumed to be proportional to the fractional volume of each constituent (air, ice, gravel) and its individual specific heat capacities. Values used for these individual coefficients are listed in Table 1. Solving the total energy budget in terms of changes in specific and latent heat caused by conductive and radiative energy transfer processes

allows us to define the change in the latent heat within a layer of ice as,

$$E_l = E_c + E_s - E_r$$

Thereby, the change in latent energy within a defined depth interval can be evaluated once the specific heat change is known and the net radiation and conductive heat fluxes are evaluated. The change in the liquid water fraction (V_w) is then evaluated as,

$$V_w = E_l L_w^{-1} \rho_w^{-1} \Delta z^{-1}$$

where L_w and ρ_w are the latent heat and density of water, respectively, and Δz is the distance between the thermocouples for which E_l was derived.

RESULTS

East Lobe Lake Bonney

Ice temperatures during the austral summers on the east lobe of Lake Bonney were isothermal at 0°C for approximately 86 days in November to February of 1993-1994 and 66 days during December to February of 1994-1995 (Figure 2). In contrast, midwinter temperature gradients were typically 6°C m^{-1}. Freezing fronts passed the first (0.5 m) thermocouple between February 4 and February 12 in each of the three years, indicating a relatively narrow time-period when seasonal freezing of the liquid water within the ice was initiated.

The rate at which the freezing fronts propagated into the ice was similar in 1994 and 1995, with the rate of freezing front progression averaging 0.030 m d^{-1} in 1994 and 0.034 m d^{-1} in 1995. Conversely the 1993 rate of freezing front progression was 0.019 m d^{-1}, approximately 1.5 to 2 times slower than in 1994 or 1995. An analysis of the annual variability in the rate of freezing front progression in context of the overall energy budget shows that the liquid water content at the end of the austral summer in 1993 was two-fold greater than in 1994 and 1995 (Table 2).

The energy budgets also indicate that the liquid water was not distributed evenly in the ice. Liquid water content was low (5 to 9%) in the uppermost portion of the ice (0 to 1 m), whereas, the liquid water content was higher (27 to 32%) between the second and third thermocouples in each of the three years. The second and third thermocouples were located at 1.0 and

TABLE 1. Notation and Description of Variables.

Variable	Variable Description	Value(s)	Units
E_T	internal energy	variable	KJ m^{-2}
E_c	conductive heat loss	variable	KJ m^{-2}
E_l	change in latent heat	variable	KJ m^{-2}
E_r	absorbed short-wave radiation	variable	KJ m^{-2}
E_s	change in specific heat	variable	KJ m^{-2}
I_d	downwelling irradiance of PAR	variable	μmol photons m^{-2} s^{-1}
K_d	attenuation coefficient for downwelling irradiances	1.5	m^{-1}
K_o	attenuation coefficient for scalar irradiances	variable	m^{-1}
k_i	thermal conductivity of lake ice	variable	W m^{-1} °C^{-1}
k_g	thermal conductivity of vapor phase	variable	W m^{-1} °C^{-1}
k_{pi}	thermal conductivity of pure ice	variable	W m^{-1} °C^{-1}
k_s	thermal conductivity of sand	variable	W m^{-1} °C^{-1}
L_w	latent heat of fusion for water	335	KJ kg^{-1}
R_o	specular reflection of PAR	0.05	%
Q_r	shortwave-radiation	variable	W m^{-2}
V_g	fractional volume of gravel	variable	%
V_{pi}	fractional volume of pure ice	variable	%
V_s	fractional volume of sand	variable	%
V_w	fractional volume of water	variable	%
z	depth	variable	m
ß	PAR conversion factor	4.2	μmol photons J^{-1}
ρ_w	density of water	1000	kg m^{-3}
ρ_i	density of pure ice	917	kg m^{-3}

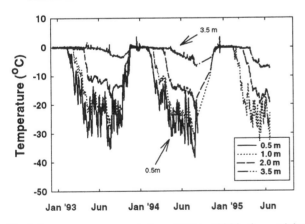

Fig. 2. Ice temperature records from 1992 to 1995 in the east lobe of Lake Bonney. Depths denote where the thermocouples were originally positioned in the ice. Note that the uppermost thermocouple was above the ice surface for a period in 1993 before being reset in the ice, and from September 1994 to the end of the record.

2.0 m in the 1993 and 1994 season; in 1995 they were estimated to be 50 cm higher in the ice at the time of the autumnal freezing. The vertical resolution of V_w (1–1.5 m) is relatively large and makes no distinction

as to the dispersion of the liquid water throughout these depth intervals. For instance, when V_w is 20% over 1 m, we cannot make the distinction between a 20 cm layer of liquid water with 80 cm of solid ice or a 1 m layer of ice with 20% of the water dispersed throughout.

Vapor bubbles with arching morphologies and sediment pockets were found from 1.8 to 2.2 m in the ice on the east lobe of Lake Bonney [Adams et al., this volume, also see color images on the CDROM that accompanies this volume]. Adams et al.[this volume] have shown these bubble morphologies to be indicative of refrozen pockets of liquid water. Bubbles indicative of pockets of liquid water were conspicuously absent both above and below these depths. The bubble distributions and morphologies [Adams et al., this volume], together with our energy budgets, lead us to conclude that the values for V_w of 20–30% in the east lobe of Lake Bonney over the 1 to 2 m depth intervals are indicative of a 30 to 40 cm layer consisting of 80 to 90% liquid water at a depth of about 2 m from the ice surface. We further contend that the ice above and below this liquid water layer remains relatively solid (with liquid water comprising 5 to 10%). Values for

TABLE 2. Energy Budget Analysis Based on the Ice Temperature Records from the East Lobe of Lake Bonney for Years 1993 to 1995. Period of Freezing Front Passage is the Time during which the Freezing Front Passed from the Top to the Bottom of the Given Depth Interval. The Average Temperature Gradient ($\partial T/\partial z$) was the Average Temperature Gradient above the Freezing Front during the Time when the Freezing Front was Passing through the Depth Interval.

Year and Thermocouple Intervals	Depth Interval (m)	Period of Freezing Front Passage	Average $\partial T/\partial z$ (°C m^{-1})	Average ∂T (°C)	E_c	E_s	E_r	E_l	V_w (%)
1993									
T # 1 – T # 2	0.5 – 1.0	12 Feb – 9 Mar	7.7	3.1	25,800	5460	16,400	14,860	9.9
T # 2 – T # 3	1.0 – 2.0	9 Mar – 26 Apr	15.2	7.0	97,700	1240	4480	94,460	33.3
T # 3 – T # 4	2.0 – 3.5	26 Apr –18 Jul	8.7	5.6	96,500	9900	0	106,400	23.8
1994									
T # 1 – T# 2	0.5 – 1.0	10 Feb – 27 Feb	12.3	7.0	28,300	4190	23,900	8590	5.7
T # 2 – T# 3	1.0 – 2.0	27 Feb – 9 Apr	13.5	9.1	74,500	16,200	9350	81,350	27.3
T # 3 – T# 4	2.0 – 3.5	9 Apr – 20 May	9.0	9.3	49,600	16,300	0	65,900	14.7
1995									
T # 1 – T # 2	in air – 0.25	N.D. – 12 Feb	N.D.	N.D.	N.D.	N.D.	N.D.	N.D.	N.D.
T # 2 – T # 3	0.25 – 1.25	12 Feb – 25 Mar	13.4	10.0	73,800	17,700	18,800	72,700	24.4
T # 3 – T # 4	1.25 – 2.75	25 Mar – 26 Apr	14.1	4.9	60,500	8540	304	69,040	15.4

N.D. = not determined because thermocouple # 1 was out of the ice during the freezing period.
E_c, E_s, E_r, and E_l units are in KJ m^{-2}.

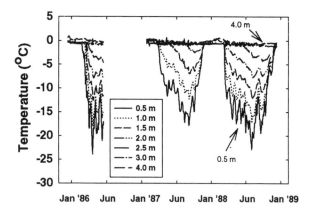

Fig. 3. Ice temperature records from 1985 to 1988 in Lake Hoare.

V_w of 24% and 15% in the lower depth intervals for years 1993 and 1995 are likely to have resulted from a portion of the liquid water pockets extending below the upper thermocouple defining the lower depth interval.

Lake Hoare

Ice temperatures in the austral summers at Lake Hoare were isothermal (at 0°C) from November 16, 1987 to February 7, 1986 (Figure 3) for a total of 111 days. Because the thermocouples were placed in the ice after the ice had become isothermal in 1985 and there

was a data recording gap in 1986, the number of days of isothermy could not be determined for these years.

Timing of the freezing front progression past the first thermocouple differed each year. In 1986, the –1°C isotherm passed the first thermocouple (T#1) on February 24, whereas it took an extra month to reach T#1 in 1987 (March 25), and passed T#1 on March 8, 1988. Interpreting this year-to-year variation in the timing of the initial freezing front progression is confounded by the movement of the thermocouples relative to the surface of the ice over time.

Unlike the thermocouples at Lake Bonney, the thermocouple string at Lake Hoare was not repositioned during the second austral summer of the time-series and the records of the depths of the thermocouples are less complete. However, by analyzing the ice temperature gradients during times when the freezing fronts had reached the bottom of the ice (i.e., late winter to early spring), we were able to estimate the depth of the thermocouples when the thickness of the ice was known. Temperature gradients in the ice were computed from temperature profiles in late winter (Figure 4). The depth at which the ice temperature was 0°C (the y-intercept) was then assumed to be the thickness of the ice, Z_i. The resultant linear equation for ice temperatures,

$$Z_t = Z_i + \frac{dZ}{dT_i} T_i(Z_t)$$

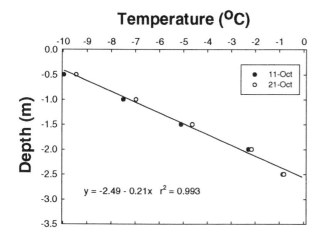

Temperature (°C)

$y = -2.49 - 0.21x \quad r^2 = 0.993$

Fig. 4. Temperature profiles in the ice cover of Lake Hoare during October 1987. Linear profiles ($r^2 = 0.993$) are found during the late-winter/early-spring, which extend from the ice surface to the ice-water interface. Assuming the y-intercept is actually at the ice-water interface then these profiles in conjunction with known ice thickness allowed us to estimate the vertical position of the thermocouples in the ice.

was then solved for the actual depth of the thermocouples, Z_t, using the temperatures measured by the in-ice thermocouples, $T_i(Z_t)$.

The ice thickness, Z_i, at Lake Hoare was 3.2 to 3.3 m in October of 1987 [*Wharton et al.*, 1993]. The average daily temperature gradients at this time were 4.7 to 5.2°C m⁻¹ and the daily average temperatures at T#1 [$T_i (Z_t)$] were –9.45 to –9.96°C. Using this information, T#1 was estimated to have moved downward from its original position of 0.5 m in 1985 to a position between 1.3 and 1.4 m in 1987. Therefore the delay in the timing of the -1°C isotherm passing to T#1 in 1987 can be explained, in part, by the thermocouples moving deeper into the ice despite of the continual upward movement of the ice (Table 3). However, the –1°C isotherm passed all the way to T#5 (2.5 m) by 29 March in 1986 and was between T#1 and T#2 (1.3 to 1.8 m) by this time in 1987, indicating that the year-to-year variation in freezing front progression did exist at Lake Hoare during the study period.

The detailed analysis of the autumnal energy budgets (Table 3) shows liquid water fractions were < 15% at all depths in the ice at the end of the 1985-1986 summer season (Table 3). In the following summers (1986-1987 and 1987-1988) liquid water contents

reached 50 to 70% within the ice interior between the depths of 1.8 to 2.8 m. Therefore year-to-year variations in the freezing front progression at Lake Hoare were largely caused from variations in the liquid water fractions in the ice cover generated in the proceeding austral summers. Again, the energy budgets make no distinction as to the dispersion of the liquid water throughout the depth intervals bounded by the thermocouples.

Observations reported by *Squyres et al.* [1991] and those of *Adams et al.* [this volume] both indicate that sediments are present at depths of 1 to 2.5 m in the ice. Furthermore arching bubble morphologies indicative of pockets of liquid water were present at these depths during 1987 to 1988 [*Squyres et al.*, 1991] and in 1995 [*Adams et al.*, this volume]. These observations are similar to those of the ice at Lake Bonney and further suggest that the high liquid water content in the ice at approximately 2 m (70% by volume) is present as a concentrated layer of water pockets in association with sediments.

Sensitivity Analysis

Variance in estimates of V_w can arise from uncertainties in the following: (1) the ability of the radiative transfer model to accurately predict Q_r; (2) the true depths of the thermocouples; (3) the physical coefficients used to calculate thermal conductivities and specific heat capacities of the ice; and (4) the temperature measurements themselves. A sensitivity analysis was conducted during which coefficients and parameters in the energy budgets were individually varied over ranges that we believe encompass the natural variability in each coefficient or parameter. Table 4 lists these parameters, the range of values tested, and the change in estimates of V_w resulting from a change in each parameter. This sensitivity analysis was performed for two depth intervals on the Lake Bonney data from 1993 to illustrate the source of potential errors in the V_w calculations. The analysis was done at two depth intervals because the calculation's sensitivities are expected to differ with depth in the ice from changes in the relative proportion of E_r to the total energy budget as a function of depth and time of freezing (early autumn versus late autumn-early winter).

As expected, variations in parameters used to calculate Q_r (i.e., depth of the thermocouples, K_d (PAR), R_o and β) produce the largest uncertainties in V_w values near the surface of the ice (0.27% to 14.6% from 0.5 to 1.0 m versus 0.3% to 2.5% from 1.0 to 2.0

TABLE 3. Energy Budget Analysis Based on the Ice Temperature Records from Lake Hoare for years 1986 to 1988. Period of Freezing Front Passage is the Time During which the Freezing Front Passed from the Top to the Bottom of the Given Depth Interval. The Average Temperature Gradient ($\partial T/\partial z$) was the Average Temperature Gradient above the Freezing Front during the Time When the Freezing Front was Passing through the Depth Interval.

Year and Thermocouple Intervals	Depth Interval (m)	Period of Freezing Front Passage	Average $\partial T/\partial z$ (°C m⁻¹)	Average ∂T (C°)	E_c	E_s	E_r	E_l	V_w (%)
1986									
T # 1 – T # 2	0.5 – 1.0	23 Feb – 9 Mar	5.1	2.6	19,200	4600	1,490	22,310	14.7
T # 2 – T # 3	1.0 – 1.5	9 Mar – 14 Mar	5.8	1.4	7820	2420	168	10,072	6.8
T # 3 – T # 4	1.5 – 2.0	14 Mar – 20 Mar	3.5	2.1	5650	3630	73	9207	6.2
T # 4 – T # 5	2.0 – 2.5	20 Mar – 29 Mar	3.1	1.6	7460	2880	31	10,309	6.9
1987									
T # 1 – T # 2	1.3 – 1.8	25 Mar – 3 Apr	24.3	2.7	22,600	4700	46.7	27,253	18.3
T # 2 – T # 3	1.8 – 2.3	3 Apr – 30 May	4.8	2.5	72,900	4430	0	77,330	51.9
T # 3 – T # 4	2.3 – 2.8	30 May – 22 Aug	4.5	2.4	102,000	4350	0	106,350	71.3
T # 4 – T # 5	2.8 – 3.3	22 Aug – N.D.*	N.D.*	N.D.*	N.D.*	N.D.*	N.D.*	N.D.*	N.D.*
1988									
T # 1 – T # 2	0.4 – 0.9	7 Mar – 8 Mar	21.5	1.6	5780	2820	1380	7220	4.8
T # 2 – T # 3	0.9 – 1.4	8 Mar – 8 Mar	0.8	< 1	< 216	< 926	154	< 988	< 0.7
T # 3 – T # 4	1.4 – 1.9	8 Mar – 14 Mar	1.5	0.9	1980	1720	82	178	0.1
T # 4 – T # 5	1.9 – 2.4	14 Mar – 24 Jun	3.4	1.9	94,600	3350	88	97,862	65.6

N.D.* = not determined because the freezing front never reached thermocouple #5.

E_c, E_s, E_r, and E_l units are in KJ m⁻².

m, Table 4). This is because E_r comprises a larger fraction of the total energy budget during the early autumn when daily insulation is still substantial and radiation fluxes near the surface of the ice are larger. Variability in the parameters predicting Q_r have relatively little effect on the calculated values for V_w deeper in the ice (because Q_r deeper in the ice during the late autumn comprises a small fraction of the total energy budget). Other parameters creating notable variations in estimates of V_w are the changes in ice temperatures, and the temperature gradients used to predict conductive heat losses. Because the thermocouples offsets were approximately 0.1 to 0.7°C the temperature measurements are likely to have been accurate only on the order of 0.5°C. Therefore we tested the sensitivity at 1°C to estimate the higher limits to our uncertainties in V_w values. The range of V_w estimates based on this level of temperature precision ranged from 2.4% to 9.1% between 0.5 and 1 m and from 4.4% to 6.2% between 1.0 to 2.0 m. The range of uncertainties in V_w in the 0.5 to 1.0 m depth interval is relatively large when compared to the standard estimate of V_w, which was 9.9%. Whereas the range of estimates in the 1 to 2.0 m depth interval are small compared to the standard estimate of 32%.

Overall the sensitivity analysis shows that the estimates of liquid water contents are rather robust deep in the ice and are less certain in the upper regions. However the uncertainties are not large enough to obscure the trend showing that the majority of the liquid water in these ice covers was present deep in the ice at depths that coincided with the sedimentary layers and arching bubble patterns.

DISCUSSION

Depth of Liquid Water

Most of the data analyzed exhibited maximum V_w values at 2 m in the ice (Figure 5). This location coincides with the depths where sedimentary matter has been shown to be at its maximum [*Wing and Priscu*, 1993; *Adams et al.*, this volume]. The association of high liquid water fractions and arching bubble morphologies in direct association with pockets of sediments all support the notion that absorption of solar radiation by the lithogenic material is the primary process governing the generation of liquid water in these ice covers. If radiation absorption by the ice itself is the primary process generating liquid water, the

TABLE 4. Sensitivity Analysis on the Estimates of Liquid Water Fractions (V_w) in the East Lobe of Lake Bonney (1993) to a Range of Uncertainties in Variables used in the Analyses. Estimates of V_w using the Standard Values for the Variables were 9.9% at 0.5 to 1.0 m and 33.3% at 1.0 to 2.0 m.

Variable	0.5–1.0 m			1.0–2.0 m		
	Standard Values	Range of Values	Range of V_w%	Standard Values	Range of Values	Range of V_w%
Thermocouple depth	0.5 and 1.0	± 0.25	7.6 – 22.3**	1.0 and 2.0	± 0.25	31.9 – 34.3
K_d (PAR)	1.5	± 0.5	10.4 – 10.7	1.5	± 0.5	32.3 – 34.4
R_o	0.05	± 0.05	9.5 – 0.5	0.05	± 0.05	33.2 – 33.5
ß	4.2	± 0.3	9.1 – 10.7	4.2	± 0.3	33.1 – 33.6
k_s	0.2	± 0.2	9.9 – 9.9	0.2	± 0.2	33.3 – 33.3
k_a	0.024	± 0.02	9.9 – 9.9	0.024	± 0.02	33.3 – 33.4
c_s	820	± 50%	9.9 – 9.9	820	± 50%	33.3 – 33.3
c_a	1000	± 50%	9.9 – 10.0	1000	± 50%	33.3 – 33.4
ρ_s	2000	± 50%	9.9 – 9.9	2000	± 50%	33.3 – 33.3
ρ_a	1.2	± 50%	9.9 – 9.9	1.2	± 50%	33.3 – 33.3
V_g (above depth interval)	1	± 100	9.2 – 9.8	1	± 100	32.1 – 33.5
V_g (in depth interval)	1	± 100	9.9 – 10.1	21.6	± 100	33.3 – 33.5
I_d(0,PAR)	135	± 10	7.5 – 10.8	59	± 10	32.8 – 33.9
Days of Freezing	25	± 3	9.2 – 10.7	47	± 3	31.5 – 35.2
Average change in temperature	3.1	± 1	8.7 – 11.1	7.0	± 1	32.7 – 33.9
Average ΔT above depth interval	7.67	± 1	5.4 – 14.5**	7.56	±1	29.1 – 37.6

** = uncertainties greater than 50% of those estimated using the standard values.

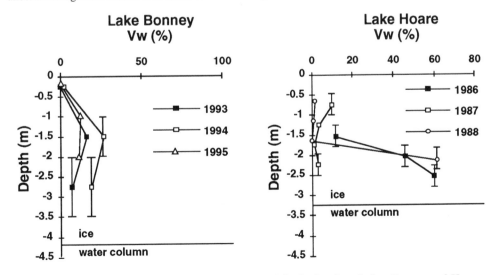

Fig. 5. Vertical profiles of liquid water content (V_w) in the ice from Lakes Bonney and Hoare. Whiskers demarcate the depth intervals overwhich the value of V_w is centered.

vertical distribution of liquid water would be expected to approximate the exponential decrease in short-wave radiation. Lake Hoare did show low V_w values exhibiting an exponential decrease during 1986 (Figure 5), implying that internal melting was not significant during this season. However in the other two years monitored at Lake Hoare and in all three years analyzed at Lake Bonney, internal melting predominated and liquid water contents were well in excess of 10% at depths associated with lithogenic material.

We have made the assertion that radiation absorption by lithogenic material is a primary process generating liquid water in the ice covers. The temperature records and the irradiance measurements before the advancement of the freezing fronts at Lake Bonney provide an independent method to assess whether or not radiation absorption by sediments in the ice can account for the high liquid water fractions at 2 m depth.

Using the previously described radiative transfer model, we have calculated the potential for liquid water generation in the ice during the summer months. For this analysis we assumed the following: (1) a simple exponential decay of I_d(PAR) to the depth of sediment inclusions; (2) that sediment inclusions absorb 100% of the energy impinging on their upper cross-sectional surface area; and (3) absorbed energy melts ice only during the period when the ice is isothermal.

Using the downwelling irradiance data at the surface of the ice at Lake Bonney [*Dana et al.*, this volume] as input for the model, we calculate that energy absorption would have been 1.2×10^8 J m^{-2} of sand cross-section during the 86 days of ice isothermy during the 1993-1994 season (Figure 2) (assuming the sediments were at 2 m). This first-order calculation translates into 0.39 m^3 H$_2$O (m^{-2} sand area) yr^{-1} (i.e., 39 cm yr^{-1}). Assuming the arching bubbles are remnant traces of liquid water pockets [*Adams et al.*, this volume], their dimensions are proxies for previous internal ice melt. In 1995 these arching bubbles extended 20 to 40 cm above sand inclusions [*Adams et al.*, this volume] and are assumed to represent a water pocket formed during the previous year with dimensions of 0.20 to 0.40 m^3 (m^{-2} sand area). These dimensions are on the same order as the 0.39 m^3 H$_2$O (m^{-2} sand area) predicted from the model. Therefore the solar radiation absorption in the ice interior by the sedimentary layers would have been sufficient to melt the pockets of liquid water within the ice interior.

Interannual Variations in Liquid Water Contents

Our previous calculations show that radiation absorption by lithogenic matter is a primary factor generating liquid water in the ice covers. It follows that annual variations in the liquid water contents should be related to factors effecting this net energy transfer. These factors include the optical properties of the ice and (if present) snow, the ice depth of the sediments, the flux of short-wave radiation, and the length of time flux of short-wave radiation, and the length of time over which absorbed radiation produces ice melt (i.e., the time over which the ice is isothermal). The length of time over which the ice is isothermal is influenced by all of the factors governing the total energy budget of the ice such as long-wave radiation, atmospheric temperatures, relative humidity, windspeed, in addition to short-wave radiation.

Differences in the liquid water contents of the Lake Hoare ice between 1986 and 1987 were likely to have been caused by interannual differences in the local weather and atmospheric conditions. During the austral summer of 1985 to 1986 the number of degree-days above freezing was 46 compared to 90 in 1986 to 1987 [*Clow et al.*, 1988]. Therefore the ice was isothermal for a longer period of time in the austral summer of 1986 to 1987, and the liquid water generation within the ice would have been greater, despite the fluxes of short-wave radiation being comparable [*Clow et al.*, 1988].

The high liquid water content at the end of the 1987 season at Lake Hoare indicates that the ice covers can reach advanced stages of deterioration. If the liquid water fractions in the upper 1.3 m were comparable to the 1986 season (i.e., 10 to 20%) we then calculate that the 3.3 m thick ice cover contained roughly 2 cubic meters of solid ice per square meter of ice cover. The ice cover had lost approximately one-third of its total mass. In order to completely melt, the ice cover would have to loose the additional two-thirds of the ice. This is unlikely to occur in one season given the current climate conditions. However two or three austral summers with a large number of degree days above freezing coupled with mild winters with lower than usual number of degree days below freezing could lead to the complete loss of the ice covers. These specific conditions which could lead to ice-out should be further explored through additional model constructions and analysis.

Ecological Considerations

Ice covers control the fluxes of energy and materials into and out of the lakes in the McMurdo Dry Valleys and, therefore, govern the energetics and biogeochemical cycles within the lakes. For example, the flux of photosynthetically active radiation through the ice is the primary path of new energy into the lakes; this flux of energy is primarily controlled by the thickness and optical properties of the ice covers [*Palmisano and Simmons*, 1987; *McKay et al.*, 1994; *Howard-Williams et al.*, this volume]. The optical properties of ice are functions of the amount and types of impurities [e.g., *Fritsen et al.*, 1992; *McKay et al.*, 1994] as well as the number, size, and shapes of vapor inclusions. Bubble morphologies and distributions are directly related to the liquid water generated in proximity to sediment inclusions [*Adams et al.*, this volume]. Liquid water may also fill bubbles (either entirely or partially) in the ice and refreeze during the winter season creating opaque disk-shaped fractures along the basal-plane of the ice crystals [*Adams et al.*, this volume]. Through these processes liquid water generation in the ice covers directly alters the distributions and morphologies of bubbles and the transparency of the ice.

The quantitative effects of the bubbles created by the process of liquid water generation and refreezing on ice optics has yet to be determined. However *McKay et al.* [1994] found it necessary to treat the ice covers as having several horizontal layers with distinct optical properties which varied independently throughout the year in order to achieve agreement between radiative transfer models and irradiance measurements. The analysis of *McKay et al.* [1994] further implies that the seasonal metamorphosis associated with melting is quantitatively important in determining variations in radiant energy fluxes through the ice covers.

The internal melting also creates liquid water habitats for microbial life co-occuring with the sediments in the ice [*Wing and Priscu*, 1993; *Fritsen and Priscu*, 1996; *Priscu and Fritsen*, 1996]. Once these microbes are exposed to liquid water they are capable of net photosynthesis and growth. New nutrients required for net growth are made available through the melting process. In 1995, inorganic nitrogen concentrations ranged from 0.61 to 4.8 mmol N m^{-3} in ice cores below the depths of the sediments in the east lobe of Lake Bonney. If one assumes that ice sediments melt approximately 0.3–0.4 m^3 m^{-2} yr^{-1} as they move downward (to balance the net annual upward ice movement) then the sediments and associated microbes encounter new nitrogenous nutrients in ice meltwater at rates of 0.18 to 1.92 mmol N m^{-2} yr^{-1} as they melt downward each summer season. The supply of nutrients to these microbes is likely to influence how fast these populations grow and accumulate.

Acknowledgments: We wish to thank members of the field teams who contributed to the installation and maintenance of the data aquisition systems on the east lobe of Bonney, and Gary Clow for his initial work on acquiring ice temperature records from Lake Hoare. This work was funded by the National Science Foundation Office of Polar Programs grant OPP-94-19423 and OPP-92-11773 to J.C.P.

REFERENCES

Adams, E.E., C.H. Fritsen, and J.C. Priscu, Permanent ice covers on lakes in the McMurdo Dry Valleys, Antarctica: Bubble Morphologies and Metamorphosis, this volume.

Anderson, D.W., R.A. Wharton, Jr., and S.W. Squyres, Terrigenous clastic sedimentation in Antarctic dry valley lakes. in *Physical and Biogeochemical Processes in Antarctic Lakes, Antarctic Research Series*, vol 59, edited by W. Green and E. Friedmann, pp. 71-81, AGU, Washington D.C., 1993.

Clow, G.D., C.P. McKay and R.A. Simmons, Climatological observations and predicted sublimation rates at Lake Hoare, Antarctica. *J. Climate*, 7: 715-728, 1988.

Craig, H., R.A. Wharton Jr., and C.P. McKay, Oxygen supersaturation in an ice-covered Antarctic lake: Biological versus physical contributions. *Science*, 255: 218-221, 1992.

Dana, G.L., R.A. Wharton Jr., R. Dubayah, Solar radiation in the McMurdo Dry Valleys, Antarctica, this volume.

Doran, P.T., R.A. Wharton, and B.W. Lyons, Paleolimnology of the McMurdo Dry Valleys, Antarctica. *J. Paleolimnology*, 10: 85-114, 1994.

Fritsen, C.H., and C.W. Sullivan, Influence of particulate matter on spectral irradiance fields and energy transfer in the Eastern Arctic Ocean. SPIE Vol 1750. *Ocean Optics* 11, 527-541, 1992.

Fritsen, C.H., and J.C. Priscu, Photosynthetic characteristics of cyanobacteria in permanent ice-covers on lakes in the McMurdo Dry Valleys, Antarctica. *Antarctic Journal of the United States*, 31, 1996, In press

Howard-Williams, C., A. Schwarz, I. Hawes, and J.C. Priscu, Optical Properties of Dry Valley Lakes, this volume.

Kirk, J.T.O., *Light and Photosynthesis in Aquatic Ecosystems*. 2nd Edition Cambridge, 1994.

Lizotte, M.P., and J.C. Priscu, Spectral irradiance and bio-optical properties in perennially ice-covered lakes of the dry valleys (McMurdo Sound, Antarctica). *Antarctic Research Series*. 57, 1-14, 1992.

McKay, C.P., G.D. Clow, D.T. Anderson, and R.A. Wharton Jr., Light transmission and reflection in perenially ice-covered Lake Hoare, Antarctica. *J. Geophys. Res.*, 99(C10), 20,427-20,444, 1994.

Morel, A., and R.C. Smith, Relation between total quanta and total energy for aquatic photosynthesis. *Limnol. Oceanogr.,* 19, 519-600, 1974.

Neddell, S.S., D.W. Anderson, S.W. Squyres, and F.G. Love, Sedimentation in ice-covered Lake Hoare, Antarctica. *Sedimentology*, 34: 1093-1106, 1987.

Palmisano, A.C., and G.M. Simmons Jr., Spectral downwelling irradiance in an Antarctic Lake. *Polar Biology*, 7, 145-151, 1987.

Priscu, J.C., Extreme supersaturation of nitrous oxide in a poorly ventilated Antarctic lake. *Limnol. Oceanogr.,* 41, 1544-1551, 1996.

Priscu, J.C., and C.H. Fritsen, Antarctic lake ice microbial consortia: origin, distribution and growth physiology. *Antarctic Journal of the United States*. 31, 1996, In press.

Roesler, C.S., and R. Iturriaga. Absorption properties of marine-derived material in Arctic sea ice. *SPIE Ocean Optics* 12, 1994.

Spigel, R.H., and J.C. Priscu, Physical limnology of the McMurdo Dry Valley Lakes, this volume.

Squyres, S.W., D.W. Anderson, S.S. Neddell, R.A. Wharton Jr., Lake Hoare, Antarctica: Sedimentation through a thick perennial ice cover. *Sedimentology,* 38, 363-379. 1991.

Wharton Jr., R.A., C.P. McKay, R.L. Mancinelli, and G.M. Simmons, Perennial N2 supersaturation in an Antarctic lake, *Nature*, 325, 343-345, 1987.

Wharton Jr., R.A., C.P. McKay, G.D. Clow and D.T. Anderson, Perennial ice covers and their influence on Antarctic lake ecosystems, in *Physical and Biogeochemical Processes in Antarctic Lakes, Antarctic Research Series*, vol 59, edited by W. Green and E. Friedmann, pp. 53-70, AGU, Washington D.C., 1993.

Wing, K.T., and J.C. Priscu. 1993. Microbial communities in the permanent ice cap of Lake Bonney, Antarctica: relationships among chlorophyll-*a*, gravel, and nutrients. *Antarctic Journal of the United States*. 28(5): 247-249.

Adams, E.E. Department of Civil Engineering, Montana State University, Bozeman, MT 59717.

Fritsen, C.H., Department of Biological Sciences, Montana State University, Bozeman, MT 59717.

McKay, C.P., Solar Systems Exploration Branch, NASA Ames Research Center, Moffett Field, CA 94035

J.C. Priscu, Department of Biological Sciences, Montana State University, Bozeman, MT 59717.

(Received December 10, 1996; accepted April 30, 1997.)

PERMANENT ICE COVERS OF THE MCMURDO DRY VALLEY LAKES, ANTARCTICA: BUBBLE FORMATION AND METAMORPHISM

Edward E. Adams

Civil Engineering Department, Montana State University, Bozeman, Montana

John C. Priscu and Christian H. Fritsen

Department of Biological Sciences, Montana State University, Bozeman, Montana

Scott R. Smith and Steven L. Brackman

Civil Engineering Department, Montana State University, Bozeman, Montana

The permanent ice covers of liquid water based lakes in the McMurdo Dry Valleys are thermodynamically active and display a well defined but transitory stratigraphy. We discuss the annual development of the physical structure of the ice based on field measurements, data gathered during the austral winter and spring of 1994 and 1995, laboratory experiments, and quantitative analysis. In general, the ice growth takes place on the bottom of the ice cover with ablation from the top. Sediment deposited on the ice surface by aeolian processes migrates downward through the ice. The migration is driven by a combination of solar absorption and seasonal warming, leaving a liquid melt trail in its path. The sediment collects in discrete pockets forming a layer with clean ice above and below. Attenuation of solar energy coupled with the brief duration of the relatively warmer portion of the summer season limit the sediment's level of descent. During the austral summer an aquifer is created in the ice, with its lower boundary marked by the sediment layer. The aquifer is connected to the lake water through conduits and the lower ice remains essentially dry. A complex ice stratigraphy is produced as the result of top down freezing of the liquid water in the ice during fall and winter. Inverted tear-drop shaped bubbles with diameter generally under 5mm are produced in the upper meter of the ice. Arching plume-like bubbles and umbrella shaped waves of small spherical bubbles develop as the liquid freezes in the vicinity immediately above the sediment pockets. These patterns are governed by the shape of the freezing front. Liquid filled cavities in the ice induce local curvature of the freezing front, the shape of which is determined by differences in thermal conductivity of the water phases. Hoar frost, produced by temperature gradients, is apparent on the upper surfaces of many bubbles. Just below the sediment, a cluster of circular horizontal fractures develop when expansion due to the phase change of liquid entrapped in cylindrical bubbles causes failure. Fracture occurs on the basal plane of the S1 (c-axis vertical) ice. The lower region of the ice cover is characterized by vertically oriented cylindrical gas bubbles that develop when water freezes to the bottom of the ice cover. The bubbles, fractures, and sediment configuration influence light transmission/absorption, heat flux, and mass transport, all of which are important to the biogeochemical processes in the lake.

281

INTRODUCTION

In the Antarctic environment, the lake ice cover is a dynamic but permanent geologic structure. The ice is an essential feature influencing biologic activity in the liquid water beneath the ice cover and within a sediment layer in the ice itself [*Wing and Priscu*, 1993; *Priscu*, 1995; *Lizotte et al.*, 1995; *Fritsen et al.*, this issue]. An understanding of the physics of the ice on temporal and spatial scales will provide new information on the influence of the physical environment on the microbial populations both within the ice and in the liquid water column beneath the ice. Most of the lakes in the McMurdo Dry Valleys of Antarctica are unique in that they maintain a perennial ice cover overlying liquid water. The geomorphology and complex stratigraphy of the ice on these lakes determines the transmission, scattering and absorption of infrared and visible radiation [*Lizotte and Priscu*, 1993; *Howard-Williams et al.*, this volume]. In addition, ice shields the underlying liquid from wind induced mixing [*Spigel and Priscu*, this volume]. Ice thickness, which varies among the lakes, ranges from 3 to 6 m and is maintained by accretion on the bottom, at the liquid-ice interface. New ice formation is balanced by sublimation [*Clow et al.*, 1988] at the upper surface and melt on the bottom during the summer. The lake water is replenished by glacial melt [*Chinn*, 1993].

As early as 1966, *Henderson et al.* [1966] noted that a liquid "water table" developed within the ice cover during summer, however little research has examined the influence of the liquid water on the ice bubble morphology. *Squyres et al.* [1991] encountered liquid water 2.1 m below the surface while constructing sampling holes in Lake Hoare during November. The ice was sufficiently permeable so that the hole could not be pumped dry. They also noted that the depth at which liquid water was encountered was coincident with the bottom of in-ice sediment. *Squyres et al.* [1991] described bubble morphology within the ice hole. Some of the structure they observed has similarities to that observed in second year arctic lake ice [*Swinzow*, 1966]. *Adams et al.* [1996] noted that in a core extracted November 17, 1994 liquid was present near a sand inclusion 250 cm below the ice surface in otherwise dry ice. Within days, liquid saturated ice was encountered approximately 100 cm below the surface in the same vicinity. Liquid water ascended to the hydrostatic level of the lake after the in-ice aquifer was penetrated. Liquid water was not present at the surface, so downward infiltration did not contribute to the aquifer at this time.

Ice grows in the direction of the temperature gradients that must exist at the solid-liquid interface for active freezing to occur. Vapor bubbles in the ice are the product of gas exsolution and occlusion during freezing and are oriented relative to the direction of ice growth [*Carte*, 1961; *Bari and Hallet*, 1974]. As the ice front advances downward, the solid phase rejects dissolved gases thereby in-

creasing the supersaturation level of the liquid at the interface until, in the presence of a nucleate, a critical value for bubble formation is reached. A consequence of bubble formation is that the supersaturation of the adjacent liquid is lowered. Size, number, and shape of bubbles are to a large degree a function of ice growth rate. As the growth rate increases, the number of bubbles also increases, but the size decreases. If the growth rate of the ice and of the supersaturation of the liquid are "balanced," cylindrical bubbles will form. Waves of bubbles will form during steady freezing when the formation reduces supersaturation sufficiently to constrict the bubble; continued freezing again increases the supersaturation level causing the process to repeat.

This paper presents a physical description of the salient characteristics of the morphology of the ice structure in the permanent ice covers of lakes in the McMurdo Dry Valleys. In addition laboratory experiments that corroborate the observed morphology of in situ lake ice are described.

FIELD SITE DESCRIPTION

The McMurdo Dry Valleys of Antarctica are located at 160°–164° E, 76°30′–78°20′ S. These valleys receive precipitation on the order of 10 mm·y-1, [*Bromley*, 1986] most of which is lost to sublimation. This low precipitation coupled with the continental ice sheet barrier formed by the Transantarctic Mountains produce the dry valley environment [*Chinn*, 1993]. The lakes, situated on the valley floor, are surrounded by a barren rocky/sandy mountainous landscape with alpine and continental glaciers entering from the mountainsides and along the valley floor.

We observed considerable variation of the ice covers among lakes in the McMurdo Dry Valleys with respect to sediment content, internal structure, and surface topography. Lake Vanda for example is virtually free of sediment and has a smooth surface while the ice cover at Lake Miers has large sand deposits over a meter high and a very rough ice topography. To offer a concise presentation, the discussion presented here will focus on Lake Bonney, which is intermediate among lakes with respect to sediment deposition and surface topography.

METHODS

Field

Ice cover morphology was determined largely by field studies performed during the austral spring 1994 and late winter 1995. The 1995 late winter (August–September) study provided an opportunity to examine the stratigraphy of the ice before being altered by seasonal warming.

Ice cores were obtained using a 3 inch SIPRE ice coring tool. Cores extracted during winter extended from the surface of the ice to within 50 cm of the ice-liquid interface (total ice thickness 4 m). The remainder of the ice was then penetrated using an ice auger and the ice thickness measured by lowering a weighted tape measure, constructed so that it hooked onto the bottom of the ice.

Immediately upon extraction, ice cores were placed in flexible polyethylene tubing and labeled for position and orientation. A number of the samples were examined in a field laboratory on Lake Bonney where a photographic record of the morphology was made. The morphology of the ice was exposed using a warm aluminum plate to melt slightly the surface of the ice core in order to eliminate scratches left by the corer.

A second method used to examine internal structure of the ice on Lake Bonney in situ during the 1995 winter study was a 3.4 m deep pit excavated using an oil free chain saw. A hot air blower was used to melt slightly or "polish" the exposed pit wall so that the internal structure of the ice cover could be readily examined. This early season pit proved to be particularly valuable for obtaining large scale information on the ice stratigraphy. Large blocks of ice were also removed from the pit for examination.

Year round air and ice temperature data at Lake Bonney were recorded using thermocouples on a Campbell 21X data logger. They were placed in the ice at 0.5 m, 1 m, 2 m and 3.5 m below the upper surface of the ice [*Fritsen et al.*, this volume].

Laboratory

Ice was grown in a cold laboratory to examine bubble morphology resulting from downward freezing of an advancing ice front. The ice was grown in a nominal 60 cm diameter by 22 cm deep cylindrical vessel. Sides and bottom of the open container were well insulated. The vessel was placed on the floor of a –20°C cold room and immediately filled with 21°C distilled water. The ice structure for the laboratory and some of the lake ice was ascertained using cross polarized light on thin cross sections, in conjunction with polyvinyl formal crystal etching [*Higuchi and Muguruma*, 1958].

In another process, crystallographic oriented clear ice samples, used for subsequent experiments, were produced using the following laboratory procedure. Distilled water was partially degassed by boiling for 3 to 5 hours, then placed in sealed containers and cooled to approximately 1°C. The insulated container described above, which had been maintained at 21°C, was placed in the freezer and filled with the prepared water. The water was kept in still air conditions to minimize mixing from air currents, and allowed to freeze. This process produced a vertical c-axis oriented crystal structure in the ice. An average a-axis length of about 2 cm resulted. In other samples, a chilled insulated cover placed over the vessel increased the a-axis to 9 cm. The upper 15 cm of the sample yielded the bubble free ice. Samples of the clear ice approximately 20 cm

long, 13 cm wide, and 10 cm deep were extracted from the top center of the larger unit.

Experiments to determine the cause of horizontal fractures observed in the lake ice were conducted using these clear oriented samples. Three 1.3 cm diameter holes were drilled to a depth of 8.5 cm into each block. The holes were used as a model of vapor cylinders observed in the Antarctic ice. Holes were drilled in the ice blocks at three different orientations: parallel with the c-axis, perpendicular to the c-axis, and oriented at 45° to the c-axis. The cylinders were filled with water using a large syringe to avoid trapping air; the sample was then frozen. Top down freezing of the liquid-filled cylinders was achieved by floating the blocks in an ice-water bath in the middle of the same insulated vessel used to grow the primary ice sample. The ensemble was then frozen in –10°C laboratory over the course of several days.

Ice specimens for studying bubble morphologies above the sediment pockets were developed by freezing water in vertically oriented, transparent polycarbonate cylinders placed on an insulated platform in a cold laboratory. Dimensions of the polycarbonate freezing containers were 8.2 cm inside diameter, 0.3 cm wall thickness, and 30 cm in height. A 9 cm PVC couple securely positioned a sheet of latex rubber on the container bottom to allow for expansion of the ice. Thermal conductivities of the polycarbonate, liquid water, and ice are 0.17 W·(m°K)$^{-1}$, 0.58 W·(m°K)$^{-1}$, and 2.17 W·(m°K)$^{-1}$ respectively.

FIELD OBSERVATIONS

A description of bubble structure, the sediment layer within the ice cover, internal fracturing, internal melting, and other features are presented in this section. This narration describes the structure from the bottom of the ice sheet upward.

In the bottom 1.5–2 m of the ice cover, vertically oriented vapor cylinders up to 20 cm long are the predominant feature. The ice in this region varies from relatively clear sections to discrete zones with a high concentration of cylindrical bubbles of varying diameter, some in excess of a centimeter (Figure 1). Larger diameter cylinders appear in conjunction with a lower concentration of bubbles.

A feature first observed during August and September 1995 is a zone of horizontal fractures. This zone is located at about 1.5 m from the bottom of the ice and consists of horizontal fractures up to 10 cm in diameter. In cross section these appear as thin horizontal planes with a string of fine vertical bubbles located near the projected center (Figure 2a). When viewed from an oblique angle the bubble strings pass through or end at the nominal centroid of a circular crack. Several of these fracture disks may be apparent along a bubble string (Figure 2b). Frequently, this line of bubbles extends upward to the longitudinal axis at the bottom of a vertical vapor cylinder (Figure 2a). Ice on the bottom of the cylinder bulges upward into the tube

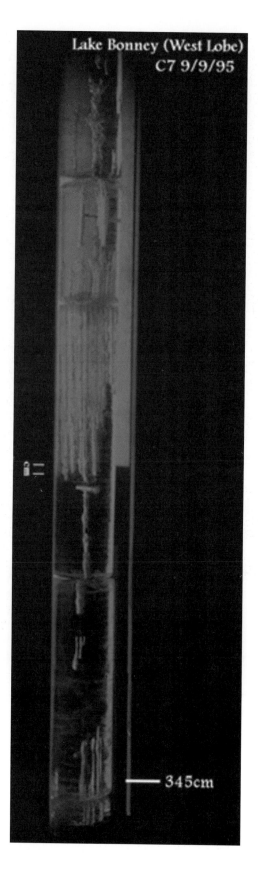

to form a convex ice surface. When viewed from above the fractures appear as circular white disks.

Located at about 1.6 to 2 m from the bottom of the ice cover, there is a zone of sediment composed of sand and gravel sized particles (Figure 3a). A number of former vapor cylinders have been filled with sediment in the lower region of the layer. Above the sediment filled cylinder layer is a zone where sediments form larger pockets several centimeters across. A lower concentration and much smaller clusters of sediment composed of only a few sand grains occurs in the upper portion of this region.

A striking change in the structure of the bubbles occurs at the sediment layer. A few of the more distinct general patterns are presented in Figures 3b and 3c. Some are elongated but not strictly vertical and tend to arc in a plume-like pattern from the sediment pocket. Another morphology consists of waves of small spherical bubbles arranged in umbrella patterns over the sediment pockets (Figure 3c). Plumes have been observed in excess of 30 cm in vertical extent (Figure 3d). Other tapering tubular bubbles grow along the surface of the sediment. Chains of larger generally vertical, inverted teardrop shaped bubbles can be seen above the center of some of the pockets (Figure 3b). (Color images of figures 3a through 3d, which show more detail, are included in the CDROM accompanying this volume.)

In the region above the sediment layer and its associated bubbles, the bubbles are generally smaller and have an inverted teardrop shape (Figure 4a). These are mixed with vertically oriented tubes. In the region extending from the sediment layer to the top of the ice sheet, bubbles were distinctly frosted on the upper surface (Figure 4b).

Figure 5 shows a sample of an ice core taken from just above the sediment layer on Lake Bonney shortly after liquid was first encountered in the ice (November 1994).

ANALYSIS AND LABORATORY RESULTS

Ice Growth and Gas Supersaturation

Thermocouple data from the east lobe of Lake Bonney during 1993 indicates that active ice growth will take place at the liquid-ice interface between August and November for a period of about 100 days per year (Figure 6). The rate of growth of the ice, dX/dt, is calculated from the energy equation (Figure 7) in the form

$$\rho_i L \frac{dX}{dt} = (k_i \frac{\partial T_i}{\partial x} - k_i \frac{\partial T_l}{\partial x}) \qquad (1)$$

Fig. 1 Section of "3 inch" ice core (10 cm diameter) showing the vertical cylindrical bubble structure. (The horizontal cylinder apparent near the scale is an artifact of sampling.)

Figure 2a

Fig. 2. Horizontal fractures and thin bubble train which occur in remnants of the vertical cylindrical vapor bubbles Fractures are the result of stresses produced by trapped liquid water expansion due to freezing. Note also the concave lower surface of the remaining portion of the vapor cylinder (a). A series in of fractures are produced due to stress relief as a result of the ice failure followed by continued freezing (b).

Figure 2b

where the subscripts i and l represent ice and liquid, t is time, k is the thermal conductivity, T is temperature, ρ is the density, and L is the heat of phase change. Assuming a temperature gradient in the liquid of 0.5°C·m-1 and an ice thickness of 3.9 m at the start of the growth period, this model predicts ice growth of about 3.5 mm d-1 (Figure 7) between August and November.

As the ice surface advances during freezing and rejects vapor into the liquid at the ice surface, a gas concentration gradient normal to the ice surface develops. The governing equation describing the ratio of gas concentration to some initial value as a function of time, t, and position, x, in the fluid ahead of the moving ice front is given as [*Carte*, 1961]

$$\frac{C(x,t)}{C_o} = 1 + \frac{1-\xi}{\xi}\left\{exp\left(-\frac{Rx}{D}\right)\ldots\right.$$

$$\left. - exp\left[-\frac{R}{D}(1-\xi)(x+\xi Rt)\right]\right\} \tag{2}$$

where C is the solute concentration, C_o is the initial solute concentration, D is the diffusivity of air in water, x is the distribution coefficient (ratio of solute air in the solid and liquid), and R is the freezing rate of the interface. To illustrate the development of air supersaturation in water at the interface we examine a constant ice growth rate of 3.5

Figure 3c

Fig 3. Sediment layer which is located about 1.5 to 2 m from the bottom of the 4 cm ice cover on Lake Bonney. It is composed of predominantly sand sized particles (a). Complicated vapor bubble morphology such as these bubble chains and plume-like bubbles develop directly above sediment pockets in the lake ice (b). Arching wave of umbrella shaped bubbles (c).

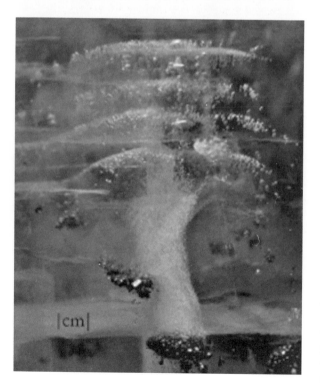

Figure 3b

$mm \cdot d^{-1}$ (based on results shown in Figure 6). The concentration ratio would increase 12 fold, as shown for 30 cm of ice growth (Figure 8), owing largely to the slow diffusion through the liquid. This increase does not account for the development of bubbles which would lower the calculated values. It should also be noted that the lake water is in an initial state of supersaturation. In addition, even slight convection with respect to oxygen and nitrogen that may occur in the lake would lower the concentration at the interface.

cm

Laboratory Bubble Formation

Insulative properties of the polycarbonate cylinders used to examine the sediment related bubble morphology caused the top of the water exposed to the air to freeze first. The combined components of vertical and horizontal heat flux then produce a curved freezing front. Alteration of thermal properties as the result of the phase change further influences the heat flux, contributing to the development of a concave downward, circular paraboloid as the freezing front. Figure 9 illustrates a progression of the freezing front in a vessel held at an ambient room temperature of $-7°C$. Note the parabolic freezing front that produces the arching bubble shapes, the vertical linear front that develops horizontal cylindrical morphologies, and the bubble chains that develop near the top center. The development of a circular paraboloid at the freezing front was common to all samples which were frozen in the 8 x 30 cm polycarbonate containers at temperatures of $0°C$, $-7°C$, and $-23°C$.

The plume-like bubble patterns developed in these experiments for ambient laboratory temperatures of $0°C$ and $-7°C$ but not at $-27°C$. The plume shape was observed to develop normal to the curved path of the advancing ice front. Nearly horizontal cylindrical bubbles developed on the sides at depth following the nearly vertical orientation of the freezing front in this region. This pattern is similar to the vertically oriented cylindrical bubbles that develop at the flat horizontal freezing front in the lake ice and in the large laboratory vessel. Some of the large arching cylindrical bubbles exhibit ridges on the top surface, a characteristic similar to that observed in some of the lake ice. The largest cylindrical and plume like bubbles developed when the ambient laboratory temperature was near $0°C$, inducing a slow growth rate.

Bubble chains in the laboratory ice were located near the top center of the sample cylinder (Figure 10a) interior to the arching and horizontal cylindrical bubbles, and were most pronounced at the slower freezing rates. Bubble chains develop when the ice front is encroaching on the center and the gas concentration increases accordingly. The mechanism of formation is presumably similar to that described by *Carte* [1961] for the waves of bubbles. Bubbles forming near the bottom of the test cylinder were observed to detach and buoyantly ascend to the vertex of the parabaloid during laboratory experiments, contributing to bubble formation at that location.

A cloud of small spherical bubbles developed when the freezing rate was increased in the presence of lower ambient laboratory temperature ($-27°C$). These clouds appear similar to those observed just above the sediment in in the lake ice (Figure 3c). Like bubble chains, cloud bursts appear in the center of the frozen sample. The higher freezing rate results in the large number of small bubbles.

Figure 3d

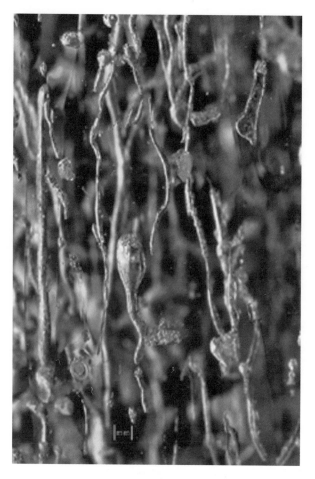

Figure 4a

Figure 4b

Fig. 4 Inverted teardrop and fine tube-like bubbles predominate in the upper regions of the ice cover . The largest bubble in the figure is about 2 mm (a). Frost on the top surfaces is common. The result of sublimation and condensation across the bubble driven by a strong temperature gradient (b).

Umbrella shaped bubble waves (Figure 10b) formed in the laboratory when the vertical walls of the polycarbonate cylinder were wrapped with additional insulation, reducing the horizontal component of freezing. As the curved freezing front advanced, a layer of small spherical bubbles formed, following the contour of the freezing front. The formation of small bubbles presumably lowered gas supersaturation and clearer ice followed until the liquid water again became supersaturated with gas resulting in another wave of bubbles.

These laboratory results are essentially qualitative and were conducted to gain insight into probable physical constraints that would cause the formation of the complex bubble structure observed in the lake ice. In this regard, the experimental results were successful. Morphologies similar to those observed in the lake ice have been reproduced and potential mechanisms that would lead to appropriate boundary conditions are identified in the following section.

Horizontal Fractures

Water frozen in the large vessels, as described in the laboratory methods above, initially forms a cap of ice across the top, trapping dissolved gasses in the liquid below. In cross section, the top 15 cm of the laboratory ice is clear but transitions below to predominantly vertical cylindrical bubbles. Interspersed among the cylinders were occasional chains of teardrop shaped bubbles with the tapered end pointing downward.

Several methods of freezing the liquid which filled the holes drilled in the clear crystal oriented ice samples were investigated to examine the consequence of the manner of freezing. Water injected into the holes of a block which had come to thermal equilibrium with the −10°C cold room froze within seconds. A large number of very small scattered bubbles developed in the ice filled hole, but no cracking occurred. Three other samples were brought to approximately 0°C, and the holes filled with water. The tops of

Fig. 5 An ice core taken when liquid water was present in the ice. This appears to be a remnant of plume-like bubble. Water has drained from the sample.

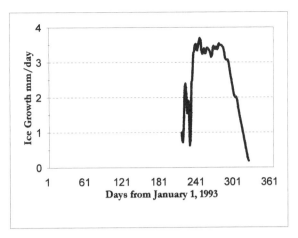

Fig. 7 Calculated ice growth to the bottom of the ice sheet based on thermocoule data for Lake Bonney, 1993.

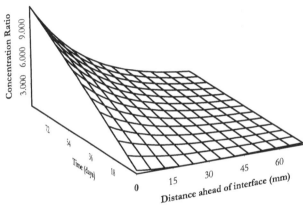

Fig. 8 An example of the supersaturation increase in the liquid at the freezing front for ice to grow 30 cm (about 86 days) at a rate of 3.5 mm·d[-1].

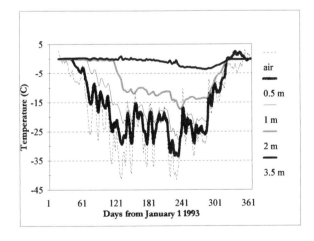

Fig. 6 Temperature profile for Lake Bonney, 1993.

the holes of two were sealed by freezing with a −10°C aluminum plate. Individual samples placed in a −10°C and a −2°C environment produced cracking in each but no orientation was apparent. The third block was placed in the −2°C environment without sealing the top. No cracking was induced in this case.

In each laboratory specimen subjected to the predominant top down freezing process described above, a well defined circular fracture occurred on the basal plane. The basal plane is orthogonal to the c-axis of the ice crystal. This was consistent for the samples composed of large and small crystals and in the case where a hole was placed in a single crystal, regardless of the orientation of the crystal axis relative to the direction of freezing. The basal plane

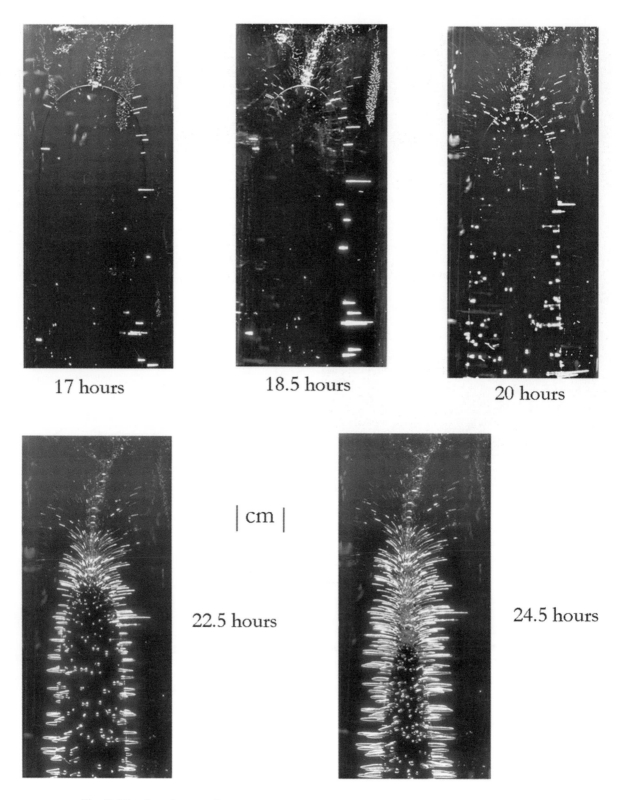

Fig. 9 Time lapse images of ice growing at –7°C in an 8 x 30 cm polycarbonate cylinder. Note the parabolic shape of the freezing front as the ice advances downward and inward. Bubble structure is oriented normal to the freezing front.

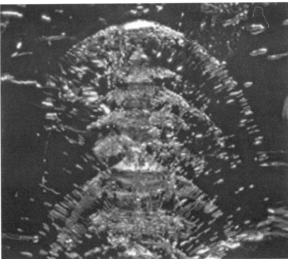

Figure 10b

Fig. 10 Laboratory grown bubble chain and plume-like sturcture (a). Laboratory grown waves of umbrella patterned small spherical bubbles (b).

Figure 10a

failure for the vertically oriented crystal structure is similar to that observed in the lake ice (see Figures 5a and 5b), i.e., a thin line of bubbles developed on the central vertical axis of each refrozen cylinder.

Horizontal fractures in the essentially vertically oriented c-axis lake ice are hypothesized to be the result of hydrostatic pressure induced by the volumetric increase from phase change during top down freezing. An analysis describing the influence on fracturing due to volumetric expansion during solidification is presented. Water expands upon solidification, therefore downward freezing in the cylindrical hole causes compression of the fluid trapped below, producing a rise in the hydrostatic pressure. This pressure then induces a stress field in the surrounding ice which may lead to failure.

Ice with a crystallographic orientation exhibits orthotropic behavior, governed by its crystal structure, whereby (depending on loading conditions) it will tend to fail along its basal plane due to shearing stress [Michel, 1978]. Michel [1978] found that a specimen of ice loaded in uniaxial tension or compression, regardless of the loading orientation with respect to the basal plane, always failed

along the basal plane of S1 ice in shear, except at the extremes when the basal plane is parallel or normal to the direction of a uniaxial load. Ashton [1986] reported a range of 0.2 to 4.0 MPa for shear strength of ice.

In the case of downward freezing of the liquid filled cylinder, the increased pressure in the entrapped fluid is examined by considering the equation for the bulk modulus, κ, of a fluid at constant temperature, T.

$$\kappa = -V \left[\frac{\partial p}{\partial V} \right]_T \qquad (3)$$

where p is pressure and V is volume. The negative sign indicates that as pressure increases volume decreases. Rearranging this equation and expressing it in terms of finite values yields,

$$\Delta P = \kappa \frac{-\Delta V}{V_o} \qquad (4)$$

where V_o is the initial fluid volume. For the sake of our discussion, if we consider 4.5 cm of the original 8.5 cm long (1.27 cm diameter) hole used for the laboratory study to have frozen, an unconfined expansion of 0.47 cm³ would result (assuming the density ratio of ice to water as 0.917). Neglecting visco-elastic deformation of the ice this gives $\Delta V = -0.47$ cm³ and an unconfined initial volume, $Vo = 5.54$ cm³. Using $\kappa = 1964$ MPa for water at 0°C, the increase in pressure is calculated to be $\Delta P = 178$ MPa. By similar analysis, if 2.5 cm in this example had frozen then the pressure would be 70 MPa; freezing 6.5 cm would yield 443 MPa.

This increase in pressure in the fluid induces stress in the surrounding ice. Principal stresses in a thick walled pressure vessel may be calculated in cylindrical coordinates [e.g., *Boresi and Sidebottom*, 1993] using the set of equations,

$$\sigma_{rr} = \frac{p_1 a^2 - p_2 b^2}{b^2 - a^2} - \frac{a^2 b^2}{r^2(b^2 - a^2)}(p_1 - p_2)$$

$$\sigma_{\theta\theta} = \frac{p_1 a^2 - p_2 b^2}{b^2 - a^2} + \frac{a^2 b^2}{r^2(b^2 - a^2)}(p_1 - p_2)$$

$$\sigma_{zz} = \frac{p_1 a^2 - p_2 b^2}{b^2 - a^2} \qquad (5)$$

where σ indicates normal stress and the subscripts imply cylindrical coordinate directions (radial, angular, vertical), a and b are, respectively, the inner and outer diameters of the cylinder, p_1 and p_2 are the internal and external pressures and r is the radial distance to the point of interest. To calculate the increase in stress, ΔP is used for p_1, and p_2 is taken as 0, $a = (1.27/2)$ cm, $b = 30$ cm (the radius of the container in which the sample is centered). Values are calculated at $r = a$ where the ice stress will be a maximum. Using the values discussed above, this yields values for stress of $\sigma_{rr} = -p = -178$ MPa (compression), $\sigma_{\theta\theta} = p = 178$ MPa (tension), and $\sigma_{zz} = 2.0$ MPa (tension). The maximum shearing stress will occur on a plane oriented at 45° to the maximum and minimum normal stresses. In this case $\sigma_{\theta\theta}$ and σ_{rr} are the maximum and minimum values so the maximum shearing stress is oriented on a plane at 45° to the $\rho-\theta$ axes. The maximum shear stress, τ_{max}, is found from

$$\tau_{max} = \frac{\sigma_{max} - \sigma_{min}}{2} \qquad (6)$$

which yields of $\tau_{max} = 178$ MPa for the example under discussion.

In the ideal case of perfectly vertical c-axis crystals and perfectly vertical cylindrical holes, no shear stress would be generated on the basal plane. However if a crystal orientation is rotated about the r coordinate relative to the vertical by an amount ϕ, the shear stress may be computed on this plane using the equation,

$$\tau_{\theta z} = \frac{\sigma_{\theta\theta} - \sigma_{zz}}{2} \sin(2\phi) + \tau_{\theta z} \cos(2\phi) \qquad (7)$$

It would require that a slight off vertical orientation of less than 1 degree of the crystal c-axis would be sufficient to induce shear stress failure in the range reported in the literature [*Ashton*, 1986]. Results from our laboratory experiments and analysis support the contention that horizontal fractures observed in the lake ice are the result of expansion caused by downward freezing of liquid water that has infiltrated cylindrical bubbles.

DISCUSSION AND CONCLUSIONS

Figure 11 presents the spatial and temporal interaction of the processes inherent to the ice covers of the McMurdo Dry Valley lakes. The relationship between mechanisms in the atmosphere, the ice cover and the liquid lake water are shown as seasonal transition between the wet summer months and the dry winter months.

The near surface bubble morphology within the ice in the ridge and troughs differ. Ice in the ridges is more porous than the ice at the trough surface that results from freezing of summer surface melt ponding in depressions. The topology of the ridges provides more surface exposure to the atmosphere, which coupled with its higher porosity ice will likely cause a higher ablation rate of the ridge ice.

During the summer months, sediments absorb solar radiation forming liquid water pockets. Although most of the vertical elongate bubbles remain dry, liquid water infiltrates those just below the sediment layer. Liquid lake water is connected to the sediment layer via vertical cracks in the ice cover. As winter approaches, top down freezing produces teardrop shaped bubbles in the upper region and plume shaped bubbles, umbrella waves, and bubble clouds above the sediment pockets. Temperature gradients across the bubbles produce hoar frost in the teardrop shaped bubbles, which greatly increases light attenuation through the ice. The liquid water that drains into and partially fills the elongate bubbles freezes, producing disk shaped horizontal fractures. The deeper elongate bubbles remain dry and unaffected by the summerwinter transition. Sediment is presumably lost from the ice cover through the vertical cracks in the ice cover. This contention is supported by large discrete sediment piles that have been observed on the bottom of the lakes [*Squyres et al.*, 1991].

It is likely that the liquid water in the summer aquifer eventually saturates the ice from the sediment layer up to the hydrostatic lake level. The morphology in the core sample in Figure 5 indicates that the liquid is penetrating a plume-like vapor inclusion characteristic of those above the sediment. Penetration of this sort increases the liquid water permeability of the ice as the in-ice aquifer develops.

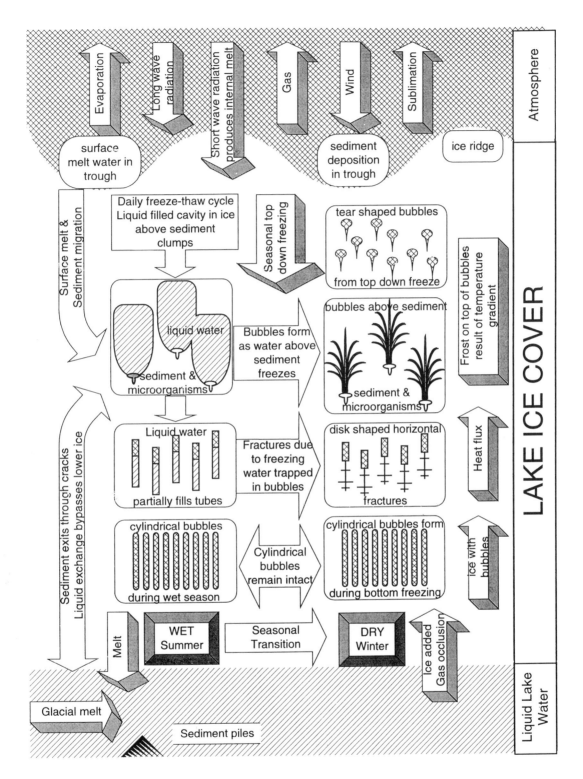

Fig. 11 A conceptual model of the temporal and spatial veriation of the ice cover. Liquid water in the ice cover which develops in the summer months plays a predominant role in the structure of the ice morphology during top down winter freezing. Cylindrical bubble develop from the freezing of lake water at the bottom.

Craig et al. [1992] contend that gas bubbles are a mechanism for the transfer of gas from lake water to the atmosphere. *Craig et al.* [1992] assumed that gas in bubbles formed at the bottom of the ice cover is carried with the ice as it gradually works upward to the ablating surface. This process has also been considered by *Priscu* [1997] and *Priscu et al.* [1996] as a mechanism for the transport of gas from lake water to the atmosphere. Our data show that the *Craig et al.* [1992] scenario is valid for the portion of the ice only below the sediment layer i.e., where the cylindrical bubbles exist, but would not account for the gas transported through the liquid water conduits, or in the region above the sediment. It is likely that when the summer ice is liquid saturated this water is in communication with the atmosphere and presumably will have equilibrium values of the dissolved gases.

Minerals, nutrients, and microbes associated with aeolean sediments migrate through the ice where a portion may enter the water column and the lake bottom. Calculations by *Simmons et al.* [1987], in an examination of the warming of subsurface ice to the melting point, indicate that solar absortion would be ineffective to cause melt for $-1°C$ ice unless the sand particles were very large. However when we consider that in the warmer months the ice becomes isothermal at $0°C$ (see Figure 6), the amount of energy required to facilitate this process is much lower. The equation used by *Simmons et al.* [1987] indicates that, if all other variables are held constant, the radius of the particle varies linearly with the temperature difference between the sediment particle and the ice, with the particle size going to zero as the temperature difference approaches zero.

The continuity of the bubbles in the lake ice just above the sediment zone implies a previous equivalent continuity of the liquid phase in the ice. We believe that a situation exists in which the large volume of liquid water in the aquifer above the sediment layer occurs in pockets formed by the downward sediment migration moving through the liquid-saturated ice, assisted in its descent by absorbing solar radiation. Melt paths result with the sediment pockets at the bottom. We demonstrated in the laboratory that neither sediment nor microbes need be present to form the over sediment bubble morphology, but a mechanism is necessary for controlling the direction of the freezing front. If a liquid filled cavity above the sediment is produced then a situation similar to the laboratory experiments we described would exist, whereby the higher conductivity of the ice would cause a curved freezing front leading to the shapes observed in the lake ice.

In addition to exsolution of gases due to freezing, microbes active when liquid water is available at the sediment layer [*Wing and Priscu*, 1993] produce gas as the byproduct of biologic processes. Although these biologically generated gasses are not required for the development of bubbles in the ice, they can contribute to the supersaturation level and thus modify the bubble morphology in their immediate proximity. Since the presence of the bubbles strongly influences the manner in which light is scattered the microorganisms can contribute to the physical structure of their niche in the ice. Although additional investigation is warranted, the bubble structure will influence the manner in which the light energy will be made available to the absorptive sediment and the microbes, many of which are photoautotrophs [*Wing and Priscu*, 1993]. This concept is particularly intriguing when coupled with the reflection from the horizontal fractures, in that it offers the possibility of additional energy for melt and photosynthesis. Hence a situation may exist where the physical structure of the ice can benefit microorganisms that have been observed to grow in association with the lake ice sediment layer.

REFERENCES

Adams, E. E., J. C. Priscu, and A. Sato, Some Metamorphic Processes in Antarctic Lake Ice. *Antarctic Journal of the United States*, 1996.

Ashton, G. D, *River and Lake Ice Engineering*. Water Resources Publications, Littleton, Colorado, U.S.A. 1986.

Bari, S. A. and J. Hallett, Nucleation and Growth of Bubbles at An Ice-Water Interface. *Journal of Glaciology*. 13(69), 489–520, 1974.

Boresi, A. P. and O. M. Sidebottom, *Advance Mechanics of Materials*, Fifth Ed., John Wiley & Sons, New York, 1993.

Bromley, A.M., Precipitation in the Wright Valley, N.Z. *Antarctic Record* , Special Supplement, 6, 60–68, 1986.

Carte, A. E., Air Bubbles in Ice. *Proceedings of the Physical Society* (London), 77(495),757–768, 1961.

Chinn, T. J., Physical Limnology of the Dry Valley Lakes in *Physical and biogeochemical processes in Antarctic lakes*. Antarctic Research Series, vol. 59, edited by W. J. Green and E. I. Friedman, 1–51, 1993.

Clow, G. D., C. P. McKay, G. M. Simmons, and R. A. Wharton Jr., Climotological Observations and Predicted Sublimation Rates at Lake Hoare, Antarctica. *Journal of Climate* , 1988

Craig, H, R.A. Wharton, and C. P. Mckay, Oxygen Supersaturation in Ice Covered Antarctic Lakes: Biological Versus Physical Contributions. *Science*, 255,318–321, 1992.

Fritsen, C. H., E. E. Adams, C. P. McKay, and J. C. Priscu, Permanent Ice Covers of the McMurdo Dry Valley Lakes, Antarctica: Liquid Water Content, this volume

Henderson, R.A., W.M. Prebble, R.A. Hoare, K.B. Popplewell, D.A. House, and A.T. Wilson, An ablation rate for Lake Fryxell, Victoria Land, Antarctica. *Journal of Glaciology*, 6, 129–133, 1966.

Higuchi, K and J. Muguruma, Etching of ice crystals by the use of plastic replica film, *Journal of the Faculty of Science*, Hokkaido University, Japan, Ser VII, 1(2),81–91 1958.

Howard-Williams, C., A.-M. Schwarz, I. Hawes, and J.C. Priscu. Optical properties of the McMurdo Dry Valley Lakes, Antarctica, this volume

Lizotte, M.P. and J.C. Priscu. Spectral irradiance and bio-optical properties in perennially ice-covered lakes of the dry valleys (McMurdo Sound, Antarctica). *Antarctic Res. Ser.* 57:1–14, 1992.

Lizotte, M.P., T.J. Sharp, and J.C. Priscu. Phytoplankton dynamics in the stratified water column of Lake Bonney, Antarctica: I. Biomass and productivity during the winter-spring transition. *Polar Biology*, In Press

Michel, B., *Ice Mechanics.* Les Presses De LíUniversite Laval, Quebec, 1978

Priscu, J.C. Phytoplankton nutrient deficiency in lakes of the McMurdo Dry Valleys, Antarctica. *Freshwater Biology,* In Press, 1995.

Priscu, J. C., M. T. Downes, and C. P. McKay. Extreme supersaturation of nitrous oxide in a poorly ventilated Antarctic lake. *Limnology and Oceanography*, 1996.

Priscu, J.C., The Biogeochemistry of nitrous oxide in permanently ice-covered lakes of the McMurdo Dry Valleys, Antartica in *Global Change Biology Special Issue, Microbially Mediated Atmospheric Change* (J. Prosser, ed.), 1997.

Simmons, G. M. Jr, R. A. Wharton Jr., C. P. McKay, S. S. Nedell, and G. Clow, Sand/ice interactions and sediment deposition in perenially ice-covered Antarctic Lakes, *Antarctic Journal of the United States*, 237–240, 1987.

Spigel, R.H. and J.C. Priscu, Physical limnology of the McMurdo Dry Valley Lakes, this volume.

Squyres, S. W., D. W. Anderson, S. S. Nedell, and R. A. Wharton, Lake Hoare, Antarctica: sedimentation through a thick perennial ice cover. *Sedimentology.* 38, 363–379, 1991.

Swinzow, G. K., Ice cover of an arctic proglacial lake. *U. S. Army Cold Regions Research and Engineering Laboratory (USA CRREL), Research Report* 155, 1966.

Wing, K. T. and J. C. Priscu, Microbial communities in the permanent ice cap of Lake Bonney, Antarctica: Relationships among chlorophyll-*a*, gravel, and nutrients. *Antarctic Journal of the United States.* 28: 246–249, 1993.

E. E. Adams, Civil Engineering Department, Montana State University, Bozeman, MT, 59717

S. L. Brackman, Civil Engineering Department, Montana State University, Bozeman, MT, 59717

C. H. Fritsen, Department of Biological Sciences, Montana State University, Bozeman, MT, 59717

J. C. Priscu, Department of Biological Sciences, Montana State University, Bozeman, MT, 59717

S. R. Smith, Civil Engineering Department, Montana State University, Bozeman, MT, 59717

(Received December 13, 1996; accepted April 14, 1997.)

THE SOIL ENVIRONMENT OF THE
MCMURDO DRY VALLEYS, ANTARCTICA

Iain B. Campbell and Graeme G.C. Claridge

Land and Soil Consultancy Services, Nelson, New Zealand

David I. Campbell and Megan R. Balks

Earth Sciences, University of Waikato, Hamilton, New Zealand

The soils in the McMurdo Dry Valley region are a key component of the polar desert ecosystem. Formed in an environment of low precipitation, severe cold and minimal biological activity, the soils have distinctive cold desert features in which the principal processes of oxidation and salinization are slowly superimposed on the regolith materials. Although climatic conditions are extreme, there is considerable variation in the soil environment. At the macro-scale five distinct soil regions are identified, broadly corresponding with temperature differences across the dry valley region. Soil moisture and chemical characteristics are important for characterizing the soils. At the microscale, appreciable variations occur over short distances as a result of parent material and site differences which affect the radiational and thermal properties of the soils. Surface albedo and air temperature are key factors influencing the soil thermal regime and moisture availability. Soil salinity is important in determining the occurrence of ice cement and, to an extent, the variation in soil temperature extremes. Because of the great age and stability of the soils and the extremely slow rate at which soil processes operate, the soils of the McMurdo Dry Valleys are very susceptible to damage from human activities.

INTRODUCTION

The Antarctic cold desert is a unique ecosystem which extends over much of ice-free Antarctica, distinguished by extreme cold and extreme aridity. One of the most important components of the ecosystem is the soil.

Throughout the Transantarctic Mountains soils are found on exposed bare ground. The environment in which these soils have formed is characterized by very low temperatures, low precipitation, low available moisture, negligible biological activity, and extraordinary landscape stability. In spite of the slow soil-forming processes, considerable diversity within the soil environment is found. The soil variability is, to a large degree, a result of the marked range of climate that occurs throughout the Transantarctic Mountains. The very great age range of the land surfaces, and the extended time scale [*Prentice et al.,* this volume] over which weathering processes have been operating has had an important influence on soil properties. However there is also much local diversity in soil and ecosystem characteristics as a result of a range of microsite factors and energy relationships.

The McMurdo Dry Valley region, while not covering the complete range of soil conditions found throughout the Antarctic cold desert, nevertheless embraces considerable variation within the soil forming environment. In this chapter, we outline the main soil environments of the dry valleys region and discuss some of the factors and processes responsible for soil and ecosystem diversity.

SOILS OF THE COLD DESERTS

The soils of the ice-free regions of Antarctica, and in particular those of the dry valley region, were described as cold desert soils by *Markov* [1956] and by *Tedrow and*

Ugolini [1966]. Soil development and soil weathering ranges from virtually nil on young land surfaces (<50,000 years), to soils with distinctive pedogenic characteristics on the older land surfaces(>10 my) [*Campbell and Claridge,* 1978, 1981, 1987]. Notable features of the soils (Figure 1) include the widespread occurrence of a pebble or boulder surface pavement, a soil form that is dominated by coarse but extremely variable textures, lack of cohesion and soil structural development, very weakly developed chemical weathering, negligible organic matter content, wide variations in salinity, and the existence of either ice-cemented or non ice-cemented permafrost at variable depth. The extent to which these properties are developed at any particular location is largely dependent on regional and local climate, on site factors such as aspect and proximity to snow or moisture sources, on the age of the land surface, and on the characteristics of the materials from which the soils are formed. In the coldest and most arid parts of the cold desert, along the inland edge of the Transantarctic Mountains, especially in the south, the soils have thick salt horizons containing a high proportion of nitrate and very little chloride [*Claridge and Campbell,* 1968]. In warmer locations to the north and near the coast, the soils are moister and contain smaller amounts of salts, which are largely chloride with a very much lower proportion of nitrate.

With increasing age, the proportion of fine particle size material increases, as does cohesion. There are more obvious changes in soil chemical properties, more particularly in the intensity and depth of oxidation, in the soil salinity, and in the extent to which coarse particles are fragmented by salt weathering. The wide range in soil environments found within the McMurdo Dry Valleys is largely a result of significant climatic, geomorphic, and geological differences within this region. The climate varies from comparatively mild in coastal regions (xerous to subxerous) [*Campbell and Claridge,* 1969], where mean annual temperatures are around −18°C, to moderately severe in the inland areas adjacent to the polar plateau (xerous) [*Campbell and Claridge,* 1969] where mean annual temperatures are estimated to be around −30°C.

The McMurdo Dry Valley region is geomorphologically diverse, with coastal lowland and marine terraces, entrenched inland and coastal valleys with steep sides and long narrow floors, upland valleys, wide cirques, broad plateaus and high mountains. The soil environment is influenced by differences in rock type and by differences in the ages and sedimentary characteristics of many of the glacial deposits. Soils derived from hard rocks such as dolerite or granodiorite, for example, have a relatively high proportion of boulders and coarse particles (Figure 1),

while sand-sized particles are more abundant in soils that are derived from less durable rocks such as sandstones, which occur widely throughout the dry valley region. The amount of energy absorbed by the soil and therefore its thermal characteristics are influenced by color and ground surface texture [*MacCulloch,* 1996].

An important feature of the dry valley region is its exceedingly long glacial history, which is recorded by widespread glacial deposits that date from at least the Miocene [*Prentice et al.,* this volume]. Numerous glacial episodes, many of which have originated from periodic expansions of the West Antarctic Ice Sheet and the alpine glaciers [*Campbell and Claridge,* 1978, 1987, 1988; *Denton et al.,* 1984, *Marchant et al.,* 1993, 1996] have left a wide variety of glacial deposits ranging from deep sandy and silty tills to shallow ice-cored bouldery moraines. The wide range in the physical characteristics of the surface forming deposits together with the prolonged periods during which they have been weathered has resulted in much variation in the soil environment.

SOIL CHARACTERISTICS OF THE MCMURDO DRY VALLEYS

The soils of the dry valleys can be separated into a number of differing groups or classes based on physiographic position. These are soils of the coastal regions, of the main valley floors, of the valley sides, of the upland valleys and of the inland and polar plateau fringes. This generalized grouping of soils largely follows a previous subdivision [*Campbell and Claridge,* 1969, 1982] which recognized that climate differences within the Transantarctic Mountains imparted significant differences to the soil properties primarily as a result of differences in available soil moisture.

Soils of the Coastal Regions

Geography and parent materials. The coastal region includes the coastal fringe of the dry valleys (Figure 2) and other scattered coastal areas in McMurdo Sound from Cape Roberts in the north to Walcott Bay, Brown Peninsula, and Black Island in the south, along with the coastal areas of western Ross Island. It is characterized by young soils with negligible soil weathering features. The soils and land surfaces, up to an elevation of about 1100 m, date from the period when the Ross Ice Sheet withdrew from McMurdo Sound at the end of the Wisconsin Glaciation. Sandy to silty bouldery till deposits are widespread and probably form the most extensive soil forming materials. The tills in the coastal region commonly have mixed

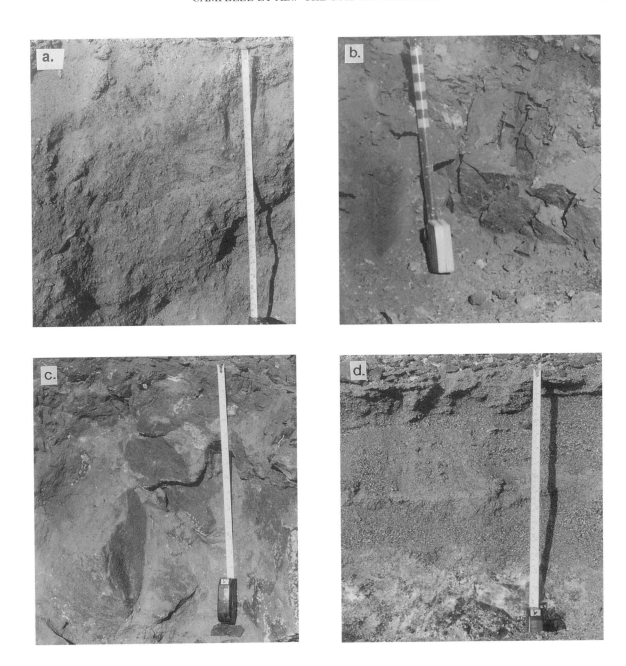

Fig. 1. Soil profiles illustrating a range of soil features and conditions. a) A weakly developed soil on sandy gravel till derived from the Rhone Glacier, upper Taylor Valley. b) A moderately weathered soil on fragmented granodiorite bedrock, Sessrumnir Valley, Asgard Range. c) A well weathered valley side soil derived from bouldery till laid down by the Taylor Glacier. Note the salt deposits throughout the profile. d) A weakly developed soil in thin till over ice, near the Rhone Glacier. Note the clear boundary between dry soil and ice, with no sign of melting.

lithologys, often containing fragments of kenyite, a rock only known to occur on Ross Island [*McCraw*, 1967], although they are usually dominated by the locally occurring rocks. On the volcanic islands, for example, volcanic rock such as scoria is the dominant component of the drift deposits, whereas on the mainland coast, basement rocks predominate along with scattered volcanics. Where tills are absent, the soils are predominantly formed on fragmented bedrock. In some coastal sites such as at Marble Point, the soils are often formed from raised beach deposits of reworked bouldery gravels.

Fig. 2. Landscapes illustrating subregions of the McMurdo Dry Valleys. a) Coastal region near the mouth of Taylor Valley. b) Valley floor and valley side regions, Wright Valley from near the Meserve Glacier. Lake Vanda in distance. c) Sessrumnir Valley, an upland valley in the Asgard Range, looking across the Wright Valley to the Olympus Range. d) Plateau fringe region, above Sessrumnir Valley, Asgard Range, showing blockfield and high precipitation environment.

The coastal region has a comparatively mild climate. Air temperatures are probably similar to those observed at Scott Base [*Bromley,* 1994] where the mean annual temperature was –19.7±1.65°C. Of greater significance, however are the summer maximum temperatures which, from November to February, are frequently above 0°C. During this period, surface soil temperatures may exceed 14°C or more for a short time each day with a diurnal amplitude greater than 12°C (Figure 3). In winter however, surface soil temperatures may fall as low as –38°C .

Soil temperature measurements that we have made, over two consecutive years at Marble Point, have shown that the surface horizon of the soil may remain unfrozen for 40 days, but that the number of freeze-thaw cycles are in fact very few. In the coastal region, the depth to which the 0°C isotherm penetrates and thawing occurs in summer is quite

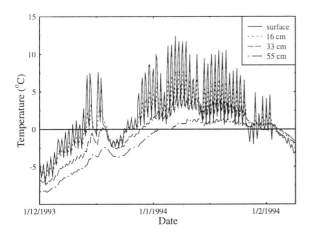

Fig. 3. Soil temperatures to 55 cm at Marble Point during the summer 1993–1994. The record covers the entire period when the surface temperatures were above freezing.

variable and ranges from about 30 to 60 cm. This variation appears to result largely from differences in site conditions and is probably a reflection of local site energy relationships.

The coastal region has an average of about seven falls of snow per month during summer. Observations have shown that summer snowfalls seldom persist on the ground surface for longer than 24 hours and, while some moistening of the soil surface may occur, most of the snow is lost through sublimation. Winter snow falls are much more persistent and are commonly concentrated into scattered permanent or semi-permanent drifts by wind. With the onset of summer, many of these drifts melt around the edges where the snow is in contact with the warming soil, giving rise to patches of moistened soil and transitory stream flows.

Moisture status. Liquid water derived from the thawing of permanent ice and semi-permanent snow patches is widespread in the coastal region during summer, resulting in numerous small stream flows and small lakes.

Seasonal water occurrences are defined as those which regularly flow in a channel, or accumulate in a depression or thaw in a lake in most years. Adjacent to the seasonal water flows, the soils are commonly moistened by capillary movement for a distance of several meters and the annual and seasonal persistence of water in these sites frequently provides suitable soil conditions for the development of moss, algae, or lichen communities.

Transitory water occurrences are most commonly found adjacent to small winter accumulations of snow which completely disappear by ablation and thawing during summer. Investigation of the soils at one of these sites (Figure 4) showed that the surface 5 cm of the wetted soil

had a gravimetric moisture content of 15% at the edge of the thawing snow patch, 12% at 1.5 m distance and 2.5% 4 m away in the capillary fringe zone. As thawing proceeded, the snow patch retreated and disappeared, the moistened soil soon dried out and the moisture content dropped to <1% over 14 days.

While liquid water occurrences are common within the coastal region, more than 95% of the ground and soil surfaces are not moistened by liquid water and the supply of moisture for these soils is primarily from the limited thawing of periodic snow falls, which may occasionally moisten the soil surface for a few hours.

Investigations of the water content of soils from about 50 sites in the coastal region [*Campbell et al.,* 1994; *Balks et al.,* 1995] have shown that the gravimetric moisture content in the surface soil horizon (approximately 3 cm) averages around 1% (Figure 5). The moisture content then increases steadily down through the soil, reaching around 10% near the base of the active zone, then rises abruptly in the permafrost. The average active layer water content for all horizons from 23 well drained sites in the coastal region was 5% and is greater than in soils from the other dry valley regions, apart from atypical sites that are wetted from sources such as thaw patches, streams, or lakes. The water content of the underlying ice-cemented permafrost in the coastal region is also higher than the other regions, at times reaching 150%.

Soil properties. The soils of the Marble Point area are formed on till, mainly of Ross I age but, especially around the northern end of the area, the tills are thin and the basement bedrock is exposed. These are a complex of gneiss, schists, and marble. The tills themselves are unweathered (weathering stage 1) [*Campbell and Claridge,* 1975], but the basement rocks show considerable evidence

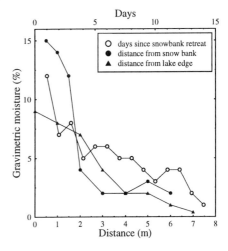

Fig. 4. Water content of soils adjacent to liquid water sources.

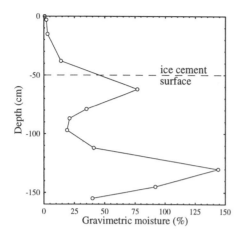

Fig. 5. Typical pattern of soil moisture distribution in the active zone and the permafrost of a well-drained coastal region soil.

of staining and physical disintegration [*Kelly and Zumberge,* 1961]. Our investigations have revealed the presence of patches of relatively strongly weathered soils in many places, especially where the underlying basement contains veins of very micaceous schist.

A soil from Marble Point (Figure 6) may be considered representative of these coastal soils on till. Less than half the soil is less than 2 mm in particle size and most of the fine earth is in the 2–0.5 mm fraction. Nevertheless 1–3% of the material is of clay size, a feature typical of glacially ground rocks such as schist or gneiss, which are naturally fine-grained. The free iron oxide content (the oxides extractable by reduction in the presence of complexing agents [*Mehra and Jackson,* 1960]) is also very low, indicative of very little weathering having taken place. Free iron oxide content is also low within the ice-cemented part of the profile, although there is an increase in the clay content which may represent slight differences in the nature of the underlying till.

Some small areas of strongly weathered soils occur in the Marble Point area, which may represent remnants of a much older surface. The apparently older soils have much stronger brown colors and are finer in texture with up to 15% clay recorded at one site and free iron oxide contents of up to 6%. The pattern of young tills and older soils is complex, but in general older soils are found on topographic highs where basement rocks are closer to the surface.

Both the "old" and the "young" soils are slightly alkaline, with pH values of up to 9.8 close to the surface and above 8 throughout most of the profile. This is attributed to the ubiquitous presence of marble fragments, which occur as boulders within the till or as individual

calcite crystals fretted off marble outcrops and widely distributed by wind. These crystals form part of the desert pavement over much of the area.

Despite the presence of marble, the soluble salts in the soils are mainly sodium chloride, with only minor quantities of other cations present. Potassium levels are high in comparison to calcium and magnesium, indicating a dominant marine influence. The dominant anion is chloride, with lesser amounts of sulphate, consistent with a direct marine origin for the salts. Nitrate can be detected in the salts, but at very low concentrations (Table 1). Carbonate and bicarbonate ions must of necessity be present because of the pH values but are present at levels too low to measure by simple techniques.

Further south, in Walcott Bay, soils are formed on thick tills of Ross I age, much of which is ice-cored. Basement rocks do not outcrop close to sea level, but are exposed on the hill slopes surrounding the bay. The particle size

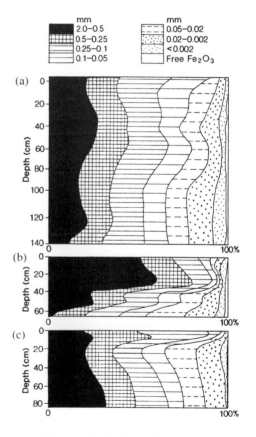

Fig. 6. Particle size distribution of the fine earth fraction for soils at a) Marble Point, b) Scott Base, and c) near Vanda Station. Note the uniformity of the deep till from Marble Point, the high proportion of coarse material near the surface of the soil from Scott Base, with increasing fine material with depth, and the sorting indicated in the surface of the water-washed till from Lake Vanda.

TABLE 1. pH, Total Soluble Salts (TSS), and Ions Present in Water Extracts of the Soil above the Ice Cemented Permafrost from a Number of Localities within the McMurdo Cold Desert Region.

Soil locality	pH	TSS %	Ca	Mg	Na	K	Cl	SO$_4$	NO$_3$
					—me/100 g—				
Coastal regions									
Marble Point	9.8	0.5	0.05	0.09	7.6	0.1	6.9	0.7	0.03
Walcott Bay	9.4	0.03	0.4	0.1	1.8	0.3	0.03	0.05	–
Scott Base	8.9	0.04	0.01	0.08	0.6	0.01	0.2	0.2	0.05
Lake Vanda	8.7	0.08	0.4	0.2	0.3	0.06	0.7	0.2	0.01
Barwick Valley	8.6	1.0	4.6	1.6	4.4	0.04	3.6	2.9	3.8
Northwind Valley	8.7	0.1	0.4	0.5	1.8	0.03	0.01	0.5	0.2
Valley walls									
Meserve Glacier	8.2	12.0	5.9	13.4	245	2.1	308	27.5	4.1
Upland valleys									
Coombs Hills	7.7	0.5	7.1	2.5	3.7	0.03	0.01	11.6	1.7
Sessrumnir Valley	6.4	0.5	0.9	2.8	3.9	0.03	2.8	2.6	0.3
Plateau fringe									
Elkhorn Ridge	8.0	0.7	0.8	0.8	3.8	0.01	0.1	2.3	1.0
Asgard Range	8.3	0.6	2.2	2.3	11.2	0.04	2.1	11.3	2.0

distribution of these tills is similar to that of the tills at Marble Point, but the clay contents are lower, probably because the tills are deeper and there is little contribution from underlying finer textured material. These tills are also alkaline, but with very low contents of soluble salts, mainly sodium sulphate and sodium chloride.

Soils on volcanic material, such as those on Hut Point Peninsula, are formed on scoria and broken rock rather than on till. Nevertheless they contain considerable amounts of fine material, probably because the scoria tends to break up very readily under the action of frost. Material collected from between the large boulders or blocks of more resistant basalt, contain about the same proportion of fine earth as do soils from till, although in general, most of the fine earth falls in the 2–0.5 mm fraction (Figure 6). There is a much higher proportion of silt and clay-size material in these soils than those on till, but as *Wellman* [1964] and *Claridge and Campbell* [1988] showed, much of the fine material is loessial in origin and is derived from the till exposures to the west.

The soils in the Hut Point Peninsula area are also alkaline, with pH values in the range 8.8–9.4. They are subjected to a much greater marine influence than those of coastal southern Victoria Land, because of their exposed situation, and therefore are subject to the input of marine salts. Fine-grained calcium carbonate in the loessial

material may also contribute to the high pH of soils in the Hut Point Peninsula area.

In most soils the salt content is generally low, probably because there is sufficient moisture available from the frequent light snowfalls to flush out the salts (Table 1). Small amounts of salt are present, with a very slight increase at the surface, where they are concentrated by evaporation, and within the ice-cement. Sodium is the most common cation, with some magnesium and lesser amounts of calcium and potassium. The anions are chloride and sulphate, with only trace amounts of nitrate. This chemistry is consistent with a marine origin for the salts, with some small contribution from rock weathering, indicated by the magnesium derived from weathering of olivine present in the basalt. The proportion of calcium is generally very low, probably because any calcium ions liberated by weathering or present in the precipitation are taken up in the formation of the calcium carbonate coatings on the underside of stones that are a feature of these soils.

Soils of the Valley Floors

Geography and parent materials. The valley floors include the inland valley floors (Figure 2) of the Taylor, Wright, and the Victoria Valley system, up to an elevation of around 800 m.

The soil environment of the valley floors largely reflects the long and complicated glacial history of these valleys. Formed and excavated initially by eastward flowing ice, probably in the Oligocene, the surfaces have been influenced by tectonic uplift which raised some parts of the valley floors from below sea level and created drainage reversals resulting in the formation of inland saline lakes. Following the uplift, there have been numerous subsequent ice invasions from both the east and the west but these do not appear to have greatly changed the shape of the valley floors. Post glacial aeolian, fluvial, and fluvioglacial activity have however modified many of the deposits and contributed to the wide variety of soil forming environments on the valley floors.

Tills form extensive valley floor deposits. They range in age from the younger Late Quaternary Ross or Alpine drifts to those of Pliocene age. The textural characteristics of the soils from tills are closely related to lithological properties of the tills. For example, the younger tills commonly contain high proportions of coarse fragments, whereas the older tills often have greater proportions of sand and silt particles. The younger deposits are almost completely unweathered, but older soils show a small amount of oxidation, some granular disintegration, and some accumulation of salts within the soil profile. Ice cement is generally very close to the surface in the youngest soils and at a depth of 50 cm or more in the older soils on till.

Fluvial and fluvioglacial deposits occur in numerous places on the valley floors and they include sediments from streams, pro-glacial lakes, fans, and lake beach deposits. Most of these deposits have formed from re-worked glacial sediments and the soils are often very sandy. At a series of 12 sites that we examined on lake beach deposits near Lake Vanda, for example, the proportion of fine (<2 mm) material averaged 78% for all horizons to 40 cm depth.

The soils of the fluvial and fluvioglacial deposits are invariably weakly developed, ranging from bouldery gravels on the deposits where fine sediments have been transported away, to sandy gravel plains where sediments have been deposited.

Aeolian deposits are not extensive on the valley floors and are mostly restricted to sandy sediments and soils near the eastern ends of the valleys. The valley floor soils in general, however, have higher amounts of sand in them than the soils at greater elevations (Figure 1), as sand, which is progressively moved down the valleys by aeolian transport, ultimately becomes incorporated into the valley floor soils through cryoturbic processes. Sand dunes stabilized by ice are found at the eastern end of the Victoria Valley [Lindsay, 1973].

Bedrock surfaces are a significant component of the valley floors, particularly in the vicinity of Lake Vanda and on Nussbaum Riegel in Taylor Valley. The soils of these areas comprise a mosaic ranging from patches of bare rock with superficial surface crumbling, to soils greater than 50 cm deep on fragmented bedrock (Figure 2). Bedrock fragmentation occurs because of thermal expansion and contraction. The soils typically consist of a stony surface horizon with about 20% sand, overlying angular fractured bedrock with about 5% sand occurring in lenses along the fracture surfaces.

Climate. The soil climate on the valley floors differs from that found in the coastal region. Mean annual temperatures are probably similar to those in the coastal areas but temperature extremes are greater. Short-term records from Vanda Station showed that mean daily maximum and minimum air temperatures from October to March were higher than for Scott Base (max >7°C) but minimum winter temperatures were much colder [Thompson et al., 1971b]. January soil temperature measurements from a number of sites near Vanda Station suggest that the 0°C isotherm may reach about 30 cm depth [Thompson et al., 1971a].

Snow fall on the dry valley floors is appreciably less than in the coastal dry valley region. Bromley [1988] reported annual snowfall totals for Vanda Station of 82, 7, and 115 mm (water equivalent) for the three years in which observations were made for a full year. Longer-term measurements made by a tube-type gauge which may collect more falling snow than blowing snow averaged 13 mm yr^{-1} over a 20 year period (Chinn, personal communication). Field observations suggest that very little moisture finds its way into the soil because of rapid sublimation. Because of the lower snowfall, there are fewer snow drifts, consequently little moisture is supplied to the soils from transitory snow accumulations. The persistent low humidity winds which dry the soils are another important influence on the soil climate of the valley floors.

Moisture status. Liquid water for the soils comes primarily from stream flows originating alongside thawing glacier margins, from thawing of a few permanent snow drifts in higher altitude gullies, or from the seasonal thawing of the frozen lakes. Adjacent to these seasonal sources of water, the soils are moistened and the wetting zone is extended by capillary flow, over distances of several meters. In these wetted margins, the soil water contents progressively diminish from saturation at the water source to very dry several meters away (Figure 4).

Where there is no obvious water source, the soils of the valley floors have very low moisture contents. The average gravimetric water content for all horizons from

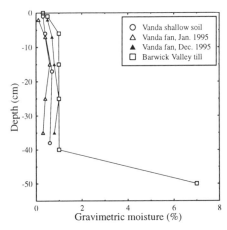

Fig. 7. Soil water content for valley floor soils from Barwick and Wright Valleys.

13 sample sites on shallow soils over bedrock at Vanda Station was 0.5% (Figure 7). At another location, the water content at 12 sites for all horizons to 40 cm depth in January also averaged 0.5%. The same sites sampled 11 months later in December had water contents averaging 0.7%, suggesting that there may be some variation within or between seasons. The water content for the surface horizons is very low (average 0.3%) and generally increases with depth. Where ice cement is present, soil water contents rise to 6% with summer thawing providing some liquid moisture.

The moisture content differences between the Barwick Valley and Vanda soils, shown in Figure 7, suggest that there may be small but biologically significant differences between each of the three main valleys [*Freckman and Virginia,* this volume]. As well as small moisture content differences between the valleys, however, it is likely that there are differences within each valley with soil moisture contents increasing towards the coast.

Soil properties. The soils of the valley floors range widely in chemical properties because of the large differences in the nature of the soil parent materials and the ages of the soils.

The soils of the lower part of the Taylor Valley are formed from tills of Ross I age and are very weakly developed. These soils were mapped and described by *Claridge* [1965] and *McCraw* [1967]. For the most part they are of weathering stage 1 of *Campbell and Claridge* [1975]. These soils are similar to the till soils found on the coast except that they contain somewhat smaller amount of fine earth. Only small amounts of free iron oxides are present, indicating limited weathering. The soils have pH values in the range 8–9 and contain small amounts of water soluble salts. The salts are mainly sodium chloride and

sulphate, but potassium and magnesium ions are also present. The salts are generally dispersed throughout the soil profile but are occasionally found as discrete nodules under stones, where they are concentrated when moisture from melting snow, which accumulates around stones, evaporates.

Minimal weathering is also indicated by the nature of the clay-size minerals. The greater part of the clay-sized fraction consists of feldspar and mica, with lesser amounts of other clay minerals. The mica is unweathered and reduced to clay-size by physical processes. The other clay minerals present, are all micas in varying degrees of hydration, indicating a slight degree of weathering. In some soils, smectites are found, especially those containing magnesium-rich minerals.

Farther inland, soils around Lake Vanda in the Wright Valley, although formed on rather different material than the Ross I drifts of the lower Taylor Valley, show some similarities. Around Lake Vanda, the drifts are local and have been reworked to some extent by recent changes in lake levels. They are characterized by relatively low contents of fine earth, with most falling in the 0.5–0.25 and 0.25–0.1 mm size range (Figure 6).

In the Barwick Valley, around Lake Vashka, the tills are somewhat older than those around Lake Vanda, and contain higher amounts of fine earth, with more material in the 2–0.5 mm size range. On the other hand, the soils contain higher amounts of silt and clay, and slightly more free iron oxides, indicating a greater degree of weathering (Table 2).

The soils in the Lake Vanda area are slightly less alkaline than the soils of the coastal region, with pH values usually below 8.5 in the upper part of the profile but increasing down the profile. Close to Lake Vanda, where the soils were submerged a few thousand years ago, salt contents are low, although they tend to rise further away from the lake. Calcium is the major cation, while nitrate ion contents are generally very low. Some soils contain free gypsum crystals, accounting for the relatively high content of calcium and sulphate.

Older soils, such as those found around Lake Vashka, in the Barwick Valley, contain higher amounts of salt, also with a high proportion of calcium. They also contain appreciable amounts of nitrate ion, an indication of their greater age.

In places on the floors of the valleys, depressions or kettleholes occur, sometimes with moist soils or even small ponds at their lowest point. The ponds or moist soils are often highly saline, having accumulated salts over a long period of time through slow migration of ions through the surrounding soils. Because these soils are unfrozen for

TABLE 2. Fine Earth (<2 mm) Content (% of whole soil), and Particle Size Distribution of the Fine Earth Fraction, Averaged over the Material above the Ice-cemented Permafrost, for a Water-Washed Till near Lake Vanda, and for a Soil on Older Till near Lake Vashka in the Barwick Valley.

Site	fine earth (%)	Particle Size Range (mm)							Iron oxide (%)
		2–0.5	0.5–0.25	0.25–0.1	0.1–0.05	0.05–0.02	0.02–0.002	<0.002	
Vanda	45	13.8	46.2	30.0	3.4	1.6	2.0	2.4	0.2
Barwick	69	31.5	26.0	20.0	4.0	5.1	6.4	6.7	0.3

much longer periods than the dry soils of the valley floors, they are weathered to a very much greater extent. The soils tend to be olive in color, rather than brownish, because of a smaller amount of free iron oxides coating the particles. They contain a much higher proportion of clay minerals, largely smectite, as a consequence of greater weathering and neoformation of clay minerals under saline, alkaline conditions. The salts contain much higher proportions of calcium and magnesium chlorides than do the soils of the valley floors, because these ions form solutions with low freezing point, which can migrate downslope into undrained hollows and accumulate.

Thus in interpreting the environment of the valley floors, attention must be paid to the nature of the material forming the soil, and minor differences in topography, because they influence moisture content and salinity.

Soils of the Valley Sides

Geography and parent materials. The valley sides form the most extensive soil landscape unit within the dry valleys, (Figure 2) but are probably the least studied part of the entire region. They range in altitude from about 200 to 1600 m and progressively steepen upslope.

The valley sides are very stable. Since the valleys were cut, there have been only minor modifications by eastward and westward ice invasions, which have reached part way up the sides of the major valleys, as well as by fluctuations of a number of alpine glaciers which extend down the valley sides. The main deposits on which the soils are formed are tills, talus, and bedrock.

Tills are for the most part patchy and more extensive on the younger and lower slopes, particularly adjacent to the alpine glaciers where distinctive moraine and soil sequences are found. As with the valley floors, the tills on the valley sides range in age from Pliocene sandy to silty tills associated with early up- and down-valley ice advances, to younger, coarse-textured drifts close to the present alpine glacier margins. The soils on the older tills are more weathered, have increased salinity levels, and are non ice-cored. The soils on the valley side tills are therefore

highly variable and reflect the considerable range in age and sedimentary characteristics of the drift deposits [*Everett,* 1971].

Talus deposits are mostly thin and typically comprise a superficial layer of bouldery detritus over bedrock. The soils are dominated by coarse particle sizes with a low proportion of fine earth. Despite their appearance, these soils are quite stable and are not subject to mass movement. The bedrock soils on the valley sides are typically restricted to small accumulations of granular and sandy detritus along bedrock fracture planes and cleavages. They are more common on higher surfaces where bedrock outcrops are more extensive.

Climate. The climate on the valley sides is probably similar to the valley floors except that soil temperature extremes may be greater because of the pronounced aspect differences between the north and south faces. Precipitation increases with increasing altitude, indicated by the thicker snow cover with altitude after snowfalls.

Moisture status. Results of soil moisture measurements from three sites on the valley side north of Lake Vanda are shown in Figure 8. The gravimetric water content of these soils is extremely low and averages only 1% for all horizons. It is only in a few gullies and the marginal areas adjacent to alpine glaciers where thawing of semi-permanent snow and glacier ice occurs that the soils are moistened by liquid water flows.

Soil properties. Many of the younger soils are very similar to the soils of the valley floors, but there is a higher proportion of older soils which have been preserved. Old Pliocene tills, such as the Peleus Till described by *Prentice et al.* [1993] or the soils around the Meserve Glacier [*Everett,* 1971] are fine-textured and sometimes contain marine diatoms and other fossils. Soils from the vicinity of the Meserve Glacier contain a higher proportion of fine sand than the soils around Vanda Station (Figure 6). Soil pH is variable, ranging from 7.7 to 9.1, but tends to be slightly above 8. These soils are highly saline, with distinct salt horizons. At one location a horizon containing almost 50% salt occurs between 5 and 6 cm depth. The salt in this horizon is almost completely sodium chloride, with some

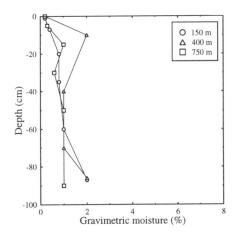

Fig. 8. Moisture content profiles of soils from valley sides, Wright Valley at altitudes of 150 m, 400 m, and 750 m..

calcium and magnesium ions present. The proportion of nitrate to chloride is low in this horizon, but is higher in the horizon below, probably a consequence of the greater solubility and mobility of nitrate salts. The maximum concentration of magnesium ions is also found below the dominant salt horizon, once again illustrating the greater solubility of magnesium salts.

These soils probably represent the most developed of the soils on the valley sides. The soils on talus tend to be much coarser, but, within the large blocks which make up the talus, there is usually much fine material. This is generally young and unweathered, but may contain material which formed part of the old valley side tills. Thus the soil matrix may be somewhat weathered and contain patches of salts.

Valley side soils can therefore be considered variable, and contain niches capable of supporting plant life, as well as patches of highly saline soils which are probably sterile.

Soils of the Upland Valleys

Geography and parent materials. The upland valleys cover a significant part of the McMurdo Dry Valley region, primarily in the Olympus and Asgard Ranges, but also to a lesser extent in the St. Johns and Convoy Ranges, the Coombs Hills, and in the head of the Taylor and the Barwick Valleys. This soil landscape unit comprises an ancient system of cirque valleys, (Figure 2) generally above an altitude of about 1500 m. The upland valleys probably formed in Oligocene or Miocene times, during the early glaciation of Antarctica, but the valleys are now largely devoid of glaciers, apart from valley head or headwall ice accumulations. Land surface ages vary greatly because of the past complex interactions between local and regional ice movements.

Glaciers have entered the upland valleys from their mouths, as major valley glaciers have expanded and flooded back into ice-free cirques, or from the head of the valley as ice has flooded through high passes from the inland ice, leaving a complex of tills of varying ages. Younger tills have been deposited from small glaciers in the heads of the valleys, however the tills are generally formed from re-sorted older tills. The oldest soils are formed on tills in the center of the valleys.

Climate. The climate of the upland valleys is cooler than that of the coastal region. We estimate that mean annual temperature probably ranges between around –25°C to –30°C, with air temperatures of –10°C to –18°C commonly being recorded in the summer. January soil temperatures, measured from short term recordings, may reach 10°C at the soil surface with diurnal fluctuations to –10°C. In contrast to the coastal region and valley floor soils however, there are a greater number of daily freeze and thaw cycles at the soil surface. Below 10–20 cm depth, however, the soils do not thaw. The position of the ice cement in the soil does not appear to relate to the maximum penetration of the 0°C isotherm (permafrost) and may first occur at more than 2 m depth in some places.

Precipitation in the upland valley region is appreciably greater than on the valley floors. Precipitation is >100 mm yr^{-1} at the Heimdall Glacier, above Vanda Station (*Chinn,* personal communication) but may be higher nearer the coast and lower farther inland, as valleys close to the coast are ice filled, while those to the west are completely ice-free. Most snowfall, however, is lost through removal by wind or ablation and the surface of the soil is only occasionally moistened from melting snow. Liquid water flows, melt pools and frozen lakes are rare and only found to the north in the Convoy Range, consequently there are very few sites where significant amounts of liquid water are present in the soils.

Moisture status. The water content of soils in the upland valley region differs from that found in the coastal region and the valley floors. At 13 sites at Beacon Heights, for example, the average gravimetric water content measured for all horizons above the ice cement was 1.5% (Figure 9), approximately twice that for the soils on the valley floors, but appreciably less than the average of 5% water contents for the soils in the coastal region. As illustrated by Figure 9, there is a negligible increase in water content with increasing depth, except in soils which are underlain by ice cement. In these soils the ice cement water content is appreciably lower than is found in the soils of the coastal region.

Similar moisture characteristics were found in soils in the Coombs Hills and Convoy Range areas (Figure 9). There, the gravimetric water content of soil horizons from

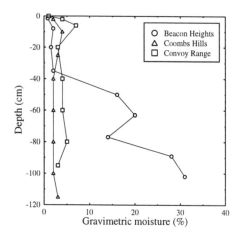

Fig. 9. Moisture content profiles of soils from the upland valleys.

15 sites was approximately 3% for Coombs Hills and Greenville Valley soils. About 55% of these soils were without ice-cement.

Soil properties. Weathering is clearly more advanced in the upland valleys and most evidence points to ages in excess of 3.5 million years for these surfaces [*Campbell and Claridge,* 1975; *Marchant et al.,* 1996]. The content of clay-size material is a little higher than in the soils of the coastal region or the valley floors. The clays could have been produced by weathering in situ, but are more likely to have been blown in [*Claridge and Campbell,* 1988]. Most of the soils are placed in weathering stages 4 and 5 of *Campbell and Claridge* [1975].

Salt content is usually high, and often form a distinct horizon, generally about 5 cm below the surface. The salt horizon may vary in thickness from a 1–2 cm thick intermittent band, up to 15 cm thick. In a few cases the salts comprise up to 10% of the horizon. Sodium is generally the dominant cation, but calcium and magnesium are present in proportions much higher than in sea water. This has been taken as an indication of release of these cations from ferromagnesian minerals weathering on the surface of dolerite boulders and releasing iron to form the wide spread iron staining commonly called desert varnish [*Claridge and Campbell,* 1984]. The proportion of potassium is, in general, very low in the absence of appreciable amounts of potassium feldspars that could release it by weathering. In soils from tills containing granite, potassium forms a higher proportion of the salts [*Claridge and Campbell,* 1977].

The dominant anion in the soils of the Convoy Range, to the north of the dry valley region, is sulphate, and only traces of chloride are present, while nitrate accumulates in the salt horizon. In the soils of the Asgard Range, more

chloride is present, but the dominant anion is still sulphate. Nitrate ion concentration is generally the same as chloride. The difference is a good example of the general pattern, described by *Claridge and Campbell* [1977], that nitrate dominates when the primary source of the precipitation is windblown snow from the interior. Conversely, chloride is derived mainly from moist air moving inland from the Ross Sea and does not reach ice-free areas at high altitudes far from the coast.

In general the soils of the upland valleys are old, well weathered by Antarctic standards, and saline, with relatively high concentrations of nitrates. These features may inhibit biological activity, except in isolated pockets where salts may be leached by melting water from rare snowfalls.

Soils of the Plateau Fringe

Geography and parent materials. The plateau fringe region includes areas, primarily above about 2000 m, on the western margin of the dry valleys at the edge of the Polar Plateau. It comprises an ancient, essentially undissected land surface (Figure 2) of mainly dolerite rocks, but with occasional occurrences of other rocks. The extreme age of the plateau fringe surfaces is indicated by the widespread occurrences of intensively pitted dolerite rocks. Till deposits are rare and the soil mantle appears to be felsenmeer formed from the disintegration of local bedrock by expansion, contraction, and wedging processes. Ground surfaces are typically undulating with dolerite outcrops interspersed with felsenmeer and snow patches. Patterned ground is sometimes extensive and indicates the widespread occurrence of ice-cemented ground. The soils typically consist of stony or bouldery gravel derived from weathering bedrock. The ice cement depth has been found at depths varying from 10 to 50 cm. In a few places, where snow cover is restricted, strongly weathered residual soils may be present along bedrock clefts and fracture planes.

Climate. Mean annual temperatures in this region are estimated to be around –30°C. Precipitation is derived from periodic snowfalls and from extensive occurrences of snow blown from the Polar Plateau. The soils are only moistened when the ground surface temperature briefly rises to thaw some of the surface snow or ice cement in the soil. While the frozen water content of these soils may be quite high, the soils are rarely thawed enough to provide significant liquid water for soil processes.

Moisture status. There is very little known about the moisture status of these soils.

Soil properties. In this area soils are extremely variable and mainly formed from the fretting of dolerite grains, which accumulate in hollows and rock crevices. They tend to be very coarse in texture, with much of the material ranging from rock fragments up to 10 mm in diameter down to individual mineral grains around 0.5 mm in diameter. Most of the smaller grains are blown away, unless they combine to form aggregates.

The chemistry of the salts contained within these soils is very similar to the soils of the upland valleys 200–300 m lower in altitude, with similar pH values, around 8.3 for the samples studied. Sodium is usually the dominant cation, while sulphate is the dominant anion, with lesser amounts of nitrates and chlorides. In upland plateaus of the Convoy Range, however, chlorides are almost completely absent.

Sometimes traces of even more weathered soils, which we consider to be remnants of an even older weathering stage are found in crevices. These soils are often highly colored, contain relatively higher amounts of clay-size material and free iron oxides (4% clay and 0.5–0.7% iron oxides) than the less weathered soils. The few examples of these soils that have been studied have pH values of around 5 and are highly saline. However, because of their limited extent, they presumably have little ecological importance.

SOIL CHEMISTRY

Introduction

Although dry valley soil formation is dominated by physical weathering processes, there is ample evidence of chemical weathering. The effects of chemical weathering are revealed mainly by differences in oxidation of the soil surface and soil profile. Environmental factors such as temperature, precipitation, and age of the soil surface influence the degree and extent of chemical weathering and result in clearly identifiable weathering differences between the soils of the different regions. Such features as color, staining and polish of the surface pavement, depth of oxidation within the soil profile, the texture of the soil matrix, the distribution and amount of salts within the soil, the nature of the salts, and the depth to ice-cement are all related to environment and have an influence on the ecology of the soil.

Soil Salinity

The salts present in the soils differ widely from place to place and very many crystalline phases have been recognized amongst them. Most of the phases that can be formed from the cations sodium, potassium, magnesium, and calcium with the anions chloride, sulphate and nitrate have been identified in the soils of the dry valleys [*Claridge and Campbell*, 1977; *Keys and Williams*, 1981]. Near the coast the salts in the soils are of direct marine origin and the ions present are in the same ratio as in sea water. Farther inland, chlorides are preferentially removed from precipitation and the proportion of sulphate found in the soil increases. The anions present are sulphate, nitrate and iodate, while chloride ions are absent [*Claridge and Campbell*, 1968]. The nitrate is considered to be ultimately derived from the sea, added to the atmosphere as protein fragments present in sea foam and transported to Antarctica. It is fully oxidized during transport and arrives as nitrate in snowfall on the inland ice sheet. It accumulates on high land surfaces along the Transantarctic Mountains which receive their snow cover, predominantly, as windblown snow from the inland ice sheet, rather than as direct precipitation, which generally arrives in moist air masses moving inland from the coast.

Another suggested source of nitrate in polar snows is fixation from the atmosphere by auroras, but this can be ruled out by the association of traces of iodate, which is concentrated in marine organisms, with the nitrate deposits. Because the soils of the inland areas are extremely old, nitrate has had time to form considerable accumulations. Nitrate can also be detected in young soils, such as those of Marble Point, indicating that traces of nitrate arrive in precipitation from local seas, but it is either leached out, because of the greater amount of moisture available in coastal regions where temperatures are warmer, or is utilized by the small amount of plant life in these soils.

The cations in the salts originate largely from sea water, as indicated by the dominance of sodium. However there are important contributions from weathering, indicated by the correlation of the salt composition with geology. In inland areas, where tills contain a high proportion of dolerite, that has had time to weather on the surface, calcium and magnesium are present in large amounts and sometimes dominate the salt, while tills containing granite generally have salts with a higher proportion of potassium than is present in sea water.

There can be separation of species during migration of salts through soils. When a solution with the composition of sea water freezes the first salt to precipitate is mirabilite (sodium sulphate decahydrate) and the residual fluid after most of the salts have precipitated contains calcium chloride [*Nelson and Thomson*, 1954]. This accounts for the presence of calcium chloride brines in such situations as Don Juan Pond in the Wright Valley or the saline lakes of the Convoy Range. Sodium sulphate efflorescences are extensive on some soil surfaces in the Hut Point Peninsula area. These have precipitated following extensive disturbance and removal of the soil surface, and subsequent evaporation of the underlying ice-cement.

Salt Migration in Soils

Salts migrate in Antarctic soils over time, as indicated by the accumulation of salt horizons, the presence of nodules of salt under some surface stones and the transient appearance of salt efflorescences on soil surfaces. Salts move as a result of moisture gradients, which are a function of moisture dynamics within the soil and previous site history.

Moisture is added to the soil surface mainly from melting snow and moves through the soil by capillary flow. Even at low temperatures, thin films of liquid will remain unfrozen if they consist of extremely concentrated salt solutions [Ugolini and Anderson, 1973]. These brines will move in response to temperature gradients, in particular towards a freezing front [Buchan, 1991]. The salts may accumulate at levels where moisture films moving from deep in the soil finally evaporate. Thus in some soils salt horizons form while in others salts concentrate in nodules under flat stones in the desert pavement. The solubility and hygroscopicity of individual salts varies, permitting some form of separation as shown by the different zones of concentration of nitrates and chlorides in the soils of the valley sides and elsewhere.

We have followed the movement of salts, experimentally, by adding salt (LiCl) to the soil surface and tracing its movement over a period of years. We have found that on dry sites of the valley floor near Lake Vanda the salts have moved less than 1 m over a three year period. In the coastal region migration of up to 3 m occurred at dry sites, while at wet sites, where moisture is available seasonally from melting snow, movement of up to 5 m in three years has taken place. Lithium chloride is a very soluble and hygroscopic salt and some movement would be expected. During the period of time equivalent to the age of the oldest soils it is expected that added lithium chloride would have moved out of the soil at periodically wet sites, although it may reach equilibrium at sites where there is little moisture available.

Similarly we have followed the movement of pollutant ions, such as Cu, Pb, and Zn, arising from human activities. In general, these metals are present as hydrous oxides, rather than salts, which are much less soluble and movement is slow. Nevertheless we have detected movement of up to 0.5 m in 30 years [Claridge et al., 1995].

Significance of Soil Chemistry for the Ecology of the Dry Valley Soils

Most of the soils are alkaline, with pH values ranging from 7.5 to 9.5. Because of the very low clay and organic matter contents, there is very little buffering capacity and high pH values are not matched by measurable amounts of bicarbonate or carbonate ion. Thus, where salinity is low organisms can easily modify the local soil environment by excreting materials which counteract the effect of high pH.

The variable salinity is a much more important factor. Young soils (<20,000 yrs) generally have very low salt content, although these soils may concentrate salt in places where moisture is lost by evaporation. Thus near Lake Vanda, where there is a pronounced wetting front close to the lake edge, salt contents are highest just inland of the wetting front. Most older soils have accumulated salts over time, either throughout the profile, concentrated in a discrete salt horizon, or as nodules, and these are likely to have an inhibiting effect on plant or animal life. The older soils generally contain a high proportion of nitrate ion, which in small amounts can stimulate biological activity, but in high amounts is toxic. Even in coastal zones, very small amounts of water-soluble nitrate can be detected, which we consider arrives in precipitation, and this must provide a nutrient source for biological activity.

THE SOIL MICROCLIMATE

Introduction

The bare ground surfaces of the dry valleys create their own distinctive climates because they absorb more solar radiation than snow or ice surfaces [Weyant, 1966]. The absorbed energy is used to heat the soil and air, melt ice and snow, and drive evaporation (sublimation) from wet (ice) surfaces. This is the reason that liquid water becomes available in the soil, or as river flow, or within the permanent ice covers of dry valley lakes [Fritsen et al., this volume] in an otherwise frigid landscape.

At a local or microscale the differences in radiational and thermal behavior of soil materials becomes very important, firstly in determining the amount of energy available at the surface, and secondly in determining the amount and depth to which the soil will warm.

The Surface Radiation Balance

The dry valleys are an extreme radiational environment. For much of the year the amount of solar radiation arriving at the surface is minimal or non-existent, however for the

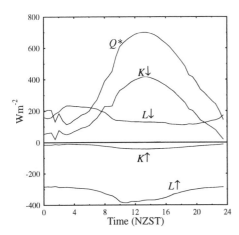

Fig. 10. Measured and estimated radiation balance components for a basalt soil near Scott Base, December 23, 1994. $L\uparrow$ is estimated from measured soil surface temperature [Oke, 1987] and $L\downarrow$ is estimated as the residual from Equation 1.

brief summer period the combination of frequent clear skies and 24-hour insolation provides the surface with amongst the largest solar radiation receipt in the world [Weyant, 1966; Dana et al., this volume].

The surface radiation balance describes the component inputs and outputs of radiant energy at the surface. Incoming solar radiation ($K\downarrow$) is either reflected ($K\uparrow$) or absorbed as net solar radiation (K^*) at the soil surface. All surfaces emit longwave radiation ($L\uparrow$) at a rate dependent on their temperature and emissivity. The surface receives longwave radiation ($L\downarrow$) from the atmosphere and clouds, however the net longwave radiation (L^*) is usually negative (describing a loss by the surface) because the surface is usually warmer than the sky or cloud base. The sum of all radiation balance components is the net radiation, Q^*,

$$\begin{aligned} Q^* &= K\downarrow + K\uparrow + L\downarrow + L\uparrow \\ &= K^* + L^* \end{aligned} \tag{1}$$

where components representing a loss of energy from the surface are defined as negative. The net radiation is the amount of energy available for heating the soil and air and driving latent heat (water phase change) processes. The most important surface property affecting the radiation balance is the albedo, or shortwave reflectivity, a. The albedo is a function of surface color, roughness and moisture content [Oke, 1987].

Figure 10 shows measured and estimated radiation balance components for a mainly clear-sky day at a site near Scott Base. The incoming solar radiation describes a bell-shaped curve, peaking at 700 W m^{-2} at solar noon,

and remaining positive through the entire 24-hour period. Most of this large radiation input is absorbed by the dark basalt soil and provides a large positive Q^*. The major radiant energy loss from the surface is via $L\uparrow$, which achieved its maximum magnitude in late morning corresponding to maximum surface temperature.

Soil albedo is the major surface factor causing differences in net radiation. Table 3 lists radiation balance components for clear-sky days at three snow-free sites: Scott Base (basalt soil), Northwind Valley, Convoy Range (sandstone soil), and in the Coombs Hills (a dolerite soil). Both the Scott Base and Coombs Hills soils are dark-colored with mean albedos of 0.06 and 0.07 respectively, while the light, polished sandstone surface at the Northwind site had the largest albedo of 0.26. The large albedo at the Northwind site results in a much smaller Q^* than the other two sites, hence there was less energy available to heat the soil and air and ablate ice.

Surface albedo is not constant; it changes as the physical condition of the soil changes and because of radiation geometry. Figure 11 shows the diurnal variation in albedo for clear-sky days at the three sites. Both the Coombs Hills and Northwind sites were shaded at night and this caused discontinuities because the reflective behavior of the surface differs for direct and diffuse radiation. At the Coombs Hills snowfall beginning around 1900 hrs caused a rapid increase in albedo. At all sites albedo decreased during the morning to reach minimum values at solar noon, which is common behavior for many surface types [Oke, 1987; Arnfield, 1975]. At low solar altitudes reflection is specular (mirror-like) while at high solar angles there is more scattering and increased opportunities for radiation trapping, hence albedo is lower. Albedo also displayed a dependence on solar azimuth, with the rate of decline in the forenoon period not matched by equivalent rates of increase in the afternoon. This is attributed to preferential polishing of desert pavement surfaces by prevailing winds at each site. Hence there are different reflective characteristics depending on the horizontal direction of the solar beam. This diurnal variation in albedo could be important for applications where radiation balance components are being modeled.

Summer snowfalls are relatively common in parts of the dry valleys, and snow cover has a dramatic effect on the radiation balance. Table 4 lists data for two days, at Scott Base and the Coombs Hills, when there was a temporary thin snow cover. At Scott Base, during this partly cloudy day, the high albedo meant that K^* was small, however Q^* of 7.4 MJ m^{-2} d^{-1} still allowed considerable soil heating and ablation processes to occur. At the Coombs Hills, despite the clear sky and large $K\downarrow$, Q^* was negligible.

TABLE 3. Radiation Flux Densities (MJ m^{-2} d^{-1}) and Albedo (1000–1400 hrs), for Clear-Sky Days at Three Snow-Free Sites. Net Longwave Flux Calculated as $L^*=Q^*-K^*$.

Site/Date	Surface Type	$K\downarrow$	K^*	L^*	Q^*	α
Scott Base December 23, 1994	basalt	35.2	32.9	–14.3	18.6	0.06
Northwind Valley January 21, 1995	sandstone	31.3	22.8	–13.1	9.7	0.26
Coombs Hills January 11, 1995	dolerite	33.0	29.9	–14.4	15.5	0.07

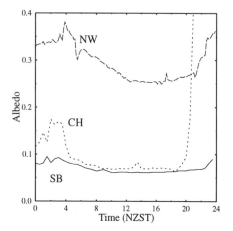

Fig. 11. Diurnal pattern of measured albedo for the three field sites on clear-sky days: Scott Base (SB), December 23, 1994; Northwind Valley (NW), January 21, 1995; Coombs Hills (CH), January 11, 1995.

While the highly reflective snowpack was partly responsible, by reducing K^*, the clear sky allowed large longwave radiation loses, hence total Q^* was negligible. Clouds generally prevent large longwave losses, as seen at Scott Base (Table 4), hence cloudy conditions in the presence of a thin snow cover can result in more energy for soil heating and ablation processes than can clear-sky conditions.

Soil Temperature

Soil surface temperature. Solar radiation absorbed at the soil surface raises its temperature. Heat energy is conducted down into the soil volume (soil heat flux) at a rate dependent upon the vertical temperature gradient and the soil thermal properties. Heat is also lost to the air (sensible and latent heat fluxes) and radiatively as longwave energy, $L\uparrow$. The rate and magnitude of soil surface temperature changes are dependent on the radiation balance, the properties of the overlying air, i.e., its temperature and the windspeed, as well as upon the properties of the soil.

Measurements of soil and rock surface temperature have been made across a range of environments in Antarctica, and large temperature variations are common, with maximum temperatures well above 0°C. For example, *Miotke* [1980] measured near-surface rock temperatures exceeding 20°C in the Darwin Mountains (80°S).

Table 5 presents a summary of measured soil surface temperatures for three surface types across a range of environments and sky conditions. Maximum surface temperatures can be well above freezing, even at the Coombs Hills site, 2200 m altitude. Mean daily soil temperature only exceeded 0°C at the low altitude Scott Base site, however. Large ranges of diurnal soil surface temperature are a feature of Antarctic soils when radiation inputs are large, with all three sites experiencing ranges in temperature greater than 22°C. Large surface–air temperature gradients are also commonly in excess of 20°C in the lower meter of the atmosphere, indicating that the turbulent sensible heat flux will be extremely large, and much of the adsorbed solar radiation may be lost this way. Figure 12 shows the diurnal variation in soil surface temperature at each site. Soil surface temperature maximums are similar at Scott Base and the Coombs Hills, however the minimum temperatures are quite different because they are more affected by air temperature. The maximum temperature a bare surface can attain is more closely linked to incoming solar radiation than to air temperature; however the latter has more control over minimum temperatures. Hence we expect that the greatest ranges of diurnal soil surface temperature will be found in environments with the coldest climates, such as the plateau fringe.

Figure 13 shows that soil surface temperature is strongly linked to $K\downarrow$ and windspeed [*Balks et al.*, 1995]. At a site near Scott Base a simple regression model

TABLE 4. Radiation Flux Densities (MJ m^{-2} d^{-1}) and Albedo (1000—1400 hrs), at Two Sites with Thin Snow Cover. Net Longwave Flux Calculated as $L*=Q*-K*$.

Site/Date	Surface	$K\downarrow$	$K*$	$L*$	$Q*$	α
Scott Base December 28, 1994	basalt	20.0	10.9	−3.5	7.4	0.46
Coombs Hills January 14, 1995	dolerite	33.1	8.9	−8.7	0.2	0.73

TABLE 5. Soil Surface and Air Temperatures (T_{air}) and Mean Windspeed (u) for Three Soils. "Snow" Refers to the Presence of a Surface Snowpack. Site Locations and Details are Listed Below.

Site/Date	Conditions	Soil Surface Temperature (°C)				T_{air} °C	u m s^{-1}
		max.	min.	range	mean		
Scott Base[1]							
December 23, 1994	clear sky	17.8	−4.7	22.5	4.5	−5.2	4.6
December 25, 1994	cloudy	11.0	−2.3	13.3	4.6	−4.8	2.9
December 28, 1994	snow	3.8	−3.6	7.4	−0.9	−3.4	8.9
Northwind Valley[2]							
January 21, 1995	clear sky	8.5	−13.5	22	−2.9	−9.6	2.2
January 20, 1995	cloudy	−1.8	−14.9	13.1	−7.9	−10.8	2.7
Coombs Hills[3]							
January 11, 1995	clear sky	14.3	−13.6	27.9	0.4	−12.9	3.0
January 13, 1995	cloudy	2.5	−13.0	15.5	−6.7	−15.6	2.1
January 14, 1995	snow	−6.4	−14.3	7.9	−10.5	−17.1	4.7

[1] Scott Base: 44 m asl; 75°50'S; basalt soil

[2] Northwind Valley: ≈1500 m asl; 76°46.3'S; sandstone soil

[3] Coombs Hills: ≈ 200 m asl; 76°48.3'S; dolerite soil

provided a good prediction of the diurnal variation of soil surface temperature.

Diurnal soil temperature regimes. Figure 14 shows the diurnal soil temperature variation with depth at Scott Base and the Northwind Valley for clear-sky conditions. Both sites displayed significant heating and cooling cycles that diminish and lag in time with increasing depth into the soil. The 0°C isotherm penetrates to greater than 18 cm at Scott Base but barely reaches 9 cm at the Northwind site. At Scott Base temperatures in the soil tend towards 0°C in the absence of solar heating, and this is evident in the early morning period (Figure 14), while at Northwind temperatures at depth are much colder and there is not the same tendency for the soil to become isothermal. Temperature variations with depth are significantly greater at the Northwind site, evident by comparing the 18 cm and 19 cm temperatures at the respective sites in Figure 14.

Figure 15 shows vertical profiles of mean soil temperature for a range of conditions at the three locations. At Scott Base the soil down to 20 cm depth was warmer than 0°C on average, while at the two high-altitude sites the soil was colder than 0°C. The Northwind soil was colder than the Coombs Hills soil because of its higher albedo (Table 3). At Scott Base cloud cover did not result in the soil cooling, as it did at the colder sites.

Soil temperature and ice cement. There are strong links between the moisture component of the soil and soil climate. In Antarctica little, is understood about the mechanisms and rates of soil moisture movement in liquid and vapor phases. Ice cement is a common feature of Antarctic soils, although an understanding of its distribution and interactions with changing climate is still lacking. Phase changes of water in the soil affect the soil thermal regime. When ice melts, energy equivalent to the latent heat of fusion (3.33×10^5 J kg^{-1}) is absorbed with no

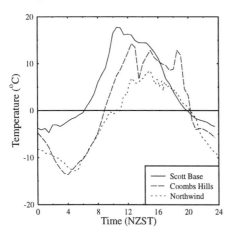

Fig. 12. Soil surface temperature for mainly clear-sky days at three sites with bare soils (see Table 5 for site details).

resulting temperature increase. An equivalent amount of energy is released to the soil when liquid soil water freezes, resulting in warming. These latent heat processes provide a strong buffer against temperature change in the soil when phase changes are occurring. *Anderson and Tice* [1989] demonstrated that for saline Antarctic soils water does not freeze at a single well defined temperature, but instead the phase change occurs across a range of temperature, with some liquid water existing at temperatures below –70°C.

The effect of this delayed phase change can be clearly seen in Figure 16. For the non-saline Scott Base soil, diurnal temperature changes damp out as the ice-cement zone is reached around 20–25 cm depth, corresponding with temperatures close to 0°C. For the saline soils at the Coombs Hills and Northwind Valley, soluble salt contents of 10.5% and 8.1% respectively cause phase changes to be delayed. Hence there is no evidence of temperature buffering, so diurnal temperature changes occur throughout the upper soil profile. This effect can be seen in Figure 14, where temperature changes in the saline Northwind Valley soil occur to much greater depths than at Scott Base, despite similar amplitudes of soil surface temperature variation and similar soil thermal properties (Table 6). At Scott Base in summer the upper level of the ice-cement remains close to 0°C, causing migration of soil moisture to the zone of phase change [*Buchan*, 1991], and, paradoxically, limiting the penetration depth of the 0°C isotherm so that the soil active layer is relatively shallow. At Scott Base under a thin snow pack (Figure 15) the soil became approximately isothermal at 0°C and ice cement was noted near the surface, while at the Coombs Hills, in similar conditions, the soil cooled dramatically with no evidence of ice cement.

Annual soil temperature. Annual cycles of soil temperature have been described at Vanda Station by *Thompson et al.* [1971b] and at Marble Point by *Balks et al.* [1995]. At Marble Point the upper soil layers are warmer than 0°C for a short period during December, January, and February (Figures 3 and 17). During this time strong diurnal cycles of temperature change occur which diminish with depth. After summer the soil cools steadily and the coldest temperatures are found near the surface. Temperature fluctuations of up to 15°C occur throughout winter, presumably driven by rapid changes in the temperature of the overlying airmass. The soil temperature cools rapidly after summer and then levels off to reach a minimum in late autumn, then warms again before declining to the coldest temperature, below –30°C, in August. This pattern has been described as the "coreless" or "kernlose" winter, a feature noted in air temperature records at both coastal and inland stations across Antarctica (e.g., at Scott Base by *Sansom* [1984]). *Wexler* [1959] and *van Loon* [1967] attributed the temporary warming in winter to an intensification of the meridional temperature gradient in autumn causing increased cyclonicity and southward advection of warm air.

Soil Thermal Properties

Table 6 lists soil physical and thermal data for three sites [*MacCulloch*, 1996]. Soil heat capacity and thermal conductivity measurements were obtained using powered probe methods described by *Bristow et al.* [1994]. Soil moisture content was low at all sites however the soil was slightly moister at Scott Base, where there was a well

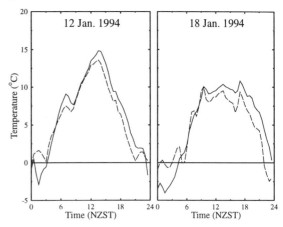

Fig. 13. Dependence of soil surface temperature (T_S) on solar radiation ($K\downarrow$), and windspeed (u). Solid line = measured; dashed line = modelled ($T_S = 0.031 K\downarrow - 1.13 u - 1.76$). Regression model was fitted for January 12, 1994 (r2 = 0.93) and used to predict surface temperature on January 18, 1994 (r2 = 0.81) [*Balks et al.* 1995].

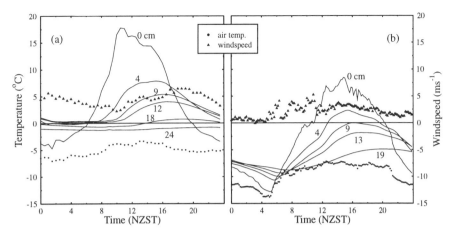

Fig. 14. Diurnal cycles of soil and air temperature, and windspeed, at (a) Scott Base (December 23, 1994) and (b) Northwind Valley (January 21, 1995) on mainly clear-sky days in summer. Depths (cm) are indicated on graphs.

defined ice cement zone, below 20 cm depth, because of its low soluble salt content. Soil heat capacity was highest at Scott Base because of its higher soil moisture content and higher bulk density. All three soils had soil thermal conductivities typical of dry sandy soils [Oke, 1987], and these did not change significantly through the range of soil moisture content present in the field. The low thermal diffusivity characterizes the total thermal response of the soil [Oke, 1987] with large variations in temperature confined to the uppermost layers.

IMPACTS OF HUMAN ACTIVITIES ON THE SOIL ENVIRONMENT

Introduction

The dry valleys are among the most accessible, attractive, and scientifically interesting areas of the Antarctic continent, and they are among the most frequently visited. Bases have been established and facilities such as air strips and helicopter landing pads have also been developed in this region. The impacts of human activities result from both the physical disturbance of the soils and from the introduction of foreign substances, sometimes accidentally (as in the case of fuel spills), and sometimes in the course of waste disposal. However field waste disposal is becoming less frequent as increased emphasis is being placed on removing all wastes, including human waste.

The land surfaces and soils in ice-free areas of Antarctica are very easily disturbed by human activities. The soils are typically protected by a desert pavement, a thin layer of gravel and coarse sand formed by the winnowing out of fine material by wind over a long period

of time until stability is achieved. If the desert pavement is disturbed, the soil surface may take a very long time to recover. Soil processes operate very slowly in the Antarctic environment because of the low temperatures, the general absence of vegetation, and scarcity of available water. Surface recovery following disturbance is, in most areas, a process driven primarily by wind.

It is generally accepted that one of the unique and most cherished features of the Antarctic dry valleys is their remoteness, and the impression that visitors have that very few, if any, people have visited the area before them. It is therefore important to endeavour to minimize human impacts, both from an aesthetic and an ecological perspective.

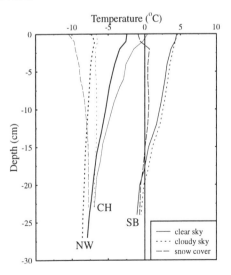

Fig. 15. Vertical profiles of mean soil temperature at three sites for a range of summer days. SB=Scott Base; NW=Northwind Valley; CH=Coombs Hills.

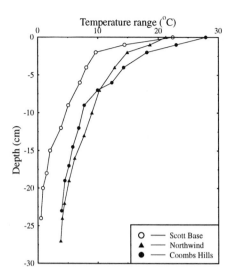

Fig. 16. Vertical profiles of diurnal temperature range (T_{max}–T_{min}) at three sites for clear-sky days.

Physical disturbance of soils in the dry valley areas can be considered at three scales. Major disturbance occurs where earth moving equipment or explosives are used to recontour a land surface for construction of facilities such as airstrips, helicopter landing pads and flat areas for base construction. Moderate disturbance occurs where activities such as camping, vehicle traffic and soil sampling have been undertaken. Low-level disturbance results from activities such as walking over the landscape. In some areas even low level disturbance can have a long-lasting impact on the landscape due to displacement of the desert pavement.

Impacts of Earth Moving Activities on Antarctic Soils

We have investigated the short term effects of earth moving activities in an experiment undertaken near Scott Base [*Balks et al.,* 1995]. Soil was removed from the active layer, using earth moving equipment, creating a "cut" area, and a "fill" area where the soil material was deposited. A comparable undisturbed area was also monitored. Soil moisture changes were recorded, following disturbance, using a neutron probe. Temperatures were measured at depths to 60 cm, using thermocouples, and surface radiation balance components were also measured.

Following removal of the active layer, the previously frozen soil, exposed in the "cut" area warmed rapidly, melting the ice within the soil. The released soil water evaporated at rates of up to 3 mm d^{-1}, precipitating salts on the soil surface. There was marked slumping and lowering of the newly formed soil surface. Following disturbance the previously dark colored soil surface (with an albedo of approximately 5%) lightened considerably (to an albedo of about 11%), but when the disturbed soil surface became moist, it darkened to an albedo similar to the undisturbed soil surface. The higher moisture content of the disturbed soil resulted in smaller diurnal temperature variations than those measured for the undisturbed soil (Figure 18). Most of the initial melt-out in the cut area occurred in the top 20 cm of soil in the twenty days following disturbance (Figure 19).

The long term effects of earth moving activities were investigated at Marble Point [*Campbell et al.,* 1994], where earthmoving work had been carried out in 1958–1959 and then abandoned. In the summers of 1989–1990 and 1991–1992 we identified areas of cut and fill, and nearby undisturbed areas, which were sampled at depths down to 2 m to assess the soil moisture, texture, and related soil properties.

In undisturbed soils the moisture content was generally very low (<10%) in the active layer, but considerably higher (generally in the range 20–80%), and more variable, within the permafrost. Where a soil was buried by addition of fill material a new active layer formed at the surface, but the

Table 6. Physical and Thermal Properties of Soil in the Active-Layer at Three Sites.
ρ_b = Dry bulk density; θ = Volumetric moisture content; C_s = Heat capacity (at field moisture content); k_s = Thermal conductivity (oven dry); κ_s = Thermal diffusivity. Errors included for C_s and k_s are 1.96 × standard error.

Site	ρ_b kgm^{-3}	θ %	soluble salt (%)	C_s MJ m^{-3}K^{-1}	k_s W m^{-1}K^{-1}	κ (m^{2}s^{-1} ×10^{-6})
Scott Base	1778	5.8	0.52	1.939 ± 0.055	0.261 ± 0.014	0.138
Northwind Valley	1638	2.9	8.08	1.549 ± 0.04	0.205 ± 0.012	0.132
Coombs Hills	1643	4.2	10.46	1.757 ± 0.04	0.217 ± 0.02	0.124

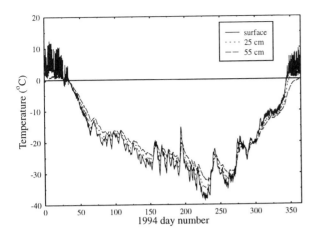

Fig. 17. Soil temperature regime at Marble Point, 1994.

high water contents within the permafrost had not been re-established in the fill material over the 30 years since disturbance occurred.

Where the soil active layer was removed (cut sites), a new active layer would have formed relatively rapidly, as illustrated in the experiment undertaken at Scott Base. The cut surfaces at Marble Point often had pronounced salt efflorescences, which were considered to have originated from the salts contained within the permafrost ice cement and were released upon thawing, probably in the year or so following excavation. The moisture profiles for cut surfaces at Marble Point were generally similar to that for undisturbed surfaces, with a low moisture content in the active layer and a higher, variable moisture content beneath.

Most land surfaces in Antarctica are dominated by patterned ground, the surface expression of a network of ice wedges in the underlying permafrost. On cut surfaces at Marble Point, unmodified since 1959, some shallow patterned ground cracks had formed, but these could be seen as part of a former polygon network where the troughs marked the position of ice wedges, which had melted out after the overlying material was removed. Occasionally, polygonal cracks were observed to extend from a scraped surface to fill material nearby, but most fill material showed little sign of new patterned ground formation, presumably because the ice content in the upper, newly formed permafrost was too low.

The impacts of earthmoving activities are generally devastating for biological communities, destroying organisms. Where salt accumulates on the surface following removal of material and melting of newly exposed permafrost, the new surface is unlikely to be readily recolonised by organisms. However in areas where water is available, algae and moss have been observed to become re-established, for instance on sites formed by

bulldozing at Marble Point 33 years earlier [*Campbell et al., 1992*].

Where the surface material is removed by human activity, as is often the case on Hut Point Peninsula, salts contained within the ice-cement are released and concentrated at the surface by evaporation. Sometimes they form a thin crust of salt which resembles a dirty snow cover. This salt crust eventually disappears over a period of years and the salts that it contains are blown away and dispersed over a wide area.

Impacts of Campsites and Other Moderate-Scale Soil Disturbance Activities

Much of the field work that is undertaken in the dry valleys is supported by helicopters and tent camps which are established for periods of up to a few weeks. When helicopters land to load and unload equipment and personnel, the resulting impacts include skid marks on the ground surface, blowing dust, and some trampling around the landing site. Once a tent site is selected, the surface stones need to be cleared from an area large enough to pitch a tent, and some ground surface smoothing may be undertaken. Other disturbances around a camp site arise from dumps of equipment and fuel, general trampling around the camp area, walking to field, toilet, and water (snow) supply sites, and accidental spills of fuel or waste waters. Depending on the nature of the scientific investigations that are undertaken, there may be a range of other impacts on a site.

A system of assessment criteria for rapid visual evaluation of terrestrial environmental impacts, at sites such as camp-sites, is described in *Campbell et al.* [1993]. The criteria that are considered include disturbed stones,

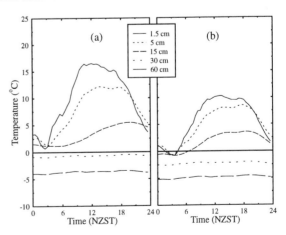

Fig. 18. Diurnal soil temperatures for (a) an undisturbed and (b) a disturbed site at Scott Base, January 1994 [*Balks et al., 1995*].

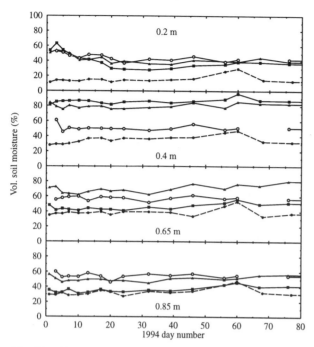

Fig. 19. Changes in soil moisture content during January to March 1994 for an undisturbed (dashed line) and three disturbed sites (solid lines), measured using a neutron probe [*Balks et al.* 1995].

stone impressions, boot imprints, visibly disturbed area, surface impressions, walking tracks, foreign objects, fuel spills, and biological disturbance. Disturbances to cobble and boulder sized stones are often highly visible because large stones are usually quite stable and develop a weathering pattern, often with staining, oxidation, abrasion, and rounding or pitting on the upper, exposed, surface, and pale colored calcite coatings on the underside. The removal, or overturning, of large stones often also leaves an imprint or indentation in the ground surface that may persist for a long time.

Many of the soil disturbances can be avoided, remedied, or at least mitigated if field parties are made fully aware of them. Completing an assessment of impacts, as one departs from a site will help raise people's awareness of their impacts and the manner in which they can be minimized.

Impacts of Human Trampling on Antarctic Soils

In areas frequently walked over, such as those adjacent to bases or experimental areas, tracks soon become evident. We measured the rate at which tracks form by walking along a set route a number of times and observing changes in the nature of the surface [*Balks et al.,* 1995] at three different

sites with different surface cover, ranging from unconsolidated sediments to coarse bouldery till.

A track forms as stones are dispersed sideways from the path that is being formed, uncovering the underlying material, which is often finer in texture and lighter in color. After as few as 20 passes, there is a distinct decrease in the number of surface stones along the line of the track and an increase in color contrast between the track and the undisturbed surface alongside. With further use the definition of the path, as well as its width, continues to increase, but at a decreasing rate. On stony surfaces the average rate of track formation is slower, as track development takes place mainly in small patches of fine material between the boulders.

The impact of trampling is progressive, but changes can become apparent even after one pass. On one particularly sensitive area, a fan surface beside Lake Vanda covered with a very loose and fine-textured soil, we were able to observe tracks made by two people several years before. After as few as 20 passes, the effects of trampling are very obvious and can be seen from long distances. The greatest visible effect occurs on soft, fine textured soil, especially when the desert pavement differs markedly in color from the finer material below. Where the surface cover consists of large rocks, the effect is less and it is possible, by choosing where to walk, to avoid almost all effects of passage.

Because significant paths form after only a very few passes and then develop further, people should be encouraged to follow an already established path, once created. Impact can be lessened by, wherever possible, following routes across large boulders or bedrock.

Soil Chemical Contamination

Because of the sometimes difficult working conditions in Antarctica and the reliance on liquid fuels for cooking, powering vehicles, aircraft, and generators, fuel spills are inevitable. Excavation of a number of sites where fuel spills have occurred, both recently and up to 30 years ago, indicates that where the soils contain ice-cemented permafrost, the fuel migrates down through the active layer to the permafrost surface and then migrates laterally. This has been observed to create a larger subsurface area of contamination than that visible at the soil surface. A large portion of the fuel evaporates, leaving behind a black deposit on the soil surface. Recent work [*Wardell,* 1995] has shown that oil degrading bacteria are active in soils on Ross Island. Therefore it may be possible to apply bioremediation to the clean-up of oil contaminated soils in the Antarctic.

Remediation strategy should be determined by the size of the spill and the nature of the hydrocarbon. The area contaminated by larger spills can be limited by early clean-up. Minor spills may be left to evaporate before later removal of residues. Consideration should be given to the environmental impacts of clean-up activities, which at some sites may exceed the impact of the initial fuel spill.

The impacts of disposal of small amounts of greywater and urine, as has occurred in the past at camp sites in the dry valleys, is largely unknown and has not been studied in any detail. However at sites that have been occupied for extended periods contaminants including nutrients, such as nitrogen and phosphorus, and a range of heavy metals have accumulated in the soils [*Sheppard et al.,* 1994]. For example, at Lake Vanda, where the former Vanda Station has been used by quite large groups for extended periods over the summer, there has been a long history of greywater disposal onto the soil surface. The level of the lake has risen over recent years, as a consequence of increased Onyx River flows, and is starting to intercept the area of contaminated soil. There are visible algae growths in the lake margins, adjacent to areas where urine and greywater have been disposed of in the past, indicating that some of the applied nutrients have migrated into the lake.

CONCLUSIONS

Soils are a key component of the cold desert ecosystem in the McMurdo Dry Valleys. In this unique environment, where climatic conditions are extreme, there is considerable variation in the processes that govern soil formation and soil properties. The result is a spatially complex and highly diverse soil environment. Formed within a very ancient land system where natural processes operate extremely slowly, the soil environment is very stable and little influenced by biological processes.

The physical properties of the soils are strongly influenced by glacial events. Large variations in the texture, thickness, lithological composition and distribution of till deposits, together with their subsequent modification by fluvial or aeolian processes, have given rise to a soil mantle with a very high degree of spatial variation. The intrinsic properties of the soil mantle have been only weakly modified by the weathering processes of oxidation and salinization. With increasing landscape age, the land surface becomes more subdued and soil reddening, oxidation, and salinity slowly increase.

The dry valley soils have extreme thermal regimes because solar radiation inputs are minimal for part of the year but very large during summer. The greatest extremes in soil temperatures are found at high altitudes and latitudes where the largest differences between air and surface temperatures occur. Most bare surfaces at any locality have temperatures above freezing for short periods if exposed to adequate solar radiation. Albedo is the most important surface factor influencing the soil thermal regime and this can vary greatly over short distances and be altered by soil disturbances resulting from human activities.

Permafrost properties are affected by the soil thermal regime, soil moisture content and salinity. In non-saline soils, energy transformations at the freezing point provide a buffer against rapid temperature changes within the soil. This effect is not evident in the saline soils that occur inland at high altitude sites because phase changes occur across a wide temperature range.

Within the McMurdo Dry Valley region, five groups of soils are differentiated based on geomorphic and pedological characteristics as well as temperature and soil moisture differences.

The coastal region is distinguished by a comparatively mild climate and soils with a relatively high moisture content. These soils are generally young, with little weathering, and contain only minor amounts of salts. Spatial variability of soil moisture content is high and in places there is considerable biological activity. The valley floors have soils that are lithologically more complex. The soil environment is much more arid, except close to streams and lakes where soil moisture is supplied by capillary flow. The soils range from very weakly to moderately weathered depending on the age of the surface on which they form, and biological activity is limited. Older soils contain a higher proportion of salts than those of the coastal region.

The soils of the valley walls are formed on a complex of old tills, screes, and colluvium. They tend to be older than the soils of the valley floor because they have escaped the more recent glacial advances. The soils on the valley walls are generally drier than those on the valley floor. Some of the soils are highly saline with thick salt horizons containing a high proportion of nitrate.

The upland valleys, above 1500 m, are part of an ancient system of cirques that contain tills ranging from young to extremely old. This region comprises a distinctly different soil environment. Temperatures are colder than on the valley floors, and although precipitation is higher, liquid water is only rarely present. On the older tills, the soils are more weathered than elsewhere and have distinct salt horizons in which nitrate and sulphate are the dominant anions. The soil environment in this region probably inhibits significant biological activity except in isolated pockets.

The plateau fringe, where the lowest temperatures and greatest precipitation occur, comprises the harshest environment of the dry valley region.

Because of the extreme slowness of dry valley ecosystem processes, the soils are easily damaged by human activities, ranging from mechanical activities, spills and, contaminations of various kinds to human trampling. Ecosystem recovery from disturbances varies according to the intensity of the initial disturbance and also the environmental conditions, with recovery time scales ranging from 1 year to greater than 100 years. Disturbances of the most arid soils generally last longer than disturbances of the moister coastal region soils.

The effects of small-scale human activities can be reduced by careful consideration of the timing and location of activities and by greater understanding of ecosystem processes. As the levels of human activity increase in the dry valleys, however the probability of disturbances also increases and there will be a need to develop guidelines

Within the McMurdo Dry Valley region, five groups of soils are differentiated based on geomorphic and pedological characteristics as well as temperature and soil moisture differences.

The coastal region is distinguished by a comparatively mild climate and soils with a relatively high moisture content. These soils are generally young, with little weathering, and contain only minor amounts of salts. Spatial variability of soil moisture content is high and in places there is considerable biological activity. The valley floors have soils that are lithologically more complex. The soil environment is much more arid, except close to streams and lakes where soil moisture is supplied by capillary flow. The soils range from very weakly to moderately weathered depending on the age of the surface on which they form, and biological activity is limited. Older soils contain a higher proportion of salts than those of the coastal region.

The soils of the valley walls are formed on a complex of old tills, screes, and colluvium. They tend to be older than the soils of the valley floor because they have escaped the more recent glacial advances. The soils on the valley walls are generally drier than those on the valley floor. Some of the soils are highly saline with thick salt horizons containing a high proportion of nitrate.

The upland valleys, above 1500 m, are part of an ancient system of cirques that contain tills ranging from young to extremely old. This region comprises a distinctly different soil environment. Temperatures are colder than on the valley floors, and although precipitation is higher, liquid water is only rarely present. On the older tills, the soils are more weathered than elsewhere and have distinct salt horizons in which nitrate and sulphate are the dominant anions. The soil environment in this region probably inhibits significant biological activity except in isolated pockets.

The plateau fringe, where the lowest temperatures and greatest precipitation occur, comprises the harshest environment of the dry valley region.

Because of the extreme slowness of dry valley ecosystem processes, the soils are easily damaged by human activities, ranging from mechanical activities, spills and, contaminations of various kinds to human trampling. Ecosystem recovery from disturbances varies according to the intensity of the initial disturbance and also the environmental conditions, with recovery time scales ranging from 1 year to greater than 100 years. Disturbances of the most arid soils generally last longer than disturbances of the moister coastal region soils.

The effects of small-scale human activities can be reduced by careful consideration of the timing and location of activities and by greater understanding of ecosystem processes. As the levels of human activity increase in the dry valleys, however the probability of disturbances also increases and there will be a need to develop guidelines and measures to minimize impacts.

Acknowledgments. The authors acknowledge the Foundation for Research, Science and Technology, the New Zealand Lottery Grants Board, and the University of Waikato for financial assistance, and the New Zealand Antarctic Programme for logistic support.

REFERENCES

Anderson, D.M. and A.R. Tice, Unfrozen water contents of six Antarctic soil materials, in *Cold Regions Engineering Proc. Fifth Int. Conf., Am. Soc. of Civil Engineers, New York,* 353–366, 1989.

Arnfield, A.J., A note on the diurnal, latitudinal and seasonal variation of the surface reflection coefficient, *J. App. Meteorol., 14,* 1603–1608, 1975.

Balks, M.R., D.I. Campbell, I.B. Campbell, and G.G.C Claridge, *Interim results of 1993/94 soil climate, active layer and permafrost investigations at Scott Base, Vanda and Beacon Heights, Antarctica,* University of Waikato, Antarctic Research Unit special report 1, 1995.

Bristow, K.L., R.D. White and G.G. Kluitenberg, Comparison of single and dual probes for measuring soil thermal properties with transient heating, *Aust. J. Soil Res., 32,* 447–464, 1994.

Bromley, A.M., Precipitation in the Wright Valley, *NZ Antarctic Record, 6, special supplement,* 60–68, 1988.

Bromley, A.M, The climate of Scott Base 1957–1992, *National Institute of Water and Atmospheric Research* NIWA/Clim/R/94-002, 1994.

Buchan, G.D., Soil temperature regime, in *Soil analysis physical methods,* edited by K.A. Smith and C.E. Mullins, Dekker, New York, 551–612, 1991.

Campbell, I.B. and G.G.C. Claridge, A classification of frigic soils: the zonal soils of the Antarctic continent, *Soil Sci., 107,* 75–85, 1969.

Campbell, I.B. and G.G.C. Claridge, Morphology and age relationships of Antarctic soils, *Quaternary Studies Roy. Soc. NZ Bull., 13,* edited by R.P. Suggate and M.M. Cresswell, 83–88, 1975.

Campbell, I.B. and G.G.C. Claridge, Soils and late Cenozoic history of the upper Wright Valley region, Antarctica, *NZ J. Geol. Geophys., 21,* 636–643, 1978.

Campbell, I.B. and G.G.C. Claridge, Soil research in the Ross Sea region, *J. Roy. Soc. NZ, 11,* 401–410, 1981.

Campbell, I.B. and G.G.C. Claridge, The influence of moisture on the development of soils in the cold deserts of Antarctica, *Geoderma, 28,* 221–238, 1982.

Campbell, I.B. and G.G.C. Claridge, *Antarctica: soils, weathering processes and environment,* Elsevier, Amsterdam, 368p, 1987.

Campbell, I.B. and G.G.C. Claridge, Landscape evolution in Antarctica, *Earth Sci. Rev., 25,* 345–353, 1988.

Campbell, I.B., G.G.C. Claridge and M.R. Balks, The properties and genesis of cryosols at Marble Point, McMurdo Sound region, Antarctica, *in Proc. 1st Int. Conf. on Cryopedology, Puschino,* edited by D.A. Gilchinsky, 59–66, 1992.

Campbell, I.B., M.R. Balks and G.G.C. Claridge, A simple visual technique for estimating the effect of fieldwork on the terrestrial environment in ice-free areas of Antarctica, *Polar Record, 29,* 321–328, 1993.

Campbell, I.B., G.G.C. Claridge, and M.R. Balks, The effect of human activities on moisture content of soils and underlying permafrost from the McMurdo Sound region, Antarctica, *Antarc. Sci., 6,* 307–316, 1994.

Claridge, G.G.C., The clay mineralogy and chemistry of some soils from the Ross Dependency, Antarctica, *NZ J. Geol. Geophys., 8,* 186–220, 1965.

Claridge, G.G.C., and I.B. Campbell, Origin of nitrate deposits, *Nature, 217,* 428–430, 1968.

Claridge, G.G.C., and I.B. Campbell, The salts in Antarctic soils, their distribution and relationship to soil processes, *Soil Sci., 123,* 377–384, 1977.

Claridge, G.G.C., and I.B. Campbell, Mineral transformations during the weathering of dolerite under cold arid conditions, *NZ J. Geol. Geophys., 27,* 533–545, 1984.

Claridge, G.G.C., and I.B. Campbell, Loess sources and aeolian deposits in Antarctica, *in Loess, its distribution, genesis and soils,* edited by D.N. Eden and R.J. Furkert, Balkema, Amsterdam, 33–45, 1988.

Claridge, G.G.C., I.B. Campbell, H.K.J. Powell, Z.H. Amin, and M.R. Balks, Heavy metal contamination in some soils of the McMurdo Sound region, Antarctica, *Antarc. Sci., 7,* 9–14, 1995.

Dana, G.L., R.A. Wharton and R. Dubayah, Solar radiation in the McMurdo Dry Valleys, Antarctica, this volume.

Denton. G.H., M.L. Prentice, D.E. Kellogg, and T.D. Kellogg, Tertiary history of the Antarctic Ice Sheet: evidence from the dry valleys, *Geology, 12,* 263–267, 1984.

Everett, K.R, Soils of the Meserve Glacier area, Wright Valley, Victoria Land, Antarctica, *Soil Sci., 112,* 425–488, 1971.

Freckman, D.W. and R.A. Virginia, Soil biodiversity and community in the McMurdo Dry Valleys, this volume.

Fritsen, C.H., E.E. Adams, C.M. McKay and J.C. Priscu, Permanent ice covers of the McMurdo Dry Valley lakes, Antarctica: liquid water content, this volume.

Kelly, W.C. and J.H. Zumberge, Weathering of a quartz diorite at Marble Point, Antarctica, *J. Geol., 69,* 430–446, 1961.

Keys, J.R. and K. Williams, Origin of crystalline, cold desert salts in the McMurdo Sound region, Antarctica, *Geochim. Cosmochim., Acta, 45,* 2299–2309, 1981.

Lindsay, J.F, Reversing barchan dunes in Wright Valley, Antarctica, *Geol. Soc. of Amer. Bull. 84,* 1791–1798, 1973.

MacCulloch, R.J, *The microclimatology of Antarctic soils,* Unpublished MSc (Hons) thesis, University of Waikato, NZ, 1996.

Marchant, D.R., G.H. Denton, G.E. Sugden and C.C. Swisher, Miocene glacial stratigraphy and landscape evolution of the western Asgard Range, Antarctica, *Geografisker Annaler, 75A,* 303–350, 1993.

Marchant, D.R., G.H. Denton, C.C. Swisher, and N. Potter, Late Cenozoic Antarctic palaeoclimate reconstructed from volcanic ashes in the dry valleys region of southern Victoria Land, *Geol. Soc. Amer. Bull., 108,* 181–194, 1996.

Markov, K.K., Some facts concerning periglacial phenomena in Antarctica, *Vestnik Moskobckogo Universitet (Geografiya), 1,* 139–148, 1956.

McCraw, J.D., Soils of the Taylor dry valley, Victoria Land, Antarctica, with notes on soils from other localities in Victoria Land, *NZ J. Geol. Geophys., 10,* 498–579, 1967.

Mehra, O.P and M.L, Jackson, Iron oxide removal from soils and clays by a dithionite-citrate system, buffered with sodium bicarbonate, *Clays Clay Mins., 7,* 317–327, 1960.

Miotke, F., Microclimate and weathering processes in the area of Darwin Mountains and Bull Pass, dry valleys, *Antarc. J. US, 15,* 14–16, 1980.

Nelson, K.H. and T.G. Thomson, Crystallisation of salts from sea water by frigid concentration, *J. Mar. Res., 13,* 166–182, 1954.

Oke, T.R., *Boundary Layer Climates,* University Press, Cambridge, U.K., 1987.

Prentice, M.L., J.G. Bockheim, S.C. Wilson, L.H. Burkle, D.A. Hodell, C. Schlüchter, and D.E. Kellogg, Late neogene Antarctic glacial history: evidence from central Wright Valley, in *The Antarctic palaeoenvironment: a perspective on global change,* edited by J.P. Kernott and D.A. Warnke, *Antarct. Res. Ser., 60,* 207–250, 1993.

Prentice, M.L., J. Kleman, and A.P. Stroeven, The composite landscape of the northern McMurdo Dry Valleys: Implication for Antarctic Tertiary glacial history, this volume.

Sansom, J., The temperature record of Scott Base, Antarctica. *N.Z. J. Sci. 27,* 21–31, 1984.

Sheppard, D.S., I.B. Campbell, G.G.C. Claridge and J.M. Deely, Contamination of soils around Vanda Station, Institute of Geological and Nuclear Sciences Science Report 94/20, 140p, 1994.

Tedrow, J.C.F., and F.C. Ugolini, Antarctic Soils, *Antarctic Soils and Soil-forming Processes*, edited by J.C.F. Tedrow, Antarct. Res. Ser., 8, 161–177, 1966.

Thompson, D.C., R.M.F. Craig and A.M. Bromley, Climate and surface heat balance in an Antarctic dry valley, *NZ J. Sci., 14,* 245–51, 1971a.

Thompson, D.C., A.M. Bromley and R.M.F. Craig, Ground temperatures in an Antarctic dry valley, *NZ J. Geol. Geophys., 14,* 477–483, 1971b.

Ugolini, F.C. and D.M. Anderson, Ionic migration and weathering in frozen Antarctic soils, *Soil Sci., 115,* 461–470, 1973.

van Loon, H., The half-yearly oscillations in the middle and high southern latitudes and the coreless winter, *J. Atmos. Sci., 24,* 472–486, 1967.

Wardell, L.J., Potential for bioremediation of fuel-contaminated soils in Antarctica, *J. Soil Contam., 41,* 111–121, 1995.

Wellman, H.W., Later geological history of Hut Point Peninsula, Antarctica, *Trans. Roy. Soc. NZ. Geol., 2,* 147–154, 1964.

Wexler, H., Seasonal and other temperature changes in the Antarctic atmosphere, *Quart. J. Roy. Meteorol. Soc., 82,* 196–208, 1959.

Weyant, W.S., The Antarctic climate, in *Antarctic soils and soil-forming processes*, edited by J.C.F. Tedrow, Antarc. Res. Ser., 8, 47–59, 1966.

M.R. Balks, Earth Sciences Dept., University of Waikato, Pvt. Bag 3105, Hamilton, New Zealand

D.I. Campbell, Earth Sciences Dept., University of Waikato, Pvt. Bag 3105, Hamilton, New Zealand

I.B. Campbell, Land and Soil Consultancy Services, 23 View Mount, Nelson, New Zealand

G.G. C. Claridge, Land and Soil Consultancy Services, 23 View Mount, Nelson New Zealand

(Received September 4, 1996; accepted February 25, 1997)

SOIL BIODIVERSITY AND COMMUNITY STRUCTURE IN THE MCMURDO DRY VALLEYS, ANTARCTICA

Diana Wall Freckman

Natural Resource Ecology Laboratory, Colorado State University, Fort Collins, Colorado

Ross A. Virginia

Environmental Studies Program, Dartmouth College, Hanover, New Hampshire

A conceptual model is proposed that defines the soil and environmental conditions determining suitable and unsuitable habitats for soil biota in the McMurdo Dry Valley ecosystem of Antarctica. We hypothesized that if dispersal of soil fauna among the dry valleys was equal, then diverse and abundant communities of soil organisms would develop in all suitable habitats. The majority of soils sampled across the valleys (65%) support up to three soil invertebrate taxa (tardigrades, rotifers, nematodes). The rest of the soils are presumed to be unsuitable habitats as none of the target organisms were found, but there appears to be no single soil property that defines a suitable or unsuitable habitat. Most soils contain only one invertebrate taxa (nematodes); two and three taxa communities are rare. There are no other soil systems known where nematodes represent the top of the food chain and where food webs appear so simple in structure. Nematodes are more abundant and more widely distributed than either tardigrades or rotifers. The species diversity of nematodes is very low (n = 3), with only *Scottnema lindsayae*, a microbial feeder, occurring throughout the dry valleys. In many locations, soil conditions may be outside the tolerances of dispersing organisms preventing community establishment, thus creating the patchy distribution of soil biota that uniquely defines the dry valley landscape. The unusually low diversity and low functional redundancy of the dry valley soils suggest that these systems will be highly disrupted by the loss or decline of even a single species that is sensitive to environmental change. We suggest that nematodes may be a useful indicator organism for detecting environmental change in the dry valley system. As more people enter the dry valleys, human-induced impacts will directly affect soil habitats and the associated biota and the ecosystem functions they perform.

INTRODUCTION

Soils are the dynamic center for a majority of terrestrial ecosystem processes, most of which are mediated by soil organisms. Soils are also a major global storage reservoir for carbon in the form of organic matter (estimates of about 1500 x 10^15 g C are stored in soils) [see *Schlesinger*, 1991]. Soil organisms are an important factor in soil formation and soil development. They control rates of organic matter decomposition and as a consequence, act to regulate the amount of carbon stored in the world's soils. Soil biota are actively linked to biogeochemical processes such as biological nitrogen fixation, methane and nitrous oxide production and soil respiration. The living microbes, fungi, and numerous phyla of invertebrates that comprise the soil food web influence ecosystem productivity by making essential elements available for plant growth, while at the same time contributing to the rate of production and consumption of radiatively active gases [*Beare et al.*, 1995; *Freckman*, 1994; *Schimel et al.*, 1994].

323

The ecological and edaphic patterns controlling the distribution and populations of soil fauna are poorly known in all terrestrial ecosystems [*Hendrix et al.,* 1986; *Freckman and Caswell,* 1985; *Freckman,* 1994]. Nonetheless we must gain this knowledge to improve our predictions of the impact of disturbance, such as global change, on soil biota and the processes they control across ecosystems [*Boag et al.,* 1991; *Heywood,* 1995; *Kennedy,* 1995; *Wynn-Williams,* 1996a; *Freckman and Virginia,* 1997]. In most terrestrial ecosystems, there is an enormous complexity of interactions among the diversity of soil organisms. The extent of soil invertebrate biodiversity is poorly studied even though these soil biota act as critical links between atmospheric, terrestrial, and aquatic systems [*Freckman,* 1994; *Franzmann,* 1996; *Courtright et al.,* 1997]. Despite the development of new technology including molecular methods, it has been difficult to determine the roles of specific taxa in affecting ecosystem processes or to determine the factors responsible for regional to global patterns of soil biodiversity. The taxonomic complexities of the majority of soil biota have not allowed species-level responses and interactions to be tested experimentally in the field. Nevertheless, field research grouping the invertebrates into functional groups and food webs, and studies using simplified laboratory microcosms, have shown that soil fauna are important determinants of biogeochemical processes [*Moore et al.,* 1996].

Nematodes are among the most numerous and diverse groups of soil organisms. Since these meso-fauna are thought to be ubiquitous in terrestrial ecosystems [*Sohlenius,* 1980; *Freckman,* 1982; *Procter,* 1990] and are a major component of soil food webs [*Petersen and Luxton,* 1982; *Freckman and Mankau,* 1986; *Freckman et al.,* 1987], comparisons of their abundance, biomass, and community structure can be made across ecosystems. Because of their key interactions in the cycling of carbon and nutrients in the soil system, nematodes have been considered to be important integrators of soil ecosystem processes and indicators of soil disturbance [*Hunt et al.,* 1987; *Bongers,* 1990; *Freckman and Ettema,* 1993; *Coleman and Crossley,* 1995; *Niles and Freckman,* in press].

Research in extreme environments having low diversity soil nematode communities, such as the Antarctic, provides an opportunity to study linkages between species diversity and ecosystem functioning that are masked by the overwhelming complexity of soil biodiversity in other terrestrial ecosystems. We have been studying the ecology and distribution of nematodes, tardigrades, and rotifers in the McMurdo Dry Valleys since the 1989–1990 austral summer. In Antarctica, there are fewer genera and species and fewer trophic groups [*Freckman and Virginia,* 1997] than in any other terrestrial ecosystem (see Figure 1). We hypothesized that the relationships between species diversity and function and the soil environment should be more apparent in the dry valleys of Antarctica than in other terrestrial ecosystems.

Our research can be characterized as moving from descriptive to more integrated soil ecosystem studies within the conceptual framework of the McMurdo National Science Foundation Long Term Ecological Research (LTER) Program [*Franklin et al.,* 1990]. Our primary efforts are to understand how nematodes, the more advanced organisms of the three invertebrate groups studied in the dry valleys, survive, disperse, colonize, and develop functional communities in Antarctic dry valley soils, and how these communities may be affected by disturbance.

Prior research on nematodes in Antarctic soil food webs and their mechanisms of survival has been concentrated in the maritime and sub-Antarctic regions where higher plants are present [*Maslen,* 1981a,b; *Block,* 1984; *Pickup,* 1990a,b; *Pickup and Rothery,* 1991] (Table 1). The soil biological systems within the McMurdo Dry Valleys are greatly simplified compared to the maritime Antarctic [*Gressitt,* 1967; *Block,* 1984; *Vishniac and Klingler,* 1988; *Vincent,* 1988; *Wynn-Williams,* 1990, 1996a]. The dry valleys lack vascular plants, and invertebrate species diversity is low [*Cameron et al.,* 1970]. Studies of nematodes in dry valley soils have been primarily restricted to the "favorable" environments of high soil moisture, high organic carbon, low salinity, and low elevation [*Cameron,* 1972]. This research has resulted in species descriptions [*Yeates,* 1970; *Timm,* 1971], biogeographic surveys [*Wharton and Brown,* 1989] and studies on mechanisms of survival [*Wharton and Brown,* 1991; *Pickup and Rothery,* 1991; *Wharton,* 1995]. Only in rare habitats of moist, moss-vegetated dry valley soils [*Vincent,* 1988; *Schwarz et al.,* 1992], are higher invertebrates, such as springtails and mites found at the top of the food chain [*Walter et al.,* 1986].

Nematodes, rotifers, and tardigrades are aquatic animals. The extreme environment of the dry valleys with its lack of moisture and cold temperatures, would appear to provide an unfavorable habitat for nematodes. *Kennedy* [1993] proposed that moisture limitations were probably more important than low temperatures for explaining organism distribution in Antarcti-

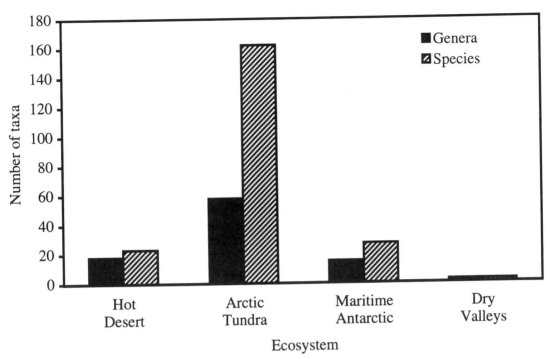

Fig. 1. Number of nematode taxa in cold or arid ecosystems. Data from [*Sohlenius*, 1980; *Maslen*, 1981a; *Freckman and Mankau*, 1986; *Freckman and Virginia*, 1997].

ca. Nematodes, tardigrades, and rotifers have remarkable survival strategies that allow them to survive for extended periods without water. Investigations in temperate and desert ecosystems have shown that at any stage in their life cycle under conditions of environmental stress, these phyla change their morphology, lose 99% of their free water, and enter an ametabolic state, anhydrobiosis [*Crowe*, 1971]. Nematodes in anhydro-biosis have been known to survive for >60 years and to revive from exposure to conditions such as 0% relative humidity, vacuum, and liquid N_2 [*Crowe and Clegg*, 1978; *Freckman*, 1978; *Freckman*, 1986]. Research has shown additional survival mechanisms such as cold-hardiness [*Pickup*, 1990a,b; *Pickup and*

TABLE 1. Nematode Species Identified from Contiguous Continental Antarctica.

Genus	Species	Reference	Habitat	Trophic level
Eudorylaimus	*antarcticus**	Steiner 1916 Yeates 1970	moss, soil, algal mat	O/P
Monhystera	*villosa*	Bütschli 1873	algal mat on soil	MF
Panagrolaimus	*davidi*	Timm 1971	algal mat on volcanic soil	MF
Plectus	*antarcticus**	de Man 1904	freshwater pond, algae	MF
	frigophilus	Kirjanova 1958	freshwater pond, algae, moss, sandy soil	MF
	globilabiatus	Kirjanova 1958	freshwater pond	MF
	murrayi	Yeates 1970	moss	MF
	parietinus	Bastian 1865	freshwater pond	MF
Scottnema	*lindsayae**	Timm 1971	volcanic soil, glacial moraine	MF

*Indicates species present in the dry valley samples.
O/P = omnivore/predator; MF = microbial feeder.

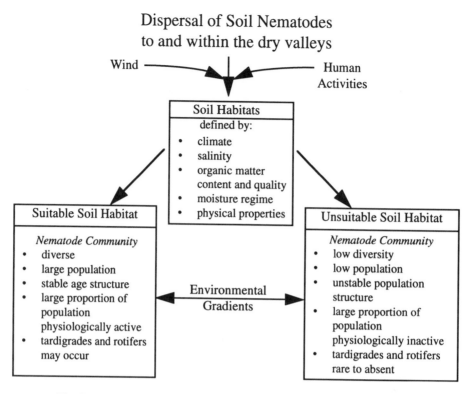

Fig. 2. Conceptual model of nematode dispersal and soil habitat suitability.

Rothery, 1991] and cryobiosis [*Wharton and Brown,* 1991; *Wharton,* 1995].

Soil nematodes in the dry valleys survive unfavorable soil conditions in anhydrobiosis (Freckman, Virginia, and Powers, unpublished data) as they do in hot deserts [*Freckman,* 1978; *Freckman et al.,* 1987]. Previous studies have shown that nematodes in anhydrobiosis are dispersed by wind, thus colonizing new soil habitats [*Orr and Newton,* 1971; *Carroll and Viglierchio,* 1981]. From these observations, we hypothesize that wind is the primary means of dispersal for nematodes in the dry valleys. Wind has been shown as the vector for spores and pollen reaching sub-Antarctic regions from South America [*Marshall,* 1996].

CONCEPTUAL MODEL

In this chapter we present and examine a conceptual model of the factors controlling nematode community structure and associated biota in Antarctic soils (Figure 2). Our conceptual framework is based on the assumption that nematodes have had the opportunity to disperse as anhydrobiotes by wind to all locations in the dry valleys [*Courtright et al.,* in press]. As a

consequence, communities of nematodes could potentially establish in all soils of the dry valleys, as they have in all other terrestrial ecosystems studied. Soil habitats can be classified as either suitable or unsuitable for nematode activity, survival and reproduction. If climate (temperature, precipitation) is similar in soils across the dry valleys, then it is the local physical (texture) and chemical soil characteristics (e.g., salinity) which would determine habitat suitability and nematode community structure and biodiversity. Habitat suitability can be defined by a variety of soil conditions, including organic matter content and quality, salinity, moisture, and physical properties. If nematodes are dispersed to a suitable soil habitat, we would expect to find established nematode communities that have numerous species, a high population density, a stable age structure, and a high percentage of living nematodes (Figure 2). If the soil habitat is unsuitable, the development of the nematode community will be limited or nematodes will be absent. We recognize that our categories of suitable and unsuitable habitats can be thought of as end points along a continuum of habitat suitability, a continuum produced by environmental gradients and factors that characterize the soil habitat. Expanding this framework to the ecosystem would suggest that suitable habitats contribute to active processing of soil carbon and nutrients through soil microbial-nematode interactions. Conversely, in sites where nematodes are in the anhydrobiotic state or are absent, we would predict an uncoupling of the soil food web interactions involving nematodes and other soil biota, resulting in lower rates of nutrient cycling.

SOIL HABITATS

In the most extensive study in continental Antarctica to date, we examined dry valley soils for nematodes and associated invertebrates. This work is continuing to examine new geographic locations and soil conditions in the McMurdo Dry Valley region and/or Ross Island. Antarctic dry valley soils are unique since they contain both large amounts of soluble salts and have permafrost [Bockheim, 1997]. Since precipitation is very low, the soils of the dry valleys are not leached and weathering products accumulate in the soil profile [Pastor and Bockheim, 1980; Bockheim, 1997]. The soil is underlain with an ice-cemented layer about 10–30 cm below the soil surface. Dry valley desert soils are generally poorly developed, coarse textured, and have low biological activity [Campbell and Claridge, 1987; Campbell et al., this volume]. Organic C and N accumulation is much lower than in hot desert soils due to the lack of plant cover and low rates of production [Cameron et al., 1970]. Most dry valley soils are classified as Pergelic Cryothents or Cryopsamments [see Bockheim, 1997].

Using sterile techniques [Powers et al., 1995], we collected soil samples across four valleys in the McMurdo Dry Valley system. Sampling locations were selected to span a wide range of environmental conditions within each valley. Moist habitats were located near streams and glaciers and generally had soil moisture contents greater than 5% (w/w). Dry habitats, defined as less than 5% soil moisture content, consisted of soils from dry polygons (sorted and non-sorted), and soils from large expanses of unstructured xeric soils [Campbell et al., this volume]. The valleys and the number of soil samples collected per valley are as follows: Garwood Valley (78° 02'S, 164° 10'E) (41 samples), Taylor Valley, site of the McMurdo Dry Valley LTER, (77° 37'S, 160° 50'E) (178 samples), Wright Valley (77° 31'S, 161° 50'E) (103 samples), and the more remote Victoria Valley (77° 23'S, 162° 00'E) (93 samples). Three phyla, nematodes, rotifers, and tardigrades were extracted by the modified sugar centrifugation technique [Freckman and Virginia,

1993]. Only one microarthropod, a mite, was found in the 415 samples.

A notable feature of dry valley soils is the high percentage that contain no invertebrates, a striking contrast to temperate ecosystems where most soils contain more than eight invertebrate phyla [Heywood, 1995]. Our results show that of the 415 soils sampled across the four valleys, about 35% had no animals (Figure 3). Of the remaining soils with soil fauna, nematodes are by far the most abundant invertebrate. Communities of either rotifers or tardigrades, or of two and three phyla are scarce, occurring in less than 5% of the samples. It is striking that soils with a complexity of greater than two phyla rarely occur in the dry valleys, which indicates that either the majority of soils are unsuitable habitats for the development of complex communities, or that organisms have not dispersed to these locations.

The soils that are presumed to be unsuitable habitats, that is lacking invertebrates, vary between dry valleys (Table 2). Victoria and Wright Valley have the highest proportion of soil samples without invertebrates (>50%), while Taylor Valley has the lowest fraction (20%). This pattern may be explained by lower soil moistures in Victoria and Wright Valley than in Taylor Valley, and the higher soil salinity in Victoria Valley, in part since soils there are older [Campbell et al., this volume]. The most frequent soil community type is one where nematodes occur alone. In Garwood and Taylor Valleys nematodes without associated invertebrates occur in about 65% of the samples. In Wright and Victoria Valleys this value is lower, 42% and 34%, respectively. Communities consisting of all three invertebrate taxa do not occur in Wright and Victoria Valleys, and even when found in Taylor Valley, the three-taxa communities exist in only about 8% of the soils. The lack of invertebrate diversity at the taxonomic level of phyla in these soils suggests that tardigrades, rotifers, and nematodes differ in their habitat requirements.

Nematodes are the most abundant group of invertebrate biota in the valleys, averaging about 700 kg^{-1} dry soil for all soils sampled (Figure 4). Even though Victoria Valley has the smallest proportion of soils with nematodes (Table 2), densities are similar to Garwood and Taylor Valleys. Nevertheless Victoria and Wright Valleys appear to be a more extreme habitat for soil invertebrates, since Victoria Valley has the lowest densities of rotifers and no tardigrades, while Wright Valley has very few nematodes. One of the challenges of our research is to determine which features

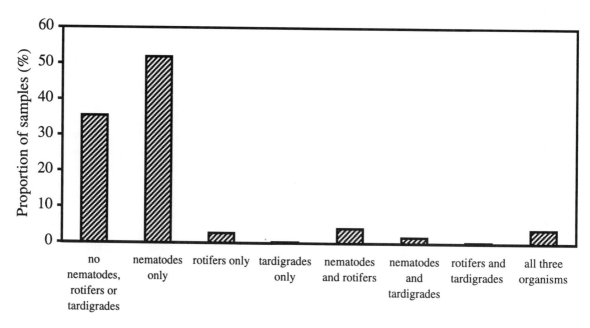

Fig. 3. Presence of invertebrate taxa in 415 soil samples collected from 4 McMurdo Dry Valleys.

of these valleys (e.g., remoteness, distance from marine or lacustrine sources of carbon [*Burkins et al.*, in press]), limits soil biodiversity and the function of soil communities.

Nematode communities in terrestrial ecosystems, from hot deserts to Arctic forests [*Sohlenius*, 1980] have a high species diversity (Figure 1), indicating wide dispersal and numerous suitable habitats for community establishment and maintenance [*Yeates*, 1970]. In hot deserts, the distribution of the nematode community is related to suitable habitats defined by plant distribution, patterns of soil organic matter accumulation, and soil properties such as depth, soil moisture, root mass, N, available P, and pH [*Freckman and Mankau*, 1986; *Freckman and Virginia*, 1989; *Virginia et al.*, 1992]. From these and other studies [*Wasilewska*, 1971; *Steinberger et al.*, 1984; *Robertson and Freckman*, 1995] we conclude that plant distribution and plant-related soil processes (e.g., decomposition) govern nematode abundance and diversity in hot deserts. Unlike hot desert soils, a notable percentage (35%) of soils sampled from the dry valleys lack nematodes [*Freckman and Virginia*, 1991; *Freckman and Virginia*, 1997], but where nematodes occur, peak densities (4000

TABLE 2. Percentage of Samples from each Valley that Contained Various Combinations of the Three Soil Invertebrate Taxa Found in Dry Valley Soils.

Combinations	Garwood Valley (n=41)	Taylor Valley (n=178)	Wright Valley (n=103)	Victoria Valley (n=93)
No Biota	24.4	19.7	52.4	51.6
Nematodes only	65.9	63.5	41.8	34.4
Rotifers only	0.0	0.6	1.9	8.6
Tardigrades only	0.0	0.6	0.0	0.0
Nematodes and rotifers	0.0	6.2	1.0	5.4
Nematodes and tardigrades	4.9	1.7	1.9	0.0
Rotifers and tardigrades	0.0	0.0	1.0	0.0
Nematodes, rotifers, and tardigrades	4.9	7.9	0.0	0.0

Sample number given in parenthesis.

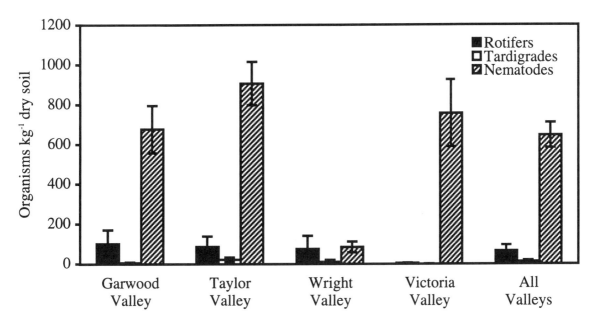

Fig. 4. Mean number (± standard error) of rotifers, tardigrades, and nematodes in four dry valleys.

kg⁻¹ dry soil) are comparable to those in other deserts [*Freckman and Mankau*, 1986].

In temperate ecosystems, nematodes interact in the soil food web at four to five trophic levels. Nematodes are generally classified into trophic/functional groups based on genus or species morphology and biological data [*Yeates et al.*, 1993]. In the dry valley soils we examined, the nematode community is reduced to the endemic species *Scottnema lindsayae, Timm* [1971], a microbial feeder, and additional species in two genera, *Plectus antarcticus*, a bacterial feeder, and *Eudorylaimus antarcticus*, an omnivore-predator [*Freckman and Virginia*, 1991]. The genera *Plectus* and *Eudorylaimus* have a global distribution. In culture, *Scottnema* can significantly reduce bacteria and yeast populations [*Overhoff et al.*, 1993] and based on other studies of related nematode species [*Ingham et al.*, 1985] we hypothesized that grazing by *Scottnema* can increase carbon and nitrogen mineralization. *Plectus* feeds on bacteria, and although the food source of the dry valley *Eudorylaimus* has not been confirmed, other species of *Eudorylaimus* are fungal or algal grazers or are predacious on other nematodes [*Yeates et al.*, 1993]. *Scottnema* dominates in both density and biomass in dry valley soils, whereas densities of *Plectus* and *Eudorylaimus* are considerably lower in most valleys and both species are rare in Victoria Valley (Figure 5). The tardigrades may be predaceous on nematodes, but as noted earlier, they occur in only a few soils with other invertebrates, so their role in influencing nema-

tode populations is unclear. Thus, unlike most ecosystems, the decomposition of soil organic matter in the dry valleys appears to be controlled by only two functional groups, microbivores and omnivore/predators and trophic interactions in the soil influencing nutrient cycling are limited to microbes (yeast and bacteria) and micro-invertebrates [*Vishniac*, 1996; *Wynn-Williams*, 1996b] (Figure 6). There are no other soil systems known where nematodes represent the top of the food chain and where food webs appear so simple in structure.

Protozoa contribute to the complexity of the soil community [*Bamforth et al.*, 1996; *Smith*, 1996] and may provide competition for the microbial-feeding nematode, *Scottnema*. The biomass of soil invertebrates in dry valley soils is exceedingly low. Nematode biomass (mg dry weight m⁻² to a depth of 10 cm) was calculated [see *Freckman*, 1982] for the three dry valley nematode species based on geometric measurement of 50–100 specimens (depending on species) (Table 3), and was compared to values for soil protozoa [see *Bamforth et al.*, 1996]. In a continuing collaborative study, *Bamforth et al.*, [1996] found amoebae densities averaged 57 organisms g⁻¹ dry soil except for a Miers Valley site (243 organisms g⁻¹ dry soil). Total flagellate and amoebae biomass (based on 18 samples) was approximately 7.3 mg dry weight m⁻² soil, which is lower than the estimates for *Scottnema* (42.3 mg dry weight m⁻² for Taylor Valley), but slightly higher than the other nematode species (Table

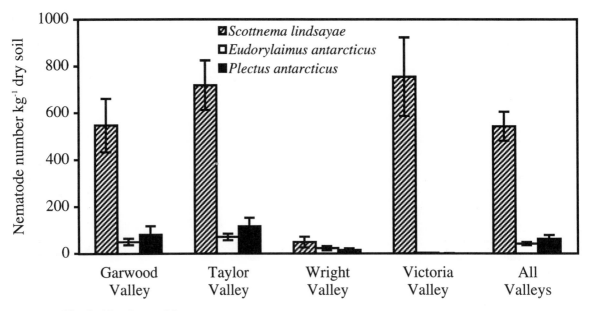

Fig. 5. Abundance of three nematode species in dry valley soils. Error bars denote standard error.

3). As a comparison, in a North American hot desert soil, nematode bacterial feeder biomass was lower than in the dry valleys, 27.4 mg dry weight m[-2] [*Freckman and Mankau*, 1986]. However, total nematode biomass in the hot desert is higher overall, since the hot desert nematode community is composed of five trophic levels instead of two. Data are not available to discern whether protozoa are present in all the Antarctic soils included in our nematode analysis.

The unusually low diversity and low functional redundancy of the dry valley soils suggest that these systems will be highly disrupted by the loss or decline of even a single species that is sensitive to environmental change [*Freckman and Virginia*, 1997]. The two

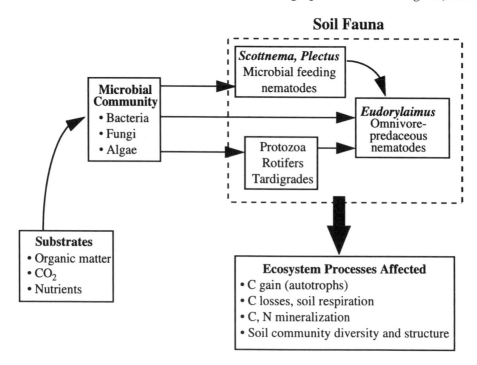

Fig. 6. Hypothesized food web model for major trophic relationships in dry valley soils.

TABLE 3. Nematode Biomass (mg dry weight m-2) to a Depth of 10 cm for
Three Species across Four of the McMurdo Dry Valleys.

Valley	Scottnema lindsayae	Eudorylaimus sp.	Plectus antarcticus	Total Nematodes
Garwood	21.2	6.8	3.4	31.4
Taylor	27.8	9.6	4.9	42.3
Wright	1.9	2.9	0.6	5.4
Victoria	29.2	0.2	0.0	29.4

nematode species found at a field experiment site near Lake Hoare, Taylor Valley, responded differently to manipulations of soil resources and climate. The density of the rarer omnivore-predator species, *Eudorylaimus*, declined in response to changes in the soil environment (increases in moisture, temperature, and carbon), while the more abundant and probable prey nematode species, *Scottnema*, generally increased. This could be due to a direct effect of our treatments altering the soil environment [*Freckman and Virginia, 1997*] and perhaps increasing the microbial food source for *Scottnema* (bottom-up effect), or an indirect effect of the reduction of the predator allowing the prey species to increase in numbers (top-down effect) [*Carpenter and Kitchell*, 1993]. The largest effect for either species occurred for the treatment combination where we increased soil temperature, moisture, and carbon. These field results show that individual species within soil communities can be impacted differentially by disturbances that alter soil resources and soil climate, resulting in changes in the soil food webs and community structure. *Campbell et al.* [this volume] have shown that the physical and chemical properties of dry valley soils are susceptible to human disturbance and that these effects are long lasting. Thus anthropogenic disturbance may have measurable effects on the diversity, structure, and distribution of Antarctic soil communities. Based on such observations we suggest that nematodes may be a useful indicator organism for detecting environmental change in the dry valley system.

We have hypothesized that nematodes disperse with wind easily between dry valleys ensuring genetic exchange, but an alternate hypothesis would be that populations are geographically isolated and might show considerable genetic drift and adaptation to local microscale climatic conditions [*Courtright et al.*, in press]. Determining the identification of nematodes is difficult because many species are similar morphologically and morphological characters can change with food source or habitat [*Anderson*, 1968; *Schiemer*,

1982; *Thomas and Wilson, 1991*]. Molecular techniques provide better resolution for studying genetic structures of nematode populations and speciation, because DNA and RNA data are less subjective than measures of variation in morphological characters or investigator interpretation. This makes molecular approaches useful for nematode species identification [*Williamson*, 1991]. Ribosomal DNA sequence data can be used to ascertain the geographic origin of nematode species which is important for determining phylogenetic relationships [*Caswell-Chen et al.*, 1992]. *Van der Knaap et al.* [1993] easily differentiated bacterial-feeding nematode species of the genera *Caenorhabditis, Acrobeloides, Cephalobus*, and *Zeldia* using arbitrarily-primed polymerase chain reaction (ap-PCR).

Courtright et al. [in press] tested our hypothesis that dispersal does not limit nematode distribution in the dry valleys by examining the nuclear and mitochondrial genome of the most widely dispersed nematode, *Scottnema*. Two segments of rDNA that encoded the D2 and D3 expansion segments of the nuclear large rRNA subunit and a section of mtDNA were sequenced to determine genetic variation within individuals of *Scottnema*. The specimens were collected from Garwood, Taylor, Wright, Victoria, and Alatna Valleys. *Courtright et al.* [in press] found the pattern of nuclear variation was most consistent for a single species defined morphologically as *Scottnema*, but mitochondrial analysis showed significant differences in the frequency of halotypes in each geographic sample. The mitochondrial results indicate that the dry valley populations are not identical and that some restriction to gene flow exists between the valleys. Thus it would appear that either dispersal of nematodes by wind is equal across the valleys and that genetic variation is occurring at small microscale habitats, or that dispersal is not equal to all valleys. As more people enter the dry valleys, movement of soil and associated nematodes will increase the rates of nematode dispersal and introductions into the remote regions, as has occurred between nematodes in agro- and other terrestrial ecosystems.

The soils of the McMurdo Dry Valleys region are described in detail by *Campbell et al.* [this volume]. Local-scale variation in soil chemistry and soil climate is high, adding to the complexity in deriving relationships between soils and biota in the dry valleys. Despite the heterogeneity of dry valley soils, relationships between soil biota and soil properties can be discerned. In an analysis of over 100 samples from three dry valleys *Freckman and Virginia* [1997], found that the three nematode genera recovered are each associated with different soil variables. For example, the density of *Scottnema*, the most abundant and widespread nematode in the dry valleys, was best related by multiple regression to soil chemical properties that are captured in measurements of soil pH and electrical conductivity (salt accumulation). However only about 20% of the population variance could be explained by these soil properties. The less abundant *Plectus* and *Eudorylaimus* were linked to soil parameters associated with soil organic matter such as total nitrogen and organic carbon. These comparisons were based on soils collected from widely separated locations in Taylor, Wright, and Garwood Valleys. When soils are examined within a local area at finer scales of resolution, and along environmental gradients such as increasing elevation [*Ho et al.*, in press], the predictive relationships between soils and nematode abundance may be higher. In a study of soils and biota with increasing elevation along the south shore of Lake Hoare, soil salinity was associated with nematode population structure [*Powers et al.*, in press]. As soil salinity increased, the percent age of nematodes recovered as living decreased.

Soil salinity is probably a significant variable in the definition of habitat suitability for nematodes. The lack of leaching, resulting from the extreme aridity of the dry valleys, along with salt inputs from the adjacent Ross Sea, leads to the accumulation of salts and other weathering materials in the soils [*Campbell and Claridge*, 1987]. Older soils in particular, commonly have a saline horizon several centimeters thick about 10–20 cm below the surface. In younger soils, salt accumulations are more intermittent and tend to occur beneath surface stones [*Campbell et al.*, this volume]. *Campbell et al.* [this volume] point out that soil salinity alters soil thermal regimes by influencing the formation of ice cement and in addition, change soil osmotic potential, thus affecting the activity of biota. Our previous work has shown nematodes to

occur across a wide range of soil organic matter and salinity concentrations [*Freckman and Virginia*, 1997]. Organisms that survive in these soils must be adapted to highly saline conditions or able to regulate water flow across their membranes [*Sømme*, 1995]. Total salt concentration was significantly higher in dry valley soils lacking nematodes than in soils collected at the same time which supported nematodes [*Freckman and Virginia*, 1997]. The soils lacking nematodes had higher concentrations of soluble cations and anions (Figure 7). The geochemistry of these soils may be influenced by the deposition of marine salts as evidenced by the high concentrations of Na^+ and Cl^- ions in the soils not supporting nematodes. The osmotic concentration of the soil solution surrounding nematodes will periodically reach very high levels as saline soils dry or begin to freeze adding to the physiological challenges facing soil invertebrates in dry valley soils [*Sømme*, 1995].

CONCLUSIONS

The McMurdo Dry Valleys support low diversity soil invertebrate communities. There are no other soil systems known where nematodes represent the top of the food chain and where food webs are so simple in structure. Here in these simple systems, the loss of a single species due to soil alteration or climate change can be detected and the associated effects on nutrient cycling can be discerned. No single soil property or factor defines the presence of a suitable habitat for soil invertebrates in dry valley soils. The wide geographic distribution of the single dominant nematode, *Scottnema lindsayae*, in the dry valleys suggests dispersal is not the main limiting factor determining the presence and diversity of soil invertebrate communities. Rather, interactions involving soil, climatic, and biological factors probably determine the biodiversity and community structure of soil invertebrates in the dry valleys. Nematodes, rotifers, and tardigrades have highly evolved survival strategies, (i.e., anhydrobiosis) that permit dispersal by wind and community establishment in some of the most extreme soil environments on earth. Local soil conditions however, can be outside the tolerances of these dispersing organisms. This would prevent community establishment and create the patchy distribution of soil biota that uniquely defines the dry valley landscape.

We find that ecosystems occurring in extreme environments are particularly attractive for study because their response to disturbance can be better

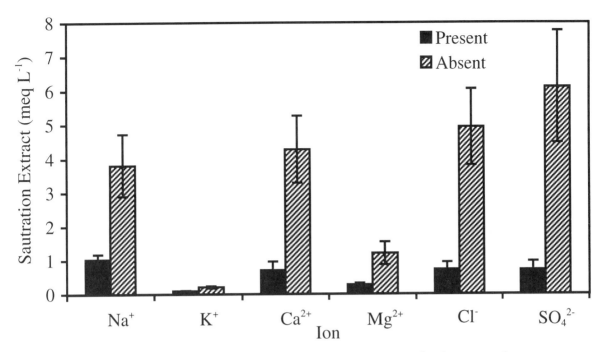

Fig. 7. Ion concentrations in soil saturation extracts from dry valley soils where nematodes were present (n = 86) or absent (n = 44). Modified from *Freckman and Virginia* [1997].

quantified than in more complex and more productive ecosystems. We predict that the low diversity and simple community structure of dry valley soils makes them highly susceptible to human influence.

Acknowledgments. We thank L. E. Powers, M. Ho, and the field teams for their assistance in sampling and laboratory analyses and S. S. Bamforth for collaborating with us and for his work on protozoa. The logistic support of the NSF McMurdo Station laboratory staff and the USA VXE-6 and RNZAF helicopter crews is gratefully acknowledged. We especially appreciate the contribution of A. N. Parsons for data analysis and assistance in preparing this manuscript. We thank J.M. Blair, I.B. Campbell, and also M.B. Burkins and A.M. Treonis for reviewing this manuscript. This research was supported by National Science Foundation Grants DEB 9115734 to D. W. Freckman, and OPP 9120123 and OPP 9421025 to D. W. Freckman and R. A. Virginia, and is a contribution to the McMurdo LTER (OPP 9211773) and the Jornada Basin LTER (DEB 9240261).

REFERENCES

Bamforth, S. S., D. W. Freckman, and R. A. Virginia, Amoebae biodiversity in the Antarctic Dry Valley Soils (abstract*), 7th International Conference on Small Freeliving Amoebae*, Adelaide, 7-12 Jan. 1996.

Bastian, H. C., II. Monograph on the Anguillulidae, or free Nematoids, marine, land and freshwater; with descriptions of 100 new species. *Trans. Linn. Soc. Lond., 25*, 73-184, 1865.

Beare, M. H., D. C. Coleman, D. A. Crossley, Jr., P. F. Hendrix, and E. P. Odum, A hierarchical approach to

evaluating the significance of soil biodiversity to biogeochemical cycling. *Plant and Soil, 170,* 5-22, 1995.

Block, W., Terrestrial microbiology, invertebrates and ecosystems, in *Antarctic Ecology*, edited by R. M. Laws, pp. 163-236, Academic Press, New York, 1984.

Boag, B., J. W. Crawford, and R. Neilson, The effect of potential climatic changes on the geographical distribution of the plant-parasitic nematodes *Xiphinema* and *Longidorus* in Europe, *Nematologica, 37*, 312–323, 1991.

Bockheim, J. G., Properties and classification of cold desert soils from Antarctica, *Soil Sci. Soc. Am. J., 61*, 224-231, 1997.

Bongers, T., The maturity index: an ecological measure of environmental disturbance based on nematode species composition. *Oecologia, 83*, 14-19, 1990.

Burkins, M. B., C. P. Chamberlain, R. A. Virginia, and D. W. Freckman, The natural abundance of carbon and nitrogen isotopes in potential sources of organic matter to soils of Taylor Valley, Antarctica. *Antarct. J. US.,* in press.

Bütschli, O., Beiträge zur Kenntnis der freilebenden Nematoden. *Nova Acta Leop. Carol., 36 (5),* 1-144, 1873.

Cameron, R. E., Microbial and ecologic investigations in Victoria Valley, Southern Victoria Land, Antarctica, *Ant. Res. Ser., 20,* 195-260, 1972.

Cameron, R. E., J. King, and C. N. David, Microbiology, ecology and microclimatology of soil sites in Dry Valleys of Southern Victoria Land, Antarctica, in *Antarctic Ecology,* edited by M. W. Holdgate, pp. 702-716, Academic Press, London, 1970.

Campbell, I. B., and G. G. C. Claridge, *Antarctica: soils, weathering processes and environment*, 368 pp., Elsevier, Amsterdam, New York, 1987.

Campbell, I. B., G. G. C. Claridge, D. I. Campbell, and M. R. Balks, The Soil Environment of the McMurdo Dry Valleys, Anarctica, this volume.

Carpenter, S. R., and J. F. Kitchell, The temporal scale of variance in limnetic primary production. *Am. Nat.* 129:417-433, 1987.

Carroll, J. J., and D. R. Viglierchio, On the transport of nematodes by the wind, *J. Nematol., 13*, 476-483, 1981.

Caswell-Chen, E. P., V. M. Williamson, and F. F. Wu, Random amplified polymorphic DNA analysis of *Heterodera cruciferae* and *H. schachtii* populations, *J. Nematol., 24*, 343–351, 1992.

Coleman, D. C., and D. A. Crossley, Jr., *Fundamentals of soil ecology.* Academic Press, Inc., London. 205 p., 1996.

Courtright, E. M., D. W. Freckman, R. A. Virginia, L. M. Frisse, J. T. Vida, and W. K. Thomas, Nuclear and mitochondrial DNA sequence diversity in the Antarctic nematode *Scottnema lindsayae. Molec. Ecol.* in press.

Crowe, J. H., Anhydrobiosis: an unsolved problem, *Amer. Naturalist, 105*, 563-571, 1971.

Crowe, J. H., and J. S. Clegg, *Dry biological systems*, 357 pp., Academic Press, New York, 1978.

De Man, J. G., Nematodes libres, *Résult. Voyage S. Y. Belgica*, Zoologie, 55 pp, 1904.

Franklin, J. F., C. S. Bledsoe, and J. T. Callahan, Contributions to the Long-Term Ecological Research program, *BioScience, 40*, 509-523, 1990.

Franzmann, P. D., Examination of Antarctic prokaryotic diversity through molecular comparisons. *Biodiversity and Conservation, 5*, 1365-1378, 1996.

Freckman, D. W., Ecology of anhydrobiotic soil nematodes, in *Dry Biological Systems,* edited by J. Crowe and J. Clegg, pp. 345-357, Academic Press, New York, 1978.

Freckman, D. W., Parameters of the nematode contribution to ecosystems, in *Nematodes in Soil Ecosystems,* edited by D. W. Freckman, pp. 81-97, Univ. of Texas Press, Austin, 1982.

Freckman, D. W., The ecology of dehydration in soil organisms, in *Membranes, metabolism and dry organisms,* edited by A. C. Leopold, pp. 157–168, Cornell University Press, Ithaca, 1986.

Freckman, D. W. (Ed.), *Life in the Soil. Soil biodiversity: its importance to ecosystem processes*, Report of a workshop held at the Natural History Museum, London, England, 26 pp., Colorado State University, Fort Collins, 1994.

Freckman, D. W., and E. P. Caswell, Ecology of nematodes in agroecosystems, *Ann. Rev. Phytopathol., 23*, 275-296, 1985.

Freckman, D. W., and C. H. Ettema, Assessing nematode communities in agroecosystems of varying human intervention, *Agric. Ecosyst. Environ., 45*, 239-261, 1993.

Freckman, D. W., and R. Mankau, Abundance, distribution, biomass and energetics of soil nematodes in a Northern Mojave desert. *Pedobiol., 29*, 129-142, 1986.

Freckman, D. W., and R. A. Virginia, Plant-feeding nematodes in deep-rooting desert ecosystems, *Ecology, 70*, 1665-1678, 1989.

Freckman, D. W., and R. A. Virginia, Nematodes in the McMurdo Dry Valleys of southern Victoria Land, *Antarct. J. US., 26*, 233-234, 1991.

Freckman, D. W., and R. A. Virginia, Extraction of nematodes from Dry Valley Antarctic soils, *Polar Biol.., 13*, 483–487, 1993.

Freckman, D. W., and R. A. Virginia, Low diversity Antarctic soil nematode communities: distribution and response to disturbance, *Ecology, 78*, 363-369, 1997.

Freckman, D. W., W. G. Whitford, and Y. Steinberger. Effect of irrigation on nematode population dynamics and activity in desert soils. *Biol. Fertil.. Soil, 3*, 3-10, 1987.

Gressitt, J. L., Entomology of Antarctica. *Antarct. Res. Ser.,* 10, 1-33, 1967.

Hendrix, P. F., R. W. Parmelee, D. A. Crossley, Jr., D. C. Coleman, E. P. Odum, and P. M. Groffman, Detritus food webs in conventional and non-tillage agroecosystems, *BioScience, 36*, 374-380, 1986.

Heywood, V. H. (Ed.), *Global Biodiversity Assessment*, UNEP, Cambridge University Press, Cambridge, UK, 1995.

Ho, M., R. A. Virginia, L. E. Powers, and D. W. Freckman, Soil chemistry along a glacial chronosequence on Andrews Ridge, Taylor Valley, *Antarct. J. US.*, in press.

Hunt, H. W., D. C. Coleman, E. R. Ingham, R. E. Ingham, E. T. Elliott, J. C. Moore, C. P. P. Reid, S. L. Rose, and C. R. Morley, The detrital food web in a shortgrass prairie, *Biol. Fert. Soils, 3*, 57-68, 1987.

Ingham, R. E., J. A. Trofymow, E. R. Ingham, and D. C. Coleman, Interactions of bacteria, fungi, and their nematode grazers: effects on nutrient cycling and plant growth, *Ecol. Monogr., 55*, 119-140, 1985.

Kennedy, A. D., Antarctic terrestrial ecosystem response to global environmental change, *Ann. Rev. Ecol. Syst., 26*, 683-704, 1995.

Kennedy, A. D., Water as a limiting factor in the Antarctic terrestrial environment: a biogeographical synthesis, *Arct. Alp. Res., 25*, 308–315, 1993.

Kirjanova, E. S., Antarkticheskie predstaviteli presnovodnykh nematod roda *Plectus* Bastian (Nematodes Plectidae) [Antarctic specimens of freshwater nematodes of the genus *Plectus* Bastian (Nematoda, Plectidae)], *Inf. Byull. Sov. Antarkt. Eksped., 3,* 101- 103, 1958.

Marshall, W. A. Biological particles over Antarctica, *Nature* 383, 680, 1996.

Maslen, N. R., The Signy Island terrestrial reference sites: XII, Population ecology of nematodes with additions to the fauna, *Br. Antarct. Surv. Bull., 53*, 57-75, 1981a.

Maslen, N. R., The Signy Island terrestrial reference sites: XIII, Population dynamics of the nematode fauna, *Br. Antarct. Surv. Bull., 54*, 33-46, 1981b.

Moore, J. C., P. C. de Ruiter, H. W. Hunt, D. C. Coleman, and D. W. Freckman, Microcosms and soil ecology: critical linkages between field studies and modeling food webs, *Ecology, 77*, 694-705, 1996.

Niles, R. K., and D. W. Freckman, From the ground up: nematode ecology in bioassessment and ecosystem health, in *Plant-Nematode Interactions*, edited by K. R. Barker, G. A. Pederson, and G. L. Windham, Agronomy Monograph, American Society of Agronomy, Crop Science Society of America and Soil Science Society of America, Madison, WI, in press.

Orr, C. C., and O. H. Newton, Distribution of nematodes by wind, *Plant Dis. Rep., 55*, 61-63, 1971.

Overhoff, A., D. W. Freckman, and R. A. Virginia, Life cycle of the microbivorous Antarctic Dry Valley nematode *Scottnema lindsayae* (Timm 1971), *Polar Biol., 13*, 151–156, 1993.

Pastor, J., and J. G. Bockheim, Soil development on moraines of Taylor Glacier, lower Taylor Valley, Antarctica, *Soil Sci. Soc. Am. J., 44*, 341-348, 1980.

Petersen, H., and M. Luxton, A comparative analysis of soil fauna populations and their role in decomposition processes, *Oikos, 39*, 287-388, 1982.

Pickup, J., Seasonal variation in the cold-hardiness of a free-living predatory Antarctic nematode, *Coomansus gerlachei* (Mononchidae), *Polar Biol., 10*, 307-315, 1990a.

Pickup, J., Strategies of cold-hardiness in three species of Antarctic dorylaimid nematodes, *Polar Biol., 10*, 167-173, 1990b.

Pickup J., and P. Rothery, Water-loss and anhydrobiotic survival in nematodes of Antarctic fellfields, *Oikos, 61*, 379-388, 1991.

Powers, L. E., D. W. Freckman, and R. A. Virginia, Spatial distribution of nematodes in polar desert soils of Antarctica, *Polar Biol., 15*, 325–333, 1995.

Powers, L. E., D. W. Freckman, M. Ho, and R. A. Virginia. Soil properties associated with nematode distribution along an elevational transect in Taylor Valley, Antarctica, *Antarct. J. U.S.*, in press.

Procter, D. L. C., Global overview of the functional roles of soil-living nematodes in terrestrial communities and ecosystems, *J. Nematol., 22*, 1–7, 1990.

Robertson, G. P., and D. W. Freckman, The spatial distribution of nematode trophic groups across a cultivated ecosystem, *Ecology, 76*, 1425–1432, 1995.

Schimel, D. S., B. H. Braswell, E. A. Holland, R. McKeown, D. S. Ojima, T. H. Painter, W. J. Parton, and A. R. Townsend, Climatic edaphic, and biotic controls over carbon and turnover of carbon in soils. *Global Biogeochemical Cycles, 8*, 279-293, 1994.

Schlesinger, W. H., *Biogeochemistry*, 443 pp., Academic Press, San Diego, 1991.

Schwarz, A. M. J., T. G. A. Green, and R. D. Seppelt, Terrestrial vegetation at Canada Glacier, Southern Victoria Land, Antarctica, *Polar Biol., 12*, 397–404, 1992.

Smith, H. G., Diversity of Antarctic terrestrial protozoa. *Biodiversity and Conservation, 5*, 1379- 1394, 1996.

Sohlenius, B., Abundance, biomass and contribution to energy flow by soil nematodes in terrestrial ecosystems, *Oikos, 34*, 21-32, 1980.

Sømme, L., *Invertebrates in hot and cold arid environments*, 275 pp., Springer-Verlag, New York, 1995.

Steinberger, Y., D. W. Freckman, L. W. Parker, and W. G. Whitford, Effects of simulated rainfall and litter quantities on desert soil biota: nematodes and microarthropods, *Pedobiol. 26*, 267-274, 1984.

Thomas, W. K., and A. C. Wilson, Mode and tempo of molecular evolution in the nematode *Caenorhabditis*: cytochrome oxidase II and calmodulin sequences, *Genetics, 128*, 269–279, 1991.

Timm, R. W., Antarctic soil and freshwater nematodes from the McMurdo Sound region, *Proc. Helminth. Soc. Wash., 38*, 42-52, 1971.

Van der Knaap, E., R. J. Rodriguez, and D. W. Freckman, Differentiation of bacterial-feeding nematodes in soil ecological studies by means of arbitrarily-primed PCR, *Soil Biol. and Biochem., 25*, 1141–1151, 1993.

Vincent, W. F., *Microbial ecosystems of Antarctica*, Cambridge University Press, New York, 1988.

Virginia, R. A., W. M. Jarrell, W. G. Whitford, and D. W. Freckman, Soil biota and soil properties associated with the surface rooting zone of mesquite (*Prosopis glandulosa*) in historical and recently desertified habitats, *Biology and Fertility of Soils, 14*, 90-98, 1992.

Vishniac, H. S. and J. M. Klingler, Yeasts in the Antarctic Deserts in *Perspectives in microbial ecology*, edited by F. Megusar and M. Gantar, pp. 46–51, Proc. 4th Int. Symp. Microb. Ecol., Ljubljana, 1986, 1988.

Vishniac, H. S., Biodiversity of yeasts and filamentous microfungi in terrestrial Antarctic ecosystems. *Biodiversity and Conservation, 5*, 1365-1378.

Walter, D. E., R. A. Hudgens, and D. W. Freckman, Consumption of nematodes by fungivorous mites, *Tyrophagus* spp. (Acarina:Astigmata:Acaridae). *Oecol., 70*, 357-361, 1986.

Wasilewska, L., Nematodes of the dunes in the Kampinos forest. II. Community structure based on numbers of individuals, state of biomass and respiratory metabolism. *Ecol. Polska, 19*, 651-688, 1971.

Wharton, D. A., Cold Tolerance Strategies in Nematodes. *Biol.. Rev., 70*, 161-185, 1995.

Wharton, D. A., and I. M. Brown, A survey of terrestrial nematodes from the McMurdo Sound region, Antarctica. *New Zealand J. Zool.., 16*, 467–470, 1989.

Wharton, D. A., and I. M. Brown, Cold-Tolerance Mechanisms of the Antarctic Nematode *Panagrolaimus davidi*, *J. Exp. Biol., 155*, 629-641, 1991.

Williamson, V. M., Molecular techniques for nematode species identification, in *Manual of Agricultural Nematology*, edited by W. R. Nickle, pp. 107–123, Marcel Dekker, New York, 1991.

Wynn-Williams, D. D., Ecological aspects of Antarctic microbiology, in *Advances in Microbial Ecology*, edited by K. C. Marshall, pp. 71–146, Plenum Press, New York, 1990.

Wynn-Williams, D. D., Response of pioneer soil microalgal colonists to environmental change in Antarctica. *Microb. Ecol., 31*, 177-188, 1996a.

Wynn-Williams, D. D., Antarctic microbial diversity: the basis of polar ecosystem processes. *Biodiversity and Conservation, 5*, 1271-1293, 1996b.

Yeates, G. W., Two terrestrial nematodes from the McMurdo Sound region Antarctica, with a note on *Anaplectus arenicola* Killick, *J. Helminthology, XLIV (1)*, 27-34, 1970.

Yeates, G. W., T. Bongers, R. G. W. de Goede, D. W. Freckman, and S. S. Georgieva, Feeding habits in soil nematode families and genera – an outline for soil ecologists, *J. Nematol., 25*, 315–331, 1993.

Diana Wall Freckman, Natural Resource Ecology Laboratory, Colorado State University, Fort Collins, CO 80523

Ross A. Virginia, Environmental Studies Program, Dartmouth College, Hanover, NH 03755

(Received October 1, 1996;
accepted April 10, 1997.)

SCIENCE AND ENVIRONMENTAL MANAGEMENT IN THE MCMURDO DRY VALLEYS, SOUTHERN VICTORIA LAND, ANTARCTICA

Colin M. Harris

International Centre for Antarctic Information and Research

Scientific research is the principal human activity in the McMurdo Dry Valleys. Concerns have been expressed recently that with an increasing level of activity, and the advent of tourism into the region, there is a need for more formal approaches to environmental management. A recent United States National Science Foundation (NSF) workshop called for developing a management plan, utilizing zoning to manage human uses, and for developing a Geographical Information System (GIS) to archive and make accessible an up-to-date record of environmental data for the area. The support of the science community for these proposals is critical to sustain the long-term scientific and environmental values of the region.

INTRODUCTION

Scientific research and its associated logistic support is the principal human activity in the McMurdo Dry Valleys. As the papers in this volume illustrate, extensive research has been undertaken in a wide range of disciplines in the dry valleys over the last 40 years, especially by United States and New Zealand scientists [*Hatherton*, 1990]. Scientists generally access the region from McMurdo Station (United States) or Scott Base (New Zealand), located on Ross Island about 80 km distant from the dry valleys (Figure 1). The level of research activity has increased since 1993 with the selection of the McMurdo Dry Valleys as a site in the United States National Science Foundation (NSF) Long Term Ecological Research (LTER) Program [*Wharton*, 1993]. In recent years Italy also has become more actively engaged in scientific research in the area, and the first visit by tourists to the dry valleys was in 1993 [*Vincent*, 1996].

The environment of the dry valleys has extremely high and internationally significant scientific, environmental, aesthetic, and wilderness values. The environment is sensitive to human impact and has a low capacity to absorb and recover from changes. Biologi-

cal growth-rates in the cold desert environment are slow, and landscapes have evolved over many thousands of years. Much of the scientific value of the McMurdo Dry Valleys derives from the fact that the environment has been relatively undisturbed by human activity. The increasing level of activity and an appreciation of the ease with which the unique environment and scientific values can be disturbed has led to new initiatives to manage better both science and the environment in the region. This heightened awareness culminated in a workshop, sponsored by the NSF and held in Santa Fe, New Mexico in March 1995, to discuss central environmental issues in the dry valleys [*Vincent*, 1996]. The Santa Fe workshop concluded that there is a need for more formal and coordinated approaches to management to ensure long-term sustainability of scientific and environmental values.

This paper aims to summarize important issues and potential problems arising from the multinational activities in the dry valleys, necessarily drawing on the findings of the Santa Fe workshop [*Vincent*, 1996]. The views expressed, however, are of the author and not necessarily those of Santa Fe workshop participants, nor those of the national programs operating in the area.

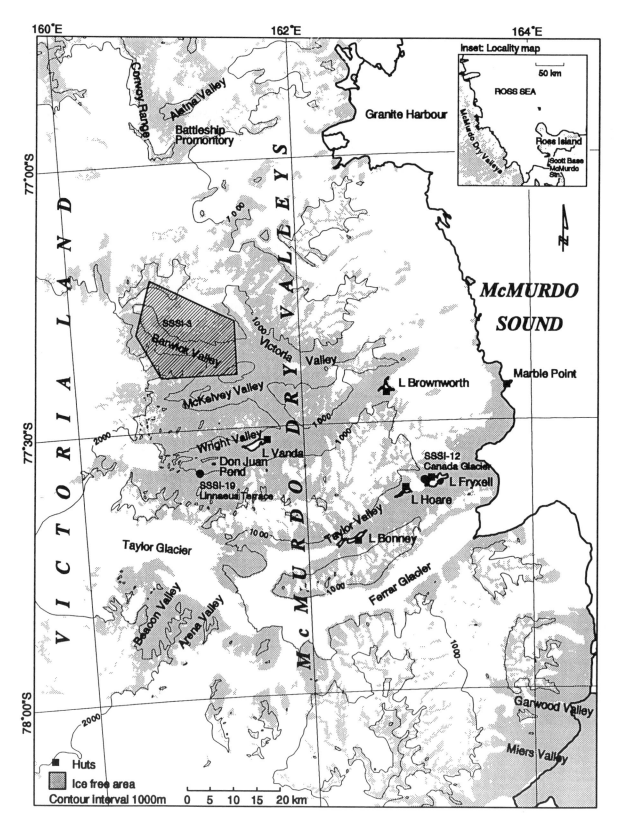

Fig. 1. The McMurdo Dry Valleys region, showing the principal ice-free area, protected areas, and research hut locations.

EXISTING MANAGEMENT INSTRUMENTS
AND MECHANISMS

In 1991 the Antarctic Treaty Parties agreed upon the Protocol on Environmental Protection to the Antarctic Treaty (the Madrid Protocol), which rationalized existing rules and provided a new framework for more comprehensive approaches to environmental management in Antarctica. As of May 1997 ratification had been completed by 24 of the 26 treaty states that agreed to the Madrid Protocol, with ratification required by the final two (Japan and Russia) for the agreement to come into full international legal effect. These two states are expected to ratify the Protocol in the near future, and most countries have already begun to implement Madrid Protocol provisions as though the agreement were in force. The Madrid Protocol provides a range of mechanisms for environmental management and those of most relevance to this paper are those on environmental impact assessment and on the protection and management of special areas.

Under the Madrid Protocol, all activities must be assessed for their potential environmental impacts before taking place. If it is concluded at the "Preliminary Stage" of assessment that the impacts will be "less than minor or transitory" then the activity can proceed. Otherwise an Initial Environmental Evaluation (IEE) must be prepared, which includes detail sufficient to assess whether impacts are likely to have more than a minor or transitory impact. If the impact is expected to be more than minor or transitory, the Madrid Protocol requires that a Comprehensive Environmental Evaluation (CEE) be prepared.

The formal mechanism in the Madrid Protocol providing for strictest control of activities is the Antarctic Specially Protected Area (ASPA), which will replace the existing Sites of Special Scientific Interest (SSSIs) and Specially Protected Areas (SPAs) that are more well-known to Antarctic scientists today. These designations provide for special protection of either sites of long-term special scientific interest or examples of unique or outstanding features and ecosystems in Antarctica. Revision of management plans for existing SSSIs and SPAs so they comply with the provisions in the Madrid Protocol has already been initiated by a number of countries [Harris, 1994a]. A second management mechanism is provided in the Madrid Protocol to assist coordination in areas where there is a risk of mutual interference or cumulative environmental impacts or where there is a need to minimize environmental impacts but not necessarily impose such stringent conditions as within an ASPA. This is called an Antarctic Specially Managed Area (ASMA) and replaces the old Multiple-Use Planning Area, which thus far has not been extensively applied. Both the ASPA and ASMA require management plans, but a key difference between them is that the former requires a permit for entry while the latter does not. ASPAs can be contained within ASMAs, but not vice versa.

ACCESS TO AND MOVEMENT WITHIN THE
MCMURDO DRY VALLEYS

Access to the region is principally by helicopter from Ross Island, although a small number of parties may travel over sea ice to the southern Victoria Land coast and then proceed on foot (Figure 1). Some wheeled vehicles were used in the region in the 1960s and 1970s, but this practice has been discontinued. The only land vehicles now in use are small all-terrain vehicles used to assist lake research programs, and these are not used on ice-free ground.

Helicopter traffic in the dry valleys over the summer period is frequent and may consist of several aircraft operating on a daily basis [Harris and Croteau, 1996]. There was a 30% increase in the number of helicopter hours flown in the region between 1974 and 1994 (from 636 to 822 hours) [Vincent, 1996]. Most flight activity is in the Taylor Valley, although flights are wide-ranging through the region to support the variety of science projects and visits by officials and the media. On-board helicopters operating from cruise ships have made tourist access possible in recent years, but thus far the number of flights has been low. Landing pads have been marked at most permanent field camps, and at two of the three designated protected areas in the dry valleys (at Canada Glacier and Linnaeus Terrace) where landings are restricted to specific sites. The current management plan of the third protected area (Barwick Valley) states that helicopter access into this area should be avoided (Figure 1). Air routes through the dry valleys are otherwise unrestricted and there are no other permanently designated landing sites, although there are a number that are informally marked.

Foot travel within range of the permanent and temporary camps and landing sites is the other main form of access around the dry valleys. In some places, foot prints have been observed to remain in soils for many years, while at others they disappear rapidly in the strong winds. At this time there are no formally designated foot trails in the region, although in a few places traffic has been sufficient to develop effectively permanent tracks over ice-free ground (e.g., the route to a stream weir from the Lake Fryxell Hut to Canada Stream, which is now being proposed as a designated route in the new management plan prepared for the Canada Glacier protected area (Figure 1) [*Harris*, 1994b; *MFAT*, 1997]).

FIELD CAMPS

Seven semipermanent scientific field camps are located in the McMurdo Dry Valleys, five of which are in Taylor Valley where the majority of LTER work is being conducted, and a permanent helicopter refuelling facility is located at Marble Point (Figure 1). The Taylor Valley camps are the most substantial, each consisting of a mess building, several research huts and accommodation/tent space for between 6 and 15 workers. Both semipermanent and temporary field camps may be established anywhere within the dry valleys outside of the protected areas subject to prior environmental impact assessment. Tourist groups have generally not yet established camps.

WASTE MANAGEMENT

Waste disposal regulations in the McMurdo Dry Valleys have always been comparatively strict, and general practice has been to remove most solid wastes since regular research began in the region in the 1960s. Some human wastes were incinerated in propane-fuelled combustors at several of the semipermanent camps, but this practice was discontinued in the dry valleys in 1996. In the past, some domestic liquid wastes were discarded at field camps, but this practice has also been discontinued. Where this had occurred at New Zealand's Lake Vanda Station, affected soil had to be removed because rising lake levels threatened to inundate the site and lead to contamination of the pristine lake waters [*NIWA*, 1993]. The old Vanda Station has been decommissioned and a smaller camp established at another site 50 to 60 meters from the lake.

Over time, rules have become more stringent, and the United States for example has had a policy of waste removal at all locations, including the dry valleys, since 1993. The United States and New Zealand have agreed to a consistent policy on waste disposal in the dry valleys based on the "Code of Conduct" presented in Table 1. The general policy is that all solid and liquid wastes, including all human wastes and water used for any human purpose (includ-

ing scientific sampling from lakes), are to be removed from the area. This includes human waste generated away from field camps, which must be containerized and returned for disposal. Because all wastes are now transported out of the region, considerable effort is made to minimize the amount generated by field parties.

FUEL AND MATERIALS MANAGEMENT

Fossil fuel is used for three main purposes in the McMurdo Dry Valleys: for aircraft, heating/cooking, and powering equipment. A light diesel (JP8) is used in aircraft and for power and heating, while gasoline is used for equipment such as generators, augers, and the all-terrain vehicles used on the lakes. Propane, gasoline ("white gas"), or kerosene may be used for cooking. As helicopters are generally refuelled outside of the dry valleys at nearby Marble Point, the amount of fuel actually transported into and stored in the dry valleys is relatively small and localized at camps. Occasionally drums may be carried aboard aircraft or cached for refuelling in the field.

The most substantial fuel spill event known in the dry valleys occurred when an air drop of JP8 to Vanda Station in 1984 ruptured on landing on the lake ice; the resulting spill was burned and contamination of the lake was thus minimized. Light diesel was used as a drilling fluid in the 1970s Dry Valley Drilling Project (DVDP), and a number of releases were reported, together with other forms of contamination [*Parker et al.*, 1978]. Significant spills are now required to be reported, but, with the exception of the records kept of the DVDP, information on past activities and effects is contained in voluminous field party reports and has yet to be researched and described. Minor spills occur in routine engine or other refuelling operations, and small quantities of fuel are ejected from fuel lines in some helicopters when engines are turned off.

Camp and scientific equipment and building materials are at risk of dispersal in the prevailing winds common in the dry valleys, and foreign materials occasionally escape into the environment. Special measures are required to ensure materials are held secure.

IMPACTS FROM SCIENCE

Any scientific activity results in impact, and the range of possible impacts is diverse. The *SCAR /COMNAP* [1996] workshops on environmental

TABLE 1. Code of Conduct for field work in the McMurdo Dry Valleys
(Source: *Antarctica New Zealand* [The New Zealand Antarctic Institute])

Environmental Code of Conduct for Field Work in the McMurdo Dry Valleys

Why are the Dry Valleys considered so important by the scientific community? The Dry Valleys ecosystem contains geological and biological features that date back thousands to millions of years. Many of these ancient features could be easily and irreversibly damaged by inadvertent human actions. Unusual communities of microscopic life forms, low biodiversity, simple food webs with limited trophic competition, severe temperature stress, aridity and nutrient limitations are other characteristics which make the Dry Valleys unique. This ancient desert landscape and its biological communities have very little natural ability to recover from disturbance. Research in such systems must always aim to minimize impacts on land, water and ice to protect them for future generations. This code suggests how you can help to ensure this.

General Conditions:

- Your visit to the valleys should have as little impact as possible. Everything taken into the valleys must be removed and returned to Scott Base. Do not dump any unwanted material on the ground.
- Activities which would result in the dispersal of foreign materials should be avoided (e.g., do not use spray paint to mark rocks) or conducted inside a hut or tent (all cutting, sawing and unpacking).
- Solar and wind power should be used as much as possible to minimize fuel usage.
- The location of any disturbance, spill, camp site, soil pit, or other sampling site should be mapped and recorded in your field report for eventual transfer to a management GIS. Where possible, the GPS coordinates of the site should be recorded.
- Water used for ANY human purpose must be removed or treated in a grey water evaporator.
- All human waste must be collected and removed.
- Do not leave any travel equipment behind (e.g. ice screws, pitons). Avoid building cairns.
- Ground vehicle usage should be restricted to snow and ice surfaces.
- When travelling on foot, stay on established trails whenever possible. Avoid walking on vegetated areas and delicate rock formations.

Field camps: location and set up

- Campsites should be located as far away as practicable from lake shores and stream beds to avoid damage or contamination. Do not camp in dry stream beds.
- Campsites should be reused to the greatest extent possible. Before entering the Dry Valleys you should attempt to determine the location of previously used campsites in the area you are visiting.
- Ensure that equipment and supplies are properly secured at all times to avoid dispersion by high winds. High velocity katabatic winds can arrive suddenly and with little warning.
- Maximize the use of fixed helicopter pads. Use markers clearly visible from the air to mark pads.

Fuel and chemicals

- Take steps to prevent the accidental release of chemicals such as laboratory reagents and isotopes (stable or radioactive). Chemicals of all kinds should be dispensed over drip trays or other containment. When permitted to use radioisotopes, precisely follow all instructions provided.
- Ensure spill kits are appropriate to the volume of fuel/chemicals and you are familiar with their use.
- Use fuel cans with spouts when refuelling generators and only refuel generators and vehicles over trays with absorbent spill pads.
- Never change vehicle oil except over a drip tray.

TABLE 1. Code of Conduct for field work in the McMurdo Dry Valleys (Continued)

Sampling and experimental sites:

- All sampling equipment should be clean before being brought into the Dry Valleys.
- Do not displace or collect specimens of any kind, including fossils, except for scientific and educational purposes; in SSSIs or SPAs the sample size will be specified in your collecting permit.
- Once you have drilled a sampling hole in lake ice or dug a soil pit, keep it clean and make sure all your sampling equipment is securely tethered.
- Backfill soil pits to prevent wind erosion and dispersal of deeper sediments.
- Avoid leaving markers (e.g. flags) and other equipment for more than one season without marking them clearly with your event number and duration of your project.

Lakes:

- Clean all sampling equipment to avoid cross-contamination between lakes.
- Retain any excess water or sediment for removal to your station.
- Never use explosives on a lake.
- Only use vehicles on lake ice when essential; park the vehicle on permanent ice rather than moat ice during the period of summer melt.
- Ensure that you leave nothing frozen into lake ice which may ablate out and cause contamination.
- Avoid swimming or diving in the lakes. These activities could contaminate the water body and physically disturb the water column, delicate microbial communities and sediments.

Streams:

- Use designated stream crossing points whenever possible.
- Avoid walking in the stream bed at any time; you may disturb the stream biota which represent several decades of slow growth.
- Avoid walking too close to stream sides as this may affect bank stability and flow patterns.

Valley floor and sides:

- Avoid disturbing mummified seals or penguins.
- Avoid sliding down screes or sand dunes; these features have taken many thousands of years to form and may also contain surface deposits of major scientific importance.

High Desert:

- Beware of causing damage to delicate rock formations. Some of the biological communities in them have taken several thousand years to develop.
- Collect only the minimum sample of endolithic community required for scientific analysis.

Glaciers:

- Minimize the use of liquid water (e.g., with hot water drills) which could contaminate the isotopic and chemical record within the glacier ice.
- Avoid the use of chemical-based fluids on the ice.
- If stakes or other markers are placed on a glacier, use the minimum number of stakes required to meet the needs of the research; where possible, label stakes with your event number and project duration.

monitoring identified those activities and "outputs" most likely to result in "significant" environmental impacts in Antarctica (Table 2), and this is a good summary of the types of concerns important to planning a science program in the McMurdo Dry Valleys.

In relation to the collection and removal of samples in the course of research, science groups working in the dry valleys have often developed specific codes of practice to avoid detrimental effects. For example, liquid lake water samples are not discarded on the lake's surface, as they once were, to avoid "contamination" of surface ice and underlying liquid water properties. In the case of soil excavations, some scientists replace soil layers in the order that they were removed so that subsurface sediments are not subsequently dispersed by wind [*Campbell et al.*, this volume]. However measures have been inconsistently applied and it is not difficult to find evidence of former camp or scientific activity.

The impacts of research activities in the dry valleys, while diverse, may still be considered as relatively localized and of limited scale and duration [*Vincent*, 1996], although as noted above a thorough assessment and inventory of actual impacts has never been undertaken. While it is true that science projects have undoubtedly had their impacts, most of the dry valley region might still be considered close to an undisturbed state. Away from the immediate environs of permanent camps signatures of global pollutants are probably more likely to be detected than local.

MANAGING TOURISM

Tourist visits to the McMurdo Dry Valleys are a recent phenomena and have been facilitated by a single helicopter-capable Russian icebreaker, Kapitan Khlebnikov. A total of 715 people have visited on seven tours to Taylor Valley since 1993 (Table 3), each tour lasting, on average, a total of about 6.6 hours. All tourist visits have been to the Taylor Valley, which is also the current region of most intensive scientific use. All visits are supervised by tour staff and by an independent observer from one of the national programs. However concerns were expressed by some scientists when initial landings by tour groups were near research sites. Informal agreement has now been reached between the tour operator and national programs on suitable landing areas in the lower Taylor Valley to avoid possible conflicts of interest. Legally, however, tour operators may visit other sites if they

wish, provided these are outside of formally protected areas where permits for entry are required.

As the length of tours is short, with individual passengers having only a few hours on the ground, there is currently little opportunity for tourists to move far beyond the initial point of landing. Tour operators and the national programs take steps before visits to educate tourists on the fragile nature of the dry valleys, the importance of the scientific programs, on specific prohibitions (e.g., waste disposal) and provide them with a summary of the adopted "Code of Conduct" presented in Table 1. Under these circumstances, waste disposal issues are generally limited to accidental loss of wrappers (e.g., film, food) and perhaps the need by some for urination. The principal impacts for this type of tour are likely to be related to foot traffic, possible disturbance of geological, biological, or scientific features of value, and emissions and spillages from helicopters. Visits are presently targeted at sites where impacts are expected to be minimal.

PROTECTED AREAS

Existing protected areas in the McMurdo Dry Valleys are summarized in Table 4 and shown in Figure 1. The Barwick Valley (SSSI-3) is the largest of the protected areas in Antarctica and was designated in 1975 on the grounds that it is one of the least disturbed and contaminated areas in the McMurdo Dry Valleys. As such it was considered valuable as a reference against which changes in other regions of the dry valleys being subjected to greater levels of activity could be compared. During a brief visit in December 1993 to assess management issues in the protected area, a number of signs of early scientific activity were observed, including evidence of old camp sites, soil pits, remains of a wooden crate, and a broken food cache partly submerged in Lake Vashka [*Harris*, 1994a]. Much of this material was subsequently removed, although the extent of contamination of Lake Vashka remains unknown. Despite this evidence of localized disturbance, there have been only a small number of visits to the area since it was designated for protection.

Linnaeus Terrace (SSSI-19) is located in the Asgard Range above the South Fork of the Wright Valley at an elevation of about 1650 m (Figure 1). The site is one of the richest localities for the cryptoendolithic communities in the Beacon Sandstone, was the site of the original detailed Antarctic cryptoendolithic descriptions, and is considered a type locality with outstand-

TABLE 2. Outputs Resulting from Human Activities in Antarctica and Principal Physical and Chemical Indicators of Their Impact (Source: SCAR/COMNAP, 1996)

Outputs	Indicators	Possible impacts
Air emissions	• SO_2, NOx, CO, PAH, heavy metals, fuel consumed • type, quantity, timing, duration	landscape, biological change
Dust	• particulates, albedo, water turbidity • type, quantity, timing, duration	landscape, biological change
Liquid waste (including brine)	• flow rate, suspended solids, BOD pH, fecal coliforms, nutrients (PO_4, NO_3, NO_2, NH_4) TKN • type, quantity, timing, duration	biological change
Solid waste (including dumps and debris)	• leachates, foreign materials • type, quantity, timing, duration	landscape, biological change
Fuel / hazardous materials (including fuel blowdown)	• PAH (air, water, land/snow), albedo, chemicals, radionuclides etc. • type, quantity, timing, duration	landscape, biological change
Noise	• type, quantity, timing, duration	biological change
Electromagnetic radiation	• type (frequency), quantity (strength) timing, duration	biological change
Mechanical actions, Constructions, (excavations, fill, explosions, compaction)	• topography, erosion, deposition, vehicle/foot traffic, albedo • type, quantity, timing, duration	landscape, biological change
Heat	• temperature, thermal regime • timing, duration	biological change
Introductions, Sampling, Extractions, Relocations	• alien biota, geological / biological specimens, snow/ice/water levels • type, quantity, timing, duration	landscape, biological change

1. Biological change covers all changes to individuals, populations and communities. Habitat disruption is covered under both landscape and biological change. Biological indicators are not included in the table. Aesthetic / wilderness disruption and changes to scientific capability are possible impacts that apply to all categories

2. Definitions (SCAR/COMNAP, 1996, p. *x*): "outputs" : "any physical change (e.g., movement of sediments by vehicle traffic, noise) or an entity (e.g., emissions, an introduced species) imposed on or released into the environment." The factors measured to assess the level of output are considered "indicators," while the "impact" is the consequence of the change (e.g., a reduction in nematode populations).

TABLE 3. Tourist visits to the Taylor Valley,
McMurdo Dry Valleys, 1993–96.

Tour Date	Passengers	Duration of visit (hours)
Feb 10, 1993	106	8
Feb 8, 1994	104	8
Jan 11, 1995	112	8
Feb 3-4, 1995	105	4
Feb 22, 1995	96	5
Jan 19, 1996	96	6.5
Feb 17, 1996	96	7
Total	715	46.5
Average per visit	102	6.6

Data source: D. Schoeling and E. Waterhouse [*Vincent*, 1996].

ing scientific values related to this ecosystem. The sandstones exhibit a range of fragile biological and physical weathering forms, and damaged rock surfaces would be slow to recolonize. Impact at Linnaeus Terrace is low, and the science groups working there have generally been meticulous in their efforts to minimize disturbance. However even here impacts are present in the form of rock surfaces broken by walking, sites affected by early waste disposal practices when urine was not retrograded for disposal, and through the release of the carbon–14 radioactive isotope as part of research experiments (Friedmann, personal communication, 1994). The radioisotopic contamination is considered insignificant in terms of impact on the environment, but it has rendered the small area affected (<100 x 100 m) unsuitable for radiocarbon dating, compromising the scientific value of the locality. This illustrates how significant impact may result even from the most careful of field studies.

Canada Glacier (SSSI–12) is located on the north shore of Lake Fryxell and an adjacent area of land was designated in 1985 to protect some of the richest plant growth (bryophytes and algae) in the Southern Victoria Land dry valleys (Figure 1). Most biological growth occurs in a flush area close to Canada Glacier and extends along the small meltwater stream (Canada Stream) draining into Lake Fryxell. Three moss species have been identified in the area [*Schwarz*, 1990]: *Bryum argenteum*, and *Pottia heimii*. Lichen growth is inconspicuous, but a number of epilithic and

chasmoendolithic species may be found [*Schwarz et al.*, 1992]. Over 37 species of freshwater algae and invertebrates from six phyla have been described at the site [*Broady*, 1982; *Schwarz*, 1990]. Weirs have been built on Canada Stream to quantify water flows through the area [*von Guerard et al.*, 1994]. The Canada Stream drainage area has been well-studied and documented, which adds to its scientific value. However the biological communities are fragile and vulnerable to disturbance by walking, sampling, pollution, or alien introductions. The site is of limited extent and has been subjected to increasing pressure from scientific and logistic activities. Evidence of human activity in the area includes footprints in the soft sediments and in moss beds, foot trails, abandoned markers, litter blown in from nearby camps, soil pits, cores extracted from moss turfs, and paint applied as markers on rocks. Ironically, sites damaged at known times in the past have been identified and provide one of the few areas in the dry valleys where the long-term effects of disturbance, and recovery rates, can be measured. A research hut facility was present within the area for more than a decade, but this was removed during the 1995–1996 austral summer and steps were taken to remediate visually obvious impacts at the same time. The new draft management plan for the region encourages scientists to locate their camps outside of the protected area and contains more stringent conditions on the conduct of scientific activities in the area, including sampling, constructions and the use of chemicals and isotopes [*MFAT*, 1997].

MONITORING

A comprehensive analysis and review of the requirements for monitoring of environmental impacts arising from science and operations in Antarctica was recently undertaken through two workshops held by *SCAR/COMNAP* [1996]. The final report concluded that environmental monitoring should be hypothesis-driven and tied to an environmental management strategy. The principal objectives of monitoring were considered to be (1) the protection of the scientific value of Antarctica, (2) the improvement of environmental management, and (3) meeting legal requirements under the Madrid Protocol and national legislation [*SCAR/COMNAP*, 1996]. More specific goals were as follows:

• Establish the present status of key values or resources;

TABLE 4. Existing Protected Areas in the McMurdo Dry Valleys

Site	Location	Approx. area (ha)	Main purpose
Sites of Special Scientific Interest			
SSSI 3	Barwick Valley, Victoria Land	30 000	Dry valley ecosystem reference site
SSSI 12	Canada Glacier, Taylor Valley	100	Preserve terrestrial plants and ecosystem
SSSI 19	Linnaeus Terrace, Asgard Range	300	Cryptoendolithic communities
Specially Protected Areas			
None.			

- Provide early warning of deterioration in key values or resources;
- Identify activities most responsible for deterioration;
- Evaluate current activities with a view to avoiding or mitigating deterioration;
- Verify the effectiveness of predicting impacts through the EIA process.

National programs have conducted specific assessments of impacts and monitored environmental performance of science and tourist groups, but at present there does not exist a formal, internationally coordinated framework and strategy for monitoring of human impacts in the McMurdo Dry Valleys.

MANAGING CUMULATIVE IMPACTS

Cumulative effects can be characterized as impacts on the natural and social environments from single or multiple sources which occur so frequently in time or so densely in space that they cannot be "assimilated", or that combine with effects of other activities in a synergistic manner [*Sonntag et al.*, 1987 cited in *Martin*, 1991]. A cumulative impact may be additive or interactive (e.g., synergism, antagonism, biomagnification). The presence of multiple research programs from a number of countries creates a context where cumulative impacts may be particularly hard to address, and may need proactive approaches to management.

INFORMATION MANAGEMENT

Access to a reliable base of information is essential to sound environmental management. Geographical Information Systems (GIS) and Global Positioning Systems (GPS) technologies, with their ability to tie multiple data sets to a common framework spatially and temporally, were viewed at the Santa Fe workshop as essential to establishing an efficient environmental information base. It was concluded a system-wide GIS for the McMurdo Dry Valleys should be established [*Vincent*, 1996]. The CDROM that accompanied this volume can be considered a seminal effort in this direction.

A project recently funded by NSF promises to expand current McMurdo Dry Valleys GIS capabilities through an investigation of recent environmental change using satellite imagery and aerial photography (Prentice, personal communication, 1996). The project will incorporate data layers on glaciology, hydrology, geology, and sedimentology into a GIS framework at several spatial scales. When completed, this database will be made accessible to the research community. It is expected that these data will also represent a valuable resource for managers and those conducting environmental impact assessments.

GIS vendors have been quick to see the potential of the World Wide Web to enable improved access, integration, and query of spatially referenced data. Several vendors now offer interfaces to the Web from the GIS software. This type of approach is particularly valuable in contexts where there are a number of groups from a range of disciplines and countries that would benefit from a mutually shared base of environmental data, such as we see in the dry valleys.

CODE OF CONDUCT FOR ACTIVITIES IN THE MCMURDO DRY VALLEYS

A draft environmental Code of Conduct for field work was developed at the Santa Fe workshop based on the experience and input of a wide range of scientists and program managers with experience in the dry valleys [*Vincent*, 1996]. Following the workshop, the United States and New Zealand national programs refined and adopted a Code of Conduct for field work in the McMurdo Dry Valleys, which was implemented in the 1996–1997 field season (the full New Zealand version is provided in Table 1) (Jatko and Waterhouse, personal communications 1996–1997). The Code formalizes and makes more accessible many of the approaches and procedures developed through experience by the national programs and the scientific groups working in the region. The Code provides guidance for different types of activities (e.g., establishing camps, handling fuels and chemicals, scientific sampling) and also for working in certain environment types (e.g., lakes, streams, ice-free valley floors and high altitude areas, and glaciers). In addition, The Code serves to raise the awareness of those working in the region of the special values or particular sensitivities of the environment. Management plans developed for ASMAs under the Madrid Protocol require the elaboration of a Code of Conduct for activities within the area designated, and the Code developed could easily integrate into a management plan.

DISCUSSION

In relation to procedures for environmental impact assessment of activities in the dry valleys, interpretation of which impacts are "minor" and "transitory" is presently made on a case-by-case basis by scientists and the national programs. How the terms are interpreted is important, because this will affect the type and perhaps quality of environmental assessment undertaken. In a multinational context such as the McMurdo Dry Valleys, a problem could emerge if there were inconsistencies in the rigor to which environmental assessments were prepared by scientists from different programs. Discussion toward consistent definitions is continuing in the context of Antarctic Treaty deliberations over liability for environmental damage. In present practice, programs exchange IEEs and CEEs to try and achieve a broad level of consistency in the application of EIA procedures. However prior circulation of IEEs among potentially affected parties is not a requirement, although the Protocol does provide for circulation of lists of IEEs after they have been completed. There is no requirement or mechanism for the exchange of Preliminary Stage assessments, although it may be impractical and bureaucratic to exchange all such assessments. In relation to work in the McMurdo Dry Valleys, circulation of all IEEs before activities take place would be valuable and practical. There would also be value in regular exchange of sample "Preliminary Stage" assessments between national programs to help ensure a broad level of consistency.

The spread of impacts could be reduced by requiring scientists, wherever possible, to reuse previously impacted camp sites. However at present there is no readily accessible record, or register, kept of where field parties have previously camped or worked, and information on such sites is generally gained by word of mouth. With an increase in activity and a more multinational context there is need for a more effective means to access such information. The information currently exists in several forms and could be obtained from the reports filed to national programs by field parties or derived from helicopter flight schedules (which record numbers of people, flight destinations, and amount of equipment transported). This information should be placed in an accessible geographical database, as discussed below.

Many of the semipermanent camps in the dry valleys have been located adjacent to lakes (Fryxell, Hoare, Bonney, Vanda, Brownworth) for practical reasons related to research priorities and local terrain. Helicopter pads are sometimes immediately adjacent to the lake edge, taking advantage of suitable flat ground. The net result is that most of the traffic, food, wastes, and other materials transported into and out of the dry valleys is through these focal points close to the lakes. Dust disturbance, engine emissions, fuel spills, and loss or spill of other materials all occur most intensively and present the greatest environmental risks at these points. Moreover many lake levels are rising, and several sites may be threatened with inundation within a few decades. Scientists and managers should question whether current camp and helicopter pad locations provide for long-term sustainability of the scientific and environmental values of the dry valleys lakes.

The network of protected areas in the McMurdo Dry Valleys has been developed partly in response to direct needs for protection from scientific pressure, in part to set aside a reasonably large area as a reference baseline and also to protect several sites for their outstanding qualities. However a systematic assessment is needed of whether examples of the key environmental values of the dry valleys are being adequately represented within these areas, and whether there are sites of special significance at risk that are currently excluded. Several additional sites in the McMurdo Dry Valleys have been suggested as meriting special protection [Keys et al., 1988]. For example, Don Juan Pond in Wright Valley was suggested because of its highly unusual hypersaline aquatic ecosystem, and Lake Vanda has also been suggested because of its unique properties, but the proposals have yet to be taken further. The only lakes currently under special protection are those within the Barwick Valley; other lakes in the dry valleys have quite distinct and unusual properties which may also warrant special protection. An outcrop of Beacon Sandstone at Battleship Promontory, near the Convoy Range and Alatna Valley (Figure 1), has been identified by Friedmann (personal communication, 1994) as richly colonized with the most diverse community of cryptoendolithic lichens known in Antarctica and exhibits unusual and fragile weathered landforms. Battleship Promontory is remote and inaccessible, and it does not appear to be under immediate threat, but proactive special protection would provide a safeguard to ensure the outstanding values of this site are maintained.

The permitting and reporting procedures required within protected areas represent important procedural tools for tracking human activities within these areas, thereby allowing more effective identification, assessment, and management of environmental impacts. The Scientific Committee on Antarctic Research (SCAR) has adopted a standard format for reporting visits to protected areas, which was recently endorsed by the Antarctic Treaty [*ATCPs*, 1995]. The format makes specific provision for reporting on, for example, sample quantities and locations, instrumentation or equipment installed or material released, observations of human effects, evaluation of whether the area is being adequately protected, and recommendations on further management measures needed. SCAR hopes that reports will help organize scientific use of the sites and provide the basis for evaluation of conservation practices and the effectiveness of management plans. Scientists should make use of this standard format when reporting their activities within protected areas.

The lack of an internationally coordinated environmental monitoring framework or strategy to monitor human impacts in the dry valleys needs to be addressed. Scientists working in the McMurdo Dry Valleys should consider how their own research could contribute to the objectives of environmental monitoring as noted above and outlined in detail in the *SCAR/COMNAP* [1996] report. More specifically, an internationally coordinated network of spatially distributed monitoring sites, both near to and distant from the permanent field camps and in relation to known environmental conditions/environments (terrestrial and marine), should be established in order to determine more precisely the nature, magnitudes, durations, and extents of impacts in the region. It would also be useful to establish a network of control sites in different environment types (e.g., glacial, ice-free valley floor, ice-free high desert, lake, and stream). Where possible, efforts should be made to integrate data from existing research sites. There may also be value in designating one or two long-term control (reference) areas where access is completely prohibited for some period, during which additional new impacts within the area(s) would be limited to those transported by more regional or global processes. To fulfill the role as a reference area effectively it may be necessary first to conduct a thorough baseline assessment.

In terms of cumulative impact, individually minor impacts may be of significance when accumulating over time or interacting synergistically. Cumulative environmental impacts are difficult to address under current models of environmental management in Antarctica, and the single-project based method of environmental impact assessment, in particular, is not designed to address such problems. The high level of coordination and information exchange required in the international context of operations in the McMurdo Dry Valleys represents a significant challenge to assess and minimize cumulative impacts. Because the range of cumulative effects can potentially be very large, and there are limited resources available for monitoring and assessment, there is a need to define appropriate study boundaries and effective indicators of environmental changes. There is also a need for appropriate biophysical characterizations against which change can be measured.

Some of the factors impeding environmental management in the dry valleys relate to the management of information. The Santa Fe workshop recommended a system-wide GIS be established for the region to marshal the environmental data needed for management. To some extent, the core of such a system already exists through efforts by the LTER, the United States Geological Survey, Land Information New Zealand, and the International Centre for Antarctic Information and Research, but this needs to be developed further. A full and up-to-date register is needed of the sites of most significant value. Moreover, there is need for an inventory of the past effects of human activities on the environment (e.g., impacts resulting from the Dry Valley Drilling Project) before this information is lost. Scientists should consider it their responsibility to plan data collection and report locations of campsites, sampling, and environmental impacts, so that additional benefit might be gained from their data when it might eventually be incorporated into such a system-wide GIS.

For efforts toward a McMurdo Dry Valleys GIS to be fully effective, it is important that current errors in the geographical reference datum in the region are corrected. A program of field work funded by the United States and New Zealand in 1996–1997 is expected to result in this correction, and the data gathered will be used to apply corrections to existing digital geographical data sets so they all sit on a reliable, GPS-compatible, reference framework. This is important if data being collected in the dry valleys using GPS are to be incorporated meaningfully into an underlying map framework. Once corrected, it is important that all scientists working in the region use the corrected base map so that all observations are made using the same consistent and accurate geographical datum. Use of such a map would simplify the process of integration of spatially referenced data from disparate sources so often necessary in multidisciplinary science and for impact assessments.

The recent advances in information and communications technology have enabled effective real-time access to remote databases, even from the McMurdo Dry Valleys. Access to GIS over the World Wide Web offers the potential to operationalize an environmental information system that could be shared internationally in support of science and management in the dry valleys. Such a system could enable access to a common information base for management by national programs, both in home countries and at stations in Antarctica, and provide scientists with access to a base of environmental data both in their institutions and in the field. This type of information system would be invaluable in support of a management plan for the region and would assist many forms of science.

CONCLUSION

Science conducted in the dry valleys, as illustrated within this volume, is vital to understanding fundamental questions related to the structure and functioning of ecosystem processes. Much of the research being conducted is critical to understanding questions related to global change and also the nature of life in an unusual environment at the extremes of existence on Earth. It is important that this science continues and that it is not unduly constrained by environmental regulation. It is equally important that the scientific values of the unique region are safeguarded, and that the region is managed so that its environmental,

aesthetic, and wilderness values are sustained for future generations.

The NSF workshop on environmental management in the McMurdo Dry Valleys [*Vincent*, 1996] concluded there is a need for more formal and coordinated approaches to management to sustain the long-term scientific and environmental values in the region. The workshop concluded that current approaches have "failed to deal with longer term degradation processes in the valleys associated with the continuing proliferation of camp and sampling sites" [*Vincent*, 1996; p.18] and that a management plan should be formulated for the region. The workshop recommended this plan incorporate appropriate zones to accommodate different activities and intensities of use. A management plan would provide a clearer basis for operations and make it easier for scientists to plan their activities with information sufficient to minimize and avoid environmental impacts. It would also assist coordination between scientific groups from different disciplines, and facilitate greater international cooperation, thus helping to minimize conflicts of interest between research groups. For the management plan to be effective a reliable and accessible base of environmental information would need to be developed and maintained. Development of the system-wide GIS for the McMurdo Dry Valleys suggested by the Santa Fe workshop would help meet these needs.

The national programs working in the region are taking steps in this direction, and the science community should give these initiatives their full and continued support.

Acknowledgements: The author wishes to thank Joyce Jatko and Daryl Moorhead for their valuable comments on this paper. Acknowledgement is also made to the United States National Science Foundation, *Antarctica New Zealand* and the Italian Antarctic Program for their support of field work in the McMurdo Dry Valleys region out of which the observations made in this paper have grown.

REFERENCES

ATCPs (Antarctic Treaty Consultative Parties) *Final report of the XIXth Antarctic Treaty Consultative Meeting*; Seoul, 9–20 May 1995. Seoul, Republic of Korea, 1995

Broady, P.A. Taxonomy and ecology of algae in a freshwater stream in Taylor Valley, Victoria Land, Antarctica. *Archiv for Hydrobiologie Supplement* 63.3: 331–49, 1982.

Friedmann, E.I. (ed) *Antarctic microbiology*. New York, Wiley-Liss Inc., 1993.

Harris, C.M. Ross Sea protected areas: 1993/94 Visit Report. Christchurch, International Centre for Antarctic Information and Research, 1994a.

Harris, C.M. Protected areas review: McMurdo Sound, Ross Sea. Polar Record 30(174): 189–92, 1994b.

Harris, C.M. and Croteau, K. Prototype system to support helicopter tracking operations (McMurdo, Antarctica). *GIS World* 9(8): 42–45, 1996.

Hatherton, T. (ed) *Antarctica: the Ross Sea region.* Wellington, DSIR Publishing, 1990.

Keys, J.R., Dingwall, P.R. and Freegard, J. (eds) *Improving the Protected Area system in the Ross Sea region, Antarctica: Central Office Technical Report Series No. 2.* Wellington, NZ Department of Conservation, 1988.

Martin, L.M. *Cumulative environmental change: case study of Cape Royds, Antarctica.* MSc Thesis in Environmental Science and Geography, University of Auckland, 1991.

MFAT (Ministry of Foreign Affairs and Trade, New Zealand) 1997. *Draft new management plan for SSSI-12 at Canada Glacier, Taylor Valley, Antarctica.* Agreed for submission to the XXI Antarctic Treaty Consultative Meeting, Christchurch, New Zealand, May 1997.

NIWA (National Institute for Water and Atmospheric Research, New Zealand) Review of the Draft Initial environmental Evaluation on: Decommissioning Vanda Station, Wright Valley, Antarctica. *New Zealand Freshwater Miscellaneous Report 136.* Christchurch, NIWA, 1993.

Parker, B.C., Howard, R.V. and Allnutt, F.C.T. Summary of environmental monitoring and impact assessment of the DVDP. In Parker, B.C. and Holliman, M.C. (eds.) *Environmental impact in Antarctica.* Blacksburg, Virginia Polytechnic Institute and State University, 1978.

SCAR *Antarctic Specially Protected Area Visit Report Form.* Cambridge, SCAR, 1995.

SCAR / COMNAP *Monitoring of environmental impacts from science and operations in Antarctica. Report of the SCAR / COMNAP Workshops on Environmental Monitoring in Antarctica held at Oslo, Norway, 17–20 October 1995 and at College Station, Texas, 25–29 March 1996.* Cambridge, SCAR, 1996.

Schwarz, A.-M., Terrestrial biological studies in Southern Victoria Land, Antarctica. An unpublished thesis submitted for the degree of MSc in Biological Science at the University of Waikato, New Zealand, 1990.

Schwarz, A.-M., Green, T.G.A., and Seppelt, R.D. Terrestrial vegetation at Canada Glacier, Southern Victoria Land, Antarctica. *Polar Biology* 12: 397–404, 1992.

Vincent, W.F. (ed) *Environmental management of a cold desert ecosystem: the McMurdo Dry Valleys. Report of a National Science Foundation Workshop held at Santa Fe, New Mexico, 14–17 March 1995.* Reno, Desert Research Institute, 1996.

von Guerard, P., McKnight, D.M., Harnish, R.A., Gartner, J.W. and Andrews, E.D. *Streamflow, water temperature and specific conductance data for selected streams draining into Lake Fryxell, Lower Taylor Valley, Victoria Land, Antarctica, 1990–92. Open-File Report 94-545,* United States Geological Survey. Denver, USGS, 1994.

Wharton, R.A. Jr. (ed). *McMurdo Dry Valleys: a cold desert ecosystem. Report of a National Science Foundation Workshop held at the Institute for Ecosystem Studies, Millbrook, New York, 5–7 October 1991.* Reno, Desert Research Institute, 1993.

Colin Harris, PO Box 14-199 Christchurch New Zealand

(Received January 21, 1997;
accepted May 2, 1997)

THE MCMURDO DRY VALLEY ECOSYSTEM: ORGANIZATION, CONTROLS, AND LINKAGES

Daryl L. Moorhead

Department of Biological Sciences, Texas Tech University, Lubbock, Texas

John C. Priscu

Department of Biological Sciences, Montana State University, Bozeman, Montana

The McMurdo Dry Valleys comprise one of the coldest and driest ecosystems on our planet. Despite these extremes, a variety of life exists in their soils, streams, lakes, glacial and lake ice meltwater pools, and rocks. The biota generally are dominated by prokaryotes, with eukaryotes restricted to the less stressful sites. Higher life forms include bryophytes, rotifers, tardigrades, and nematodes; vascular plants, insects, and vertebrates are lacking. Key conditions limiting life are liquid water and energy. In the presence of liquid water, radiant energy drives photoautotrophic production that provides heterotrophic communities with a carbon and energy supply. Spatio-temporal linkages between landscape components augment productivity of some communities and permit others to exist in places and at times that otherwise would be impossible. For example, organic matter supporting some soil food webs originates in lakes and streams. This organic matter is then transported to the soils by wind (spatial link among present ecosystem components) or made available by wind erosion of ancient lake and stream beds (temporal link to past ecosystems). Owing to tight spatio-temporal linkages, which we believe are necessary for the existence of life in extreme environments, ecological research in the McMurdo Dry Valleys must extend beyond isolated communities and focus on integrating system components. Herein we summarize existing knowledge of the McMurdo Dry Valley ecosystem emphasizing results from other papers in this volume. We focus on linkages among ecosystem components that augment or enable community existence.

INTRODUCTION

From the time of its formal definition, the term "ecosystem" has explicitly emphasized dynamic interactions among biotic and abiotic components of the environment [*Tansley,* 1987]. Indeed Stephen Forbes provided such a context in his earlier presentation to the Illinois Natural History Society in 1887 [*Forbes,* 1887]. Since then, studies have yielded insights to ecosystem structure and energy flow [*Odum,* 1957; *Teal,* 1962; *Odum,* 1969], element transfers [*Schulze and Zwölfer,* 1987], responses to disturbance [*Pickett and White,* 1985; *Reynolds and Tenhunen,* 1996] and functional linkages among landscape elements [*Turner and Gardner,* 1991].

Because of the complexity of ecosystems, their study requires a multidisciplinary approach. Owing to their scalar dimensions in time and space, ecosystem studies must focus on relatively large areas over long periods. One of the broadest sets of ecosystem studies ever designed is the Long-Term Ecological Research (LTER) program, currently funded by the US National Science Foundation. At present, there exist 18 LTER sites ranging from the tropical rain forest of Puerto Rico through the northern Chihuahaun Desert of New Mexico, USA, and including prairie, forest, and tundra biomes [*Van Cleve and Martin,* 1991]. The McMurdo Dry Valleys LTER site presents an end-member ecosystem of this group in that the Antarctic dry valleys are among the driest and coldest deserts on this planet. Even so, biological communities exist in the soils, lakes, and streams of the dry valleys.

The biological communities of the McMurdo Dry Valleys are relatively simple, being dominated by microbiota, lacking higher plants and animals, and consisting of relatively short food chains. However the basic ecological processes of primary and secondary production, decomposition, nutrient cycling and energy flux within these ecosystems demonstrate complex interactions among biological entities, as well as between biota and abiotic factors of the environment. Thus the McMurdo Dry Valleys offer a natural laboratory for study of basic ecological processes without the complications introduced by a plethora of higher organisms.

The presence of liquid water defines the primary limiting condition for life in this extreme environment. Liquid water is present in streams and soils only during the austral summer and even then freeze-thaw cycles are frequent. Prokaryotes dominate many of these ecosystems with eukaryotes occurring in the less stressful habitats. The extreme environment, coupled with a paucity of higher trophic groups, implies strong "bottom-up" controls on ecosystem structure with minimal influence of "top-down" controls exerted by herbivory or predation common to other environments [e.g., *Carpenter*, 1988]. In addition, the transport of water, nutrients, and organic carbon between glaciers, streams, lakes, and soils defines functional linkages that enhance biological activities in the dry valleys. While such linkages among landscape elements have been shown to exert considerable influence on ecosystem structure and function [*Turner and Gardner*, 1991], their impacts are particularly evident in extreme environments, such as deserts [*Schlesinger et al.*, 1990] and tundra [*Shaver et al.*, 1991; *Reynolds and Tenhunen*, 1996].

The purpose of this manuscript is to present a conceptual model of the major components of the dry valley ecosystems, with particular reference to the studies presented by other chapters within this volume. Owing to space constraints and the current status of our knowledge of certain system components, we cannot address all aspects of the dry valley system. Hence we focus on processes affecting primary production and linkages defining the transfer of organic carbon among system components. As in all ecosystems, and of particular importance in desert systems, the production of organic carbon initiates a cascade of biotic and abiotic transformations. This chapter, in conjunction with the other chapters in this book, provides a framework for future research initiatives within the McMurdo Dry Valleys of Antarctica. Insights to ecosystem processes gained from these studies will broaden our understanding of origins and persistence of life in extreme environments.

CONCEPTUAL MODEL OF THE DRY VALLEY SYSTEM

The climate of the McMurdo Dry Valleys is extremely arid, with most of the total annual precipitation (< 10 cm) falling as snow during the winter [*Clow et al.*, 1988]. Ambient air temperatures hover near freezing during the austral summer, so that most of the heat required for warming substrates and melting ice is provided by radiant energy [Figure 1; *Fountain et al.*, this volume]. Life exists in the soils, rocks (endoliths), streams, and lakes of the McMurdo Dry Valleys, but is faced with a paradox: sufficient heat is required to provide liquid water and adequate solar energy is needed to drive photosynthesis; however exposure to radiant energy in this low humidity environment also means exposure to dessication and rapid freeze-thaw cycles. With air temperatures rarely exceeding 0°C, large differences in the energy balance of microsites result from modest differences in orientation to the sun [*Dana et al.*, this volume; *Prentice et al.*, this volume]. Thus topographic features play a significant role in determining the spatial and temporal distribution of radiant energy in the dry valleys (Figure 1). Similarly the orientation of glacial faces providing most of the water entering dry valley streams and lakes strongly affects the amount and timing of ice melt [*Fountain et al.*, this volume] and flow patterns of streams [*Conovitz et al.*, this volume].

The spatial and temporal distributions of liquid water and photosynthetically active radiation (PAR) do not always overlap, nor do they necessarily correspond to the availability of adequate nutrients to support primary production, or organic carbon to support heterotrophs. However the glaciers, streams, lakes, and soils of the dry valleys are linked by aeolian and hydrological transfers of materials that compensate for some site-specific limitations (Figure 1). For example glaciers and streambed soils are sources of many solutes that increase in concentration as water flows from glaciers to streams to lakes, with important mechanisms of concentration being the evaporation of water [*Lyons et al.*, this volume; *Green et al.*, this

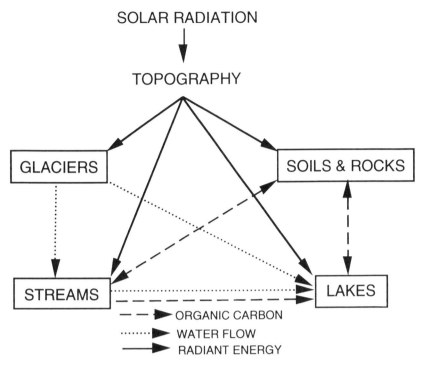

Fig. 1. Conceptual model of topographic effects on the distribution of incident radiation and the principle flow of organic carbon and water among habitats of the McMurdo Dry Valleys.

volume], and freeze-concentration [*Howard-Williams et al.*, 1989]. In addition to solutes, hydrologic processes move sediment and organic matter within streams and from streams to lakes [*McKnight et al.*, this volume]. Aeolian connections also are important linkages between dry valley components, e.g., winds deposit sediments and organic matter on glaciers and lake ice surfaces and exchange organic matter between lakes, streams and soils throughout the dry valleys [*Adams et al.*, this volume; *Lyons et al.*, this volume].

In summary, biological activity within the dry valley ecosystem is controlled by the spatial-temporal distribution of radiant energy and liquid water, with additional constraints imposed by the availability of inorganic nutrients and organic carbon. Moreover linkages between components of the dry valley landscape may ameliorate certain limitations to biota on some sites. Within this conceptual framework, similarities and differences in the controls on the ecological communities of the dry valleys can be elucidated with reference to a generic model of biomass dynamics.

GENERIC BIOMASS MODEL

Biological activity generally occurs where conditions permit liquid water to exist [*Kennedy*, 1993] and supplies of both nutrients and energy are adequate (Figure 2; see also *Vincent et al.*, [1993a]). Organic matter may accumulate through on-site photosynthesis or allochthonous inputs, and be lost through on-site respiration, release of dissolved organic carbon or mechanical erosion via wind and water. However ecosystems within the dry valleys differ with respect to the importance of various environmental controls. Herein we compare and contrast the primary controls on biological activities in the major ecosystems of the dry valleys: the microbial mats found in stream and lake beds, lake phytoplankton communities, cyanobacterial communities growing within permanent lake ice, and soil communities. Terrestrial bryophyte and endolithic communities also exist in the dry valleys but are discussed elsewhere [e.g., *Friedmann*, 1982].

Microbial Mats

Filamentous cyanobacteria (e.g., *Phormidium* spp., *Nostoc* spp.; *McKnight et al.*, this volume; *Hawes et al.*, this volume] usually dominate the algal component of benthic microbial mats found in dry valley streams and lakes. Herbivory appears to be low or absent in both environments and many of the same cyanobacterial species comprise a significant fraction of the

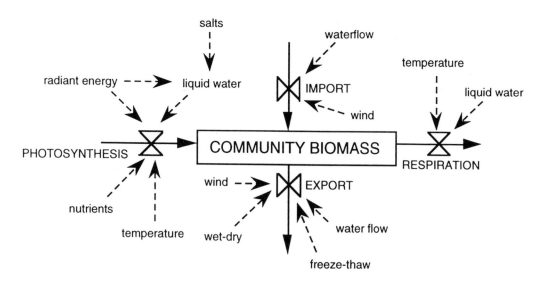

Fig. 2. General model of biomass dynamics in the McMurdo Dry Valley ecosystem, including inputs, outputs, and major factors controlling losses and gains of biomass.

communities in both environments. However the primary controls on biomass accumulation differ between streams and lakes.

In stream ecosystems liquid water is available only in summer and is provided by glacial melt, with timing, amount, and duration of flow controlled by the orientation of glacier faces to the sun, geometry of the stream, and extent of the hyporheic zone [Fountain et al., this volume; Conovitz et al., this volume]. The composition, distribution, and abundance of mats largely correlate to the stability of the streambed, with mats being most abundant in reaches with moderate gradients and stable stone pavements [Howard-Williams et al., 1986; Alger et al., 1996; McKnight et al., this volume]. Mats dominated by Nostoc spp. occupy moist areas of stream margins and in slow-moving current, while Phormidium spp. are more common in channels with faster current [Alger et al., 1996]. Sedimentation and abrasion appear to cause the major losses of mat in streams, and frequent freeze-thaw, wetting-drying, and flooding events probably compromise the structural integrity of mats [Howard-Williams et al., 1986]. Wind erosion of exposed, freeze-dried mats also may occur during winter. Thus frequent dramatic changes in water flow and temperature regimes may physically disrupt the microbial mats and enhance material export.

Losses and gains of organic matter by microbial mats may occur through transport of organic material, photosynthesis, and respiration. Vincent et al. [1993b] speculated that movements of dissolved and particulate

organic matter constituted major sources and sinks for mat biomass in Antarctic streams, which is consistent with reported patterns of mat distributions. However few observations of organic matter export have been reported; Howard-Williams et al. [1989] found substantial losses of dissolved and particulate organic N and P from streams in Taylor Valley, and Vincent and Howard-Williams [1986] noted significant downstream movement and accumulation of mat materials. In contrast, metabolic activities of mats have received much attention. Photosynthesis of stream mats generally saturates at low light intensities (< 200 µmol photons m^{-2} s^{-1}), and thus rarely is limited by the radiant energy regime when liquid water is present and the mats are active [Howard-Williams and Vincent, 1989; Hawes et al., this volume]. Although microenvironmental measurements imply that light intensity becomes limiting just beneath the surface of these mats due to heavy pigmentation [Vincent et al., 1993a, 1993b], intact mats show little negative response to light intensities exceeding saturation. For this reason Hawes et al. [this volume] argue that water temperature is the major control on net primary production of mat communities through its effects on light-saturated photosynthesis and respiration [Vincent and Howard-Williams, 1989; Hawes, 1993].

Nutrient availability could limit stream mat production, but concentrations of nitrogen and phosphorus are relatively high in stream waters [Howard-Williams et al., 1989; Alger et al., 1996]. Studies have shown significant reductions in nitrate and urea concentrations along the lengths of dry valley streams [*Howard-Williams et al.*, 1989; *Hawes and Brazier*, 1991], suggesting significant nutrient uptake by mat communities [*Moorhead et al.*, this volume]. However in situ nutrient enrichment experiments demonstrated no limitation of nitrogen or phosphorus availability on photosynthesis or pigment content of stream mats [*Howard-Williams and Vincent*, 1989].

Benthic mats within the dry valley lakes, in contrast to streams, are not subject to freeze-thaw or wet-dry cycles. The only exceptions are mats in shallow water at the lake margins where the water freezes to the bottom of the water column during winter and thaws during summer. Otherwise the permanent ice covers (3–5 m thick) on the lakes provide a physical barrier that, in combination with salt gradients, leads to stable stratification of the underlying water column [*Spigel and Priscu*, this volume]. Water input arrives from glacial melt, either directly or via inputs from meltwater streams. Thus benthic mats are permanently hydrated and ambient temperature varies little over time.

Also in contrast to stream cyanobacterial communities, light intensity probably restricts photosynthesis of mats throughout much of the lake. This is because only a small fraction of the ambient radiation is transmitted through lake ice and further attenuation occurs within the water column [*Howard-Williams et al.*, this volume]. Considerable variation in light transmission to the underlying liquid water column exists among lakes, being controlled by physical characteristics of the ice [*McKay et al.*, 1994]. However transmission seldom exceeds 5% of incident intensities for lakes at the McMurdo Dry Valley LTER site [*Howard-Williams et al.*, this volume]. In addition, long wavelengths of light (> 600 nm) are more severely attenuated than short wavelengths [*Lizotte and Priscu*, 1992]. Thus the light environment rather than temperature within dry valley lakes is presumably the major factor limiting primary production of benthic mats [*Moorhead et al.*, in press].

Nutrient concentrations clearly affect phytoplankton productivity in Antarctic lakes [*Vincent*, 1981; *Priscu et al.*, 1989; *Priscu*, 1995], but relatively little is known of nutrient controls on benthic mats. Studies have shown that concentrations of nutrients (e.g., soluble reactive phosphorus and ammonium) within microbial mats are much greater than in the overlying water column, suggesting a rapid, internal cycling of nutrients [*Vincent et al.*, 1993b]. Moreover *Hawes et al.* [1993] proposed that benthic mats in seasonal

ponds on the McMurdo Ice Shelf may serve as nutrient sinks, accumulating biomass and nutrients over time due to the absence of grazing losses or mass export. A similar paradigm presumably exists in the benthic mats of the dry valley lakes.

Benthic mats in lakes probably experience little disturbance, except for those in the shallow moat regions (previously mentioned). In deeper water mat liftoff due to the accumulation of entrained gas bubbles and occasional burial by sedimentation may represent the major disturbances experienced by these communities [*Wharton*, 1994]. However liftoff is limited to the shallower portions of the lake where light intensity is high (relative to deeper zones), and sedimentation occurs when a crack in the overlying ice permits sediments on the ice surface or entrained within the ice to pass into the water column [*Adams et al.*, this volume]. Thus burial of benthic mats in this manner appears to be localized and infrequent.

Losses of mat materials from the lake may occur via a process described by *Parker et al.* [1982]. In brief, mats in shallow water receive sufficient radiant energy to produce bubbles of oxygen within the mats. This provides sufficient buoyancy to float portions of mats to the bottom of the overlying ice cover and into open moat water, where they freeze into the ice during winter. Annual ablation of ice from the top of the ice layer and freezing of water to the bottom results in the upward movement of entrained mat materials until they are exposed on the ice surface. This mechanism is responsible for accumulations of mat materials along the shores of the seasonally melted margins of the lakes where the exposed material is readily distributed by wind.

Lake Phytoplankton

Phytoplankton communities of dry valley lakes exist in an environment characterized by low light intensity and stable stratification of the water column, both attributes resulting from the permanent ice covers on these lakes (previously discussed). While light intensity shows daily and seasonal variations, many other physical and chemical factors vary little over time. However vertical variations in environmental parameters may be large, with large gradients of salinity, temperature, and ionic concentrations occurring over depth [*Angino et al.*, 1964; *Hawes*, 1985; *Spigel and Priscu*, 1996; *Spigel and Priscu*, this volume]. The lack of wind-induced vertical mixing allows these gradients to persist over long periods, possibly since the evolution of the lakes [*Spigel and Priscu*, 1996; *Priscu*, 1995; *Priscu*, in press].

Light intensity and spectral composition also vary with depth in dry valley lakes [*Lizotte and Priscu*, 1992; *Howard-Williams et al.*, this volume; *Lizotte and Priscu*, this volume], and phytoplankton occur in relatively distinct, stratified layers [*Vincent*, 1981; *Lizotte and Priscu*, 1992; *Lizotte and Priscu*, this volume]. Phytoplankton show a considerable degree of shade adaptation, and response to temperature, with differences in photosynthetic capacity and efficiency between depths [*Lizotte and Priscu*, 1992; *Neale and Priscu*, 1995; *Neale and Priscu*, this volume]. Little evidence of photosaturation exists, so primary production is limited virtually always by light availability. In addition to light limitation, *Priscu* [1995] demonstrated that the addition of phosphorus, or nitrogen plus phosphorus, stimulated photosynthesis of phytoplankton in most lakes of Taylor Valley, particularly in the upper water column. Internal nutrient regeneration has a major role in regulating phytoplankton production in the upper water column during the austral summer [*Priscu et al.*, 1989]. Vertical nutrient profiles suggest that deep primary production maxima often are driven by diffusion of nutrients from underlying, nutrient-rich layers of water. A similar situation exists in Lake Vanda, located in the adjacent Wright Valley, wherein a deep layer of maximum chlorophyll concentration exists just above a hypersaline, nutrient-rich zone [*Vincent*, 1981; *Vincent and Vincent*, 1982; *Priscu*, 1995]. Light intensity also has been shown to regulate inorganic nitrogen uptake in these lakes [*Priscu et al.* 1987; *Priscu et al.* 1988; *Priscu*, 1989; *Priscu and Woolston*, in press]. Thus concentrations of phytoplankton and maximum rates of production are determined by the juxtaposition of favorable light, nutrient, and salinity regimes in these stratified lakes [*Priscu*, 1995]. Changes in nutrient availability, temperature, and light regime may be responsible for seasonal changes in the location, density, succession, and productivity of phytoplankton communities.

The inflow of water from melting glaciers and streams provides inputs of nutrients and organic matter to the dry valley lakes [*Green et al.*, 1988; *Howard-Williams et al.*, 1989; *Lyons et al.*, this volume]. Much of this input water appears to disperse across the lake just beneath the ice cover. This supposition is supported by observations of relatively high nutrient concentrations just beneath the ice cover, the lack of appreciable mixing between layers of water, and movements of ions released in tracer experiments

[*Lizotte and Priscu*, 1992; *McKnight et al.*, 1993; *Spaulding et al.*, 1994; *Spigel and Priscu*, this volume]. Aeolian deposits of sediments and organic matter from microbial mats are present on the surfaces of lake ice and originate from surrounding land surfaces and seasonally dry stream beds. Some of these materials enter the water column via the formation of cracks or conduits through the ice cover. Surface deposits of dark colored materials also may absorb sufficient radiant energy to melt a significant portion of the ice [*Adams et al.*, this volume; *Fritsen and Priscu*, this volume; see below], allowing direct entry to the underlying water column.

The overall biomass of phytoplankton communities in these lakes should decline with cellular respiration and potential grazing [*Laybourn-Parry et al.*, 1997; *James et al.*, this volume] during the continuously dark winter months. However photoadaptation has been demonstrated for phytoplankton in this [*Lizotte and Priscu*, 1992; *Neale and Priscu*, 1995; *Neale and Priscu*, this volume] and other systems [e.g., *Geider et al.*, 1996], and may include reductions in respiration rates with declining light intensity [*Prezelin and Sweeney*, 1978]. Degradation rates of phytoplankton are slow, especially over winter [*Priscu*, 1992; *Lizotte and Priscu*, this volume], and preliminary sediment trap collections suggest significant settling of cells from the water column during winter (Priscu, unpublished data). An unknown portion of the organic nitrogen associated with sinking organic matter is regenerated to ammonium within the water column [*Priscu et al.*, 1989] which then can be oxidized to nitrous oxide, nitrite or nitrate by nitrifying bacteria [e.g., *Voytek et al.*, this volume; *Priscu et al.*, 1996; *Priscu*, in press]. Under anaerobic conditions, present in the bottom waters of some of the lakes, the oxidized nitrogen can be reduced to nitrogen gas through denitrification [*Ward and Priscu*, in press; *Priscu*, in press].

Grazing by microzooplankton and protozooplankton may affect phytoplankton communities [*James et al.*, this volume], although these systems lack many of the phytoplankton grazers found in other aquatic communities. Mixotrophic phytoplankton species recently have been identified and phagotrophy observed [*Laybourn-Parry et al.*, 1997; *James et al.*, this volume; Laybourn-Parry, pers. communication]. These processes would allow the phytoplankton to utilize heterotrophic metabolism during low light conditions (i.e., in deep water or during winter). Fish are completely absent from the lakes eliminating predation on the lower trophic levels.

Ice Communities

A recent discovery is that biological activity is associated with sediments entrained within the ice covers on lakes in the Taylor Valley [*Wing and Priscu*, 1993; *Adams et al.*, this volume; *Fritsen and Priscu*, this volume]. Owing to the mineral nature of ice, these organisms can be referred to as cryptoendolithic in the same manner as those found in dry valley sandstone [*Friedmann*, 1982]. Although the mechanisms of establishment and perpetuation of these novel communities are just being determined, it appears that sediments, with associated cyanobacterial and bacterial cells, are deposited on the surface of lake ice by aeolian transport. These materials absorb sufficient radiant energy during the summer to allow sinking into the ice cover through meltwater channels, eventually attaining a depth at which the absorption of radiant energy is insufficient to drive further melting (usually mid-depth in the ice-covers of most lakes). These channels freeze during autumn, sealing each microcosm within solid ice. Ablation occurs primarily from the upper surface of the ice covers on lakes, with accumulation resulting from freezing of water to the undersides. This moves the entrained microcosms upward during the winter, with subsequent summer melting moving the microcosms downward through meltwater channels, the result being constant vertical positioning within the ice cover. Biotic and abiotic losses and gains have shown that biomass accumulates primarily through biological growth [*Fritsen and Priscu*, in press]. Photosynthetic studies have shown that the cyanobacterial communities saturate at relatively low light levels [<50 μmol photons m^{-2} s^{-1}; *Fritsen and Priscu*, in press] and show temperature optima near 18°C (Priscu and Fritsen, unpublished). These results together with 16S ribosomal DNA signatures imply that the microbial assemblages are of terrestrial origin [*Gordon et al.*, in press] and are comprised of complex cyanobacterial /bacterial consortia.

Soil Communities

Although soils of the dry valleys once were considered to be nearly devoid of life, recent studies have demonstrated a wide distribution of microbiota [*Powers et al.*, 1995; *Freckman and Virginia*, this volume]. Soil communities in the McMurdo Dry Valleys tend to be simple, have few trophic groups, exist in films of water on soil particles, and often are dominated by nematodes, which fill key trophic positions often

including that of top predator [*Overhoff et al.*, 1993; *Freckman and Virginia*, 1997; *Freckman and Virginia*, this volume]. Studies of nematodes in other Antarctic systems have revealed considerable cold-hardiness and freezing tolerance in the group [e.g., *Pickup*, 1990a, 1990b, 1990c; *Wharton and Block*, 1993], as well as the ability to enter an anhydrobiotic state upon desiccation [*Pickup*, 1988]. Soil nematodes of warm desert ecosytems also can enter anhydrobiotic states upon desiccation [*Freckman*, 1986; *Freckman and Mankau*, 1986; *Freckman et al.*, 1987], suggesting a common response to water-limiting conditions under hot or cold conditions. In fact mechanisms underlying freeze-tolerance and desiccation-resistance are similar [*Crowe and Madin*, 1975], and at least some species of nematode demonstrate greater freeze-tolerance when they are in an anhydrobiotic state [*Pickup and Rothery*, 1991]. Thus the frequent dominance of nematodes in soil trophic webs may result from their ability to survive the cold, dry conditions that characterize dry valley soils.

Factors controlling the spatial distribution of biota in the McMurdo Dry Valleys are uncertain. Although *Kennedy* [1993] concluded that the abundances and distributions of organisms in Antarctica were closely correlated to moisture availability, *Freckman and Virginia* [1997] reported no correlation between nematodes and soil moisture in the McMurdo Dry Valleys. Furthermore neither soil carbon nor nitrogen content were correlated to nematode distributions in this region [*Freckman and Virginia*, 1997]. However field studies demonstrated that increasing soil water, carbon, and temperature tended to increase abundances of microbivorous species while decreasing abundance of omnivore-predator species [*Freckman and Virginia*, 1997]. *Powers et al.* [1994] noted that nematode distributions in Taylor Valley appeared to be influenced by soil pH and salinity, but relationships between distributions, diversity, and abundance of nematodes and environmental factors remain equivocal.

The origins of the organic matter driving soil communities are largely unknown [*Burkins et al.*, in press]. Cyanobacteria are present in soils, but moisture conditions at the soil surface are seldom favorable for photosynthesis or growth. Nematode densities are highest at 2.5–10 cm depth, where moisture conditions are more favorable [*Powers et al.*, 1995], but light intensities are insufficient to drive photosynthesis. Allochthonous inputs of organic matter from lakes, streams and the distant ocean may provide the base of these detrital food chains, and buried sediments from

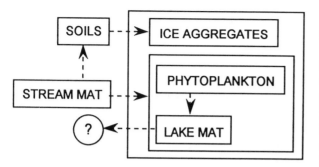

Fig. 3. Flow diagram illustrating major transfers of organic carbon between biological communities in the McMurdo Dry Valleys.

ancient lake beds also may supply soil communities [*Burkins et al.*, in press; *Freckman and Virginia*, this volume].

In summary, the existence and complexity of soil communities in the McMurdo Dry Valleys are strongly limited by a variety of environmental factors, such as low temperature and moisture availability, high pH and salinity, and low organic carbon content. Soil biota demonstrate a very patchy distribution throughout the McMurdo Dry Valleys, probably reflecting a patchwork of microenvironmental conditions. Soils are cold, dry and saline [*Campbell et al.*, this volume], and the simple food webs appear to be based largely on allochthonous inputs of organic matter from other times or places. Freeze-thaw cycles are frequent in surface soils and deeper soils remain frozen; more agreeable conditions probably exist at intermediate depths [*Freckman and Virginia*, this volume].

LINKAGES

Linkages for the transport of organisms and materials between components of the dry valley landscape enable the existence and augment production of various biological communities (Figure 1). Two major transport mechanisms, fluvial and aeolian, move three major currencies throughout the dry valleys: water, nutrients, and organic matter (including organisms). Inorganic sediments also are moved by wind and water, but travel in much the same manner as organic matter (Figure 3).

Water flows from glaciers through streams to lakes, hydrating the stream beds and benthic mats, carrying solutes and organic matter from the glaciers into the streams, and from the streams into the lakes [*Lyons et al.*, this volume]. Mat communities in streams are metabolically active within minutes of hydration

[*Vincent and Howard-Williams*, 1986], utilize inorganic nutrients and urea from the water column [*Howard-Williams et al.*, 1989], and fix molecular nitrogen from the atmosphere [*Howard-Williams et al.*, 1989]. Concentrations of other ions and both dissolved and particulate organic compounds increase along the length of streams, providing inputs of these materials to recipient lakes [*Howard-Williams et al.*, 1989; *Lyons et al.*, this volume]. These nutrient inputs may stimulate production of phytoplankton communities, especially near the ice-water interface [*Lizotte and Priscu*, 1992; *Spaulding et al.*, 1994; *Priscu*, 1995].

Dryfall and wetfall (via snow) add nutrients to land and ice surfaces in the dry valleys [*Lyons et al.*, this volume]. Fluvial movements carry these nutrients into streams and lakes, as previously mentioned. However aeolian transport is responsible for distributing sediments and organic materials around the dry valleys and probably is responsible for placing most of these materials found on surfaces of glaciers and lake ice. Perhaps the redistribution of organic matter by wind also is an important determinant of soil community development and wind erosion may expose buried sediments from ancient lake beds. Despite this uncertainty, the establishment of lake ice communities described by *Adams et al.* [this volume] and *Fritsen et al.* [this volume], and the structure of soil trophic webs [*Freckman and Virginia*, this volume], depend upon the availability of organic matter that probably originates elsewhere.

CONCLUSIONS

The availability of liquid water provides a necessary condition for life, but the spatial-temporal distribution of liquid water is discontinuous in the McMurdo Dry Valleys. The lakes in Taylor Valley are permanently hydrated, but of limited spatial extent. The streams are hydrated only during the brief, austral summer, and experience frequent freeze-thaw and wet-dry cycles even during this period. Much of the soil in the dry valleys rarely or never has adequate moisture to support metabolically active organisms, but the margins of streams, lakes, and melting glaciers, in variety of favorable microenvironments for soil communities.

Within this spatial-temporal mosaic of suitable moisture conditions, life is affected by other conditions and resources. Gradients of light, temperature, nutrient concentration, and salinity control the abundance and distribution of phytoplankton and benthic mats in the dry valley lakes. Data support the over-riding importance of streambed characteristics in the establishment and persistence of mat communities in streams. Distributions of soil organisms and development of food webs are related to moisture, pH, salinity, and temperature regimes in soils, as well as the availability of organic carbon.

Linkages between system components are responsible for the establishment of some biological communities in the dry valleys, and augment the productivities of others. For example, aeolian inputs of organic matter appear to represent the source of cells establishing communities entrained within lake ice, serve as a source of organic carbon important to the development of soil food webs, and add to the accumulation of mat materials on lake bottoms. Nutrients accompanying movements of water, sediments, and organic matter are utilized by microbial mats and plankton communities in streams and lakes. Finally the movement of water from glaciers to lakes defines the presence of both lakes and streams.

A common characteristic of the biological communities in the dry valleys is the dominance of prokaryotes as the principle autotrophs and decomposers. Cyanobacteria often dominate photoautotrophic production under extremely stressful conditions, such as in alkaline or saline waters, and geothermal streams. In the dry valleys cyanobacteria are an important component of microbial mats in streams and lakes, phytoplankton communities in certain lakes, lichen symbioses, endolithic communities, and soils. Heterotrophic bacteria are the primary decomposers in all of these systems. The close association between cyanobacteria and bacteria in many of the dry valley communities may represent a consortial arrangement promoting survival under stressful conditions.

FUTURE RESEARCH DIRECTIONS

Perhaps the most important aspects of the McMurdo Dry Valley system are the transport mechanisms that move critical materials between sites. Although fluvial processes have been studied in some detail [e.g., *Prentice et al.*, this volume; *Lyons et al.*, this volume; *Conovitz et al.*, this volume], the transport of organic matter, sediments, and solutes requires further investigation. This is particularly important with regard to nutrient dynamics along streams and the impacts of stream chemistry on recipient lakes [e.g., *Green et al.*, this volume]. Aeolian transport mechanisms have been identified as being important in a qualitative manner, but have yet to be quantified with respect to the timing, amounts, sources and sinks of materials being moved.

Of the major habitats of the dry valleys, the soil ecosystems are perhaps least understood. Only recently have detailed investigations begun to identify the types and amounts of life distributed in soils throughout the dry valleys. Correlations between distributions of key groups of soil invertebrates and physical-chemical characteristics of soils are revealing likely controls on the structure of soil communities. However little information on the metabolic characteristics, population dynamics, or material cycling within soil trophic webs has been obtained.

Microbial mat communities are common throughout the streams and lakes of the dry valleys and considerable effort has been directed toward evaluating their composition and metabolic features. However internal structure, carbon dynamics, and nutrient cycling processes in these mats are not well-known. Moreover their contributions of organic matter to other dry valley ecosystems appear to be substantial, although the type, amount, and origin of mat materials found in various habitats has not been determined, nor have the losses of mat types from various habitats been quantified. Preliminary colonization studies have begun, but mechanisms of accumulation and factors controlling structure of these communities are not well-known.

Lake systems pose a number of questions for study. The influence of the permanent ice covers as potential barriers to the flux of metabolic and abiotic gases is only beginning to be understood [e.g., *Priscu et al.*, 1996; *Priscu*, 1997]. The role of lake sediments in oxygen dynamics, nutrient cycling, and organic matter accumulation have not been examined. Vertical movements of ions in these stratified lakes appears to be by molecular diffusion of soluble compounds and settling of particulate materials. The role that microorganisms have in metal geochemistry in the saline bottom waters of certain lakes is still uncertain, though it is currently under study [e.g., *Green et al.*, this volume]. Organic carbon dynamics in lakes, particularly the vertical flux of particulate organic carbon, is just beginning to be quantified. Grazing of phytoplankton is low and dominated by microzooplankton and protozooplankton, many exhibiting a wide range of feeding strategies suggesting bottom-up controls on community structure. Phagocytotic phytoplankton species have recently been identified, although their role in the microbial loop, particularly with respect to the regulation of bacterial plankton numbers, has yet to be quantified. The role of bacterioplankton as sources or sinks of organic carbon and the relationship between net primary productivity and bacterial respiration still remains to be determined. Finally no estimates of energy or nutrient flux through these communities have been made, nor have the population dynamics of planktonic species been characterized.

In a larger context, the primary sources, sinks, and transformations of major elements, such as carbon, nitrogen, and phosphorus, have not been elucidated completely either within particular communities or between communities. Although considerable evidence suggests that transport of these materials between locations may be an important factor contributing to the overall structure and function of the McMurdo Dry Valley system, few data are available to quantify these flows or their consequences to recipient systems. Broadscale studies of material fluxes within and between components of the dry valley system are sorely needed. Most of what is currently known about ecosystem processes within the McMurdo Dry Valleys has resulted from data collections and experimental manipulations made during the austral summer. Logistics have not yet advanced to the level that would support winter research either through on-site data collection or remote sensing. It is imperative that future research address winter processes. For example, lake phytoplankton and bacterioplankton biomass is low at the end of winter and generally increases through the summer [*Lizotte et al.*, 1996; Priscu, unpublished data]. However little is know about growth and loss rates during autumn and winter. Many ecosystem process may never be delineated completely without annual data collection and experimental manipulation.

The compendium of chapters published within this book volume represents a first attempt at bringing together much of what is known about processes and linkages within the McMurdo Dry Valleys. These studies clearly show that the presence of liquid water produces a cascade of ecosystem events which are tightly coupled. We believe that tight temporal and spatial coupling among ecosystem components is the key to metabolic activity and ultimate survival of organisms within the McMurdo Dry Valleys.

In closing it is important to note that although the McMurdo Dry Valleys are among the most isolated locations on earth, they are experiencing increasing levels of anthropogenic stress and disturbance [*Harris*, this volume]. Aside from the obvious contributions of human industries to the reduction in stratospheric ozone

concentrations over Antarctica, increasing levels of sulfur, nitrogen, and other compounds in the atmosphere, and global warming, humans are also visiting Antarctica in ever-increasing numbers. Conditions supporting scientific research have improved dramatically, increasing both the number and duration of scientific visits. In addition recent entrepreneurial ventures are promoting tourism and providing access to the dry valleys via helicopter. The impacts of human activities, including scientists and tourists, only now are being explored.

Acknowledgments. We thank W. Vincent for many useful comments on the manuscript. Funding for this work was provided by USA National Science Foundation Office of Polar Programs research grants OPP-9211773 and OPP-9419423.

REFERENCES

Adams, E. E., J. C. Priscu, C. H. Fritsen, S. R. Smith, and S. L. Brackman, Permanent ice covers of the McMurdo Dry Valley lakes, Antarctica, this volume.

Alger, A. S., D. M. McKnight, S. A. Spaulding, C. M. Tate, G. H. Shupe, K. A. Welch, R. Edwards, E. D. Andrews, and H. R. House, *Ecological processes in a cold desert ecosystem: The abundance and species distribution of algal mats in glacial meltwater streams in Taylor Valley, Antarctica.* 102 pp., United States Geological Survey. Boulder, Colorado, 1996.

Angino, E. E., K. B. Armitage, and J. C. Tash, Physiochemical limnology of Lake Bonney, Antarctica. *Limnol. Oceanogr.*, 9, 207-217, 1964.

Burkins, M. B., C. P. Chamberlain, R. A. Virginia, and D. W. Freckman, Natural abundance of carbon and nitrogen isotopes in potential sources of organic matter to soils of Taylor Valley, Antarctica. *Antarct. J. U. S.*, (in press).

Campbell, I, G. G. C. Claridge, D. I. Campbell, and M. R. Balks, The soil environment of the McMurdo Dry Valleys, Antarctica, this volume.

Carpenter, S. R., *Complex Interactions in Lake Communities.* Springer-Verlag, New York, 1988.

Clow, G. D., C. P. McKay, G. M. Simmons, Jr., and R. A. Wharton, Jr., Climatological observations and predicted sublimation rates at Lake Hoare, Antarctica. *J. Clim.*, 1, 715-728, 1988.

Conovitz, P. A., D. M. McKnight, L. M. McDonald, A. Fountain, and H. R. House, Hydrologic processes influencing streamflow variation in Fryxell Basin, Antarctica, this volume.

Crowe, J. H., and K. A. C. Madin, Anhydrobiosis in nematodes: Evaporative water loss and survival. *J. Exp. Zool.* 193, 323-334, 1975.

Dana, G. L, R. A. Wharton, Jr., and R. Dubayah, Solar radiation in the McMurdo Dry Valleys, Antarctica, this volume.

Forbes, S. A., The lake as a microcosm. *Bull. Peoria Sci. Association*, 77-87, 1887, reprinted in *Bull. Illinois State Natural History Survey*, 15, 537-550, 1925.

Fountain, A. G., K. J. Lewis, and G. L. Dana, Glaciers of the McMurdo Dry Valleys, southern Victoria Land, Antarctica, this volume.

Freckman, D. W., The ecology of dehydration in soil

organisms, in *Membranes, Metabolism and Dry Organisms*, edited by A. C. Leopold, pp. 157-168, Cornell University Press, New York, 1986.

Freckman, D. W., and R. Mankau, Abundance, distribution, biomass and energetics of soil nematodes in a northern Mojave desert ecosystem. *Pedobiologia*, 29, 129-142, 1986.

Freckman, D. W., and R. A. Virginia, Low-diversity Antarctic soil nematode communities: Distribution and response to disturbance. *Ecology*, 78, 363-369, 1997.

Freckman, D. W., and R. A. Virginia, Soil biodiversity and community structure in the McMurdo Dry Valleys, Antarctica, this volume.

Freckman, D. W., W. G. Whitford, and Y. Steinberger, Nematode population dynamics and activity in desert soils: Effect of irrigation. *Biol. Fertil. Soils*, 3, 3-10, 1987.

Friedmann, E. I., Endolithic microorganisms in the Antarctic cold desert. *Science*, 215, 1045-1053, 1982.

Fritsen, C. H., and J. C. Priscu, Photosynthetic characteristics of cyanobacteria in permanent ice-covers on lakes in the McMurdo Dry Valleys, Antarctica. *Antarct. J. U. S.*, (in press)

Fritsen, C. H., E. E. Adams, C. M. McKay, and J. C. Priscu, Permanent ice covers of the McMurdo Dry Valley lakes, Antarctica: Liquid water content. this volume.

Geider, R. J., H. L. MacIntyre, and T. M. Kana, A dynamic model of photoadaptation in phytoplankton. *Limnol. Oceanogr.* 41, 1-15, 1996.

Green, W. J., M. P. Angle, and K. E. Chave, The geochemistry of Antarctic streams and their role in the evolution of four lakes in the McMurdo Dry Valleys. *Geochim. Cosmochim. Acta*, 52, 1265-1274, 1988.

Green, W. J., D. E. Canfield, and P. Nixon, Cobalt cycling and fate in Lake Vanda, this volume.

Gordon, D., B. Lanoil, S. Giovanonni, and J. C. Priscu, Cyanobacterial communities associated with mineral particles in Antarctic lake ice. *Antarct. J. U. S.*, (in press)

Harris, C. M., Science and management in the McMurdo Dry Valleys: Maintaining scientific and environmental values for the long term, this volume.

Hawes, I., Light climate and phytoplankton photosynthesis in maritime Antarctic lakes. *Hydrobiologia*, 123, 69-79, 1985.

Hawes, I., Photosynthesis in thick cyanobacterial films: A comparison of annual and perennial Antarctic mat communities. *Hydrobiologia*, 252, 203-209, 1993.

Hawes, I., and P. Brazier, Freshwater stream ecosystems of James Ross Island, Antarctica. *Antarctic Sci.*, 3, 365-271, 1991.

Hawes, I., and C. Howard-Williams, Primary production processes in streams of the McMurdo Dry Valley region, Antarctica, this volume.

Hawes, I., C. Howard-Williams, and R. D. Pridmore, Environmental control of microbial biomass in the ponds of the McMurdo Ice Shelf, Antarctica. *Arch. Hydrobiol Antarct.*, 127, 271-287, 1993.

Howard-Williams, C., C. L. Vincent, P. A. Broady, and W. F. Vincent, Antarctic stream ecosystems: Variability in environmental properties and algal community structure. *Int. Revue ges. Hydrobiol.*, 71, 511-544, 1986.

Howard-Williams, C., J. C. Priscu, and W. F. Vincent, Nitrogen dynamics in two Antarctic streams. *Hydrobiologia*, 172, 51-61, 1989.

Howard-Williams, C., A.-M. Schwarz, I. Hawes, and J. C. Priscu, Optical properties of dry valley lakes, this volume.

Howard-Williams, C., and W. F. Vincent, Microbial communities in southern Victoria Land streams (Antarctica) I. Photosynthesis. *Hydrobiologia*, 172, 27-38, 1989.

James, M., J. A. Hall, and J. Laybourn-Parry, Protozooplankton and microzooplankton ecology in lakes of the dry valleys, southern Victoria Land, this volume.

Kennedy, A. D., Water as a limiting factor in the Antarctic terrestrial environment: A biogeographical synthesis. *Arct. Alpine Res.*, 25, 308-315, 1993.

Laybourn-Parry, J., M. R. James, D. M. Mcknight, J. C. Priscu, S. A. Spaulding, and R. Shiel, The microbial plankton of Lake Fryxell, southern Victoria Land, Antarctica during the summers of 1992 and 1994. *Polar Biol.*, 17, 54-61, 1997.

Lizotte, M. P., and J. C. Priscu, Photosynthesis-irradiance relationships in phytoplankton from the physically stable water column of a perennially ice-covered lake (Lake Bonney, Antarctica). *J. Phycol.*, 28, 179-185, 1992.

Lizotte, M. P., and J. C. Priscu, Pigment Analysis of the Distribution, succession, and fate of phytoplankton in the McMurdo Dry Valley lakes of Antarctica, this volume.

Lizotte, M. P., T. R. Sharp, and J. C. Priscu, Phytoplankton dynamics in the stratified water column of Lake Bonney, Antarctica I. Biomass and productivity during the winter-spring transition. *Polar Biol.*, 16,155-162, 1996.

Lyons, W. B., K. A. Welch, K. Neumann, J. K. Toxey, R. McArthur, C. Williams, D. McKnight, and D. Moorhead, Geochemistry linkages among glaciers, streams and lakes within the Taylor Valley, Antarctica, this volume.

McKay, C.P., G.D. Clow, D.T. Andersen, and R.A. Wharton, Jr., Light transmission and reflection in perennially ice-covered Lake Hoare, Antarctica. *J. Geophys. Res.*, 99, 427-444, 1994.

McKnight, D. M., G. R. Aiken, E. D. Andrews, E. C. Bowles, and R. A. Harnish, Dissolved organic material in dry valley lakes: A comparison of Lake Fryxell, Lake Hoare, and Lake Vanda. *Am. Geophys. Union, Antarct. Res. Series*, 59, 119-133, 1993.

McKnight, D., A. Alger, C. Tate, G. Shupe, and S. Spaulding, Longitudinal patterns in algal abundance and species distribution in meltwater streams in Taylor Valley, southern Victoria Land, Antarctica, this volume.

Moorhead, D. L., W. S. Davis, and R. A. Wharton, Jr., Carbon dynamics of aquatic microbial mats in the Antarctic dry valleys: A modelling synthesis. *Proc. of the Polar Desert Confr.* Christchurch, New Zealand. Balkema Publications (in press)

Moorhead, D. , D. McKnight, and C. Tate, Modeling nitrogen transformations in Antarctic streams, this volume.

Neale, P. J., and J. C. Priscu, The photosynthetic apparatus of phytoplankton from a perennially ice-covered Antarctic lake: Acclimation to an extreme shade environment. *Plant Cell Physiol.*, 36, 253-263, 1995.

Neale, P. J., and J. C. Priscu, Fluorescence quenching in phytoplankton of the McMurdo Dry Valley lakes (Antarctica): Implications for the structure and function of the photosynthetic apparatus, this volume.

Odum, E. P., The strategy of ecosystem development. *Science*, 164, 262-270, 1969.

Odum, H. T., Trophic structure and productivity of Silver Springs. *Ecol. Monogr.*, 27, 55-112, 1957.

Overhoff, A., D. W. Freckman, and R. A. Virginia, Life cycle of the microbivorous Antarctic dry valley nematode *Scottnema lindsayae* (Timm 1971). *Polar Biol.*, 13, 151-156, 1993.

Parker, B. C., G. M. Simmons, Jr., R. A. Wharton, Jr., K. G. Seaburg, and F. G. Love, Removal of salts and nutrients from Antarctic lakes by aerial escape of bluegreen algal mats. *J. Phycol.*, 18, 72-78, 1982.

Pickett, S. T. A., and P. S. White, *The Ecology of Natural Disturbance as Patch Dynamics*. Academic Press, New York, 1985.

Pickup, J., Ecophysiological studies of terrestrial free-living nematodes on Signy Island. *Br. Antarct. Surv. Bull.*, 81, 77-81, 1988.

Pickup, J., Seasonal variation in the cold hardiness of three species of free-living Antarctic nematodes. *Funct. Ecol.*, 4, 257-264, 1990a.

Pickup, J., Seasonal variation in the cold-hardiness of a free-living predatory Antarctic nematode, *Coomanusus gerlachei* (Mononchidae). *Polar Biol.*, 10, 307-315, 1990b.

Pickup, J., Strategies of cold-hardiness in three species of antarctic dorylaimid nematodes. *J. Comp. Physiol. B.*, 160, 167-173. 1990c.

Pickup, J., and P. Rothery, Water-loss and anhydrobiotic survival in nematodes of Antarctic fellfields. *Oikos*, 61, 379-388, 1991.

Powers, L. E., D. W. Freckman, M. Ho, and R. A. Virginia, Soil properties associated with nematode distribution along an elevational transect in Taylor Valley, Antarctica. *Antarct. J. U. S.*, 29, 228-229, 1994.

Powers, L. E., D. W. Freckman, and R. A. Virginia, Spatial distribution of nematodes in polar desert soils of Antarctica. *Polar Biol.*, 15, 325-333, 1995.

Prentice, M. L., J. Kelman, and A. P. Stroeven, The composite glacial erosional landscape of the northern McMurdo Dry Valleys: Implications for Antarctic tertiary glacial history, this volume.

Prezelin, B. B., and B. M. Sweeney, Photoadaptation of photosynthesis in *Gonyaulax polyedra*. *Marine Biol.*, 48, 27-35, 1978.

Priscu, J. C., Particulate organic matter decomposition in the water column of Lake Bonney, Taylor Valley, Antarctica. *Antarct. J. U. S.*, 27, 260-262, 1992.

Priscu, J. C., Phytoplankton nutrient deficiency in lakes of the McMurdo Dry Valleys, Antarctica. *Freshwat. Biol.*, 34, 215-227, 1995.

Priscu, J. C., Biogeochemistry of nitrous oxide in permanently ice-covered lakes of the McMurdo Dry Valleys, Antarctica. In *Microbially Mediated Atmospheric Change* (J. Prosser, ed.), Special issue of *Global Change Biol.*, in press.

Priscu, J. C. and C. D. Woolston, Phytoplankton dynamics in the stratified water column of Lake Bonney, Antarctica II. Irradiance requirements for inorganic nitrogen uptake. *Polar Biol.*, in press.

Priscu, J. C., L. R. Priscu, W. F. Vincent, and C. Howard-Williams, Photosynthate distribution by microplankton in permanently ice-covered Antarctic lakes. *Limnol. Oceanogr.*, 32, 260-270, 1987.

Priscu, J. C., L. R. Priscu, C. Howard-Williams, and W. F. Vincent, Diel patterns of photosynthate biosynthesis by phytoplankton in permanently ice-covered Antarctic lakes under continuous sunlight. *J. Plank. Res.*, 10, 333-340, 1988.

Priscu, J. C., W. F. Vincent, and C. Howard-Williams, Inorganic nitrogen uptake and regeneration in perennially ice-covered Lakes Fryxell and Vanda, Antarctica. *J. Plank. Res.*, 11:335-351, 1989.

Priscu, J. C., M. T. Downes, and C. P. McKay, Extreme supersaturation of nitrous oxide in a poorly ventilated Antarctic lake. *Limnol. Oceanogr.*, 41:1544-1551, 1996.

Priscu, J. C., W. F. Vincent, and C. Howard-Williams, Inorganic nitrogen uptake and regeneration in perennially ice-covered Lake Fryxell and Vanda, Antarctica. *J. Plank. Res.*, 11, 335-251, 1989.

Priscu, J. C., Photon dependence of inorganic nitrogen transport by phytoplankton in perennially ice-covered antarctic lakes. *Hydrobiologia*, 172, 173-182, 1989.

Reynolds, J. F., and J. D. Tenhunen, *Landscape Function: Implications for Ecosystem Response to Disturbance, A Case Study in Arctic Tundra*. Springer-Verlag, Ecological Studies Series Vol. 120, New York, 1996.

Schlesinger, W. H., J. F. Reynolds, G. L. Cunningham, L. F. Huenneke, W. M. Jarrell, R. A. Virginia, and W. G. Whitford, Biological feedbacks in global desertification. *Science*, 247, 1043-1048, 1990.

Schulze, E.-D., and H. Zwölfer, *Potentials and Limitations of Ecosystem Analysis*. Springer-Verlag, Ecological Studies Series Vol. 61, New York, 1987.

Shaver, G. R., Nadelhoffer, K. H., and A. E. Giblin, Biogeochemical diversity and element transport in a heterogeneous landscape, the north slope of Alaska, in *Quantitative Methods in Landscape Ecology*. edited by M. G. Turner and R. H. Gardner, pp. 105-125, Springer-Verlag, Ecological Studies 82, New York, 1990.

Spaulding, S. A., D. M. McKnight, R. L. Smith, and R. Dufford, Phytoplankton population dynamics in perennially ice-covered Lake Fryxell, Antarctica. *J. Plankton Res.*, 16, 527-541, 1994.

Spigel, R. H., and J. C. Priscu, Evolution of temperature and salt structure of Lake Bonney, A chemically stratified Antarctic lake. *Hydrobiologia*, 321, 177-190, 1996.

Spigel, R. H., and J. C. Priscu, Physical limnology of the McMurdo Dry Valley lakes, this volume.

Tansley, A., What is ecology? *Biol. J. Linn. Soc.*, 32, 5-16, 1987.

Teal, J. M., Energy flow in the salt marsh ecosystem of Georgia. *Ecology*, 43, 614-624, 1962.

Turner, M. G., and R. H. Gardner, *Quantitative Methods in Landscape Ecology*. Springer, New York, 1991.

Van Cleve, K., and S. Martin, Long-term ecological research in the United States: a network of research sites. LTER Publication No. 11, Long-Term Ecological Research Network Office, Seattle, Washington, 1991.

Vincent, W. F., Production strategies in antarctic inland waters: Phytoplankton eco-physiology in a permanently ice-covered lake. *Ecology*, 62, 1215-1224, 1981.

Vincent, W. F., C. Howard-Williams, and P. A. Broady, Microbial communities and processes in Antarctic flowing waters, in *Antarctic Microbiology*, edited by E. I. Friedman, pp. 543-569, Wiley-Liss, Inc., 1993a.

Vincent, W. F., R. W. Castenholz, M. T. Downes, and C. Howard-Williams, Antarctic cyanobacteria: Light, nutrients, and photosynthesis in the microbial mat environment. *J. Phycol.*, 29,745-755, 1993b.

Vincent, W. F., and C. Howard-Williams, Antarctic stream ecosytems: physiological ecology of a blue-green algal epilithon. *Freshwater Biol.*, 16, 219-233, 1986.

Vincent, W. F., and C. Howard-Williams, Microbial communities in southern Victoria Land streams (Antarctica) II. The effects of low temperature. *Hydrobiologia*, 172, 39-49, 1989.

Vincent, W. F., and C. L. Vincent, Factors controlling phytoplankton production in Lake Vanda (77 0S). *Can. J. Fish. Aquat. Sci.*, 39, 1602-1609, 1982.

Voytek, M. A., B. B. Ward, and J. C. Priscu, The abundance of ammonia-oxidizing bacteria in Lake Bonney, Antarctica determined by immunofluorescence, PCR and in situ hybridization, this volume.

Ward, B. B., and J. C. Priscu, Detection and characterization of denitrifying bacteria from a permanently ice-covered Antarctic lake, *Hydrobiologia*, in press.

Wharton, Jr., R. A., Stromatolitic mats in Antarctic lakes, in *Phanerozoic Stromatolites II*. edited by J. Bertrand-Sarfati and C. Monty, pp. 53-70, Kluwer Academic, Netherlands. 1994.

Wharton, D. A., and W. Block, Freezing tolerance in some antarctic nematodes. *Funct. Ecol*., 7, 578-584, 1993.

Wing, K.T. and, J.C. Priscu, Microbial communities in the permanent ice cap of Lake Bonney, Antarctica: Relationships among chlorophyll-*a*, gravel and nutrients. *Antarct. J. U. S*., 28, 246-249, 1993.

D. Moorhead, Department of Biological Sciences, Texas Tech University, Lubbock, TX, 79409-3131

J.C. Priscu, Department of Biology, Montana State University, Bozeman, MT, 59717

(Received April 16, 1997;
accepted May2, 1997)

DIGITAL GEOSPATIAL DATASETS PERTAINING TO
THE MCMURDO DRY VALLEYS OF ANTARCTICA:
THE SOLA/AGU CDROM

Jordan Towner Hastings

University of Nevada System

The CDROM included with this volume contains a selection of maps and images, remote-sensing data, aerial and ground photography, GIS-based simulation model output, tabular data and descriptive documents, all related to the McMurdo Dry Valleys of Antarctica. A Windows® PC running a Web browser such as Netscape™ Navigator is all that is required to display these materials. Optionally, three browser "helper" applications ESRI ArcView®, MS® Excel and MS® Word can provide full access to the underlying data items. All the basic cartographic materials on the CDROM are in the public domain, having been obtained from government sources, both United States and international. Several authors in the present AGU volume have kindly supplied original research materials to the CDROM, as noted in their contributions.

INTRODUCTION

As research activities in the McMurdo Dry Valleys have grown steadily over the past fifty years, a sizable corpus of maps and images, remote-sensing data, aerial and ground photography, GIS-based simulation model output, tabular data, and descriptive documents has been produced. Many of these materials are eclectic and scattered; only a small fraction has been published. Generally, it would be desirable if historical and on-going research work could be more easily inter-related and inter-compared. With this objective, in part, for about five years the National Science Foundation's Office of Polar Programs (NSF/OPP) has been encouraging efforts to develop geographic information systems (GIS) and associated databases in the support of the U.S. Antarctic Program (USAP). GIS has proved useful in numerous other fields of research [*Tomlinson,* 1990] by its emphases on formal description, database management, and ultimately spatial integration of diverse materials over a common landscape.

The CDROM which accompanies this volume is one result of an NSF/OPP-supported activity, the SOLA Planning Workshop, which was organized specifically to explore requirements for GIS in the McMurdo Dry

Valleys region. At that workshop, the author was invited by the general editor of this volume, to develop a "sampler" of the geospatial datasets which had been discussed as suitable for GIS treatment. This sampler was to be supplied as a CDROM, made accessible to readers of the AGU volume as conveniently as possible. Given the constraints of a stand-alone CDROM with only freely-available software, it was recognized that an operational GIS could not be demonstrated; instead the focus would be on the datasets themselves. This interim project, and its eventual work product, became known as the SOLA/AGU CDROM.

The digital geospatial datasets on the SOLA/ AGU CDROM fall generally into two categories: 1) so-called framework datasets, the scanned paper maps and satellite images, which serve as de facto basemaps for the dry valleys; and 2) thematic datasets, the field measurements or observations of primary scientific interest, related to the basemaps. Usually the framework datasets are public record, while the thematic datasets are proprietary to individual investigators. In some cases datasets may incorporate both framework and thematic material: a solar radiation model based on both topography (framework) and meteorology (thematic), for example. The following primary authors in this volume have kindly contributed one or more of their original research datasets to the CDROM: Adams, Conovitz, Dana, and Prentice.

ACCESS TO CDROM

The SOLA/AGU CDROM contains a selection of geospatial data and information—maps and images, remote-sensing data, aerial and ground photography, GIS-based simulation model output, tabular data and descriptive documents—all related to the McMurdo Dry Valleys, and collectively termed resources. These resources are organized in a three-level hierarchy: catalogs, datasets, and data items.

The dataset is the fundamental organizational unit on the CDROM. Each dataset is identified by a "call number" of the form **SAnnnn** (where **nnnn** is a unique 4-digit number), similar to a library accession number. In addition to its call number, a dataset is described by author, title, publication source (institutional and/or individual), simple geospatial information (place and time), and optionally a short abstract.

Within each dataset, a variable number of separately accessible items of data and/or information are listed. Each of these items is assigned a one- or two-character identifier (ID), a name/caption, a type—map, photograph, document, etc.— and optionally a more detailed sub-description.

An overarching system of catalogs classifies and gathers the datasets together according to the type(s) of items which they contain. Each dataset appears in at least one, and often several, catalogs. Thus the datasets are cross-indexed by type in a matrix of catalogs.

A World Wide Web browser such as Netscape™ Navigator3 is the only application software required to display the resources on the CDROM. While generally used on the Web, a browser is also capable of accessing non-Web resident files, here on CDROM; the browser must provide full support for "frame sets" and "Java scripting." The browser displays a main screen, which may be divided into a number of non-overlapping panels. In each panel, an individual document is shown and may be manipulated by keyboard and/or mouse inputs from the computer user. The progression of documents shown in panels is controlled by these user actions interacting with built in "hyperlinks" interconnecting the documents themselves, as specified by their author(s).

In addition to its display capabilities, the Web browser also facilitates printing of documents, includ-

ing full-color graphics, and/or transferring the underlying data items in electronic form to local storage for use in other software packages. Further the browsers can be configured to auto- matically open selected data items in these other packages, which are then referred to as "helper" applications.

Together with the Web browser, three helper applications can provide access to, and manipulation of, all resources on the CDROM: ESRI ArcView3®, a premier desktop GIS package; and MS® Excel and MS® Word, spreadsheet and word processing packages, respectively. The contents of the CDROM were prepared using these packages. Free, limited versions of the Excel and Word packages, capable of viewing but not authoring documents, are also provided on the CDROM.

From a Windows® PC with the Web browser installed, the CDROM is accessed simply by inserting it in the reader and clicking on its top-level "Web index" file, `index.htm`. In some systems the browser starts automatically when the CDROM is inserted in the reader. After some introductory material, the Home screen appears (Figure 1); this contains the CDROM's signature graphic, the McMurdo Dry Valleys satellite image-map in its large main panel. Running down the left-hand side of this screen is the menu panel, which provides the ability to navigate the CDROM. At the top of the screen is a heading panel, which dynamically tracks the user's current location in the browser menus.

(Note: The remainder of this section is best read with the CDROM open in the browser.)

Clicking on the Resources entry in the main menu brings up a sub-menu with the matrix of catalogs described earlier; clicking again on any of the sub-menu entries displays the dataset contents of the corresponding catalog in the main panel (Figure 2). Here, each row in the dataset table is divided into three columns: Call Number, Title, and Source; all three are active Web browser links. Clicking on either Call Number or Title displays the dataset description; clicking on Source links out of the CDROM to the Web site of the institution which supplied the dataset.

Linked from the dataset/catalog system, the primary descriptive material for each dataset is organized in two sections: the Description per se, and the Item List. Optionally, a third Abstract section may be present.

The dataset Description (Figure 3) is a simple form, with standardized fields for such information as author, date and place of acquisition, and various geospatial parameters, where appropriate. All fields are, of course, read-only. Embedded in the Description panel

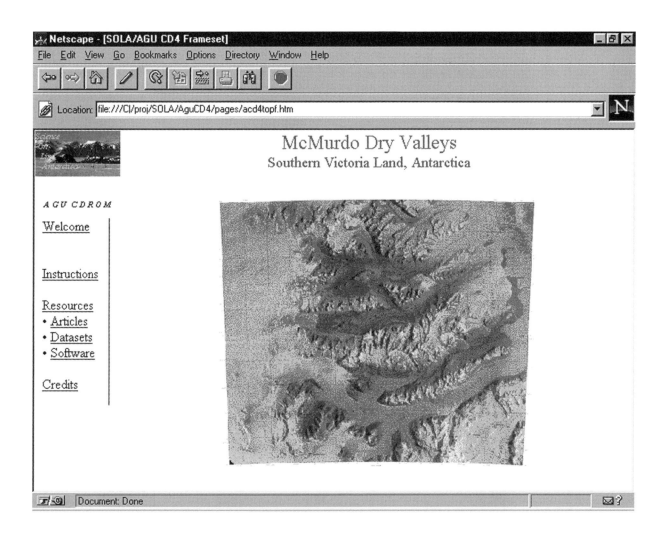

Fig. 1 Home Screen

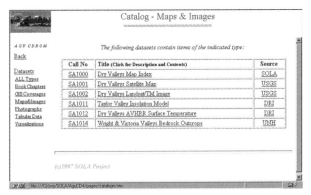

Fig. 2 Dataset/Catalog Screen

Fig. 3 Dataset Description Screen

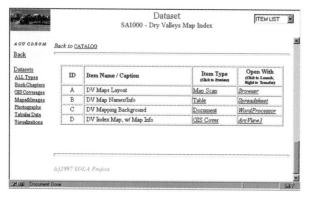

Fig. 4 Dataset Item List Screen

is a thumbnail image of one selected/typical item in the dataset. Clicking on this image brings up a scrollable list of graphical previews for all the items in the dataset; clicking further on any of these previews expands it to fill the main panel (replacing Description).

The dataset Item List (Figure 4) appears as a subsidiary table, enumerating the computer files which comprise the dataset, and which are available to be previewed, opened, or transferred individually. Each row in the item list table is again divided into four columns: ID (a letter code; sometimes a letter-digit combination), Item Name/Caption, Item Type, and Open-With (the software package most "knowledge-able" about this item). Only the last two columns contain active links. Clicking on the Item Type displays a partial-screen preview of the item (in the main panel, identical to navigating from the thumbnail in the Description panel), while clicking on Open-With launches a full-screen preview of the item in a free-standing window. By contrast, right-clicking on Open-With, initiates a transfer of the item from CDROM to local storage media. (The browser automatically presents further "pop-up" menu/navigation panels for this purpose.)

In addition to these programmed features, highlight-ed and underlined text links in the various panels also provide navigation between them, as is customary in browser documents. A Catalog link to the top-level catalog menu is always provided. Also, whenever a dataset is being explored, a "drop-down" selector appears in the upper right-hand corner of the browser screen. The highlighted menu bar in this selector confirms the active section; clicking on any other selection—Abstract, Description, or Item List—will switch to the appropriate panel.

To shortcut exploration of datasets, or in event of difficulty, clicking on any of the menu or sub-menu links will refresh the main panel. Clicking on the Back, then Home links will return to the home screen of the CDROM browser application at any time.

SUMMARY

Altogether, 13 datasets, comprising 42 items and occupying ~150+ megabytes, appear on the CDROM. All datasets are explictly geospatial; some are, in fact, manipulable GIS "coverages." Coverage is the GIS-specific term for geospatial dataset. The first six datasets (**SA1000** through **SA1005**) are believed to be the best currently available basemaps for the McMurdo Dry Valleys region; these are SOLA and/or USGS-provided resources, in the public domain, and may be utilized freely. The remaining datasets (**SA1011** through **SA1017**), are more-or-less proprietary research resources, kindly provided by contributors to this volume in Web-browsable formats. Readers are encouraged to contact the contributors directly (via the dataset Description), before making any further use or distribution of these datasets.

Acknowledgments. A publication of any sort, a CDROM in particular, represents the work of many people with diverse talents. In alphabetical order, contributors to the SOLA/AGU CDROM have been: E. Adams, J. Calkins, N. Cardinal, P. Conovitz, J. D'Angelo, G. Dana, R. Dubayah, C. Harris, J.T. Hastings, H. House, M. Prentice, J. Priscu, D. Stott, J-C. Thomas.

SOLA is an acronym for Science On Line Antarctica. The SOLA initiative envisions a GIS-based, Internet-accessible clearing-house for Antarctic research data and information, the majority of which is geospatial.

The SOLA Planning workshop, the fourth in a series of related NSF-supported workshops, was organized by the Desert Research Institute (DRI), and held 24-26 September 1996 at Granlibakken Conference Center, Tahoe City, California. NSF's Data Base Activities Program, in the Division of Biological Infrastructure (formerly Biological Instrumentation and Resources), was the primary sponsor.The SOLA initiative has been consistently supported by the National Science Foundation, under a succession of grants. The most recent grant, which primarily covers this work, is NSF/BIR95-07857.

ESRI and ArcView are registered trademarks of Environ-mental Systems Research Institute, Inc., Redlands, CA.

MS and Windows are registered trademarks of Microsoft Corp., Redmond, WA.

Netscape and Netscape Navigator are trademarks of Netscape Communications Corp., Mountain View, CA.

REFERENCE

Tomlinson, Roger F., Current and potential uses of geographic information systems: the North American experience in *Introductory Readings in Geographic Information Systems*, Donna J. Peuquet and Duane F. Marble, eds., Taylor & Francis, Ltd., 142–158, 1990.

J. T. Hastings, Department of Computer Science / 171, University of Nevada, Reno, Reno, NV 89557

(Received June 23, 1997
accepted July 5, 1997)